AF358794

# Advances in Remote Sensing for Global Forest Monitoring

# Advances in Remote Sensing for Global Forest Monitoring

Editors

**Erkki Tomppo**
**Jaan Praks**
**Guangxing Wang**
**Lars T. Waser**

MDPI • Basel • Beijing • Wuhan • Barcelona • Belgrade • Manchester • Tokyo • Cluj • Tianjin

*Editors*
Erkki Tomppo
University of Helsinki
Aalto University
Finland

Jaan Praks
Aalto University
Finland

Guangxing Wang
Southern Illinois University Carbondale
USA

Lars T. Waser
Swiss Federal Institute for Forest, Snow and Landscape Research WSL
National Forest Inventory
Switzerland

*Editorial Office*
MDPI
St. Alban-Anlage 66
4052 Basel, Switzerland

This is a reprint of articles from the Special Issue published online in the open access journal *Remote Sensing* (ISSN 2072-4292) (available at: https://www.mdpi.com/journal/remotesensing/special_issues/Global_Forest_Monitoring).

For citation purposes, cite each article independently as indicated on the article page online and as indicated below:

LastName, A.A.; LastName, B.B.; LastName, C.C. Article Title. *Journal Name* **Year**, *Volume Number*, Page Range.

**ISBN 978-3-0365-1252-5 (Hbk)**
**ISBN 978-3-0365-1253-2 (PDF)**

# Contents

# About the Editors

**Erkki Tomppo**, professor, completed his Lic. Phil (Mathematics) and PhD in statistics (thesis in Forest Biometrics) at University of Helsinki, and has studied forestry at Helsinki University. He was the professor of forest inventory at the current Natural Resources Finland, 1990–2015, and has worked with forest inventory research since 1984, leading the Finnish National Forest inventory in 1990–2004 and its method of research development 1990–2015 (until his retirement), including carbon pool assessment methods, and the operative satellite image-aided multi-source inventory. Since then, he has been a visiting scientist at the Department of Electronics and Nanoengineering, Aalto University, Finland, and is now and adjunct professor in Remote Sensing at the University of Helsinki. He was responsible for producing UN/LULUCF greenhouse gas statistics for Finland from 1998–2005. He was working for UN/FAO in Rome in 2014, establishing forest inventories and REDD+ monitoring for countries in development. He has led several projects in forest inventory, forest biodiversity and forest carbon pool estimation since 1990. The current research interests include forest inventory methods, non-parametric estimation methods for multi-source forest inventory purposes, updating of NFI information by means of multi-temporal remote sensing data, the dynamics of forest biodiversity and methods for its assessment, the estimation of carbon pools of forests and the use of dense pulse laser data in forest inventories. He was the chairman of the management committee of COST Action E43 (2004–08), with about 30 countries. He has developed the first National Forest Inventory for Tanzania in 2009–2011 and assisted several countries in revising their natural resource monitoring systems. The List of publications involve more than 200 scientific items; among these are two books, one edited book and several book chapters, mainly in forest inventory, forest remote sensing and forest biodiversity. The main scientific contribution involves the development of a multi-source forest inventory method for which he was awarded The 14th Marcus Wallenberg Prize of 2 million Swedish crowns in 1997.

**Jaan Praks**, professor, D.Sc (Tech), is an assistant professor of electrical engineering with the Department of Electronics and Nanoengineering, School of Electrical Engineering, Aalto University, Finland. He completed his BSc degree in Physics at University of Tartu, Estonia, in 1996, and his DSc (Tech.) degree in space technology and remote sensing from Aalto University, Espoo, Finland, in 2012. He has been working with microwave remote sensing, scattering modeling, microwave radiometry, hyperspectral imaging and with advanced SAR techniques, such as polarimetry, interferometry, polarimetric interferometry and tomography. One of most visited topics in his research is the remote sensing of boreal forests. Since 2009, Jaan Praks has been interested in emerging nanosatellite technology. He led a project which produced the first two Finnish satellites. Jaan Praks has also been involved in spinning off several companies in the field of satellite remote sensing. His research team is a member of Finnish Centre of Excellence in Research of Sustainable Space and he is a Principal Investigator of several small satellite missions. He is an active member of the scientific community, a member of the Finnish National Committee of COSPAR, chairman of Finnish National Committee of URSI, and chair and co-chair of many national and international conferences.

**Guangxing Wang** is a full professor of Remote Sensing and GIS with Geography and Environmental Resources, School of Earth Systems and Sustainability, Southern Illinois University Carbondale, Illinois, USA. He received his BS degree in Forestry and MS degree in Biometrics from the Central South University of Forestry and Technology, Changsha, Hunan of China, in 1982 and 1985, respectively, and the PhD in Remote Sensing of Forest Resources from the University of Helsinki, Finland, in 1996. He has research interests in remote sensing, GIS, spatial statistics and their applications to geography, natural and environmental resources with concentrations in modeling and change detection and monitoring of human activity-induced vegetation disturbance and forest biomass/carbon sequestration, quality assessment and the uncertainty analysis of data and products. He is an author and co-author of more than 130 peer-reviewed journal articles and the editor of the book Remote Sensing of Natural Resources, published in 2013. Currently, he is an editor of the Forest Remote Sensing Section of the journal Remote Sensing and an associate editor of *American Journal of Environmental Sciences*.

**Lars T. Waser**, Ph.D., is a Senior Researcher and remote sensing specialist at Swiss Federal Institute for Forest, Snow and Landscape Research WSL. He works for or leads several national, international projects and programs which are mostly related to forestry, biodiversity and land cover change. He has a large publication record and is an internationally recognized expert for remote sensing and forest research. He is the co-editor of two Q1 journals, i.e. Remote Sensing (MDPI), European Journal of Remote Sensing (Taylor and Francis), guest editor and permanent member of scientific committees of conferences. His specific research interests are related to assessing the current state and changes of forests, using novel modelling techniques and multi-source remotely sensed data. With his team, he focusses on the development of novel approaches on case study, countrywide and cross-border level and building bridges between research and practitioners.

*Editorial*

# Editorial Summary, *Remote Sensing* Special Issue "Advances in Remote Sensing for Global Forest Monitoring"

Erkki Tomppo [1,2,*], Guangxing Wang [3], Jaan Praks [1], Ronald E. McRoberts [4] and Lars T. Waser [5]

[1] Department of Electronics and Nanoengineering, Aalto University, P.O. Box 15500, 00076 Aalto, Finland; jaan.praks@aalto.fi

[2] Department of Forest Sciences, University of Helsinki, Latokartanonkaari 7, P.O. Box 27, 00014 Helsinki, Finland

[3] Department of Geography and Environmental Resources, Southern Illinois University Carbondale, 1263 Lincoln Drive, Carbondale, IL 62901-6899, USA; gxwang@siu.edu

[4] Department of Forest Resources, University of Minnesota, 1530 Cleveland Ave N, Saint Paul, MN 55108, USA; mcrob001@umn.edu

[5] Swiss Federal Institute for Forest, Snow and Landscape Research WSL, Zuercherstrasse 111, 8903 Birmensdorf, Switzerland; lars.waser@wsl.ch

* Correspondence: erkki.tomppo@aalto.fi

Citation: Tomppo, E.; Wang, G.; Praks, J.; McRoberts, R.E.; Waser, L.T. Editorial Summary, *Remote Sensing* Special Issue "Advances in Remote Sensing for Global Forest Monitoring". *Remote Sens.* **2021**, *13*, 597. https://doi.org/10.3390/rs13040597

Received: 4 January 2021
Accepted: 4 February 2021
Published: 8 February 2021

**Publisher's Note:** MDPI stays neutral with regard to jurisdictional claims in published maps and institutional affiliations.

## Editorial Summary

The need for timely, spatially, and thematically accurate information regarding forests is increasing because of the key role of forests in the global carbon balance and sustainable social, economic, ecological, and cultural development. While an increasing number of countries in the world already are conducting statistically sound forest inventories, a few for 100 years, inventories in other countries are still lacking, which makes the global information about forest statistics inaccurate.

The total global forest area based on the Forest Resource Assessment report of the United Nations FAO from 2020 was 4.06 billion hectares, which is 31 percent of the total land area. The rate of net forest loss declined from 7.8 million ha per year in the decade 1990–2000 to 4.7 million ha per year in 2010–2020. The net forest area loss in 2010–2020 was the largest in Africa, while the largest increases were in Asia and Europe. The world's total growing stock of trees decreased slightly, from 560 billion cubic metres in 1990 to 557 billion cubic metres in 2020, due to a net decrease in forest area while the stock per unit area increased from 132 in 1990 to 137 cubic metres per ha in 2020.

Remote sensing techniques have been utilized for National Forest Inventories (NFI) for many decades, first using mainly airborne sensors such as photography, but increasingly, space-borne sensors such as Landsat. Remotely sensed data have also been used at a global level by FAO for the purpose of comparison and as complementary information, since 1980. Some institutes have attempted to conduct global forest inventories, sometimes providing contradicting information. This implies that remote sensing-based approaches are vulnerable to misclassification and inaccurate estimation of forest parameters and changes, such as the recognition of temporarily unstocked forest areas, forest degradation, and species composition, as well as in strictly following the globally adopted definitions.

Active remote sensing technologies, such as Airborne Laser Scanning (ALS) and, in particular, Synthetic Aperture Radar (SAR), which are becoming increasingly available, provide new opportunities for large-area and global forest inventory, and sufficient repeat monitoring in a cost-efficient way. Technically and statistically sound methods are still being developed. Some gaps exist in global forest monitoring using remote sensing, such as difficulties in estimating subtle changes as well as the lack of statistically sound uncertainty estimates.

This Special Issue includes papers that attempt to overcome the gaps and describe state-of-the-art of remote sensing for forest parameter estimation and change monitoring at national, continental, or global scales.

A short summary of the articles published is given below.

(1) Detection of Forest Windstorm Damages with Multitemporal SAR Data—A Case Study: Finland. Multitemporal Sentinel-1 data together with other geo-referenced data were used to develop methods to localize forest windstorm damages, assess their severity and estimate the total area damaged, preliminarily without new training data. The study was the first step towards an operational system for near-real-time windstorm damage monitoring. The improved k-NN method, multinomial logistic regression and support vector machine classification methods were fine-tuned and their predictions were evaluated. A method to estimate the confidence intervals of the probabilities of the predicted categories was proposed.

(2) Forest Drought Response Index (ForDRI): A New Combined Model to Monitor Forest Drought in the Eastern United States. A new forest soil drought response index (ForDRI) for long term spatial drought monitoring was developed. It uses space-borne remotely sensed data (MODIS) and other geo-referenced data. The new index identified extreme drought periods that are compatible with those calculated from forest flux-tower data. The tree ring analyses showed the impact of the drought on the tree growth.

(3) Large Uncertainty on Forest Area Change in the Early 21st Century among Widely Used Global Land Cover Datasets. The study shows large variation and inconsistency in the estimates of forest area changes that are based on the widely used global data sets and calls for the development of a more accurate database to support forest policies and contribute to global actions against climate change.

(4) A Model-Based Volume Estimator that Accounts for Both Land Cover Misclassification and Model Prediction Uncertainty. The study estimated the effects of the uncertainty of forest species maps used in sampling and on the volume estimation. The conclusions drawn were: (1) the effects of uncertainty in the forest species map on the uncertainty of large area volume estimates are not negligible; (2) overall, the effects of uncertainty in the forest species map on area estimates were greater than the effects of uncertainty in the map on the selection of field plots used to calibrate the random forest volume prediction model; (3) the effects of the forest species map uncertainty increased for open forest species or less representative forest species; (4) bootstrapping estimates demonstrated the suitability of this technique to accommodate the effects of uncertainty from more than one source; and, (5) the results are relevant for countries that use a remote sensing-based forest/non-forest map to guide the establishment of field plots.

(5) An End-to-End Deep Fusion Model for Mapping Forests at Tree Species Levels with High Spatial Resolution Satellite Imagery. A new end-to-end deep learning fusion method using high spatial resolution remote sensing images was developed by combining the advantageous properties of multi-modality representations and the powerful features of post-processing step to optimize the classification accuracy of the dominant tree species in a highly automated way. The accuracy of the method was tested for several plantation tree species in two test sites in China, such as Chinese pine and *Larix principis* in the northern test area, and Eucalyptus in the southern test area.

(6) Wide-Area Near-Real-Time Monitoring of Tropical Forest Degradation and Deforestation Using Sentinel-1. The system introduced combined time-series analysis of small objects that were identified in Sentinel-1 data, which is, segments containing linear features and apparent small-scale disturbances. A physical model was introduced for quantifying the size of small (upper-) canopy gaps. Deforestation detection was evaluated for several forest landscapes in the Amazon and Borneo. In peat swamp forests, narrow linear canopy gaps (road and canal systems) could be detected, including many gaps that are barely visible on hi-res SPOT-6/7 images. When compared to optical data, subtle degradation signals are easier to detect and they are not quickly lost over time due to fast re-vegetation. The method looks promising in recognizing relatively small changes in forest canopy cover, such as those caused by ditching or small scale logging.

(7) A Near Real-Time Method for Forest Change Detection Based on a Structural Time Series Model and the Kalman Filter. A stochastic modelling method that combines

a structural time series model with the Kalman filter was developed for near-real-time monitoring of forest changes, caused, e.g., by damages. The method was demonstrated while using Sentinel-2 data and test sites in Malawi (dry tropical forest) and Austria (temperate deciduous, coniferous, and mixed forests). The method looks promising for an automated REDD+ (Reducing Emissions from Deforestation and Forest Degradation) services in the tropics and windthrow damage assessment or bark beetle monitoring in Central Europe.

(8) Characterizing the Error and Bias of Remotely Sensed LAI Products: An Example for Tropical and Subtropical Evergreen Forests in South China. The study used nearly 8000 in situ measurements of leaf area index (LAI) from six forest environments in southern China to evaluate the magnitude, uncertainty, and dynamics of three widely used EO LAI products. The finer spatial resolution GEOV3 PROBA-V 300 m LAI product gave the most accurae LAI estimates with a multi-site dataset and captured canopy dynamics well. The MODIS 500 m product did not capture the temporal dynamics observed in situ across southern China. The uncertainties estimated for each of the EO products are substantially smaller (3–5 times) than the observed bias of the EO products when compared to the in situ measurements, showing that uncertainties are substantially underestimated and do not fully account for their total uncertainty.

(9) Analysis of the Spatial Differences in Canopy Height Models from UAV LiDAR and Photogrammetry. Six data sets, including one LiDAR data set and five photogrammetry data sets captured from an unmanned aerial vehicle (UAV), were used to estimate the forest canopy heights. Three spatial distribution descriptors, the effective cell ratio, point cloud homogeneity, and point cloud redundancy, were developed to assess the LiDAR and photogrammetry point clouds in the grid. Large negative and positive variations were observed between the LiDAR and photogrammetry canopy heights. The stratified mean difference in canopy heights gradually increased from negative to positive when the canopy heights were greater than 3 m, which means that photogrammetry tends to overestimate low canopy heights and underestimate high canopy heights.

(10) Prediction of Individual Tree Diameter and Height to Crown Base Using Nonlinear Simultaneous Regression and Airborne LiDAR Data. A compatible simultaneous equation system of diameter at breast height (DBH) and height to crown base (HCB) error-in-variable (EIV) models were developed using LiDAR-derived data and ground-measurements for 510 *Picea crassifolia* Kom. trees in northwest China. Four versatile algorithms were evaluated for their estimating efficiencies and precision for a simultaneous equation system of DBH and HCB EIV models. The simultaneous equation system could illustrate the effect of errors that are associated with the regressors on the response variables (DBH and HCB) and guaranteed the compatibility between the DBH and HCB models at an individual level level.

(11) Land Use/Land Cover Mapping Using Multitemporal Sentinel-2 Imagery and Four Classification Methods—A Case Study from Dak Nong, Vietnam. A parametric classifier (logistic regression) and three non-parametric machine learning classifiers (improved k-nearest neighbors, random forests, and support vector machine) for the classification of multi-temporal Sentinel-2 images into LULC categories in Dak Nong province, Vietnam, were studied. A total of 446 images, 235 from the year 2017 and 211 from the year 2018, were pre-processed to gain high quality images for mapping LULC in the 6516 km$^2$ study area. The Sentinel 2 images were tested and classified separately for four temporal periods: (i) dry season, (ii) rainy season, (iii) the entirety of the year 2017, and (iv) the combination of dry and rainy seasons. Eleven different LULC classes were discriminated. The greatest accuracies were achieved for the composite IMC 4 obtained by combining dry and rainy season image sets while using the SVM classifier. The research showed the utility of combining Sentinel-2, multi-spectral, and dry and rainy season band data when mapping LULCs.

(12) Predicting Forest Cover in Distinct Ecosystems: The Potential of Multi-Source Sentinel-1 and -2 Data Fusion. The study investigated: (i) the ability of the individual sensors and (ii) their joint potential to delineate forest cover for study sites in two highly

varied landscapes that were located in Germany (temperate dense mixed forests) and South Africa (open Savanna woody vegetation and forest plantations). Multi-temporal Sentinel-1 and single time steps of Sentinel-2 data in combination to derive accurate forest/non-forest (FNF) information via machine-learning classifiers were used. The results indicated that optical sensors are capable of detecting homogeneous tree aggregations with high accuracies while failing at locating large portions of tree cover in open Savannas. The addition of multi-temporal microwave information to this data set showed multiple advantages. These are the correction of falsely classified cloud pixels, as well as an improved delineation of small forests in the Savanna ecosystem.

(13) Estimation of Changes of Forest Structural Attributes at Three Different Spatial Aggregation Levels in Northern California using Multitemporal LiDAR. Two modeling strategies to estimate changes in the stem volume (V), basal area (BA) and above ground biomass (AGB), of trees were developed using auxiliary information from two light detection and ranging (LiDAR) flights in the Black Mountains Experimental Forest, Northern California from two time points. The analyzed strategies consisted of (1) directly modeling the observed changes as a function of the LiDAR auxiliary information (sigma-modeling methods) and (2) modeling V, BA, and AGB at two different points in time, including a term to account for the temporal correlation, and then computing the changes as the difference between the predicted values of V, BA, and AGB for time two and time one. The predictions and measures of uncertainty at three different level of aggregation were evaluated.

(14) Remote Sensing Support for the Gain-Loss Approach for Greenhouse Gas Inventories. For tropical countries that do not have extensive ground sampling programs, such as national forest inventories, the gain–loss approach for greenhouse gas inventories is often used. With the gain–loss approach, emissions and removals are estimated as the product of activity data defined as the areas of human-caused emissions and removals and emissions factors defined as the per unit area responses of carbon stocks for those activities. Remotely sensed imagery and remote sensing-based land use and land use change maps have emerged as crucial information sources in facilitating the statistically rigorous estimation of activity data. Similarly, remote sensing-based biomass maps have been used as sources of auxiliary data for enhancing the estimates of emissions and removal factors and as sources of biomass data for remote and inaccessible regions. The current status of statistically rigorous methods for combining ground and remotely sensed data that comply with the good practice guidelines for greenhouse gas inventories of the Intergovernmental Panel on Climate Change was reviewed.

**Author Contributions:** Conceptualization, E.T., L.T.W., G.W., R.E.M. and J.P.; methodology, E.T., L.T.W., G.W. and R.E.M.; software, E.T. and L.T.W.; validation, E.T., L.T.W., G.W., R.E.M. and J.P.; formal analysis, E.T. and L.T.W.; investigation, E.T. and L.T.W.; resources, E.T.; data curation, E.T., L.T.W., G.W. and R.E.M.; writing—original draft preparation, E.T., L.T.W., G.W., R.E.M.; writing—review and editing, E.T., L.T.W., G.W., R.E.M. and J.P.; project administration, E.T. and L.T.W. All authors have read and agreed to the published version of the manuscript.

**Funding:** This research received no external funding.

**Institutional Review Board Statement:** Not applicable.

**Informed Consent Statement:** Not applicable.

**Data Availability Statement:** For the data of the articles of the this Special Issue, please contact the authors.

**Conflicts of Interest:** The authors declare no conflict of interest.

*Article*

# Detection of Forest Windstorm Damages with Multitemporal SAR Data—A Case Study: Finland

Erkki Tomppo [1,2]*, Ghasem Ronoud [1], Oleg Antropov [3], Harri Hytönen [4] and Jaan Praks [1]

[1] Department of Electronics and Nanoengineering, Aalto University, P.O. Box 11000, 00076 Aalto, Finland; ghasem.ronoud@aalto.fi (G.R.); jaan.praks@aalto.fi (J.P.)
[2] Department of Forest Sciences, University of Helsinki, Latokartanonkaari 7, P.O. Box 27, 00014 Helsinki, Finland
[3] VTT Technical Research Centre of Finland, P.O. Box 1000, 00076 Espoo, Finland; oleg.antropov@vtt.fi
[4] Finnish Forest Centre, Kauppakatu 19 B, 40100 Jyväskylä, Finland; harri.hytonen@metsakeskus.fi
* Correspondence: erkki.tomppo@aalto.fi or erkki.tomppo@helsinki.fi

**Abstract:** The purpose of this study was to develop methods to localize forest windstorm damages, assess their severity and estimate the total damaged area using space-borne SAR data. The development of the methods is the first step towards an operational system for near-real-time windstorm damage monitoring, with a latency of only a few days after the storm event in the best case. Windstorm detection using SAR data is not trivial, particularly at C-band. It can be expected that a large-area and severe windstorm damage may affect backscatter similar to clear cutting operation, that is, decrease the backscatter intensity, while a small area damage may increase the backscatter of the neighboring area, due to various scattering mechanisms. The remaining debris and temporal variation in the weather conditions and possible freeze–thaw transitions also affect observed backscatter changes. Three candidate windstorm detection methods were suggested, based on the improved k-nn method, multinomial logistic regression and support vector machine classification. The approaches use multitemporal ESA Sentinel-1 C-band SAR data and were evaluated in Southern Finland using wind damage data from the summer 2017, together with 27 Sentinel-1 scenes acquired in 2017 and other geo-referenced data. The stands correctly predicted severity category corresponded to 79% of the number of the stands in the validation data, and already 75% when only one Sentinel-1 scene after the damage was used. Thus, the damaged forests can potentially be localized with proposed tools within less than one week after the storm damage. In this study, the achieved latency was only two days. Our preliminary results also indicate that the damages can be localized even without separate training data.

**Keywords:** boreal forest; windstorm damage; synthetic aperture radar; C-band; Sentinel-1; support vector machine; improved k-NN; genetic algorithm; multinomial logistic regression



**Citation:** Tomppo, E.; Ronoud, G.; Antropov, O.; Hytönen, H.; Praks, J. Detection of Forest Windstorm Damages with Multitemporal SAR Data—A Case Study: Finland. *Remote Sens.* **2021**, *13*, 383. https://doi.org/10.3390/rs13030383

Received: 11 December 2020
Accepted: 18 January 2021
Published: 22 January 2021



## 1. Introduction

### 1.1. Background and Objectives of the Study

Windstorm damages have become more common in the past decades [1,2]. Windstorms cause noticeable large area forest damages in Europe, including Scandinavia and Finland. For example, in southern Sweden, approximately 4.5 million cubic meters of timber was damaged in 1999 in a single storm [3], and in 2005 and 2007 approximately 70 and 12 million cubic meters of timber fell down in similar disastrous events, respectively [4]. The forest area reported to have been cut due to damages was over 30,000 ha on more than 20,000 forest stands in Northern Finland in 2014, and more than 6000 ha in Eastern Finland in July 2020.

A rapid localization of the forest damages and removal of the fallen trees is the key for not only assessing the losses, but also avoiding further damage, caused, e.g., by insects. Severe storms require earlier sanitary cuttings (compared to original forest plans) to prevent such insect outbreaks. These ad-hoc cuttings naturally increase harvesting and removal costs, cause losses in revenue and lower the future cutting possibilities [5]. The volume of damaged trees in windstorm has exceeded the volume of the normal annual cut in some countries in Europe, e.g., Germany, Poland and Sweden [1,6]. Timely detection and mapping of a damaged forest allows additionally to optimize efforts in clearing potentially blocked roads and damaged power-lines in rural areas.

A common method to localize the damage areas for operational forest regime purposes and obtain a rough overview of the damages has been monitoring with airplane using either visual assessment or optical sensors, e.g., video camera or airborne laser scanner (ALS). Most large-area studies with space-borne data have been conducted using optical satellite instruments [7]. A recent study by Rüetschi et al. [8] presents a summary of several demonstrated approaches in mapping windthrown forest areas. Our further in-depth analysis with SAR data is given in Section 1.2.

A key prerequisite for successful operational forest management after a storm is a rapid, near-real-time localization of the damages. Damages are often large in area wherefore methods using space-borne data are appealing and cost-efficient alternatives. A central requirement is the timely availability of the remotely sensed data. SAR data are the only possibility for rapid monitoring due to their independence of light conditions and cloud cover.

SAR backscatter depends on the forest structure and biomass, the environment and weather conditions such as moisture and temperature and sensor properties. From the current operative SAR satellites, EU Copernicus program's two C-band Sentinel-1 satellites probably have the best potential for rapid monitoring, primarily due to a frequent data acquisition and a free of charge data policy [9]. The only drawback of C-band data in forest application is the low penetration to forest volume due to short wavelength, which could restrict detectability of minor damages. The ALOS PALSAR-2 with L-band SAR, with deeper penetration depth than C-band and fully polarimetric capability, would likely better suit forest applications [10,11], but the operational use is restricted by data availability [10,11]. ESA's coming forest specific P-band BIOMASS mission may provide information for monitoring aboveground biomass and its change over large areas, but will not be operated over Europe [12]. The data availability and price also restrict the usability of high resolution X-band SAR data that could enable spatial texture analysis of SAR backscatter for forest disturbance [10] (a further detailed analysis of possible scenarios is given in Section 1.2) In the future, new satellites and constellations, such as NISAR [13] and ICEYE [14,15] as well as planned DLR TanDEM-L [16] and ESA ROSE-L [17] may improve the situation significantly.

Thus, Sentinel-1 presently and in the near future seems to be the most suitable tool for forest damage assessment in Europe at operational level.

The overall goals of this study were to develop methods to localize the forest windstorm damages, assess the severity and area of damaged forests and quantify the uncertainties in forest damage prediction when using space-borne SAR data.

The detailed objectives were:
1. to study the potential of Sentinel-1 SAR data in localizing the forest windstorm damages;
2. to assess the accuracy of the developed methods; and
3. to assess the time lag from the damage to the damage detection and a ready product.

*1.2. Windstorm Damage Studies with SAR*

Several windstorm studies with SAR are shortly reviewed in this section, as well as studies using airborne SAR instruments. It is expected that the number of SAR-based studies will increase with the increasing data availability.

Green [18] investigated the sensitivity of SAR backscatter to forest windstorm damage gaps using multi-polarization C, L and P band data acquired by the NASA/JPL AIRSAR in August 1991. The study showed that changes in backscatter due to the presence of windstorm damage gaps were evident in each polarization channel used, especially with C-band HH polarization in a coniferous plantation. It is suggested that backscatter is sensitive not only to the presence but also to the shape and geometry of the windthrow gaps.

Dwyer et al. [19] used ESA's C-band ERS-1/ERS-2 interferometric image pairs and found them to be effective in differentiating between damaged and undamaged forests when the damaged areas were larger than or equal to 2–3 ha. The damage happened in Jura mountain in France in December 1999. Ready software by ESA made fast data processing possible.

Fransson et al. [3] studied the potential of CARABAS-II long wavelength SAR imagery for high spatial resolution mapping of windstorm damage forests. The results of this research show that the backscattering amplitude, at a given stem volume, is considerably higher for windstorm damage thrown forests than for unaffected forests. In addition, the backscattering from fully harvested storm-damaged areas was, as expected, significantly lower than from unaffected stands. These findings imply that VHF SAR imagery has potential for mapping windthrown forests.

Another study by Fransson et al. [4] investigated simulated wind-thrown forest mapping (controlled experiment with felling of trees) using multitemporal ALOS PALSAR (L-band), RADARSAT-2 (C-band) and TerraSAR-X (X-band) imagery. The detection methodology was based on bitemporal change detection and visual interpretation of scenes acquired before and after a simulated windthrow event. Stripmap ALOS PALSAR images were found less suitable for a damage area detection, likely due to a coarse spatial resolution. The windthrown forests were well visible when the RADARSAT-2 and TerraSAR-X HH polarization images were used.

Ulander et al. [20] used space- and airborne SAR data to map windthrown forests in southern Sweden. Analysis of the Space- and Airborne C-band SAR images including Envisat and Radarsat showed that they are unable to detect forest storm damage. The CARABAS VHF-band SAR, on the other hand, showed that these data can detect most storm-damaged forests as well as power lines, and sometimes better than the aerial photographs.

Thiele et al. [21] used TerraSAR-X data and focused first on the border line extraction of forest areas to enables a fast estimation of windthrown areas, whereby the pre-event forest border is derived from multi-spectral data. Second, clean-up operations were monitored in the affected forest areas by applying a change detection operator. They presented a method to extract the border of forest areas by fusing multi-aspect SAR images. They found that this extraction of multi-temporal changes and displacements of the forest border enables a rapid damage estimation, which is very useful to plan first clean-up operations. In addition, their intensity-based change detection showed good results to highlight small areas especially with hard to analyze data, even for human operators.

Eriksson et al. [6] showed that, when trees are felled, the backscattered signal from TerraSAR-X (X-band) increases by about 1.5 dB, while for ALOS PALSAR (L-band) a decrease with the same amount is observed. Radar images with fine spatial resolution also showed shadowing effects that should be possible to use for identification of storm felled forest.

Tanase et al. [22] applied L-band space-borne SAR data to windthrow and insect outbreak detection in temperate forests. The results show that changes in backscatter relate to the damages caused by the wind and insect outbreaks. In this case, an overall accuracy of 69–84% was achieved for the delineation of areas affected by the wind damage. The study showed that L-band space-borne SAR data can be employed over larger areas and ecosystem types in the temperate and boreal regions to delineate and detect damaged areas.

Rüetschi et al. [8] developed a straightforward approach for a rapid windthrow detection in mixed temperate forests using Sentinel-1 C-band VV and VH polarization data. Following radiometric correction of Sentinel-1 scenes acquired approximately 10 days before and 30 days after the storm event, a SAR composite images of before and after the storm were generated. The differences in backscatter before the storm and after the storm in windthrown and in intact forest were studied. A change detection method was developed.

Locations of windthrown areas of a minimum extent of 0.5 ha was suggested. The detection was based on user-defined parameters. While the results from the independent study area in Germany indicate that the method is very promising for detecting areal windthrow with a producer's accuracy of 0.88, its performance was less satisfactory at detecting scattered windthrown trees. Moreover, the rate of false positives was low, with a user's accuracy of 0.85 for (combined) areal and scattered windthrown areas. These results underscore that C-band backscatter data have a great potential to rapidly detect the locations of windthrow in mixed temperate forests within approximately two weeks after a storm event.

Other methods potentially suitable for mapping windthrown forests with SAR data include approaches demonstrated in other studies of natural and/or anthropogenic forest disturbance. These include mapping snow-damaged forest areas [23], monitoring selective logging and thinning operations in boreal and tropical forest biomes [11,24–28], forest clear cutting and other forest changes [29–34].

A common observation is that, while at L-band direct pixel-wise (or area-based) change detection using averaged-backscatter can be attempted, due to a better sensitivity to forest structure and volume, this does not really work at shorter wavelength such as C-band. Especially in the absence of fully polarimetric SAR capability. At C-band, texture analysis/extraction and subsequent image segmentation should be attempted after speckle is reduced (e.g., using image aggregation of scenes acquired before and after the forest disturbance event). At C-band, felling of trees does not change strongly the total backscatter, since the needles and smaller branches still create "bright enough" random-volume layer. Thus, texture from shadows in standing and fallen trees is the key feature to rely upon in the analysis. At even shorter wavelength and even higher resolution, such as X-band, texture analysis becomes the central way to proceed with the change detection, in addition to single pass interferometry with X-band data. Interestingly, most of the studies cited above rely on some kind of bitemporal change detection (even if "before" and "after" scenes are aggregated in two composite images). This does not really allow analyzing the added value of incorporating additional scenes of temporal dimension into the analysis. However, the idea of using textural features, even at stand level, and follow-up image segmentation appears most fruitful and is adopted and elaborated in our further analysis and methodology development.

## 2. Material

### 2.1. Test Site

The study area was selected in a collaboration with the Finnish Forest Centre. It is a forested landscape in Southern Finland in which a severe windstorm damage occurred on 12 August 2017. The fastest speed of the wind in the inland was near 30 m/s and on the sea outside the capital 32 m/s [35]. The area reaches from Helsinki capital region towards the towns Kouvola and Lappeenranta east and northeast of Helsinki region (Figure 1). The area of the forestry land in the study area, covered by Sentinel-1 scenes, is 830,000 ha. Forestry land includes three land categories: (1) forest land; (2) poorly productive forest land; and (3) unproductive land [36]. The two commonest stand level dominant tree species are Norway spruce (*Picea abies* Karst. L.) and Scots pine (*Pinus sylvestris* L.) (see also Tables 1 and 2).

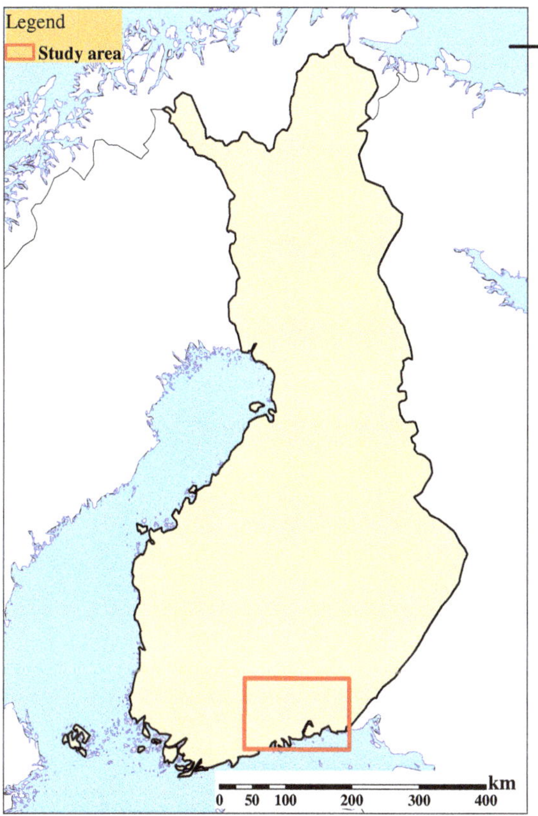


**Figure 1.** The location of the study area in Southern Finland.

*2.2. Field Data*

The training data and validation data were also selected in a collaboration with the Finnish Forest Centre and extracted from their forest database. The harvesting operation in Finnish forests presume an advance announcement and acceptance by the Forest Centre wherefore the Forest Centre has good overview of stands with wind damage but only after a longer reporting period. Two types of stands were selected for study: (a) stands in which harvesting had been planned to be carried out due to the windstorm damage; and (b) stands in which harvesting had not been reported. We call these stands damaged and non-damaged, respectively, in the following analysis. The total number of stands in reference database falling inside of all images was 977 after screening checks and removal of some doubtful stands, e.g., when stand characteristics seemed to be out-of-date in 2017. The number of damaged stand records was 313 and the number of non-damaged stand records 664. From the damaged stands, 195 stands were severely damaged and the rest slightly damaged. The severity category is assessed in the field by the forestry experts in charge. A severe damage presumes stand regeneration, while slight damage requires only removal of fallen or broken trees. The areas and growing stock characteristics are shown in Table 1 and similar statistics when the volume of growing stock is larger than 75 m$^3$/ha in Table 2.

Damaged and non-damaged stands displayed on Sentinel-1 scene from 19 August 2017, VV polarization and VH polarization are shown in Figure 2 (see also the zoomed figures with stand boundaries in Figures 3 and 4).

**Table 1.** Average stand-level areas and forest characteristics in training data, separately for damaged and non-damaged stands.

| | Damaged Stands | | | Non-Damaged Stands | | |
| | 313 Stands | | | 664 Stands | | |
| Characteristics | Minimum | Average | Maximum | Minimum | Average | Maximum |
| --- | --- | --- | --- | --- | --- | --- |
| Area, ha | 0.05 | 1.89 | 33.96 | 0.03 | 0.91 | 7.24 |
| Diameter, cm | 0.00 | 19.57 | 29.25 | 0.00 | 17.79 | 27.79 |
| Height, dm | 2.00 | 169.32 | 253.00 | 0.78 | 159.98 | 247.35 |
| Age, years | 1.00 | 56.20 | 96.61 | 0.63 | 49.27 | 98.92 |
| Basal area, $m^2$/ha | 0.00 | 18.79 | 30.00 | 0.00 | 18.27 | 30.23 |
| Volume, $m^3$/ha | 0.00 | 167.60 | 337.83 | 0.71 | 159.75 | 330.87 |
| Volume, pine | 0.00 | 52.92 | 145.50 | 0.71 | 41.13 | 182.43 |
| Volume, spruce | 0.00 | 84.31 | 304.67 | 0.00 | 68.43 | 284.36 |
| Volume, birch spp. | 0.00 | 21.89 | 110.80 | 0.00 | 34.87 | 107.63 |
| Volume, other br. l. | 0.00 | 8.48 | 103.00 | 0.00 | 15.32 | 67.60 |

**Table 2.** Average stand-level areas and forest characteristics in training data, separately for damaged and non-damaged stands with volume of growing stock greater than 75 $m^3$/ha.

| | Damaged Stands | | | Non-Damaged Stands | | |
| | 281 Stands | | | 631 Stands | | |
| Characteristics | Minimum | Average | Maximum | Minimum | Average | Maximum |
| --- | --- | --- | --- | --- | --- | --- |
| Area, ha | 0.05 | 1.89 | 33.96 | 0.03 | 0.91 | 7.24 |
| Diameter, cm | 9.17 | 20.97 | 29.25 | 6.28 | 18.38 | 27.79 |
| Height, dm | 87.83 | 181.34 | 253.00 | 64.49 | 165.06 | 247.35 |
| Age, years | 22.17 | 60.11 | 96.61 | 17.92 | 50.84 | 98.92 |
| Basal area, $m^2$/ha | 7.60 | 20.3 | 30.00 | 8.57 | 18.92 | 30.23 |
| Volume, $m^3$/ha | 77.88 | 181.4 | 337.83 | 76.89 | 165.62 | 330.87 |
| Volume, pine | 5.00 | 57.06 | 145.50 | 1.27 | 42.77 | 182.43 |
| Volume, spruce | 5.54 | 92.17 | 304.67 | 0.03 | 71.35 | 284.36 |
| Volume, birch spp. | 1.00 | 23.27 | 110.80 | 1.70 | 35.88 | 107.63 |
| Volume, other br. l. | 0.00 | 8.88 | 103.00 | 0.00 | 15.62 | 67.60 |

Selection of the Training Data and Validation Data

The field observation data were ordered based on the east and north coordinates and split into training and validation data using a systematic sampling. The goal was a representative variation both in training data and validation data. Three-fourths of the observations were selected into the training data and the remaining one-fourth into the validation data. The entire data were 977 stands, of which 732 were used as the training samples and the other 245 as validation samples (Table 3).

Windstorm damages usually occur in more advanced stands: in thinning forests stands or mature forests stands. The analyses were therefore done also with the forest stands in which the volume of the growing stock on the basis of the Finnish multi-source national forest inventory (MS-NFI) was larger than 75 $m^3$/ha, a value leaving out young forests (Table 3) (see Section 2.4 and the work of Tomppo et al. [36]). The number of those stands was 912 from, of which 683 stands belonged to the training data and 229 stands to the validation data.

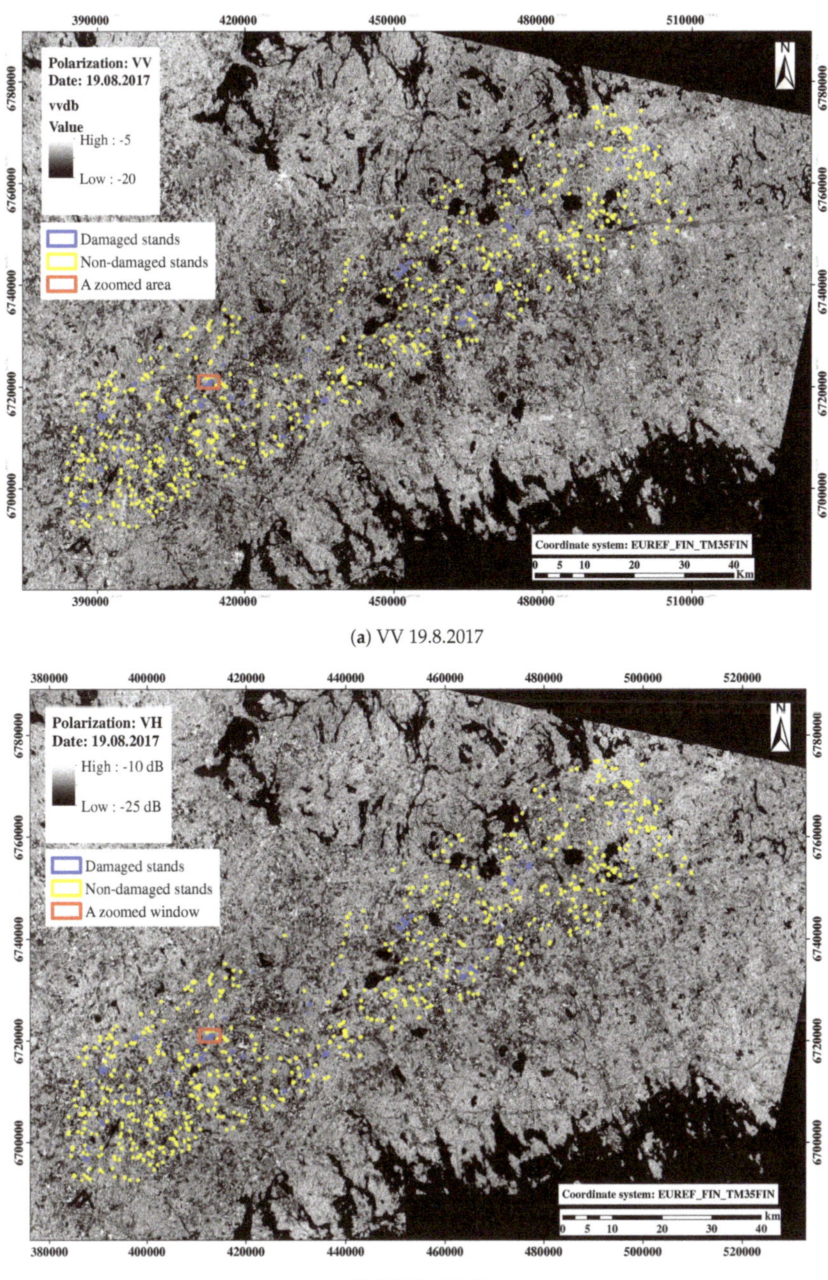

(**a**) VV 19.8.2017

(**b**) VH 19.8.2017

**Figure 2.** Damaged and non-damaged stands displayed on Sentinel-1 from 19 August 2017. VV polarization (**a**); and VH polarization (**b**) (EPSG:3067).

**Table 3.** The numbers of forest stands in the training data and validation data classified also based on the severity of the damage in the entire dataset and when the volume of growing stock is larger than 75 m$^3$/ha.

| Data | Entire Dataset | | | | Volume > 75 m$^3$/ha | | | |
|---|---|---|---|---|---|---|---|---|
| | Damage | | | | Damage | | | |
| | No | Severe | Slight | Total | No | Severe | Slight | Total |
| Training | 492 | 142 | 98 | 732 | 474 | 120 | 89 | 683 |
| Validation | 172 | 49 | 24 | 245 | 57 | 48 | 24 | 229 |
| Total | 664 | 191 | 122 | 977 | 631 | 168 | 113 | 912 |

### 2.3. Sentinel-1 SAR Data

In total, 40 Sentinel-1 GRD (ground range detected) images, Level-1 data with VV and VH polarizations, were downloaded from ESA Open Access Hub. Some of these images covered the test site only partly. In total, 27 image layers were constructed (mosaiced) from the images. The images were acquired in interferometric wide (IW) mode between 4 January and 25 September 2017. Multi-temporal data are necessary for change detection, but aggregating the data also reduces the effect of the random scattering (speckle) on the estimates and error estimates. It makes it possible to utilize the variation of the data acquisition conditions in the estimation using multifaceted information. The dates of the images are shown in Table 4.

The pixel spacing of orthorectified scenes was set to 10 m. The local digital elevation model (DEM) available from National land Survey was used (see Section 2.5). Scenes were aggregated in azimuth and range to obtain images with pixel dimensions approximately corresponding to the 10 m grid spacing. Bi-linear interpolation method was used for resampling in connection with the orthorectification. Radiometric normalization of intensity was done using a projected pixel area-based approach to minimize the effect of the topography. The scenes with a pixel size of about 13.5 m were further re-projected to the ERTS89/ETRS-TM35FIN projection (EPSG:3067) and resampled to a final pixel size of 10 m.

In total, 27 Sentinel-1 mosaics were constructed from the 40 original Sentinel-1 scenes and thus the final data stack includes 54 backscatter intensities layers of VH and VV polarizations. Examples of intensity variations before and after the damages are shown in Figures 3 and 4 with stand boundaries displayed on the intensities.

**Table 4.** The 27 Sentinel-1 data mosaics from 2017 used and their acquisition dates.

| Mosaic | Date | Mosaic | Date | Mosaic | Date |
|---|---|---|---|---|---|
| 1 | 4 January | 10 | 14 July | 19 | 20 August |
| 2 | 16 January | 11 | 15 July | 20 | 1 September |
| 3 | 28 January | 12 | 21 July | 21 | 7 September |
| 4 | 9 February | 13 | 26 July | 22 | 12 September |
| 5 | 21 February | 14 | 2 August | 23 | 13 September |
| 6 | 2 July | 15 | 7 August | 24 | 18 September |
| 7 | 3 July | 16 | 8 August | 25 | 19 September |
| 8 | 8 July | 17 | 14 August | 26 | 24 September |
| 9 | 9 July | 18 | 19 August | 27 | 25 September |

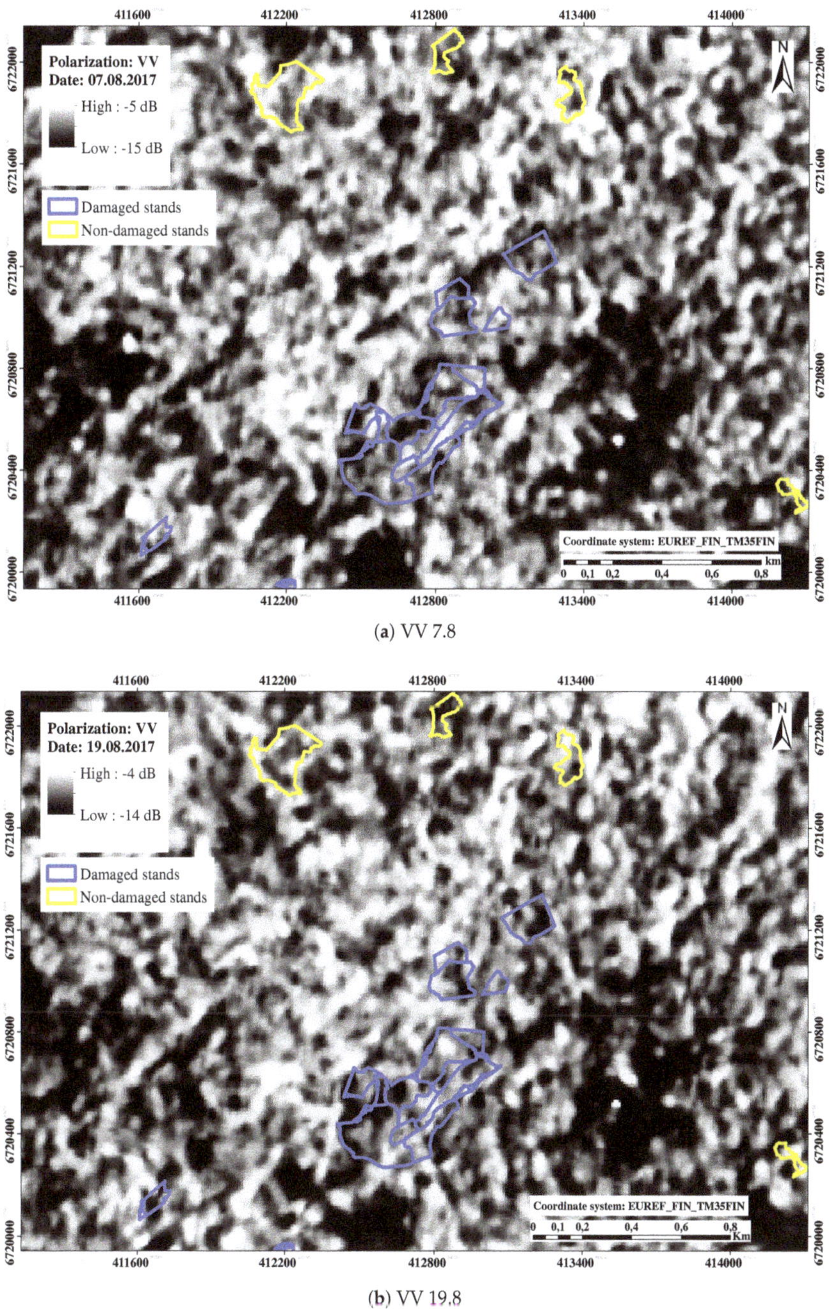



(**a**) VV 7.8



(**b**) VV 19.8

**Figure 3.** Examples of damaged and non-damaged stands displayed on Sentinel-1 scenes: (**a**) before windstorm damage, 7 August 2017; and (**b**) after windstorm damage, 19 August 2017 (**b**) (VV polarization, EPSG:3067).

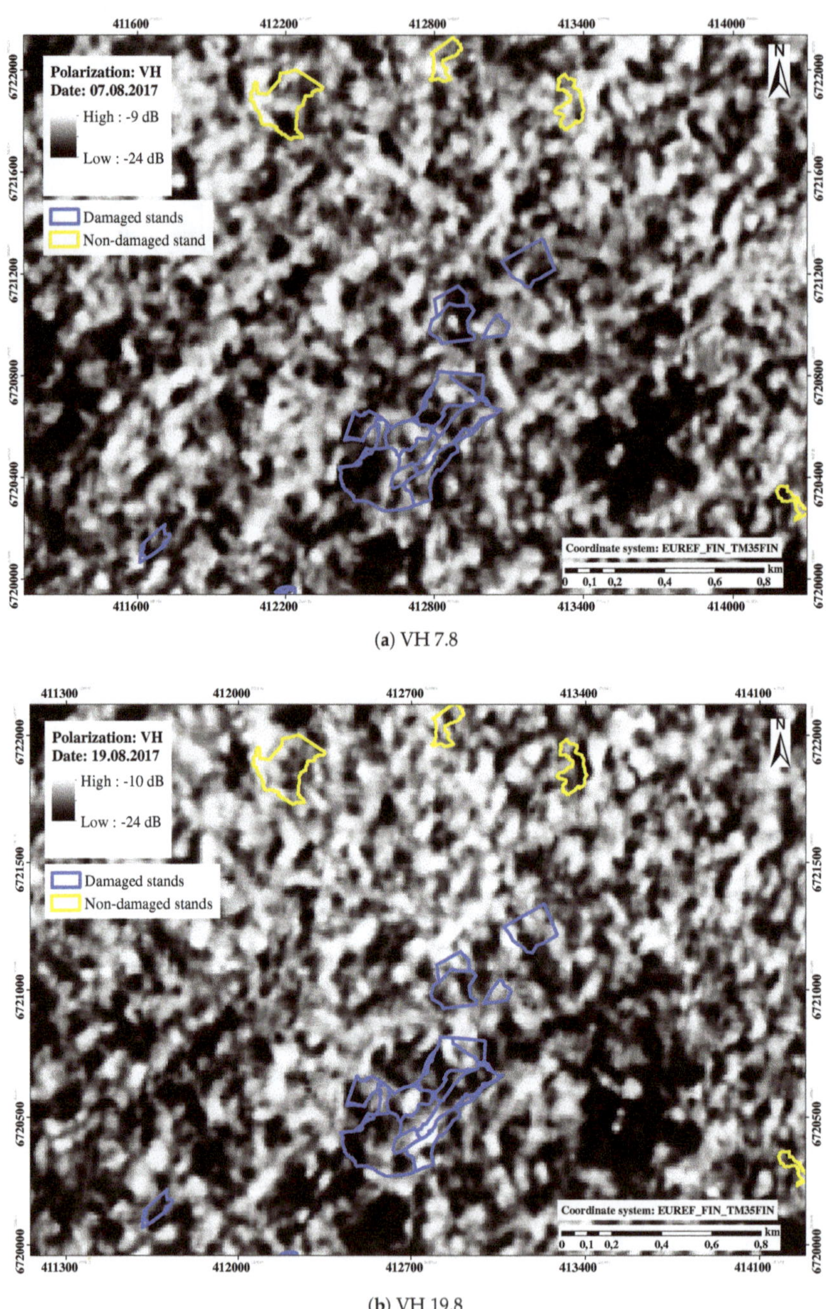


(**a**) VH 7.8

(**b**) VH 19.8

**Figure 4.** Examples of damaged and non-damaged stands displayed on Sentinel-1 scenes: (**a**) before windstorm damage, 7 August 2017; and (**b**) after windstorm damage, 19 August 2017 (**b**) (VH polarization, EPSG:3067).

*2.4. Multi-Source National Forest Inventory Data*

The raster form data from the Finnish multi-source national forest inventory (MS-NFI) were used as additional information in estimating the models to predict the windstorm damages, their severity and uncertainty [36,37]. The data have been projected to correspond to the 31 July 2017 situation and cover all forest ownership groups [38].

The following variables were used in the analyses: mean diameter of the trees, mean height of the trees, mean age of the trees and basal area of trees as well as the volume of the growing stock by tree species groups. The groups were Scots pine (*Pinus sylvestris* L.), Norway spruce (*Picea abies* Karst. L.), birch (*Betula* spp) and other broad leaved trees, mainly aspen (*Populus tremula* L.) and alder (*Alnus* spp.). The first three variables were calculated from the tree-level field measurements as weighted averages, the weight being the basal area of trees. A similar method was used when calculating stand-level averages from pixel-level estimates or measurements. A variation of volume of the growing stock is shown in Figure 5 and a zoom from a sub-area in Figure 6.

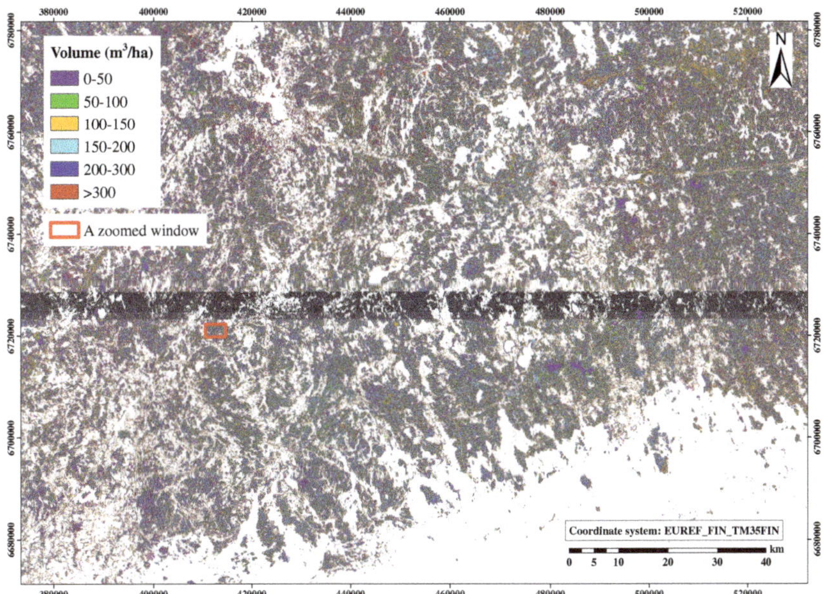


**Figure 5.** Volume of the growing stock on 31 July 2017 on the study area based on MS-NFI (EPS:3067).

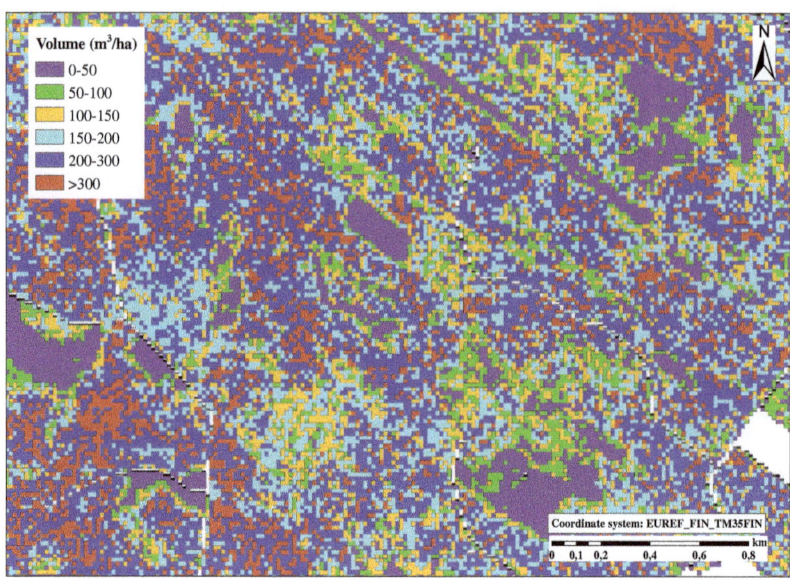

**Figure 6.** Volume of the growing stock on 31 July 2017 on the sub-area of the study area based on multi-source national forest inventory (EPS:3067).

### 2.5. Other Geo-Referenced Datasets

A digital elevation model from Land Survey of Finland was used in orthorectification and radiometric normalization of SAR images, as well as to calculate the average elevation, slope and aspect for each stand and pixel. The original pixel size is 10 m × 10 m and elevation resolution 10 cm [39].

The reason for also using these datasets as explanatory variables in the models is our assumption that the windstorm damages vary also by the aspect and steepness of the slopes of the hills.

### 3. Methods

Three different classification methods, the improved k-NN (ik-NN), multinomial logistic regression (MLR) and support vector machine classifier (SVM), were tailored for windstorm damage detection to reach a desired detection accuracy level. The observations units in the models were forest stand level averages of the Sentinel-1 intensities or other stand level quantities of the input variables. Forest patches identified and derived using a segmentation algorithm were tentatively tested as optional observation units. The reason was that the damages do not necessarily follow the stand boundaries given by the Forest Centre. The segmentation-based units were tested only with the SVM classifier. Figure 7 shows the logic and processing phases of the classification models training and windstorm-map production. The methods are described in detail in the following sections.

The main software tools used for analyses were as follows: (1) SNAP software by European Space Agency [40] was used for Sentinel-1 image pre-processing; (2) GDAL [41] was used for other image raster data pre-processing; (3) statistical computing package R [42] with own codes was used for other data handling including the field data, as well as for MLR, SVM and statistical analyses (Sections 3.4–3.6); and (4) own algorithms for ik-NN (Section 3.3), segmentation and adaptive filtering, written mainly in GNU Fortran [43] (Section 3.2).

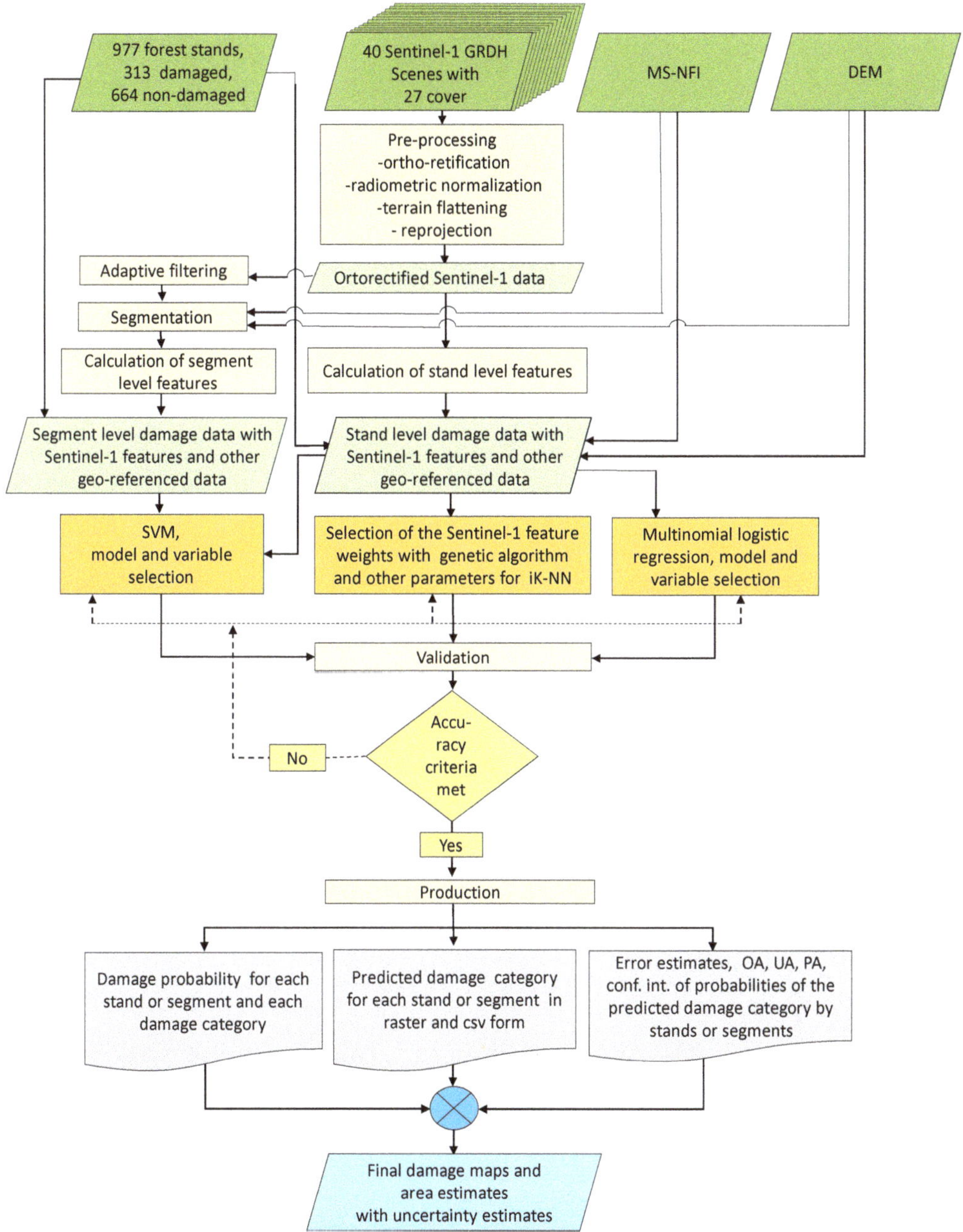



**Figure 7.** The flowchart of the windstorm detection methodology using Sentinel-1 time series: model training and damage-map production phases.

### 3.1. SAR Metrics Used in the Damage Assessment

The basic SAR metrics were stand level variables calculated from pixel level variables for both polarizations. Using stand level metrics instead of pixel level metrics reduces the the speckle effect and the radiometric variation within homogeneous stands. From the pixel level intensities ($I_{s,i}^{k,p}$), the following features were calculated for each stand $s$, for both polarizations $p$ and for each image $k$:

(a) averages,

$$10\log_{10}\overline{I_s^{k,p}} = 10\log_{10}\frac{\sum_{i=1}^{n_s} I_{s,i}^{k,p}}{n_s}, k = 1,...,27, p \in \{VV, VH\}, \tag{1}$$

where $n_s$ is the number of the pixels on stand $s$; (b) standard deviations

$$\sqrt{\sum_{i=1}^{n_s}(I_{s,i}^{k,p} - \overline{I_s^{k,p}})^2/(n_s - 1)}, k = 1,...,27, p \in \{VH, VV\}; \tag{2}$$

and (c) intensity-ratios

$$1/n_s \sum_{i=1}^{n_s} I_{s,i}^{k_1,p}/I_{s,i}^{k_2,p}, k_1 = 1...,26 \text{ and } k_2 = 2,...,27. \tag{3}$$

These predictor variables, stand-averaged intensities (Equation (1)), standard deviations (Equation (2)) and different ratio features (Equation (3)), were evaluated and, optionally, instead of the averages, also median and mode stand level intensities.

The training data and the validation data and their 'stand' boundaries concerning the damages consisted of the cutting reports rather than the forest patches really damaged by the storm. A natural question is whether it is relevant to use SAR features calculated from the areas reported to be cut because the cutting area can be larger than the damaged area. The characteristics such as median and those calculated from the quantiles SAR features inside the cutting stands were also tested, in addition to the differences of the features from SAR data before and after the damage. Median is not as sensitive to the outliers as is, e.g., the average.

### 3.2. Segmentation-Based Observation Units

The areal units provided by the Finnish Forest Centre represent forest areas to be treated by sanitary cuts due to the windstorm damage. They may be larger than the real damage area. We tested whether it was possible to delineate the entire forest area into sub-areas of which a part of the sub-areas are changed due to the damage. A common practice is to use segmentation for the delineation of the area of interest. The input for segmentation should include the information of the changes. A speckle removal before segmentation could improve the quality of segmentation.

#### 3.2.1. Adaptive Filtering

An adaptive edge-preserving filtering in further stand-level processing was tested. An own heuristic algorithm was implemented and used. It employs a set of alternative image windows (kernels) and selects the one with the smallest variance. The windows were selected inside a window of $5 \times 5$ pixels. The line and column coordinates for a set of 26 different windows are given in Table 5 in the groups of three pixels. For example, ($-2$, $-1$, 0) (2, 1, 0) means the set of pixels (line, col) ($-2$, 2), ($-2$, 1), ($-2$, 0), ($-1$, 2), ($-1$, 1), ($-1$, 0), (0, 2), (0, 1) and (0, 0), the center pixel of the window, that is the pixel for which the average value is calculated and attached. Usually 3–5 consecutive runs, depending on data, are needed to identify homogeneous forest patches.

**Table 5.** The 26 image widows (W) for adaptive filtering.

| W | $\text{lin}_{1,2,3}\ \text{col}_{1,2,3}$ | W | $\text{lin}_{1,2,3}\ \text{col}_{1,2,3}$ | W | $\text{lin}_{1,2,3}\ \text{col}_{1,2,3}$ |
|---|---|---|---|---|---|
| 1 | $(-2, -1, 0)\ (0, 0, 0)$ | 2 | $(-2, -1, 0)\ (1, 0, 0)$ | 3 | $(-2, -1, 0)\ (1, 1, 0)$ |
| 4 | $(-2, -1, 0)\ (2, 1, 0)$ | 5 | $(-1, -1, 0)\ (1, 2, 0)$ | 6 | $(-1, 0, 0)\ (2, 0, 1)$ |
| 7 | $(0, 0, 0)\ (0, 1, 2)$ | 8 | $(0, 0, 1)\ (0, 1, 2)$ | 9 | $(0, 1, 1)\ (0, 1, 2)$ |
| 10 | $(0, 1, 2)\ (0, 1, 2)$ | 11 | $(0, 1, 2)\ (0, 1, 1)$ | 12 | $(0, 1, 2)\ (0, 0, 1)$ |
| 13 | $(0, 1, 2)\ (0, 0, 0)$ | 14 | $(0, 1, 2)\ (0, 0, -1)$ | 15 | $(0, 1, 2)\ (0, -1, -1)$ |
| 16 | $(0, 1, 2)\ (0, -1, -2)$ | 17 | $(0, 1, 1)\ (0, -2, -1)$ | 18 | $(0, 0, 1)\ (-1, 0, -2)$ |
| 19 | $(0, 0, 0)\ (-2, -1, 0)$ | 20 | $(-1, 0, 0)\ (-2, -1, 0)$ | 21 | $(-1, -1, 0)\ (-2, -1, 0)$ |
| 22 | $(-2, -1, 0)\ (-2, -1, 0)$ | 23 | $(-2, -1, 0)\ (-1, -1, 0)$ | 24 | $(-2, -1, 0)\ (-1, 0, 0)$ |
| 25 | $(0, 0, 0)\ (-1, 0, 1)$ | 26 | $(-1, 0, 1)\ (0, 0, 0)$ | | |

The window with the smallest variance was selected and the average value was attached to the center, that is, the pixel in question. The windows allow the detection of narrow linear form structures in the target and preserves those structures.

### 3.2.2. Segmentation Algorithm

We decided to test whether an algorithm implemented by us could be fine-tuned just for the damage detection with SAR data. The directed tree algorithm by Narendra and Goldberg [44] was modified to use SAR data (see also [45,46]). In this simple test, we used only the difference of the intensities of two scenes, one before and one after the damage and separately VV and VH polarizations.

The algorithm utilizes an edge gradient calculated from an edge image. Any edge operator can be used to construct the edge image. Simplified steps of the segmentation algorithm are as follows. (1) Plateau points are calculated using the inverted edge image and a selected sensitivity (threshold) parameter $c$. (2) For each pixel $(i, j)$ that is not a plateau point, a parent is determined. The parent of $(i, j)$, $P(i, j)$, is the neighbor that gives the highest positive value of the inverted edge gradient among the neighbors of $(i, j)$. Ties are resolved arbitrarily. If no such neighbor exists, $(i, j)$ has no parent and is therefore a root node. The parents of each plateau points are thus determined. All points on a uni-modal plateau will belong to the same directed tree with one point on the plateau that will be called a root. (3) Once the parent of each point is determined, the points can be labeled by the directed trees (segment) they belong to. The root pixels are first labeled. (4) Once the roots have been labeled, the label of each pixel is determined by tracing of the chains to link the pixels to their root pixels (for details, see [44]). The final segmentation result does not depend on the processing order of the image.

### 3.3. ik-NN Method in Storm Damage Recognition

The well-known k-NN estimation method was tailored for and employed in storm damage recognition. The weights for the features were calculated with a genetic algorithm [36,47] and its variant for categorical variables [48]. This k-NN method is called the ik-NN method (improved k-NN) here. The advantage of the ik-NN method is the weighting of the explanatory variables based on their importance in prediction and thus smaller prediction errors compared to the ordinary k-NN method, wherefore it is called improved. Other advantages of the k-NN method are that all variables can be estimated simultaneously. It preserves thus the natural dependencies of the variables in the estimates, e.g., among stand age, mean height, mean diameter of the trees and the volume of the growing stock. It is non-parametric, and no model is needed. When the weights for training observations are collected for the calculation units, it also avoids a tendency towards the mean in the areal level estimates that is typical for many other methods (see, e.g., [36]). The k-NN estimation method became popular in forest applications when it was taken into into the operational Finnish multi-source forest inventory (e.g., [49,50]). It is very well suited for calculation of the areal level estimates.

Let us recall the main features of the ik-NN estimation with the genetic algorithm in the feature weighting. Denote the $k$ nearest feasible stands by $i_1(p), \ldots, i_k(p)$ when the distance is calculated in the feature space. The weight $w_{i,p}$ of stand $i$ to stand $p$ is defined as

$$
w_{i,p} = \frac{1}{d^t_{p_i,p}} \Bigg/ \sum_{j \in \{i_1(p),\ldots,i_k(p)\}} \frac{1}{d^t_{p_j,p}}, \quad \text{if and only if } i \in \{i_1(p),\ldots,i_k(p)\}
$$
$$
= 0 \quad \text{otherwise.}
$$
(4)

The value of $k$ was fixed to be 5 after preliminary tests using the overall accuracy as the criterion. The distance weighting power $t$ is a real number, usually $t \in [0,2]$. The value $t$=1 was used here. A small quantity, greater than zero, is added to $d$ when $d = 0$ and $i \in \{i_1(p),\ldots,i_k(p)\}$.

The distance metric $d$ employed was

$$
d^2_{p_j,p} = \sum_{l=1}^{n_f} \omega_l{}^2 (f_{l,p_j} - f_{l,p})^2,
$$
(5)

where $f_{l,p}$ is the $l$th SAR feature variable of stand $p$, $f_{l,p_j}$ is the $l$th SAR feature variable of the nearest neighbor $j$ of stand $p$, $n_f$ the number of SAR feature variables and $\omega_l$ the weight for the $l$th SAR feature variable.

The values of the elements $\omega_l$ of the weight vector $\omega$ were selected with a genetic algorithm. The details of the genetic algorithm employed are given in [47] and the modification to categorical variables in [48].

The fitness function for the categorical variables to be minimized with respect to $\omega$ vector was

$$
f[\omega, \gamma, B(\mathbf{X}) = \sum_{j=1}^{n_m} \gamma_j [1 - B_j(\mathbf{X}_j)],
$$
(6)

where $\gamma_j > 0$ is a user defined coefficient, $\mathbf{X}_j$ an error matrix, $B_j$ is the accuracy measure with response variable $j$ whose classes are to be predicted, $n_m$ is the number of response variables to be considered in the optimization procedure and $\omega$ is the weight vector to be optimized (Equation (5)). The number generations in the genetic algorithm optimization was selected to be 40 after the tests.

For categorical variables, the mode or median of the predicted classes for the nearest neighbors can be used as a prediction instead of a weighted average as is used for continuous variables. For this study, the mode gave more accurate results than the median, consistent with the earlier investigations [48]. The predicted category is the category that has the greatest sum of the weights, $\omega_{i,p}$, when summed up by classes over the k nearest neighbors. In theory, equal sums are very rare when real value weights are used; in fact, the probability is zero if rounding is not considered. In cases of equal sums for two or more classes, one class is selected randomly from among those with the greatest sum. This method was used for predicting the categorical variable obtaining the values, the value being the damage category.

### 3.4. Multinomial Logistic Regression Method

Multinomial logistic regression was tested as one optional estimation method. The probability of the damage category $k$ on stand $p$ was estimated using the model

$$
P(\text{damage category on stand } p = k | \mathbf{x}_p) = \frac{e^{\beta_k \mathbf{x}_p}}{1 + \sum_{l=1}^{L-1} e^{\beta_k \mathbf{x}_p}}, k = 1, \ldots, L-1 \text{ and}
$$
$$
= \frac{1}{1 + \sum_{l=1}^{L-1} e^{\beta_k \mathbf{x}_p}}, k = L,
$$
(7)

where $f(k, p) = \beta_k \mathbf{x}_p$ is a linear predictor function, $\beta_k$ is the vector of the regression coefficient associated with damage category $k$, $\mathbf{x}_p$ is a vector the set of the explanatory variables associated with observation (stand) $p$ and $L$ is the number of the damage categories, here 3.

*3.5. Support Vector Machine Method*

Support Vector Machine method (SVM) is a machine learning technique presently actively adopted in remote sensing [23,51–54]. SVMs are supervised learning models with associated learning algorithms that analyze data used for classification or regression. Given a set of training examples, each marked as belonging to one or the other of two categories, an SVM training algorithm builds a model that assigns new examples to one category or the other, making it a non-probabilistic binary linear classifier [55].

SVMs are based on statistical learning theory and have the aim of determining the location of decision boundaries that produce the optimal separation of classes. [56]. In the case of a two-class pattern recognition problem with linearly separable classes, the SVM selects from among the infinite number of linear decision boundaries the one that minimizes the generalization error. Thus, the selected decision boundary will be the one that leaves the greatest margin between the two classes, where the margin is defined as the sum of the distances to the hyperplane from the closest points of the two classes [56]. The margin maximization is achieved using standard quadratic programming optimization techniques. The data points that are closest to the hyperplane are used to measure the margin and are referred to as support vectors.

If the two classes are not linearly separable, the SVM tries to find the hyperplane that maximizes the margin while, at the same time, minimizing a quantity proportional to the number of misclassification errors. The trade-off between margin and misclassification error is controlled by a user-defined constant [55]. SVM can also be extended to handle nonlinear decision surfaces by projecting the input data onto a high-dimensional feature space using kernel functions [56]. Radial basis functions with accordingly selected parameters are a typical choice to serve as kernel functions [51,52]. The gamma value varied here between 0.1 and 0.005 depending on the dataset.

As SVMs are designed for binary classification, this method appears to be an ideal fit for outlier detection problems, i.e. separating damaged forest class against intact using temporal dynamics of SAR backscatter. However, for estimating severity of damage (evaluating "change magnitude"), the approach is less suitable.

*3.6. Area Estimates and Error Estimates*

We used poststratified estimators for the area and area error estimators [57], as derived and suggested by Olofsson et al. [58]. The estimators use the confusion matrix and the area estimates of the categories based on an output map, that is, the pixel level estimates of the categories. The stratified estimators of the proportion of a category $k$ is

$$\hat{p}_{.k} = \sum_{i=1}^{L} W_i \frac{n_{ik}}{n_{i.}}, \tag{8}$$

where $n_{ik}$ is the $(i, k)$ element of the confusion matrix, observed counts on the columns and classified on the rows; $n_{i.}$ is the row sum of the row $i$; $L$ is the number of the categories; and $W_i$ is the proportion of the area mapped as category $i$. The area estimate of category $k$ is

$$\hat{A}_k = A \times \hat{p}_{.k}, \tag{9}$$

where $A$ is the total area mapped.

The standard error for the poststratified estimator of the proportion of area (Equation (8)) is estimated by

$$S(\hat{p}_k) = \sqrt{\sum_{i=1}^{L} \frac{W_i \hat{p}_{ik} - \hat{p}_{ik}^2}{n_{i.} - 1}}, \tag{10}$$

where $\hat{p}_{ik} = W_i \frac{n_{ik}}{n_{i.}}$, and standard error of the the area by

$$S(\hat{A}_k) = A \times S(\hat{p}_k). \tag{11}$$

An approximate 95% confidence interval is obtained as $A_k \pm 1.96 \times S(\hat{A}_k)$.

### 3.7. Confidence Intervals of Probabilities for Individual Observations Using ik-NN

The following procedure can be used to assess the uncertainty of the prediction of the damage and non-damage of individual stands or forest areas. The k-NN estimation and its improved version ik-NN produce probabilities for the predicted category on stand $p$. These probabilities can be calculated using the weights $w_{i,p}$ (Equation (4)) as follows

$$\widetilde{prob(k)}_p = P(\text{category}(p) = k) = \sum_{i \in I_p} w_{i,p} Ind_{(cat(i)=k)}, \tag{12}$$

where $k$ is the mode category based on the largest sum of the weights $w_{i,p}$ by categories on stand $i \in I_p$ and $Ind_{cat(i)=k}$ is an indicator function of the category in stand $i$. The confidence intervals for the probabilities of the mode for the individual stands were calculated using a linear model

$$\widetilde{prob(k)}_p = a + \sum_{l=1}^{n_f} b_l f_{l,p} + c \cdot k + \varepsilon, \tag{13}$$

where $f_{l,p}$ are the SAR features (Equation (5)); $k$ is the predicted damage category, a categorical variable (factor); $a$, $b$ and $c$ are the regression coefficients to be estimated ($b$ being a vector); and $\varepsilon$ is a normally distributed random error.

The confidence intervals of the predictions were calculated in a normal way using the estimator

$$\hat{V}_f = s^2 \, x_0 \, (X'X)^{-1} \, x_0' + s^2 \tag{14}$$

for the variance for the individual prediction with a predictor vector of $x_0$, residual sum of $s^2$ and design matrix $X$ consisting of the feature vectors $f$ and predicted categories $k$.

## 4. Results

### 4.1. Selection of the Data, Variables and Methods

The capability of the different features in damage area recognition was first studied (see Section 3.1). Other variables were some traditional stand level forest characteristics (Table 1) as well as slope, aspect and altitude calculated from the digital elevation model. The stand level variables tested were mean diameter, mean height, age, basal area and the volume of growing stock as well as volumes by tree species groups (Table 1).

The methods studied in windstorm damage recognition were the improved k-NN with feature optimization based on a genetic algorithm, called ik-NN method, multinomial logistic regression (MLR) and support vector machine (SVM) (Sections 3.3–3.5). The three models were used to classify the training data into the three categories, non-damaged, severely damaged and slightly damaged. The accuracies were validated using the validation data. The explanatory variables in the models were the SAR metrics calculated from the SAR images before and after the damage as well as the other characteristics mentioned above.

Each method has several parameters to control the performance of the method. Furthermore, the combinations of the explanatory variables is large. The number of the optional methods and variable combinations is thus really large. It was not possible to carry out all possible experiments. Only the results from a few tested combinations are reported here.

### 4.2. Selection of Basic Sentinel-1 Features

Selection between the intensity variables, average, median and quantiles were done using the SVM classifier and the Sentinel-1 scenes starting on 4 January 2017 and reaching until 20 August, in total 19 image layers (see Table 4). The date of 20 August was selected close enough after the damage of 12 August. In total, three images after the damage were available. The averages of the intensities, more precisely, $\sigma^0$ (1), gave the largest OA in the validation data, 0.771 (Table 6). The averages also worked well in the cases of UA and PA. Only the results based on the averages are therefore reported here as the main results.

**Table 6.** The overall accuracies (OA) in training and validation data and user's accuracies (UA) and producer's accuracies (PA) by damage by categories in the validation data when using 19 Sentinel-1 cover (1 January 2017–20 August 2017) and the SVM classifier. A volume threshold of 75 m$^3$/ha is used in the data.

| Intensity | OA | | UA | | | PA | | |
|---|---|---|---|---|---|---|---|---|
| method | Training | Test | 1 [1] | 2 [2] | 3 [3] | 1 [1] | 2 [2] | 3 [3] |
| Average | 1.0 | 0.771 | 0.805 | 0.619 | 0.750 | 0.913 | 0.531 | 0.250 |
| Median | 1.0 | 0.747 | 0.840 | 0.511 | 0.412 | 0.913 | 0.531 | 0.250 |
| Quantiles | 1.0 | 0.763 | 0.788 | 0.647 | 0.625 | 0.930 | 0.449 | 0.208 |

[1] No damage, [2] Severe damage, [3] Slight damage.

### 4.3. Selecting a Time-Frame of Sentinel-1 Scenes

One goal of the study was to examine the minimum number of the scenes and the shortest time after the damage to achieve a feasible uncertainty level in windstorm detection. For this purpose: (a) only images until 20 August were used (19 images instead of the original 27 image layers); and (b) only images until 14 August were used, further limiting the number of images to 17. We also studied the importance of the scenes before the damage, that is, the achieved accuracies when some of the images acquired before the damage were left out of analysis. It turned out that 10 scenes gave almost as large accuracies as the entire set of the scenes from 4 January to 20 August 2017. The accuracies were thus calculated using the following numbers of the images: 27 (all scenes), 19 (all scenes until 20 August), 17 (all scenes until 14 August), the most important scenes until 20 August (10 scenes) and the most important scenes until 14 August (8 scenes) (see Section 4.4). Recall that the damage occurred on 12 August.

The importance of the Sentinel-1 scenes, acquired before the damage, was studied with all three methods. The date of these 10 scenes were: 16 January, 28 January, 9 February, 2 July, 14 July, 21 July, 26 July, 2 August, 19 August and 20 August. The dates of all available scenes in the period 4 January–20 August, in total 19 scenes, were 4 January, 16 January, 28 January, 9 February, 21 February, 2 July, 3 July, 8 July, 9 July, 14 July, 15 July, 21 July, 26 July, 2 August, 7 August, 8 August, 14 August, 19 August and 20 August (Tables 4 and 7).

When using all Sentinel-1 scenes from the period 4 January–14 August, as well as adding scenes cumulatively after 14 August, it was noticed that the overall accuracy (OA) increased only slightly. The maximum accuracy was obtained with the training and validation data with the scenes until either 12, 13 or 18 September, depending on the method. The OA in the validation data was 0.79 with SVM, 0.73 with ik-NN and 0.71 with MLR (Figure 8 and Table 7).

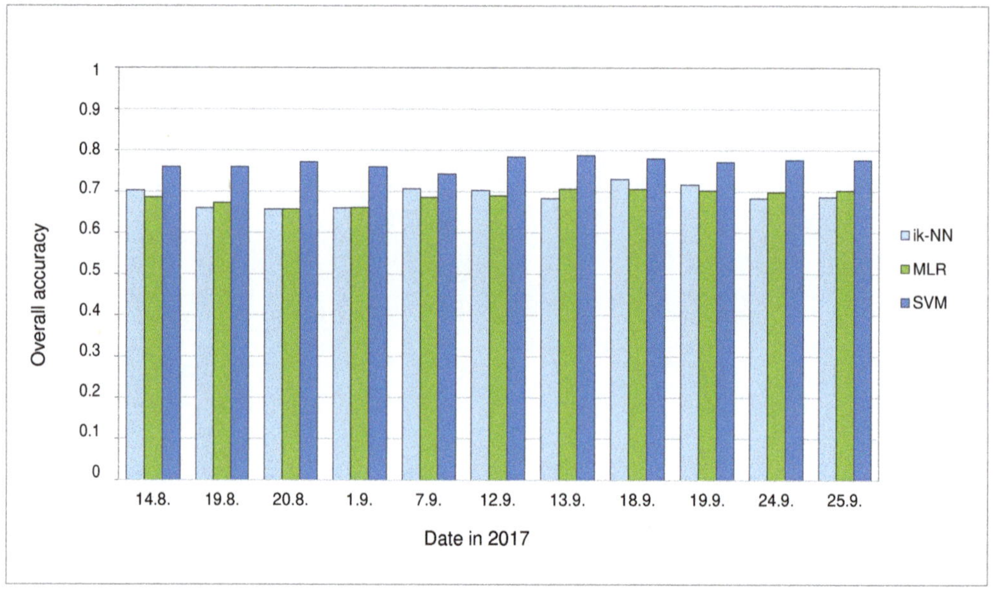


**Figure 8.** The overall accuracy with three different methods, ik-NN, MLR and SVM as a function of the acquisition date of the latest Sentinel-1 scene. The scenes were used until the date in the horizontal axis.

**Table 7.** The overall accuracy (OA), user's accuracy (UA) and producer's accuracy (PA) in the validation data using support vector machine (SVM), improved k-NN method (ik-NN) and multinomial logistic regression with four different Sentinel-1 datasets. The latest scene after the damage was from 20 August 2017, except for 27 scenes in which all Sentinel-1 scenes were used. Segmentation-based results are indicated with 'S'. A volume threshold of 75 m$^3$/ha was used for the data.

| Method and the Number of the Scenes | OA | UA by Category | | | PA by Category | | |
|---|---|---|---|---|---|---|---|
| | | 1 [1] | 2 [2] | 3 [3] | 1 [1] | 2 [2] | 3 [3] |
| SVM 8 scenes | 0.729 | 0.778 | 0.618 | 0.400 | 0.892 | 0.438 | 0.250 |
| SVM 10 scenes | 0.720 | 0.754 | 0.562 | 0.500 | 0.917 | 0.375 | 0.125 |
| SVM 17 scenes | 0.759 | 0.795 | 0.610 | 0.667 | 0.901 | 0.510 | 0.250 |
| SVM 19 scenes | 0.771 | 0.805 | 0.619 | 0.750 | 0.913 | 0.531 | 0.250 |
| SVM 27 scenes | 0.769 | 0.781 | 0.688 | 0.800 | 0.955 | 0.458 | 0.167 |
| SVM 8 scenes, S | 0.735 | 0.792 | 0.535 | 0.500 | 0.884 | 0.469 | 0.208 |
| SVM 10 scenes, S | 0.755 | 0.816 | 0.571 | 0.462 | 0.901 | 0.490 | 0.250 |
| SVM 19 scenes, S | 0.784 | 0.807 | 0.686 | 0.625 | 0.948 | 0.490 | 0.208 |
| SVM 27 scenes, S | 0.788 | 0.832 | 0.683 | 0.500 | 0.919 | 0.571 | 0.292 |
| ik-NN 8 scenes | 0.700 | 0.775 | 0.415 | 0.435 | 0.871 | 0.327 | 0.263 |
| ik-NN 10 scenes | 0.690 | 0.762 | 0.467 | 0.250 | 0.867 | 0.404 | 0.105 |
| Ik-NN 17 scenes | 0.703 | 0.781 | 0.432 | 0.434 | 0.867 | 0.365 | 0.263 |
| ik-NN 19 scenes | 0.630 | 0.747 | 0.286 | 0.133 | 0.814 | 0.308 | 0.053 |
| MLR 8 scenes | 0.703 | 0.742 | 0.435 | 0.583 | 0.917 | 0.208 | 0.292 |
| MLR 10 scenes | 0.712 | 0.763 | 0.464 | 0.533 | 0.904 | 0.271 | 0.333 |
| MLR 17 scenes | 0.686 | 0.771 | 0.407 | 0.348 | 0.879 | 0.229 | 0.333 |
| MLR 19 scenes | 0.657 | 0.794 | 0.326 | 0.296 | 0.808 | 0.286 | 0.333 |

[1] Damage, [2] Severe damage, [3] Slight damage.

## 4.4. The Accuracies with Different Methods

The overall accuracy of the classifications increases only slightly when adding the Sentinel-1 scenes after 1 or 2 scenes after the damage of 12 August (Figure 8) (see also Table 7). The latest date here is 20 August, except for 27 scenes in which all Sentinel-1 scenes were used. SVM gave the largest OA and ik-NN slightly larger than MLR. MLR is the only parametric method of the three methods tested. All methods presume the selection of the estimation parameters. The OAs are slightly larger when the backscattering-based features are calculated using segmentation-based $\sigma^0$ instead of stand boundaries-based ones (Table 7, notation 'S'). The proposed segmentation approach (tested only with SVM) increased the OA particularly when the number of the scenes is small (e.g., 10), as well as the PAs for the damage categories with 10 scenes. There are many possible combinations, therefore not all were tested here. The limited improvement can also be attributed to the fact that segmentation used the SAR data only now. A larger accuracy could be obtained when using also the inventory and other auxiliary data, similar to set of features used in the classification method.

*4.5. The Damage Map Along with Area and Area Error Estimates*

An example of the predicted damages in a map form was made using the ik-NN method. The damage was predicted only to forests in which the volume of growing stock exceeded 75 m$^3$/ha. Ten Sentinel-1 scenes with the dates of 16 January, 28 January, 9 February, 2 July, 14 July, 21 July, 26 July, 2 August, 19 August and 20 August were used.

A good practice with poststratified estimators was used in estimating the area of the damage categories (Section 3.6). The area of the classified forestry land (forest land, poorly productive forest land and unproductive land) was 799,000 ha. The severe damage was predicted for an area of 97,600 ha that corresponds to 12% of the mapped area and slight damages for an area of 76,200 ha (9.5% of the mapped forestry land area). Some individual pixels with wrong damage predictions increase the area of the damages. The large-scale damage patterns generally follow the windstorm patterns on 12 August 2017 (Figures 9 and 10). The poststratified estimates of the relative errors were 9% and 10%, respectively.

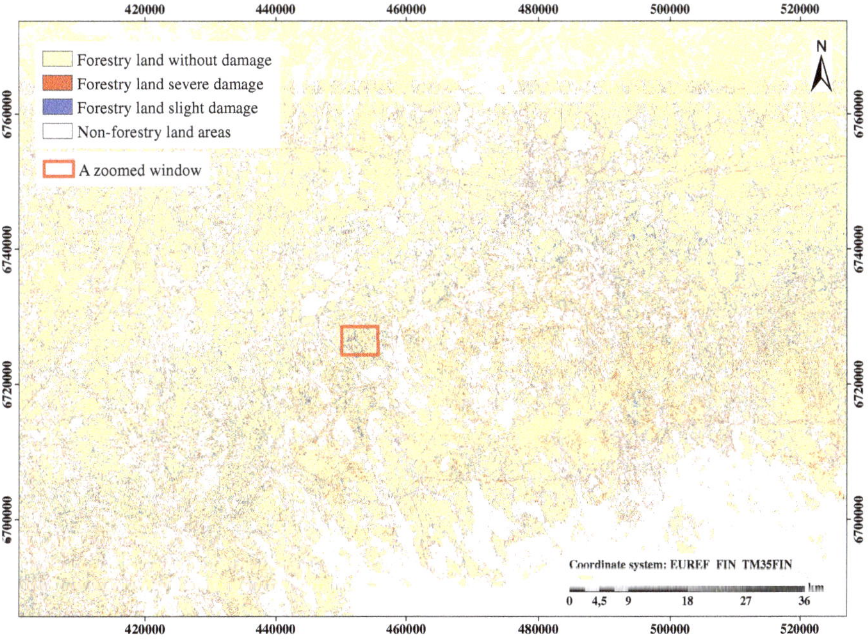


**Figure 9.** The predicted forest damages in the study area (EPGS:3067).

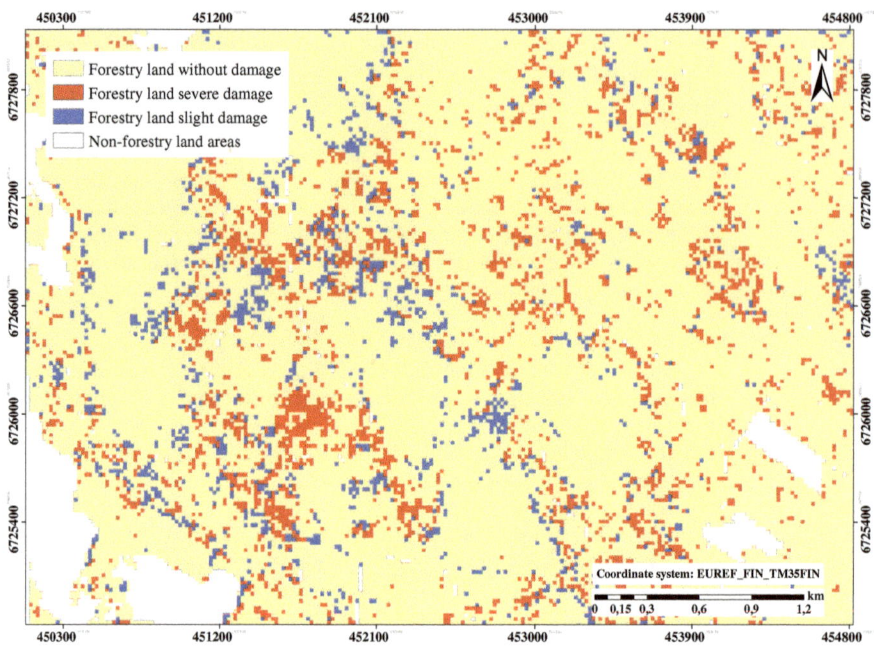

**Figure 10.** The predicted forest damages in a sub-area of the study area (EPGS:3067).

*4.6. Confidence Intervals for Individual Stand Predictions*

The probabilities of the predicted category in the validation datasets were assessed as given in Equation (12). The 95% confidence intervals of the predicted probabilities were calculated using Equation (14) and the Sentinel-1 datasets from the period 16 January–20 August, that is, in total 10 scenes (see Section 4.3). Basic statistics of the widths of the confidence intervals by the windstorm category are shown in Table 8. The median and mean of the intervals vary from 11% to 13%.

**Table 8.** Examples of the statistics of the 95% confidence intervals of the predictions of the probability of the category with the largest probability when ten Sentinel-1 scenes from the period  16 January–20 August 2017 were used. The statistics are shown by the threshold of the minimum of the stand area.

| Damage Category | Mean of Predictions | Std of Predictions | Min of Intervals | Median of Intervals | Mean of Intervals | Max of Intervals | Std of Intervals |
|---|---|---|---|---|---|---|---|
| 1 | 0.787 | 0.077 | 0.052 | 0.105 | 0.109 | 0.686 | 0.036 |
| 2 | 0.678 | 0.110 | 0.055 | 0.113 | 0.120 | 0.429 | 0.041 |
| 3 | 0.708 | 0.114 | 0.061 | 0.131 | 0.132 | 0.223 | 0.033 |

Figure 11 shows statistics of the widths of predictions intervals by the damage categories for the stands in the validation dataset together with the outliers. The statistics are the median, 25th and 75th percentiles and 1.5 times the interquartile range.

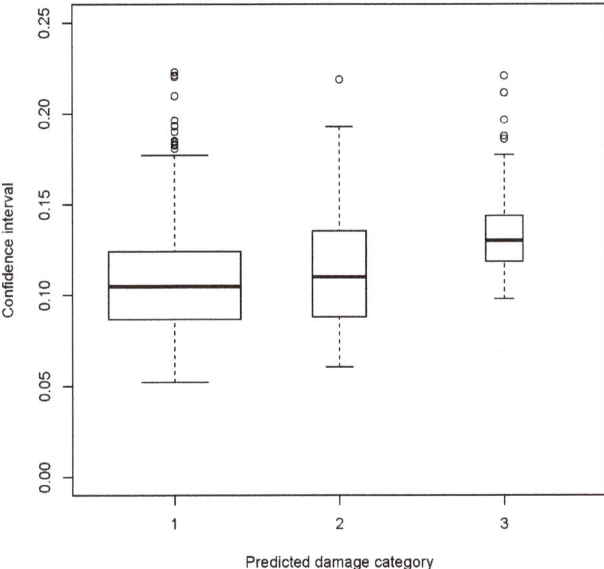

**Figure 11.** Distributions of the confidence intervals of the prediction probability of the category with the largest probability by the damage category, shown as boxplots, that is, median, 25th and 75th percentiles, 1.5 times of the interquartile ranges and individual outliers. The width of the box is proportional to the square root of the number of the observations of the damage category.

The accuracy of the individual prediction, that is, the confidence interval of the probability of the predicted category, also depends on the area of an individual damaged stand. The confidence interval becomes generally narrower when the area of a stand increases up to about 1–2 ha (Figure 12).

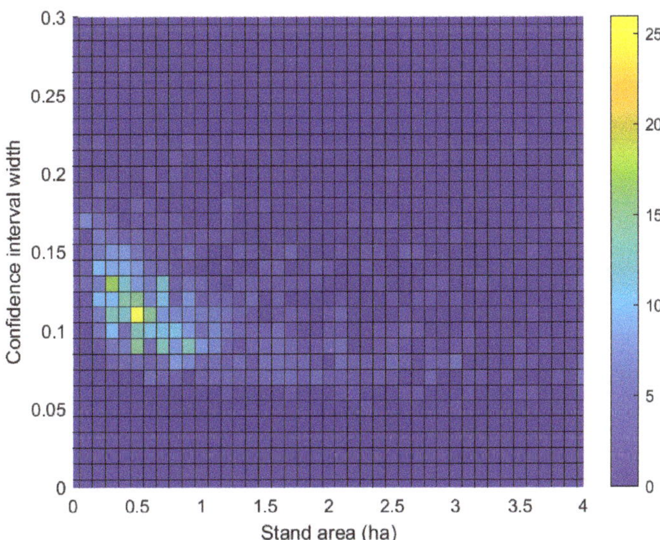

**Figure 12.** The width of the confidence interval of the probability of an individual prediction as a function of the area of the damaged stand.

*4.7. The Accuracies of the Methods without a Separate Training Data*

The use of specific training data is not necessarily possible in operational applications due to a need for very rapid or almost real-time localization of the damages. A final goal should be development of a method and a model that predict the locations of the damages without new training data and using only the existing information before the damage and SAR data (1–6 scenes) after the damage.

This possibility was tentatively tested in a simple way by selecting a set of the Sentinel-1 scenes and another distinct set for prediction. The cover numbers of the scenes for modeling were 2, 3, 4, 6, 7, 11, 12, 15, 16, 17, 18, 24, 25 and 26, and in prediction 1, 5, 8, 9, 10, 13, 14, 19, 20, 21, 22, 23 and 27 (see Table 4). The proportion of correctly predicted stands was 0.73. This simple test could already indicate if it is possible to develop an unsupervised damage prediction method.

## 5. Discussions and Future Work

The windstorm analyzed in this study took place during the Finnish summer, on 12 August 2017, wherefore the most important Sentinel-1 scenes were also from the summer. Earlier and other ongoing studies have revealed that late autumn or early winter is the best season for forest parameter estimation from SAR images in the boreal region (e.g., [59,60]). The results of this study show that the summer SAR images could also be applicable in forest change detection caused by a windstorm damage. One challenge when developing an operational damage monitoring method is that the windstorms occur all around the year, including during winter in Scandinavia and other boreal region. The weather conditions and the temperature can also rapidly change from above to below 0 °C. Under frozen conditions, a normalized radar cross-section decreases (e.g., [61]). Generally, freeze–thaw environmental transitions affect the classification methods and accuracies if they happen just after or during the damage and if a rapid assessment is needed.

This study provides some alternative methods to be developed further to be part of an operational windstorm damage monitoring system. Three different classification algorithm were tested to classify the forest observations on the three categories, non-damaged, severely damaged and slightly damaged forest patches. In total, 27 Sentinel-1 image covers were acquired and originally used, 16 before the damages and 11 after damages, in addition to the field observations and other geo-referenced data. The explanatory variables were derived from the intensities of the two polarizations VV and VH of the Sentinel-1 images and from the quantities of the other geo-referenced data. Two alternative analysis units were tested: (1) the forest stands or forest areas to be cut due to the windstorm damage; and (2) the forest patches constructed using a segmentation algorithm. The main analyses were carried out with the Alternative 1. The data were split to training data and validation for assessing the uncertainties of the results of the different methods. A statistical method was developed to construct the confidence intervals of the probabilities of the estimated damage categories. One goal was to find the minimum number of the images after the damages for a rapid operational monitoring method.

Using calculation units that are derived with a segmentation algorithm, that is, the units that are homogeneous with respect to backscattering coefficient, slightly increased the overall accuracies (OAs), and in some cases also the user's accuracies (UAs). In some cases, the UAs and producer's accuracies (PAs) were smaller than with the given boundaries. This may imply that the given boundaries followed the damaged areas or the effect of variation in the image conditions on the results is significant or the segmentation-based approach needs further development.

Although some windstorm damage studies have been published so far with space-borne SAR data, quantitative uncertainty assessments are generally lacking. Many comparisons with accuracy figures are thus not possible. Our results are competitive with, e.g., those of Thiele et al. [21]. Our study showed that the damages could be identified even a few days after the damage, which is quite unique. On the other hand, we should keep in mind that windstorm damages vary by the areal extent of the damaged forest patches and also by severity. Furthermore, forest structures and imaging conditions vary wherefore uncertainty quantities are not necessarily comparable.

Preliminary tests showed that it could also be possible to develop an unsupervised method for windstorm damage monitoring, that is, to detect the changes without a specific training data.

Detection of the windstorm damage is a demanding task. Severity of the damages changes within the area of one storm, even in a relatively small area, e.g., the one in this study, 100 km × 100 km, and depends on many factors, e.g., the structure of the growing stock, the soil properties, the terrain elevation variation and the small scale spatial variation of growing stock. Forests next to an open area or a young forest are more vulnerable to damages than the forests surrounded by mature forests. Furthermore, in addition to the changes in the growing stock, many other factors affect the changes of backscatter and also in a short time interval, e.g., changes in the moisture of the tree canopies and soil. The possibility to frequently acquire SAR data is thus important.

It should be recognized that further work is needed for a near-real-time operational monitoring system.

In the continuation work, the accuracy of the estimates will be improved by further method development and additionally using interferometric SAR data as well as meteorological data. The use of other geo-referenced data, such as land-use data, forest age and soil data and forest data from the surrounding areas, may improve the classification accuracy because the windstorm damages occur often on the borders of open areas, newly constructed roads and power-lines as well as next to young forests or forest regeneration areas.

Potential of interferometric SAR coherence, possibly combined with backscatter intensity information, should be studied using Sentinel-1 multitemporal imagery, even though temporal decorrelation can limit its utility [62]. Further, time series of bistatic TanDEM-X scenes can be used for mapping forest change, due to high sensitivity to the vertical structure of the forests [63,64]. For the latter, the limiting factor is data availability over large areas with small latency.

Coming and existing satellite missions and constellations with frequent and tailored data acquisition increase the availability of data. It is also important that data providers adopt a systemic data acquisition strategy similar to Sentinel-1 and ALOS/ALOS-2 missions in connection with the hazard monitoring, particularly windstorm detection. A background mission with at least seasonal global coverage can be suitable.

## 6. Conclusions

The methods to localize the forest damages caused by windstorms using space-borne SAR data were developed and possibilities to an operative system investigated. Multitemporal Sentinel-1 time series were used.

Support vector machine (SVM) gave the largest overall accuracies among the three methods tested, improved k-NN (ik-NN), multiple logistic regression (MLR) and SVM. The proportion of correctly classified stands (OA) in a separate validation data was 79%, and 75% if only one Sentinel-1 scene after the damage was used. The user's accuracy (UA) for severe damages was 62%, and 75% for slight damages. The producer's accuracies (PAs) were somewhat lower. The accuracy of 75% was achieved using only one Sentinel-1 scene after the damage, here two days after the damage, in addition to the data before the damage.

Using segmentation-based calculation units only slightly increased the OA, implying that this approach may presume further work. Most likely, not only SAR data, but also inventory and other auxiliary data should be used in the segmentation methodology.

The study indicates that the damages could be localized using only one Sentinel-1 scene after the damage implying a time-lag of potential satellite SAR-based assessment method would be just a few days after the damage. This gives promises that a SAR-based near-real-time semi-automatic operative system to monitor windstorm damages is feasible.

**Author Contributions:** Conceptualization, E.T., J.P., O.A., H.H. and G.R.; methodology, E.T.; validation, E.T. and H.H.; formal analysis, E.T.; investigation, E.T., O.A. and J.P.; data curation, G.R., H.H. and E.T.; writing—original draft preparation, E.T.; writing—review and editing, E.T., O.A. and J.P.; visualization, E.T., G.R and O.A.; and project administration, E.T. and J.P. All authors have read and agreed to the published version of the manuscript.

**Funding:** The study was funded by Business Finland as a part of the project conducted by the Finnish Forest Centre, grant number 4480/31/2018. O.A. was supported by Business Finland MULTICO project under grant 6501/31/2019. All authors were also supported by their Institutes.

**Institutional Review Board Statement:** Not applicable.

**Informed Consent Statement:** Not applicable.

**Data Availability Statement:** Sentinel-1 satellite data are available from ESA via Copernicus Open Access Hub at no cost. Derived data supporting the findings of this study are available from the corresponding author E.T. on request.

**Acknowledgments:** The damage and non-damage stand data with stand boundaries were provided by the Finnish Forest Centre. Four anonymous reviewers provided constructive comments that improved the text. We thank all institutions and individuals who have contributed to the article.

**Conflicts of Interest:** The authors declare no conflict of interest.

## References


1. Gregow, H.; Laaksonen, A.; Alper, M. Increasing large scale windstorm damage in Western, Central and Northern European forests, 1951–2010. *Sci. Rep.* **2017**, *7*, 46397. [CrossRef] [PubMed]
2. Usbeck, T.; Wohlgemuth, T.; Dobbertin, M.; Pfister, C.; Bürgi, A.; Rebetez, M. Increasing storm damage to forests in Switzerland from 1858 to 2007. *Agric. For. Meteorol.* **2010**, *150*, 47–55. [CrossRef]
3. Fransson, J.E.; Walter, F.; Blennow, K.; Gustavsson, A.; Ulander, L.M. Detection of storm-damaged forested areas using airborne CARABAS-II VHF SAR image data. *IEEE Trans. Geosci. Remote. Sens.* **2002**, *40*, 2170–2175. [CrossRef]
4. Fransson, J.E.S.; Pantze, A.; Eriksson, L.E.B.; Soja, M.J.; Santoro, M. Mapping of wind-thrown forests using satellite SAR images. In Proceedings of the 2010 IEEE International Geoscience and Remote Sensing Symposium, Honolulu, HI, USA, 25–30 July 2010; pp. 1242–1245. [CrossRef]
5. Gardiner, B.; Blennow, K.; Carnus, J.M.; Fleischer, P.; Ingemarson, F.; Landmann, G.; Lindner, M.; Marzano, M.; Nicoll, B.; Orazio, C.; et al. *Destructive Storms in European Forests: Past and Forthcoming Impacts*; Final Report to European Commission—DG Environment; European Forest Institue, Atlantic European Regional Office—EFIAtalantic: Joensuu, Finland, 2013; pp. 1–138.
6. Eriksson, L.E.B.; Fransson, J.E.S.; Soja, M.J.; Santoro, M. Backscatter signatures of wind-thrown forest in satellite SAR images. In Proceedings of the 2012 IEEE International Geoscience and Remote Sensing Symposium, Munich, Germany, 22–27 July 2012; pp. 6435–6438. [CrossRef]
7. Frolking, S.; Palace, M.W.; Clark, D.B.; Chambers, J.Q.; Shugart, H.H.; Hurtt, G.C. Forest disturbance and recovery: A general review in the context of spaceborne remote sensing of impacts on aboveground biomass and canopy structure. *J. Geophys. Res. Biogeosci.* **2009**, *114*, [CrossRef]
8. Rüetschi, M.; Small, D.; Waser, L.T. Rapid Detection of Windthrows Using Sentinel-1 C-Band SAR Data. *Remote Sens.* **2019**, *11*, 115. [CrossRef]
9. Torres, R.; Snoeij, P.; Geudtner, D.; Bibby, D.; Davidson, M.; Attema, E.; Potin, P.; Rommen, B.; Floury, N.; Brown, M.; et al. GMES Sentinel-1 mission. *Remote Sens. Environ.* **2012**, *120*, 9–24. [CrossRef]
10. GFOI. *Integrating Remote-Sensing and Ground-Based Observations for Estimation of Emissions and Removals of Greenhouse Gases in Forests: Methods and Guidance from the Global Forest Observations Initiative*; Group on Earth Observations: Geneva, Switzerland, 2014.
11. Antropov, O.; Rauste, Y.; Häme, T.; Praks, J. Polarimetric ALOS PALSAR Time Series in Mapping Biomass of Boreal Forests. *Remote Sens.* **2017**, *9*, 999. [CrossRef]
12. Carreiras, J.; Quegan, S.; Le Toan, T.; Ho Tong Minh, D.; Saatchi, S.; Carvalhais, N.; Reichstein, M.; Scipal, K. Coverage of high biomass forests by the ESA BIOMASS mission under defense restrictions. *Remote Sens. Environ.* **2017**, *196*, 154–162.
13. Kellogg, K.; Hoffman, P.; Standley, S.; Shaffer, S.; Rosen, P.; Edelstein, W.; Dunn, C.; Baker, C.; Barela, P.; Shen, Y.; et al. NASA-ISRO Synthetic Aperture Radar (NISAR) Mission. In Proceedings of the 2020 IEEE Aerospace Conference, Big Sky, MT, USA, 7–14 March 2020; pp. 1–21. [CrossRef]
14. Antropov, O.; Praks, J.; Kauppinen, M.; Laurila, P.; Ignatenko, V.; Modrzewski, R. Assessment of Operational Microsatellite Based SAR for Earth Observation Applications. In Proceedings of the 2018 2nd URSI Atlantic Radio Science Meeting, AT-RASC 2018, Meloneras, Spain, 28 May–1 June 2018. [CrossRef]
15. Ignatenko, V.; Laurila, P.; Radius, A.; Lamentowski, L.; Antropov, O.; Muff, D. ICEYE Microsatellite SAR Constellation Status Update: Evaluation of first commercial imaging modes. In Proceedings of the 2020 IEEE International Geoscience and Remote Sensing Symposium, 26 September–2 October 2020.

16. Moreira, A.; Krieger, G.; Hajnsek, I.; Papathanassiou, K.; Younis, M.; Lopez-Dekker, P.; Huber, S.; Villano, M.; Pardini, M.; Eineder, M.; et al. Tandem-L: A Highly Innovative Bistatic SAR Mission for Global Observation of Dynamic Processes on the Earth's Surface. *IEEE Geosci. Remote Sens. Mag.* **2015**, *3*, 8–23. [CrossRef]

17. Torres, R.; Davidson, M. Overview of Copernicus SAR Space Component and its Evolution. In Proceedings of the IGARSS 2019—2019 IEEE International Geoscience and Remote Sensing Symposium, Yokohama, Japan, 28 July–2 August 2019; pp. 5381–5384. [CrossRef]

18. Green, R.M. The sensitivity of SAR backscatter to forest windthrow gaps. *Int. J. Remote Sens.* **1998**, *19*, 2419–2425. [CrossRef]

19. Dwyer, N.; Pasquali, P.; Holecz, F.; Arino, O. Mapping forest damage caused by the 1999 lothar storm in Jura (France), using SAR interferometry. *Earth Obs. Q.* **2000**, *12*, 28–29.

20. Ulander, L.M.H.; Smith, G.; Eriksson, L.; Folkesson, K.; Fransson, J.E.S.; Gustavsson, A.; Hallberg, B.; Joyce, S.; Magnusson, M.; Olsson, H.; et al. Mapping of wind-thrown forests in Southern Sweden using space- and airborne SAR. In Proceedings of the 2005 IEEE International Geoscience and Remote Sensing Symposium, 2005, IGARSS'05, Seoul, Korea, 29 July 2005; Volume 5, pp. 3619–3622.

21. Thiele, A.; Boldt, M.; Hinz, S. Automated detection of storm damage in forest areas by analyzing TerraSAR-X data. In Proceedings of the 2012 IEEE International Geoscience and Remote Sensing Symposium, Munich, Germany, 22–27 July 2012; pp. 1672–1675. [CrossRef]

22. Tanase, M.A.; Aponte, C.; Mermoz, S.; Bouvet, A.; Toan, T.L.; Heurich, M. Detection of windthrows and insect outbreaks by L-band SAR: A case study in the Bavarian Forest National Park. *Remote Sens. Environ.* **2018**, *209*, 700–711. [CrossRef]

23. Tomppo, E.; Antropov, O.; Praks, J. Boreal Forest Snow Damage Mapping Using Multi-Temporal Sentinel-1 Data. *Remote Sens.* **2019**, *11*, 384. [CrossRef]

24. Rauste, Y.; Antropov, O.; Hame, T.; Ramminger, G.; Gomez, S.; Seifert, F.M. Mapping Selective Logging in Tropical Forest with Space-Borne SAR Data. In Proceedings of the ESA Living Planet Symposium, Edinburgh, UK, 9–13 September 2013; Volume 722, p. 168.

25. Antropov, O.; Rauste, Y.; Seifert, F.M.; Häme, T. Selective logging of tropical forests observed using L- and C-band SAR satellite data. In Proceedings of the 2015 IEEE International Geoscience and Remote Sensing Symposium (IGARSS), Milan, Italy, 26–31 July 2015; pp. 3870–3873. [CrossRef]

26. Hoekman, D.; Kooij, B.; Quiñones, M.; Vellekoop, S.; Carolita, I.; Budhiman, S.; Arief, R.; Roswintiarti, O. Wide-Area Near-Real-Time Monitoring of Tropical Forest Degradation and Deforestation Using Sentinel-1. *Remote Sens.* **2020**, *12*, 3263. [CrossRef]

27. Hethcoat, M.; Carreiras, J.; Edwards, D.; Bryant, R.; Quegan, S. Detecting tropical selective logging with SAR data requires a time series approach. *bioRxiv* **2020**. [CrossRef]

28. Antropov, O.; Rauste, Y.; Praks, J.; Seifert, F.; Häme, T. Mapping forest disturbance due to selective logging in the Congo Basin with RADARSAT-2 time series. 2020, submitted.

29. Fransson, J.E.S.; Magnusson, M.; Olsson, H.; Eriksson, L.E.B.; Sandberg, G.; Smith-Jonforsen, G.; Ulander, L.M.H. Detection of forest changes using ALOS PALSAR satellite images. In Proceedings of the 2007 IEEE International Geoscience and Remote Sensing Symposium, Barcelona, Spain, 23–28 July 2007; pp. 2330–2333. [CrossRef]

30. Santoro, M.; Fransson, J.; Eriksson, L.; Ulander, L. Clear-Cut Detection in Swedish Boreal Forest Using Multi-Temporal ALOS PALSAR Backscatter Data. *IEEE J. Sel. Top. Appl. Earth Obs. Remote Sens.* **2010**, *3*, 618–631. [CrossRef]

31. Rauste, Y.; Antropov, O.; Mutanen, T.; Häme, T. On Clear-Cut Mapping with Time-Series of Sentinel-1 Data in Boreal Forest. In Proceedings of the Living Planet Symposium 2016, Prague, Czech Republic, 9–13 May 2016; Volume SP-740.

32. Olesk, A.; Voormansik, K.; Põhjala, M.; Noorma, M. Forest change detection from Sentinel-1 and ALOS-2 satellite images. In Proceedings of the 2015 IEEE 5th Asia-Pacific Conference on Synthetic Aperture Radar (APSAR), Singapore, 1–4 September 2015; pp. 522–527. [CrossRef]

33. Tanase, M.; Ismail, I.; Lowell, K.; Karyanto, O.; Santoro, M. Detecting and quantifying forest change: The potential of existing C- and X-band radar datasets. *PLoS ONE* **2015**, *10*, e0131079. [CrossRef]

34. Pantze, A.; Santoro, M.; Fransson, J.E.S. Change detection of boreal forest using bi-temporal ALOS PALSAR backscatter data. *Remote Sens. Environ.* **2014**, *155*, 120–128. [CrossRef]

35. Finnish Meterological Institute. Windspeed Observations. Available online: https://en.ilmatieteenlaitos.fi/ (accessed on 5 January 2021).

36. Tomppo, E.; Haakana, M.; Katila, M.; Peräsaari, J. *Multi-Source National Forest Inventory—Methods and Applications. Managing Forest Ecosystems*; Springer: Dordrecht, The Netherlands, 2008.

37. Mäkisara, K.; Katila, M.; Peräsaari, J. *The Multi-Source National Forest Inventory of Finland—Methods and Results 2017*; Publications of the Natural Resources Institute Finland: Helsinki, Finland, 2019; Volume 8.

38. LUKE Natural Resources Institute Finland. *MS-NFI Products from Year 2017*; Data Download Service; LUKE Natural Resources Institute Finland: Helsinki, Finland, 2019. Available online: https://kartta.luke.fi/index-en.html (accessed on 1 December 2020).

39. Land Survey Finland. Elevation Model 10 m. In *Maps and Spatial Data*; Land Survey Finland: Helsinki, Finland, 2018. Available online: https://www.maanmittauslaitos.fi/en/maps-and-spatial-data/expert-users/product-descriptions/elevation-model-10-m (accessed on 1 December 2020).

40. European Space Agency (ESA). *SNAP-ESA Sentinel Application Platform v7.0.0*; European Space Agency (ESA): Vienna, Austria, 2020.

41. GDAL/OGR Contributors. *GDAL/OGR Geospatial Data Abstraction Software Library*. Open Source Geospatial Foundation. Available online: https://gdal.org/ (accessed on 1 December 2020).
42. R Core Team. *R: A Language and Environment for Statistical Computing*; R Foundation for Statistical Computing: Vienna, Austria, 2020.
43. GNU Fortran Project. GNU Compiler Collection. Available online: https://gcc.gnu.org/ (accessed on 1 December 2020).
44. Narendra, P.M.; Goldberg, M. Image Segmentation with Directed Trees. *IEEE Trans. Pattern Anal. Mach. Intell.* **1980**, *PAMI-2*, 185–191. [CrossRef]
45. Parmes, E. Segmentation of Spot and Landsat satellite imagery. *Photogramm. J. Finl.* **1992**, *13*, 52–58.
46. Tomppo, E. An application of a segmentation method to the forest stand delineation and estimation of stand variates from satellite images. In *Image Analyses, Proceedings of the 5th Scandinavian Conference on Image Analysis, Stockholm, Sweden, June 1987*; Springer: Berlin, Germany, 1987; Volume 1, pp. 253–260.
47. Tomppo, E.; Halme, M. Using coarse scale forest variables as ancillary information and weighting of variables in k-NN estimation: A genetic algorithm approach. *Remote Sens. Environ.* **2004**, *92*, 1–20. [CrossRef]
48. Tomppo, E.O.; Gagliano, C.; Natale, F.D.; Katila, M.; McRoberts, R.E. Predicting categorical forest variables using an improved k-Nearest Neighbour estimator and Landsat imagery. *Remote Sens. Environ.* **2009**, *113*, 500–517. [CrossRef]
49. Chirici, G.; Mura, M.; McInerney, D.; Py, N.; Tomppo, E.O.; Waser, L.T.; Travaglini, D.; McRoberts, R.E. A meta-analysis and review of the literature on the k-Nearest Neighbors technique for forestry applications that use remotely sensed data. *Remote Sens. Environ.* **2016**, *176*, 282–294. [CrossRef]
50. Tomppo, E.; Olsson, H.; Ståhl, G.; Nilsson, M.; Hagner, O.; Katila, M. Combining national forest inventory field plots and remote sensing data for forest databases. *Remote Sens. Environ.* **2008**, *112*, 1982–1999.
51. Huang, C.; Davis, L.S.; Townshend, J.R.G. An assessment of support vector machines for land cover classification. *Int. J. Remote Sens.* **2002**, *23*, 725–749. [CrossRef]
52. Camps-Valls, G.; Gomez-Chova, L.; Munoz-Mari, J.; Rojo-Alvarez, J.L.; Martinez-Ramon, M. Kernel-Based Framework for Multitemporal and Multisource Remote Sensing Data Classification and Change Detection. *IEEE Trans. Geosci. Remote Sens.* **2008**, *46*, 1822–1835. [CrossRef]
53. Pal, M.; Mather, P.M. Support vector machines for classification in remote sensing. *Int. J. Remote Sens.* **2005**, *26*, 1007–1011. [CrossRef]
54. Kuo, B.; Ho, H.; Li, C.; Hung, C.; Taur, J. A Kernel-Based Feature Selection Method for SVM With RBF Kernel for Hyperspectral Image Classification. *IEEE J. Sel. Top. Appl. Earth Obs. Remote Sens.* **2014**, *7*, 317–326. [CrossRef]
55. Cortes, C.; Vapnik, V. Support-Vector Networks. *Mach. Learn.* **1995**, *20*, 273–297. [CrossRef]
56. Vapnik, V.N. *The Nature of Statistical Learning Theory*; Springer: New York, NY, USA, 1995.
57. Cochran, W.G. *Sampling Techniques*, 3rd ed.; Wiley: New York, NY, USA, 1977.
58. Olofsson, P.; Foody, G.M.; Herold, M.; Stehman, S.; Woodcock, C.; Wulder, M. Good practices for estimating area and assessing accuracy of land change. *Remote Sens. Environ.* **2014**, *148*, 42–47. [CrossRef]
59. Kurvonen, L.; Pulliainen, J.; Hallikainen, M. Retrieval of biomass in boreal forests from multitemporal ERS-1 and JERS-1 SAR images. *IEEE Trans. Geosci. Remote Sens.* **1999**, *37*, 198–205. [CrossRef]
60. Ge, S.; Tomppo, E.; Rauste, Y.; Praks, J.; McRoberts, R.; Gu, H.; Su, W.; Antropov, O. Hypertemporal Sentinel-1 data in boreal forest growing stock prediction. 2020, submitted.
61. Liudmila, N.; Zakharova, A.I.Z. Seasonal Variations of Vegetation Cover Scattering Properties. *Radiolocation* **2019**, *11*, 49–56.
62. Jacob, A.W.; Vicente-Guijalba, F.; Lopez-Martinez, C.; Lopez-Sanchez, J.M.; Litzinger, M.; Kristen, H.; Mestre-Quereda, A.; Ziółkowski, D.; Lavalle, M.; Notarnicola, C.; et al. Sentinel-1 InSAR Coherence for Land Cover Mapping: A Comparison of Multiple Feature-Based Classifiers. *IEEE J. Sel. Top. Appl. Earth Obs. Remote Sens.* **2020**, *13*, 535–552. [CrossRef]
63. Kugler, F.; Schulze, D.; Hajnsek, I.; Pretzsch, H.; Papathanassiou, K.P. TanDEM-X Pol-InSAR Performance for Forest Height Estimation. *IEEE Trans. Geosci. Remote Sens.* **2014**, *52*, 6404–6422. [CrossRef]
64. Olesk, A.; Praks, J.; Antropov, O.; Zalite, K.; Arumäe, T.; Voormansik, K. Interferometric SAR Coherence Models for Characterization of Hemiboreal Forests Using TanDEM-X Data. *Remote Sens.* **2016**, *8*, 700. [CrossRef]

*Article*

# Forest Drought Response Index (ForDRI): A New Combined Model to Monitor Forest Drought in the Eastern United States

Tsegaye Tadesse [1,*], David Y. Hollinger [2], Yared A. Bayissa [1,3], Mark Svoboda [1], Brian Fuchs [1], Beichen Zhang [1], Getachew Demissie [1], Brian D. Wardlow [4], Gil Bohrer [5], Kenneth L. Clark [6], Ankur R. Desai [7], Lianhong Gu [8], Asko Noormets [9], Kimberly A. Novick [10] and Andrew D. Richardson [11,12]

[1] National Drought Mitigation Center, University of Nebraska-Lincoln, Lincoln, NE 68583-0749, USA; ybayissa@fiu.edu (Y.A.B.); msvoboda2@unl.edu (M.S.); bfuchs2@unl.edu (B.F.); beichen@huskers.unl.edu (B.Z.); gdemisse2@unl.edu (G.D.)

[2] USDA Forest Service, Northern Research Station, Durham, NH 03824, USA; david.hollinger@usda.gov

[3] Department of Earth and Environment, Florida International University, Miami, FL 33199, USA

[4] Center for Advanced Land Management Information Technologies, School of Natural Resources University of Nebraska-Lincoln, Lincoln, NE 68583-0749, USA; bwardlow2@unl.edu

[5] Department of Civil, Environmental & Geodetic Engineering, The Ohio State University, Columbus, OH 43210, USA; bohrer.17@osu.edu

[6] USDA Forest Service, Northern Research Station, New Lisbon, NJ 08064, USA; kennethclark@fs.fed.us

[7] Department. of Atmospheric and Oceanic Sciences, University of Wisconsin-Madison, Madison, WI 53706, USA; desai@aos.wisc.edu

[8] Climate Change Science Institute & Environmental Sciences Division, Oak Ridge National Laboratory, Oak Ridge, TN 37830, USA; lianhong-gu@ornl.gov

[9] Department of Ecology and Conservation Biology, Texas A&M University, College Station, TX 77843, USA; noormets@tamu.edu

[10] O'Neill School of Public and Environmental Affairs, Indiana University, Bloomington, IN 47405, USA; knovick@indiana.edu

[11] School of Informatics, Computing & Cyber Systems, Northern Arizona University, Flagstaff, AZ 86011, USA; Andrew.Richardson@nau.edu

[12] Center for Ecosystem Science and Society, Northern Arizona University, Flagstaff, AZ 86011, USA

* Correspondence: ttadesse2@unl.edu; Tel.: +1-402-472-3383

Received: 24 September 2020; Accepted: 28 October 2020; Published: 3 November 2020

**Abstract:** Monitoring drought impacts in forest ecosystems is a complex process because forest ecosystems are composed of different species with heterogeneous structural compositions. Even though forest drought status is a key control on the carbon cycle, very few indices exist to monitor and predict forest drought stress. The Forest Drought Indicator (ForDRI) is a new monitoring tool developed by the National Drought Mitigation Center (NDMC) to identify forest drought stress. ForDRI integrates 12 types of data, including satellite, climate, evaporative demand, ground water, and soil moisture, into a single hybrid index to estimate tree stress. The model uses Principal Component Analysis (PCA) to determine the contribution of each input variable based on its covariance in the historical records (2003–2017). A 15-year time series of 780 ForDRI maps at a weekly interval were produced. The ForDRI values at a 12.5km spatial resolution were compared with normalized weekly Bowen ratio data, a biophysically based indicator of stress, from nine AmeriFlux sites. There were strong and significant correlations between Bowen ratio data and ForDRI at sites that had experienced intense drought. In addition, tree ring annual increment data at eight sites in four eastern U.S. national parks were compared with ForDRI values at the corresponding sites. The correlation between ForDRI and tree ring increments at the selected eight sites during the summer season ranged between 0.46 and 0.75. Generally, the correlation between the ForDRI and normalized Bowen ratio or tree ring increment are reasonably good and indicate the usefulness of the ForDRI model for estimating drought stress and providing decision support on forest drought management.

**Keywords:** forest monitoring; drought; time series satellite data; Bowen ratio; carbon flux

---

## 1. Introduction

Drought has multiple direct and indirect impacts on forests. High evaporative demand from high temperature and low humidity, in isolation and especially when combined with limited soil moisture supply, can induce plant water stress [1]. To reduce water loss and prevent the development of excessively low water potentials, water-stressed plants typically close stomata. This can lead to carbon stress, reduced growth, and greater susceptibility to insects and disease. Under extreme conditions, drought stress can result in depleted carbon reserves, loss of hydraulic function, and mortality [2].

Monitoring drought impacts in forest ecosystems is complex because forest ecosystems are composed of different species with heterogeneous structural compositions [3]. In a given ecosystem, different tree species can also physiologically respond differently to drought stress [4–7]. Extreme and intense droughts can induce irreversible growth and vigor loss, resulting in tree death [8–11], which may lead to accumulation of fuel in a forest and increased fire danger. Drought conditions can also result in decreases in forest Live Fuel Moisture Content (LFMC), the mass of water contained within living vegetation in relation to the dry mass. LFMC has been identified as a factor relating to fire ignition, behavior, and severity [12].

Traditionally, climate-based drought indices such as the Keetch–Byram Drought Index (KBDI) or satellite-based indices have separately been used to monitor drought. In this study, these two complementary approaches for monitoring forest drought have been combined.

The climate-based drought monitoring approach [13–19] characterizes forest drought status indirectly (i.e., the climate-based drought indices indicate moisture deficit, but do not show levels of physiological stress or damage in forests). Thus, most climate-based indices (e.g., KBDI) infer impacts of the climatic parameters (e.g., rainfall and temperature) rather than measuring changes in forest condition directly.

The remote sensing drought monitoring approach [20–25] enables a near-real-time monitoring of forest condition at high resolution. However, an approach based on reflectance values also has limits [22]. Remote sensing data alone are insufficient to demonstrate that drought is the causal agent of a particular change in reflectance values. In addition to this, remote sensing of forest drought and its interpretations can be complex due to technical aspects of the sensor technologies and interconnections of underlying ecological processes in forested areas [26]. There is a need for an integrated wide-area drought monitoring system that focuses specifically on drought stress in forested ecosystems [27]. Most forests in the eastern U.S. are composed of different tree species with different levels of drought tolerance, which makes monitoring forest drought challenging when solely using climatic or satellite data. Both climate- and satellite-based data are powerful sources for depicting and describing drought conditions and impacts. However, they could be more powerful when merged together.

In this study, we present the Forest Drought Response Index (ForDRI), a new 'hybrid' drought tool developed to monitor and assess forest drought conditions through the integration of satellite-based observations of vegetation conditions, evapotranspiration (ET) estimates from satellite, root-zone soil moisture (satellite-estimated or modeled), climate-based drought indices, and biophysical characteristics of the environment. These input variables are combined based on their contribution (weight) determined by covariance (principal component analysis) to provide the ForDRI value at each grid point. The overarching goal of ForDRI research is to develop an integrated forest drought monitoring tool for decision makers using satellite, climate, and biophysical parameters to address the need and challenges of forest drought monitoring on the order of weeks to months and years.

The main objective of this study is to identify and monitor drought impacts on forests to help users, such as the U.S. Drought Monitor (USDM) map authors (drought experts), in characterizing drought across forested areas of the U.S. The USDM relies on experts to synthesize the climate- and satellite-based data and work with local observers to interpret the information. The USDM also incorporates ground-truthing and information about how drought is affecting people, via a network of more than 450 observers across the country, including state climatologists, National Weather Service staff, extension agents, and hydrologists [28]. The USDM map is used by policy makers (e.g., legislative and congressional offices, state forestry commissions); water supply managers; irrigation associations; agricultural trade organizations; public land managers; federal, state and local fire managers; others in the U.S. [28,29]. However, trees are likely to be more resilient to water limitation than annual plants due to their generally deeper roots and woody stems, thus the need for a forest-specific product.

## 2. Materials and Methods

### 2.1. Study Area Forest Group Type Coverage by Climate Region

The study area for the experimental analysis is the eastern U.S. (Figure 1). The predominant land cover in this region is forest cover consisting of more than 80 tree species [30]. Figure 1 shows the study area and the forest type groups based on the national forest type dataset produced by the United States Forest Service (USFS) Forest Inventory and Analysis (FIA) program and the Remote Sensing Applications Center (RSAC). The national forest type dataset was created by modeling several biophysical layers, including digital elevation models (DEM), Moderate Resolution Spectroradiometer (MODIS) multi-date composites, vegetation indices and vegetation continuous fields, class summaries from the 1992 National Land Cover Dataset (NLCD), various ecologic zones, and summarized PRISM climate data [31]. The national forest types were classified into 28 groups to portray broad distribution patterns of forest cover in the U.S. [30,32]. Our study area includes 10 major forest type groups (Figure 1).

The study area was divided into Central, East North-Central, Northeastern, and Southeastern forest/climate regions [31] (Figure 1). The Oak/Hickory (38%), Loblolly/Shortleaf Pine (17%), and Maple/Beech/Birch (15%) forest type groups dominate the study area. However, each forest/climatic region has its own characteristic and areal extent of forest group types as well as species composition. For example, the highest percent area coverage of the Northeast Climate Region is the Maple/Beech/Birch Group (about 66%), followed by the Oak/Hickory Group (about 22%). In contrast, the highest percent cover of the forest group in the Southeast Climate Region is the Oak/Hickory Group (about 40%), followed by Loblolly/Shortleaf Pine Group (about 28%). Detailed information and the data for the U.S. is available at USDA's Forest Service website at [31].

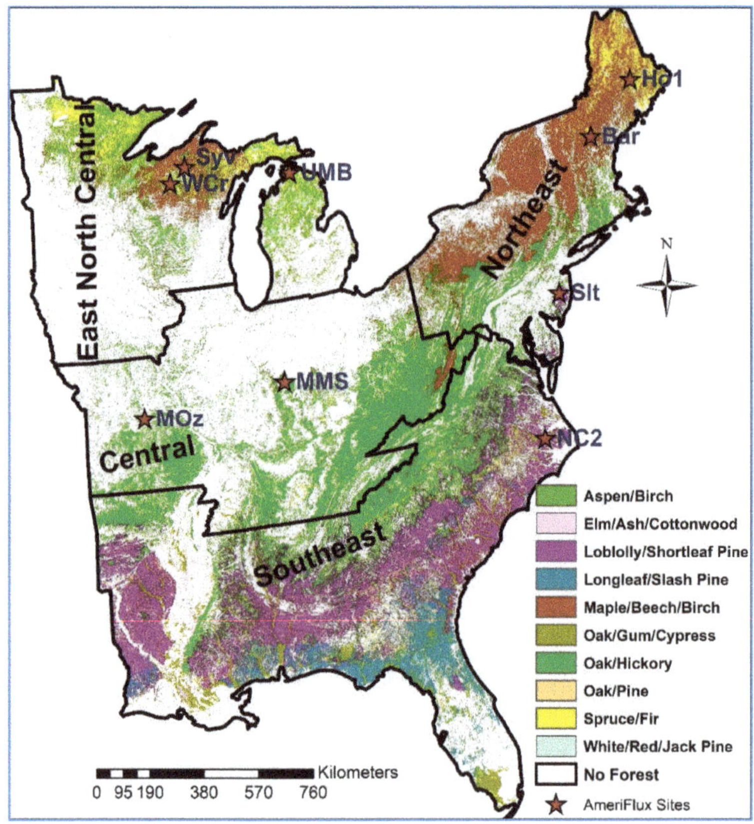



**Figure 1.** Study area for the Forest Drought Response Index (ForDRI). The map shows the ten major forest group types in the study area based on the USFS National Forest Type dataset [31].

## 2.2. Data Used in ForDRI Model Development

The ForDRI model includes water cycle variables (precipitation, temperature, evaporation, soil moisture, and vapor pressure deficit) that influence short- and long-term drought conditions that are combined with satellite-derived vegetation reflectances (NDVI) that characterize forest condition. The input variables are described in additional detail below.

### 2.2.1. MODIS-based Normalized Difference Vegetation Index (NDVI)

The normalized difference vegetation index (NDVI) information at 250-meter (m) spatial resolution is based on Moderate Resolution Imaging Spectroradiometer (MODIS) data acquired by the National Aeronautics and Space Administration's (NASA) Earth Observing System (EOS). The MODIS-based 7-day data from 2003–2017 were acquired from USGS [33] and resampled to a 1 km grid, and each dataset was standardized (Z-score) to be consistent with the other input variables. The Z-score was calculated using the formula: weekly-observed value minus weekly-mean value divided by the standard deviation. This dataset can be accessed at USGS Earth Explorer [33].

### 2.2.2. Standardized Precipitation Index (SPI)

The SPI was calculated to quantify the precipitation anomaly for three specified time-scales (the previous 12, 24, and 60 months) based on the long-term precipitation record over that specific time interval [12,34]. Since the SPI values are calculated by fitting the long-term record of precipitation over a specific time step to a probability distribution to standardize the values, we have used these three SPI values to represent different time scales of the rainfall conditions that would affect forest health. The three SPIs are selected to represent the long-term precipitation impact (from 1 year to 5 years) on tree stress. The rainfall data used to generate the time series of SPI were obtained from Applied Climate Information System (ACIS) meteorological station data across the study region. We used the available daily long-term record of each station to generate SPI at 12-, 24-, and 60-month aggregate periods and interpolated using the inverse-distance weight (IDW) method to produce 1 km resolution SPI maps.

### 2.2.3. Standardized Precipitation Evapotranspiration Index (SPEI)

Unlike the SPI, which depends only on rainfall, the SPEI is designed to take into account both precipitation and temperature. The time series of the SPEI were generated based on daily rainfall and temperature data acquired from ACIS meteorological station data. The SPEI were generated at 24- and 60-month aggregate periods and interpolated (using the IDW method) to 12.5 km spatial resolution. With the temperature input, potential evapotranspiration (PET) is calculated and a historical time series of the simple water balance (precipitation—PET) is used in determining drought. Thus, the SPEI captures the main impact of increased temperatures on water demand [35]. Two specified time periods of SPEI historical records (i.e., the previous 24 and 60 months) that represent the temperature impact on water demand (rainfall) were used in building the ForDRI model to monitor forest drought response.

### 2.2.4. Evaporative Demand Drought Index (EDDI)

The EDDI indicates the anomalous condition of the atmospheric evaporative demand (also known as "the thirst of the atmosphere") for a given location and across a time period of interest [36,37]. The EDDI is expressed as atmospheric evaporative demand (Eo) anomalies. The Eo is calculated using the Penman–Monteith FAO56 reference evapotranspiration formulation driven by temperature, humidity, wind speed, and incoming solar radiation from the North American Land Data Assimilation System datasets (NLDAS-2). EDDI is multi-scalar (i.e., captures drying dynamics that themselves operate at different timescales). We combined 12-month aggregated EDDI values with the other variables to monitor evaporative demand during forest drought.

### 2.2.5. Ground Water Storage (GWS)

GWS anomalies are calculated from Gravity Recovery and Climate Experiment (GRACE) observations [38,39]. Data from the Global Land Data Assimilation System (GLDAS), including Terrestrial Water Storage (TWS), Root Zone Soil Moisture (RZSM) at 1-meter depth, and Snow Water Equivalence (SWE), were used to convert GRACE observations into a series of GWS anomalies (i.e., GWS = TWS – RZSM – SWE). NASA provided the data (2003 to 2017) at 12.5 km resolution for the U.S. The ground water product at 1-meter depth represents deeper soil condition that can be accessed by deeper-rooted tree species. The global GRACE data (2003–2020) are also available online from the NASA GSFC Hydrological Sciences Laboratory at the NASA GESDISC data archive [40].

### 2.2.6. Palmer Drought Severity Index (PDSI) and Palmer Z Index (PZI)

The PDSI has been one of the most widely used climate-based drought indices in the U.S. [41]. The PDSI is calculated based on a simple supply-and-demand model of a water balance equation using historical records of precipitation and temperature as well as available water-holding capacity of the soil at a given location [14,15]. The PDSI is calculated using a combination of current and previous climatic conditions. In contrast to the PDSI, the Palmer Z-Index (PZI) corresponds to monthly drought conditions with no memory of previous monthly deficits or surpluses [14,15]. Thus, in this study, we have used the PDSI and 60-month PZI historical datasets to represent the short- and long-term drought conditions that impact forests.

### 2.2.7. Noah Soil Moisture (SM)

The Noah soil moisture dataset used in this study is produced using a land surface model that forms a component of the GLDAS [42–44]. The Noah soil moisture represents shallow soil depth conditions that can be accessed by shallow-rooted species. Compared to other NLDAS-2 soil moisture products (e.g., VIC), Noah soil moisture shows the best performance in simulating shallow depth soil moisture [45]. The Noah model uses a four-layered soil description with a 10-cm thick top layer and takes into account the fractions of sand and clay. Soil moisture dynamics of the top layer are governed by infiltration, surface and sub-surface runoff, gradient diffusion, gravity, and evapotranspiration [46]. The model was forced by a combination of NOAA/GLDAS atmospheric analysis fields, spatially and temporally disaggregated NOAA Climate Prediction Center Merged Analysis of Precipitation (CMAP) fields, and observation-based downward shortwave and longwave radiation fields derived using a method of the Air Force Weather Agency's agricultural meteorological system [40]. The historical data (available since 2000) has a 25 km resolution (resampled to 1 km for combining with other model inputs). This dataset is also available as NOAA's NLDAS Drought Monitor Soil Moisture [47].

### 2.2.8. Vapor Pressure Deficit

The vapor pressure deficit (VPD) represents the difference between the actual water vapor pressure in the air and the vapor pressure when the air at that temperature is saturated [48]. The VPD is one of the critical variables that control photosynthesis and water use efficiency of plants. The photosynthetic rates in leaves and canopies is inversely proportional to the atmospheric VPD [49]. Thus, it is important for forest ecosystem structure and function [50]. Average daily VPD data using the PRISM model at 4 km resolution were retrieved from the PRISM Climate Group, Oregon State University [51–53].

### 2.2.9. National Forest Groups and Types

The national forest types and forest groups geospatial dataset (1 km spatial resolution) used in this study was created by the USFS Forest Inventory and Analysis (FIA) program and the Remote Sensing Applications Center (RSAC) to show the extent, distribution, and forest type composition of the nation's forests. The dataset was created by modeling forest type from FIA plot data as a function of more than 100 geospatially continuous predictor layers. This process results in a view of forest type distribution in greater detail than is possible with the FIA plot data alone. The ForDRI model is calculated for forest areas based on this national forest type dataset acquired from the USDA Forest Service [31].

### 2.2.10. Bowen Ratio Data to Compare with ForDRI at Nine AmeriFlux Sites

Plant water stress is typically characterized by the water potential ($\psi$), which represents the tension in the water column and reflects the balance of free energy between atmospheric demand and soil water supply, modulated by leaf stomatal and hydraulic resistances [54]. Plant water potentials can be measured via pressure chamber [55] or in-situ hygrometer [56], but long-term observations across a range of sites are not available.

Energy balance considerations mean that net radiation ($R_n$) at a forest site is balanced by the energy of sensible heat ($H$) and evaporation ($\lambda E$) plus any change in storage ($S$):

$$R_n = H + \lambda E + S \tag{1}$$

The change in energy storage associated with ground or canopy heat flux is small compared to the other terms and averages over time to zero. Evaporation from a canopy in energy terms (W m$^{-2}$) is described by the Penman–Monteith equation [57]:

$$\lambda E = \frac{\Delta(R_n - S) + c_p \rho \delta_e g_a}{\Delta + \gamma(1 + g_a/g_s)} \tag{2}$$

where $R_n$ and $S$ are as above, $\delta_e$ is the vapor pressure deficit, $g_a$ and $g_s$ are boundary layer and stomatal conductances to water vapor, and $\Delta$, $c_p$, $\rho$, and $\gamma$ are thermodynamic parameters that are weak functions of temperature. The stomatal conductance, $g_s$, plays an important but not unique role in limiting $\lambda E$. If $\lambda E$ is reduced because of a change in conductance, then $H$ (and to a lesser extent, $S$) will rise because of energy balance considerations. This makes the Bowen ratio ($\beta$), defined as $H/\lambda E$, especially sensitive to changes in conductance. Stomatal conductance in turn is a function of incoming solar radiation, the vapor pressure deficit ($\delta_e$), temperature, (internal) $CO_2$ concentration, and water stress ($\psi$) [58,59]. During drought, higher temperatures and increased vapor pressure deficits can combine with soil water stress to severely limit $g_s$ and increase $H$ at the expense of $\lambda E$.

We assessed forest water stress by using sensible ($H$) and latent heat ($\lambda E$, evaporation) flux data measured at AmeriFlux network sites to calculate an integrated Bowen ratio ($\beta_i$):

$$\beta_i = \frac{\sum H}{\sum \lambda E} \tag{3}$$

Measured 30-minute $H$ and $\lambda E$ fluxes (no gap filled values) were summed over 7 days, when both were >50 W m$^{-2}$. The 7-day integration period was chosen to match the weekly timestep of ForDRI. The Bowen ratio in this context thus represents the weekly partitioning of the site net radiation. When a tree canopy is fully developed and water is passing through foliage on its way to the atmosphere, $\lambda E$ is generally greater than $H$, and $\beta < 1$. When water stress occurs, evaporation from a canopy is limited by stomatal closure and potentially, reduced foliage area. These limits result in more of the incoming energy being converted to sensible heat causing the Bowen ratio to increase.

Sensible ($H$) and latent ($\lambda E$) heat data from nine forested AmeriFlux eddy covariance sites in the eastern U.S. were used to calculate the weekly Bowen ratio ($\beta_i$). These represented all forested sites in the eastern U.S. with 12 or more years of $H$ and $\lambda E$ data (Table 1). Because there are seasonal as well as site-to-site variations in $\beta$, we normalized weekly, log-transformed integrated Bowen ratios ($\log_{10}\beta_i$) by their standard deviations ($\sigma$) from the weekly mean over the full record ($\overline{\log_{10}\beta_i}$, where a negative value indicates a higher than average $\beta_i$ and more drought-stressed conditions). This normalization (also referred to as a Z-score) occurs for each week of the growing season and helps highlight unusual behavior in the weekly $\beta_i$ values consistently across sites.

$$Z\text{--}score(\beta_i) = \frac{\overline{\log_{10}\beta_i} - \log_{10}\beta_i}{\sigma} \tag{4}$$

This normalization also means that in a long enough record there is a direct, probabilistic interpretation of values based on characteristics of the normal distribution (e.g., a 2$\sigma$ result has a single-tailed probability of ~2.27%, a 3$\sigma$ result has $p < 0.2\%$, etc.).

**Table 1.** Characteristics of AmeriFlux sites used in this analysis. DBF indicates deciduous broadleaf forest, ENF indicates evergreen needle-leaf forest, and MF indicates mixed forest. In the Köppen climate classification, Cfa = humid subtropical climate, Dfa = hot-summer humid continental climate, and Dfb = warm-summer humid continental climate.

| Site Id | Name | Lat. | Long. | Elev. (m) | Veg. | Climate | MAT (°C) | MAP (mm) | Start | End | Site ref. |
|---|---|---|---|---|---|---|---|---|---|---|---|
| US-Bar | Bartlett Experimental Forest | 44.065 | −71.288 | 272 | DBF | Dfb | 5.61 | 1246 | 2004 | 2017 | [60] |
| US-Ho1 | Howland Forest (main tower) | 45.204 | −68.740 | 60 | ENF | Dfb | 5.27 | 1070 | 1996 | 2018 | [61] |
| US-MMS | Morgan Monroe State Forest | 39.323 | −86.413 | 275 | DBF | Cfa | 10.85 | 1032 | 1999 | 2020 | [6] |
| US-MOz | Missouri Ozark Site | 38.744 | −92.2 | 219 | DBF | Cfa | 12.11 | 986 | 2004 | 2017 | [62] |
| US-NC2 | NC Loblolly Plantation | 35.803 | −76.669 | 5 | ENF | Cfa | 16.6 | 1320 | 2005 | 2019 | [63] |
| US-Slt | Silas Little Forest | 39.914 | −74.596 | 30 | DBF | Dfa | 11.04 | 1138 | 2005 | 2017 | [64,65] |
| US-Syv | Sylvania Wilderness Area | 46.242 | −89.348 | 540 | MF | Dfb | 3.81 | 826 | 2001 | 2020 | [66] |
| US-UMB | Univ. of Mich. Biological Station | 45.560 | −84.714 | 234 | DBF | Dfb | 5.83 | 803 | 2000 | 2019 | [67] |
| US-WCr | Willow Creek | 45.806 | −90.080 | 520 | DBF | Dfb | 4.02 | 787 | 1998 | 2020 | [68] |

### 2.2.11. Tree Ring Data for Evaluation

Landsat-based Phenology and Tree Ring data (1984–2013) for Eastern US Forests were acquired for evaluation of ForDRI from the Oak Ridge National Laboratory Distributed Active Archive Center (ORNL DAAC). This dataset provides a 30-year record of forest phenology and annual tree ring data at several selected forested sites in the eastern U.S. [69]. These selected sites are located in four national parks—Harpers Ferry National Historical Park (HAFE), Prince William Forest Park (PRWI), Great Smoky Mountains National Park (GRSM), and Catoctin Mountain Park (CATO). Details of sample preparation and dendrochronological analyses are presented in [70]. We have used eight sites from the four parks (two sites per park) to compare tree ring increment with ForDRI values during the summer season (June to September).

### 2.3. Methods

#### 2.3.1. ForDRI Model Development

To develop a proof-of-concept ForDRI model, we used 12 selected variables (described above) that contribute to forest drought (Figure 2). The input variables include MODIS-based NDVI, GRACE-based ground water storage, three SPI timescales (12-, 24-, and 60-month SPI), two SPEIs (12- and 24-month SPEI), PDSI, PZI, Noah soil moisture, 12-month EDDI, and VPD. To determine the contribution of each input variables objectively, we have used the principal component analysis (PCA) method. Using the PCA approach, the weights of each variable are determined based on their historical data and the covariance of all input variables (Figure 2; Step 2). This approach helps in limiting the redundant information that could influence the combined ForDRI model. In addition, the PCA-based process is automatic (using scripts), which allows us to produce a separate model for each week in a year using several inputs at a higher spatial resolution [71,72]. Figure 2 shows the method and steps to develop the ForDRI model and the process of producing maps for the forest regions. The process includes six steps from data processing to product dissemination. As shown in Figure 2, the main steps are (i) standardizing all the input variables to be consistent in combining them, (ii) determining the percent contribution (weight) of each input variable based on the covariance of the variables using the PCA method, (iii) multiplying each input variable with the proportion (weight) determined by PCA, (iv) adding the weighted input variables and standardizing the output using long historical records and generating the ForDRI maps for the selected forest regions (we generated the ForDRI maps for the four forest regions of the eastern U.S. to demonstrate and evaluate ForDRI, Figure 3), (v) evaluating the ForDRI maps using tree ring increment (dendrology) data and forest flux data (i.e., Bowen Ratio), and (vi) disseminating the ForDRI maps. In this study, Steps 1 to 5 (Figure 2) were used. For Step 4, the historical data were used in hindsight as "Near-real Time data" to demonstrate the ForDRI model's capability. The last step (i.e., Step 6, Internet portal for data access and distribution) is the potential delivery of the operational ForDRI maps to the public in the future. An operational ForDRI model is planned to be developed after expanding the model to the western U.S. and evaluating the final national ForDRI model for the continental U.S. (CONUS).

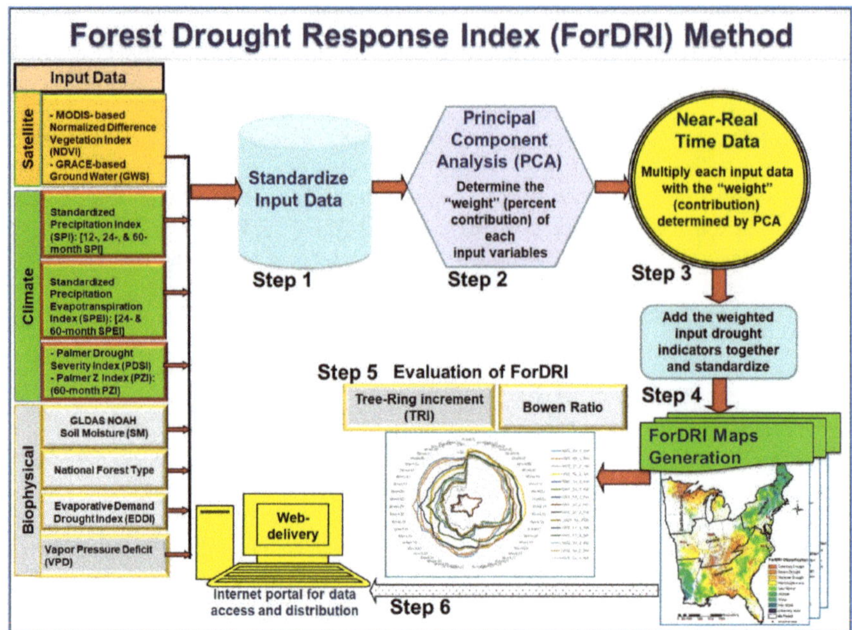


**Figure 2.** Conceptual method and steps to develop the Forest Drought Response Index (ForDRI).

## 2.3.2. Evaluation Method/Approaches for ForDRI (Both Qualitative and Quantitative Approaches)

For this initial version of the ForDRI model, we used climatic, satellite, and biophysical data for the eastern U.S. (east of 100°W) at a weekly timestep. Forests in the eastern U.S. experience occasional drought, but they tend to be shorter and more random than the seasonal droughts of the West [73]. To evaluate the ForDRI model, we needed long-term measures of forest physiological stress from a variety of sites sufficient to capture a number of significant drought events. As described earlier, our approach was to evaluate ForDRI by assessing forest water stress using sensible and latent heat (evapotranspiration) flux data measured at AmeriFlux network sites to calculate an integrated Bowen ratio and by comparison with estimates of forest growth. It is well known that drought is a primary limit on tree growth and its effects can be seen in tree ring increments [74]. We also evaluated the ForDRI model by qualitatively comparing the spatial patterns and intensity of the drought conditions depicted on the U.S. Drought Monitor (USDM) maps during selected drought years. The USDM is a hybrid product, developed using several sources of ground observation and remote-sensed data including the SPI, PDSI, NDVI, streamflow values, and other drought indicators used by the agriculture, forest, and water management sectors as well as expert feedback from regional and national climatologists.

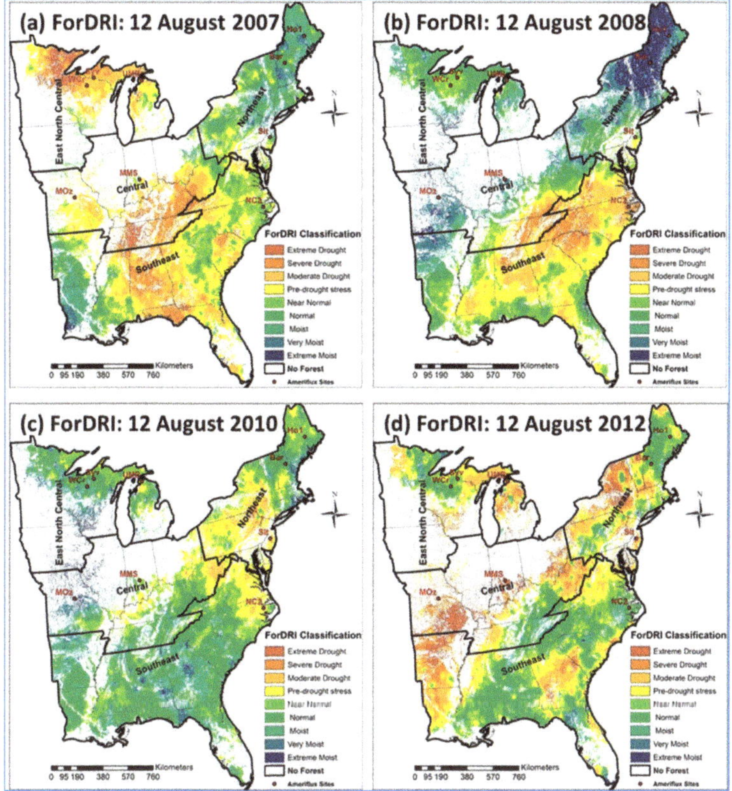



**Figure 3.** Example of the Forest Drought Response Index (ForDRI), showing maps of eastern U.S. Forest Service regions for week 32 (August 12) for selected years: (**a**) 2007, (**b**) 2008, (**c**) 2010, and (**d**) 2012.

## 3. Results

### 3.1. ForDRI Maps for Selected Drought Years

Historical ForDRI maps (780 maps at a weekly interval) were produced from 2003 to 2017. The same weeks (ending August 12) in 2007, 2008, 2010, and 2012 (Figure 3a–d) are shown below to demonstrate and evaluate the ForDRI model and products. The selection of these drought years was based upon the general long-term drought conditions of the eastern U.S. depicted by the USDM (Figure 4). Even though 2010 was not a drought year over most parts of the U.S., the Northeastern region experienced drought, as shown in Figure 3c.

### 3.2. Comparison of ForDRI with U.S. Drought Monitor (USDM)

The drought intensity estimates of ForDRI broadly agree with those for the same time period produced by the USDM (compare Figure 3; Figure 4). Note that ForDRI masks out non-forested (e.g., agricultural, rangelands, water, and urban) lands that are a focus of the USDM. In mid-August 2007 (Panel "a"), for example, both reach their most severe categories in Alabama-Tennessee and both capture intense drought west of Lake Superior. Details of the patterns differ because of differences in inputs and weighting. In mid-August 2008, for example, ForDRI indicates forest drought stress stretching well into Virginia while the USDM localizes the worst effects in a smaller region (Panel "b"). Both products agree that only mild drought is present in mid-August 2010 (Panel "c"). However, ForDRI does not indicate stress for forests in northern Louisiana while the USDM at that time is indicating short-term (e.g., agricultural) impacts are present. The extreme drought across much of the Midwest in August of 2012 [11] is clearly visible in both products (Figures 3d and 4d).

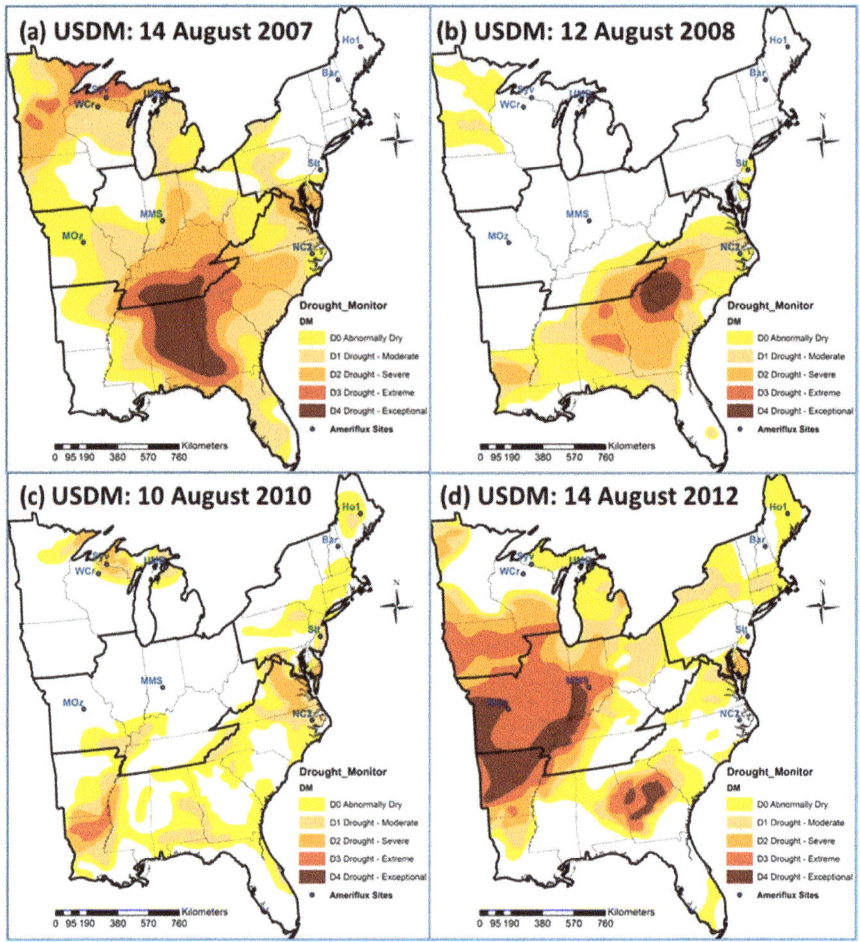


**Figure 4.** The U.S. Drought Monitor (USDM) maps for mid-August: (**a**) 2007, (**b**) 2008, (**c**) 2010, and (**d**) 2012 for qualitative comparisons.

### 3.3. Evaluating ForDRI with Bowen Ratio

Figure 5 shows the time-series comparison of the historical records of Bowen Ratio at nine AmeriFlux sites and ForDRI. During the assessment period, two of the flux tower sites, Morgan Monroe ("MMS", Monroe County, Indiana) and the Missouri Ozarks ("MOz", Boone County, Missouri) experienced "Exceptional" (D4) drought as defined by the U.S. Drought Monitor (Table 2). The North Carolina Pine site ("NC2", Washington County) experienced "Extreme" (D3) drought, while four sites experienced at least one "Severe" (D2) drought (Table 2). Two sites experienced at most "Moderate" (D1) growing season drought in the period between 2003 and 2017. Both Willow Creek ("WCr") and the Sylvania Wilderness ("Syv") sites experienced D3 events in the period between 2007 and 2010 or 2011 when they were offline (no observations available).

The Midwest drought of 2012 is easily seen in the normalized Bowen ratio flux data from both the MMS and MOz sites and is well captured by the ForDRI model (Figure 5). The 2012 drought reached D4 at both sites in August, and both model and data reached a minimum during this event. The normalized Bowen ratio reached $-2.89\sigma$ at the MOz site and $-3.26\sigma$ at MMS, consistent with single-tailed probabilities of <1% and <0.1%, indicating the severity of the drought. At both sites, the ForDRI model output is significantly correlated over the entire assessment period with the normalized Bowen ratio data (Z-score $\beta_t$) ($p < 0.001$, $r = 0.56$ at Morgan Monroe and $r = 0.76$ at the Missouri site). A late-summer D2 event at Morgan Monroe in 2010 is also well resolved in both the data and by ForDRI, as is a late summer D1 event in 2007 at both sites. However, a drought classified as D2 by the USDM at the Missouri Ozarks site in 2006 is less clear in the Bowen ratio data and ForDRI model. The ForDRI model and normalized Bowen ratio flux data disagree noticeably at Morgan Monroe in 2014 and at the Missouri Ozarks site in 2015. In both cases, the data suggest $\sim 1\sigma$ drier than normal conditions (higher Bowen ratios) while ForDRI indicated wetter than normal. This may be related to tree mortality attributable to 2012 drought that occurred in subsequent years; this delayed effect of drought [62] might complicate the Bowen ratio comparison.

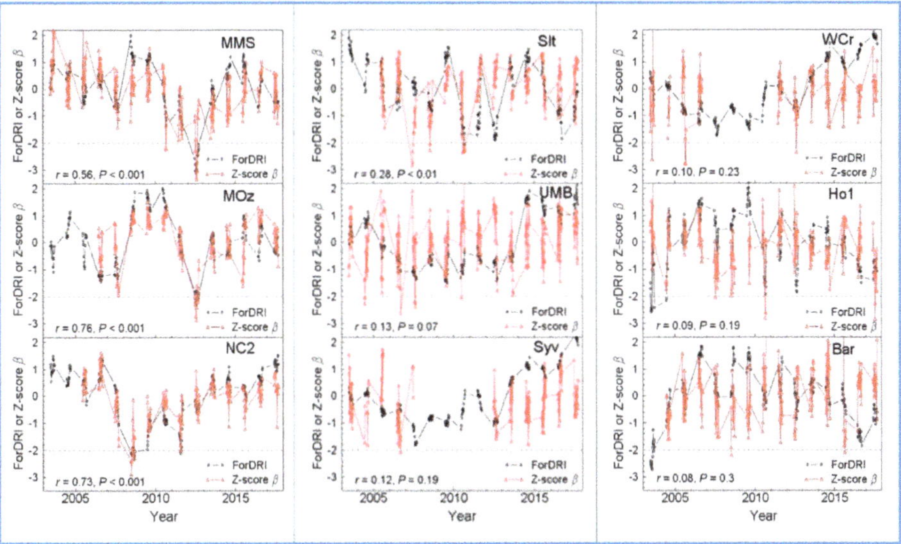

**Figure 5.** Comparison of the historical records of ForDRI values and normalized Bowen Ratio (Z-Score $\beta_t$) at nine AmeriFlux sites that include Bartlett Experimental Forest (Bar), Howland Forest (Ho1), Morgan Monroe State Forest (MMS), Missouri Ozark Site (MOz), North Carolina pine forest (NC2), Silas Little Forest (Slt), Sylvania Wilderness Area (Syv), Univ. of Mich. Biological Station (UMB), and Willow Creek (WCr).

The ForDRI model and Z-score $\beta_t$ are also well-correlated ($p < 0.001$, $r = 0.73$) at a North Carolina pine forest (NC2) site (Figure 5). The NC2 flux site experienced D2 in the fall of 2007 which worsened to D3 in the spring of 2008. This site also experienced a D2 drought throughout the summer of 2011. All of these events and their relative severity are clearly identified in both ForDRI and the normalized Bowen ratio.

The Silas Little Forest (Slt) in the New Jersey Pine Barrens is characterized by sandy soils with low water holding capacity and drought-tolerant species. The record drought in this time period was September 2010, when the USDM classified Burlington County as D2 for several weeks. The normalized Bowen ratio shows this as a $-2\sigma$ event and ForDRI identifies it as the most extreme in the interval (Figure 5). However, model and data disagree sharply at this site in the early spring of 2007 when

ForDRI was indicating normal moisture conditions while the Z-score $\beta_l$ showed this as an extreme stress departure of $-2.85\sigma$. ForDRI and the normalized Bowen ratio then came into better agreement as the growing season progressed. The difference can be accounted for by a gypsy moth caterpillar (*Lymantria dispar* L.) outbreak that removed most foliage from the forest in spring of 2007 [64]. Following the peak of herbivory in mid-June, a second, partial leaf-out occurred and resulted in a canopy with roughly half of the normal summer leaf area [64]. A secondary, lesser defoliation occurred at Silas Little in 2008. With little or no foliage, evaporation was severely constrained, and this resulted in most of the incoming energy being converted to sensible heat and a high Bowen ratio.

ForDRI identified the 2007–2009 drought at Willow Creek and the Sylvania Wilderness that reached D3 when flux data were not available, as well as lesser events. The normalized Bowen ratio data (Z-score $\beta_l$) reached a minimum of $-2\sigma$ at lesser (D2) events at these sites. However, ForDRI and Z-score $\beta_l$ were not significantly correlated at either site over the full data record (Willow Creek, r = 0.10, $p$ = 0.23; Sylvania r = 0.12, $p$ = 0.19). At UMB, the USDM reached D2 in 2005 and 2007, but these periods were poorly resolved by both ForDRI and Z-score $\beta_l$. Both Howland and Bartlett recorded only minor (D1) growing season events during the assessment period, and ForDRI and Z-score $\beta_l$ were not significantly correlated at these sites.

**Table 2.** Historic drought at AmeriFlux sites during the ForDRI assessment period based on the U.S. Drought Monitor.

| Site | County | State | Year | Dates | Intensity |
|---|---|---|---|---|---|
| MMS | Monroe | Indiana | 2012 | 26 June–4 Sept | D2 |
| | | | | 17 July–28 Aug | D3 |
| | | | | 24 July–7 Aug | D4 |
| | | | 2010 | 21 Sept–23 November | D2 |
| | | | 2007 | 21 Aug–26 Oct | D2 |
| MOz | Boone | Missouri | 2012 | 3 July–end of year | D2 |
| | | | | 17 July–16 Oct | D3 |
| | | | | 14 Aug–28 Aug | D4 |
| | | | 2006 | 8 Aug–22 Aug | D2 |
| | | | 2007 | 21 Aug–16 Oct | D1 |
| NC2 | Washington | North Carolina | 2011 | 31 May–23 Aug | D2 |
| | | | | 20 Nov–4 Mar 2012 | D2 |
| | | | 2008 | 1 Jan–26 Aug | D2 |
| | | | | 29 Jan–12 Feb, 26 Aug (one week) | D3 |
| | | | 2007 | 4 Sept–23 Oct | D2 |
| Slt | Burlington | New Jersey | 2010 | 7 Sept–28 Sept | D2 |
| | | | 2007 | June, Gypsy moth outbreak | none |
| UMB | Cheboygan | Michigan | 2011 | 29 Mar–26 Apr | D1 |
| | | | 2010 | 6 April–17 Aug | D1 |
| | | | 2007 | 28 Aug–4 Sept | D2 |
| | | | 2005 | 19 July–16 Aug | D2 |
| | | | 2003 | 7 Jan–1 April, 23 Sept | D1 |
| Syv | Gogebic | Michigan | * 2010 | 1-29 June | D3 |
| | | | | 13 April–17 Aug | D2 |
| | | | * 2009 | 22 Sept–20 Oct | D2 |
| | | | * 2008 | 26 Aug–12 May 2009 | D1 |
| | | | * 2007 | 14 Aug–4 Sep | D3 |
| | | | | 10 July–16 Oct | D2 |
| | | | 2006 | 11 July–25 July | D2 |

**Table 2.** *Cont.*

| Site | County | State | Year | Dates | Intensity |
|------|--------|-------|------|-------|-----------|
| WCr | Price | Wisconsin | 2012 | 9–23 Oct | D2 |
| | | | * 2010 | 13 April–22 June | D2 |
| | | | * 2009 | 4–18 Aug | D3 |
| | | | | 25 Jan–Aug | D2 |
| | | | * 2008 | 21 Oct–end of year | D2 |
| | | | * 2007 | 12–18 Sept | D2 |
| | | | 2005 | 6 Sept–4 Oct | D2 |
| | | | 2003 | 18–25 Mar, 22–29 July, 2 Sept–end of year | D1 |
| Ho1 | Penobscot | Maine | 2016 | 15 Nov–20 Dec | D2 |
| | | | 2016/17 | 27 Sept–7 Feb 2017 | D1 |
| | | | 2010 | 10 Aug–28 Sept | D1 |
| Bar | Carrol | New Hampshire | 2016/17 | 27 Sept–7 Feb 2017 | D1 |

* means data not available from flux site for that specific period.

### 3.4. Evaluating ForDRI with Tree Ring Increments

Tree ring increment (TRI) data from eight sites were used to assess ForDRI values at the four national parks (i.e., HAFE, PRWI, GRSM, and CATO). The tree ring increment is the width of a tree ring that shows the amount of growth taken place over one year and thus indicates the growing conditions for that year. The data that we used include 38 trees in CATO park (two per each site), 115 trees in GRSM (two per each site except one site with only one tree), 24 trees in HAFE (two per each site), and 44 trees in PRWI (two trees per each site). To analyze the correlation of the ForDRI and TRI, two sites from each national park were selected (Figure 6). Three species including American tulip tree (*Liriodendron tulipifera*), northern red oak (*Quercus rubra*), and white oak (*Quercus alba*) were selected for tree ring increment data analysis. Niinemets and Valladares [75] considered *Liriodendron tulipifera* and *Quercus rubra* moderately susceptible to drought and *Quercus alba* moderately tolerant [76]. At each of the selected park sites, the individual tree ID and species type are shown in Figure 6.

Figure 7 shows the correlation between annual tree-ring increment data and ForDRI weekly values during the summer season (June to September). The ForDRI values at a weekly interval were compared with the tree ring annual data at each site between 2003 and 2017 to identify the best period to monitor drought stress on trees using the ForDRI model. The results showed that four sites at GRSM and PRWI have higher correlations (between 0.61 and 0.82) with ForDRI during all weeks of summer (Figure 7) than the other park sites. The correlation peaked when compared with ForDRI values from mid-August. Tree ring increment at the two CATO sites also showed relatively good correlation (0.35 < r < 0.73) with ForDRI. At this site, the highest correlation (0.73) was found in July. Tree ring increments recorded at two HAFE sites showed relatively lower correlations (0.22 < r < 0.63) with ForDRI. This could be because the dominant tree species in the park (oak) are drought-tolerant. In addition, differences in the strength of these relationships may depend upon tree site specifics (ridgetop vs. valley), soils, or other factors. In addition, the frequency and intensity of drought at these four national historic parks over this relatively short interval were not identical. Generally, however, the comparison revealed that the ForDRI values showed reasonable correlation with the tree ring increment, so ForDRI maps may help decision-makers monitor tree drought stress in these parks.

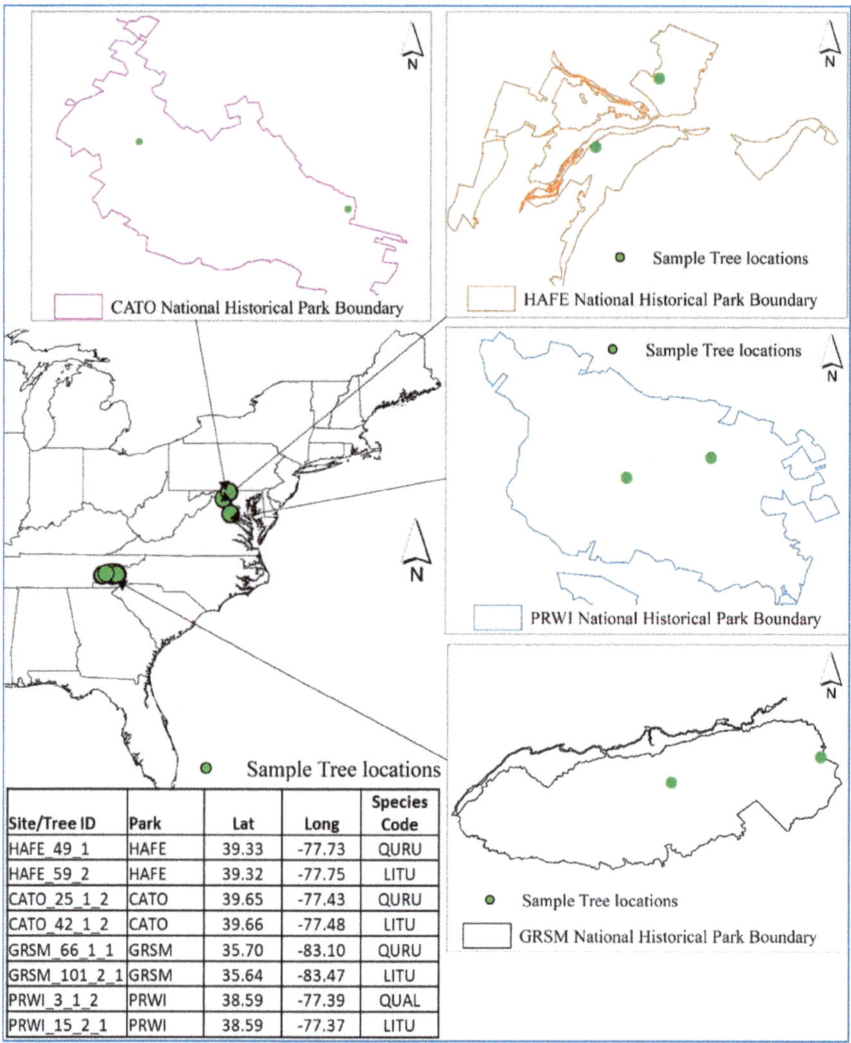



| Site/Tree ID | Park | Lat | Long | Species Code |
|---|---|---|---|---|
| HAFE_49_1 | HAFE | 39.33 | -77.73 | QURU |
| HAFE_59_2 | HAFE | 39.32 | -77.75 | LITU |
| CATO_25_1_2 | CATO | 39.65 | -77.43 | QURU |
| CATO_42_1_2 | CATO | 39.66 | -77.48 | LITU |
| GRSM_66_1_1 | GRSM | 35.70 | -83.10 | QURU |
| GRSM_101_2_1 | GRSM | 35.64 | -83.47 | LITU |
| PRWI_3_1_2 | PRWI | 38.59 | -77.39 | QUAL |
| PRWI_15_2_1 | PRWI | 38.59 | -77.37 | LITU |

**Figure 6.** Locations of the tree ring sites and their species types at the selected four national historical parks. The table in the lower left side of the figure shows the species type of each individual tree, indicating the tree species: *Quercus alba* (QUAL), *Liriodendron tulipifera* (LITU), and *Quercus rubra* (QURU).

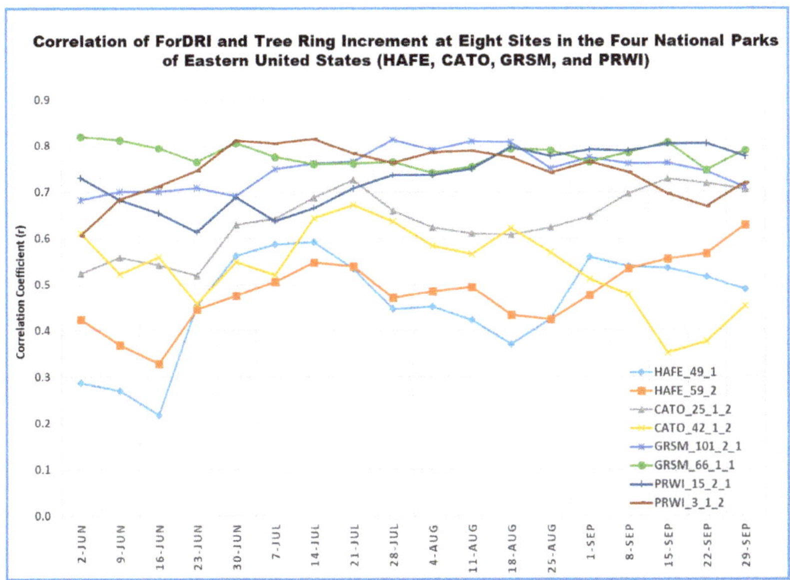



**Figure 7.** Time series correlation of ForDRI and tree ring increment data during summer season (June to September) at eight sites across four national parks in the eastern U.S.

Table 3 shows the maximum, minimum, and average correlation between ForDRI and tree ring increment data at eight sites of the four national parks in the eastern U.S. during the summer season (June to September). The correlation between ForDRI and tree ring increments at the selected eight sites during summer ranged between 0.46 (minimum) and 0.78 (maximum). The two GRSM sites had higher average correlations (0.75 and 0.78) than the PRWI (0.73 and 0.75), or other sites. Using average summer values of ForDRI accounted for over half the variance in tree ring increment at the GRSM and PRWI sites. Correlations may have been strongest at these two sites because they were impacted by the 2008 Southeast drought (Figure 3b) while the CATO and HAFE sites were not.

**Table 3.** Maximum, minimum, and average correlation of ForDRI and tree ring increment (TRI) data at eight sites during summer season (June to September) at four parks in the eastern United States.

| Site/Tree ID | Min | Max | Average |
| --- | --- | --- | --- |
| HAFE_49_1 | 0.22 | 0.59 | 0.46 |
| HAFE_59_2 | 0.33 | 0.63 | 0.48 |
| CATO_25_1_2 | 0.52 | 0.73 | 0.64 |
| CATO_42_1_2 | 0.35 | 0.67 | 0.54 |
| GRSM_101_2_1 | 0.68 | 0.81 | 0.75 |
| GRSM_66_1_1 | 0.74 | 0.82 | 0.78 |
| PRWI_15_2_1 | 0.61 | 0.81 | 0.73 |
| PRWI_3_1_2 | 0.61 | 0.81 | 0.75 |

## 4. Discussion

The ForDRI model reaches minimum values at the same times as the normalized Bowen ratio (Z-score of $\beta_L$), a relative measure of physiological water stress. Both of these measures reach minimum values at times when the USDM suggests these forested sites experienced extreme (D3) or exceptional drought (D4). Overall, ForDRI was significantly correlated with the normalized Bowen ratio. At the site level, this correlation was significant at four of the nine sites and can account for over half the variance in the flux-derived quantity. At the sites with lesser (e.g. D2) events in the record, both the normalized

Bowen ratio measurements and ForDRI tend to reach at least local minima during the drought event(s) but the correlation between these indicators across the entire time period drops. This lack of correlation at these sites is to be expected when there is little or no drought stress signal to measure. We would expect that other factors such as herbivory and other causes of foliage loss are contributing "noise" to the signals during these non-drought periods and that ForDRI and the normalized Bowen ratio are differentially sensitive to these other factors (the "noise" is uncorrelated). As mentioned earlier, stomatal conductance and $\beta$ are sensitive to a number of factors in addition to plant (or soil) water stress. These include solar radiation, temperature, and vapor pressure deficit. When significant droughts are absent at a site during the comparison periods (e.g., Bartlett Forest), our normalization scheme will highlight this other variation and magnify disagreement with ForDRI. Bowen ratio data from the Silas Little Forest supports this argument. In 2007, Bowen ratio values at Silas Little Forest reached a minimum, indicating extreme physiological stress, while ForDRI suggested no stress was present. Researchers at the forest, however, report that insects had consumed almost all of the canopy foliage at this time [64]. Without foliage to transpire water, incoming energy was converted to sensible heat and $\beta$ soared. The stress was real; it just was not caused by drought. Even so, lesser droughts (D2) are easily visible in the normalized Bowen ratio record.

Tree ring increment data were similarly significantly correlated with ForDRI, with higher correlations evident at sites that had experienced more significant drought. The long timespan of developing intense drought (drought serial autocorrelation) was observed in the correlation of annual ring increment with ForDRI estimates across the summer.

A multiyear pattern of drought stress is clearly visible in ForDRI and the normalized Bowen ratio at a number of sites, and critically, in all those that reached D3 or D4. This is an important result as it implies that serious forest drought, the kind that we are most concerned about, takes a long time to develop. It also indicates that ForDRI has a certain capacity to predict the likelihood of extreme (D3) or exceptional drought (D4) prior to, or early in, the growing season. Extreme or exceptional drought conditions seem very unlikely to develop if ForDRI is indicating average or wetter than average conditions at the beginning of the growing season. Conversely, seasons with enhanced likelihood of significant forest drought stress can also be identified. This suggests the possibility of forecasting potential drought maximum severity at the beginning of the growing season, which would be useful to fire managers and many others.

## 5. Conclusions

We have described ForDRI, a new and non-subjective indicator of forest drought. Weekly values of ForDRI have been calculated since 2003, and in that period, these values readily identify extreme (D3) or exceptional (D4) drought in several research forests. Severe (D2) and less intense droughts are also identified, but at a lower probability of success. A novel and independent measure of forest water stress calculated from forest flux-tower data, weekly, log-transformed integrated Bowen ratios ($\log_{10} \beta_i$) transformed to Z-scores from the weekly mean over the full record, similarly identifies extreme drought periods over the same record. At the sites that have experienced extreme or exceptional drought, these measures are significantly correlated, providing strong evidence for the utility of ForDRI.

The tree ring analysis also showed that the ForDRI values are correlated at the eight sites of the four national parks in the eastern U.S., indicating the drought/water stress impact on tree growth during the drought years. The results showed the potential usefulness of the ForDRI tool for decision making to monitor drought stress on trees in the eastern U.S. and suggest the model can be readily expanded to other parts of the continental U.S.

**Author Contributions:** T.T., D.Y.H., G.D., M.S., B.F., and B.D.W. conceived and designed the research; T.T., D.Y.H., Y.A.B., and B.Z. performed the data collection; T.T., D.Y.H., and Y.A.B. analyzed the results; T.T. and D.Y.H. wrote the original manuscript and T.T., D.Y.H., Y.A.B., M.S., B.F., B.D.W., G.B., K.L.C., A.R.D., L.G., A.N., K.A.N., and A.D.R. validated, revised and edited the manuscript. All authors have read and agreed to the published version of the manuscript.

**Funding:** This research was funded by USDA Cooperative Agreement, Federal Award Identification Number 58-0111-16-013.

**Acknowledgments:** The authors would like to thank the USDA, U.S. Forest Service, NASA, and USGS for providing satellite and model products, and the Department of Energy AmeriFlux Network Management Project for support of US-WCr, US-Syv, US-Ho1, US-NC2, US-MMF, and US-UMB. Research at US-Ho1, US-Bar, and US-Slt is supported by the USDA Forest Service's Northern Research Station. The authors also thank Deborah Wood of the NDMC for her editorial comments.

**Conflicts of Interest:** The authors declare no conflict of interest.

## References


1. Rigden, A.J.; Mueller, N.D.; Holbrook, N.M.; Pillai, N.; Huybers, P. Combined influence of soil moisture and atmospheric evaporative demand is important for accurately predicting US maize yields. *Nat. Food* **2020**, *1*, 127–133. [CrossRef]

2. Manzoni, S.; Katul, G.; Porporato, A. A dynamical system perspective on plant hydraulic failure. *Water Resour. Res.* **2014**, *50*, 5170–5183. [CrossRef]

3. Camarero, J.J.; Gazol, A.; Sangüesa-Barreda, G.; Cantero, A.; Sánchez-Salguero, R.; Sánchez-Miranda, A.; Granda, E.; Serra-Maluquer, X.; Ibáñez, R. Forest growth responses to drought at short-and long-term scales in Spain: Squeezing the stress memory from tree rings. *Front. Ecol. Evol.* **2018**, *6*, 1–11. [CrossRef]

4. Yin, J.; Bauerle, T.L. A global analysis of plant recovery performance from water stress. *Oikos* **2017**, *126*, 1377–1388. [CrossRef]

5. Matheny, A.M.; Fiorella, R.P.; Bohrer, G.; Poulsen, C.J.; Morin, T.H.; Wunderlich, A.; Vogel, C.S.; Curtis, P.S. Contrasting strategies of hydraulic control in two codominant temperate tree species. *Ecohydrology* **2017**, *10*, e1815. [CrossRef]

6. Roman, D.T.; Novick, K.A.; Brzostek, E.R.; Dragoni, D.; Rahman, F.; Phillips, R.P. The role of isohydric and anisohydric species in determining ecosystem-scale response to severe drought. *Oecologia* **2015**, *179*, 641–654. [CrossRef]

7. Plaut, J.A.; Yepez, E.A.; Hill, J.; Pangle, R.; Sperry, J.S.; Pockman, W.T.; Mcdowell, N.G. Hydraulic limits preceding mortality in a piñon–juniper woodland under experimental drought. *PlantCell Environ.* **2012**, *35*, 1601–1617. [CrossRef]

8. Sanchez-Salguero, R.; Camarero, J.J.; Dobbertin, M.; Fernández-Cancio, A.; Vila-Cabrera, A.; Manzanedo, R.D.; Zavala, M.A.; Navarro-Cerrillo, R.M. Contrasting vulnerability and resilience to drought-induced decline of densely planted vs. natural rear-edge Pinus nigra forests. *For. Ecol. Manag.* **2013**, *310*, 956–967. [CrossRef]

9. Camarero, J.J.; Gazol, A.; Sanguesa-Barreda, G.; Oliva, J.; Vicente-Serrano, S.M. To die or not to die: Early warnings of tree dieback in response to a severe drought. *J. Ecol.* **2015**, *103*, 44–57. [CrossRef]

10. Cailleret, M.; Jansen, S.; Robert, E.M.; Desoto, L.; Aakala, T.; Antos, J.A.; Beikircher, B.; Bigler, C.; Bugmann, H.; Caccianiga, M.; et al. A synthesis of radial growth patterns preceding tree mortality. *Glob. Chang. Biol.* **2017**, *23*, 1675–1690. [CrossRef]

11. Wolf, S.; Keenan, T.F.; Fisher, J.B.; Baldocchi, D.D.; Desai, A.R.; Richardson, A.D.; Scott, R.L.; Law, B.E.; Litvak, M.E.; Brunsell, N.A.; et al. Warm spring reduced carbon cycle impact of the 2012 US summer drought. *Proc. Natl. Acad. Sci. USA* **2016**, *113*, 5880–5885. [CrossRef]

12. Ruffault, J.; Martin-StPaul, N.; Pimont, F.; Dupuy, J.L. How well do meteorological drought indices predict live fuel moisture content (LFMC)? An assessment for wildfire research and operations in Mediterranean ecosystems. *Agric. For. Meteorol.* **2018**, *262*, 391–401. [CrossRef]

13. McKee, T.B.; Doesken, N.J.; Kleist, J. Drought Monitoring with Multiple Time Scales. In Proceedings of the 9th Conference on Applied Climatology, Dallas, TX, USA, 15–20 January 1995; American Meteorological Society: Boston, MA, USA.

14. Wells, N.; Goddard, S.; Hayes, M.J. A self-calibrating Palmer drought severity index. *J. Clim.* **2004**, *17*, 2335–2351. [CrossRef]

15. Palmer, W.C. *Meteorological Drought*; Research Paper, No. 45; US Department of Commerce, Weather Bureau: Washington, DC, USA, 1965; Volume 30, p. 58.

16. Keetch, J.J.; Byram, G.M. *A Drought Index for Forest Fire Control*; US Department of Agriculture, Forest Service, Southeastern Forest Experiment Station: Asheville, NC, USA, 1968; Volume 38.

17.	Koch, F.H.; Smith, W.D.; Coulston, J.W. An improved method for standardized mapping of drought conditions. In *Forest Health Monitoring: National Status, Trends, and Analysis 2010. Gen. Tech. Rep. SRS-GTR-176*; Potter, K.M., Conkling, B.L., Eds.; US Department of Agriculture, Forest Service, Southern Research Station: Asheville, NC, USA, 2013; pp. 67–83.

18.	Koch, F.H.; Smith, W.D.; Coulston, J.W. Drought patterns in the conterminous United States and Hawaii. In *Forest Health Monitoring: National Status, Trends, and Analysis 2012. Gen. Tech. Rep. SRSGTR-198*; Potter, K.M., Conkling, B.L., Eds.; US Department of Agriculture, Forest Service, Southern Research Station: Asheville, NC, USA, 2014; pp. 49–72.

19.	Koch, F.H.; Smith, W.D.; Coulston, J.W. Drought patterns in the conterminous United States, 2012. In *Forest Health Monitoring: National Status, Trends, and Analysis 2013. Gen. Tech. Rep. SRSGTR-207*; Potter, K.M., Conkling, B.L., Eds.; US Department of Agriculture, Forest Service, Southern Research Station: Asheville, NC, USA, 2015; pp. 55–69.

20.	Saleska, S.R.; Didan, K.; Huete, A.R.; da Rocha, H.R. Amazon forests green-up during 2005 drought. *Science* **2007**, *318*, 612. [CrossRef]

21.	Anderson, M.C.; Hain, C.; Wardlow, B.; Pimstein, A.; Mecikalski, J.R.; Kustas, W.P. Evaluation of drought indices based on thermal remote sensing of evapotranspiration over the continental United States. *J. Clim.* **2011**, *24*, 2025–2044. [CrossRef]

22.	Asner, G.P.; Alencar, A. Drought impacts on the Amazon forest: The remote sensing perspective. *New Phytol.* **2010**, *187*, 569–578. [CrossRef]

23.	Pasho, E.; Camarero, J.J.; de Luis, M.; Vicente-Serrano, S.M. Impacts of drought at different time scales on forest growth across a wide climatic gradient in north-eastern Spain. *Agric. For. Meteorol.* **2011**, *151*, 1800–1811. [CrossRef]

24.	Samanta, A.; Ganguly, S.; Myneni, R.B. MODIS enhanced vegetation index data do not show greening of Amazon forests during the 2005 drought. *New Phytol.* **2011**, *189*, 11–15. [CrossRef] [PubMed]

25.	Zhang, Y.; Peng, C.; Li, W.; Fang, X.; Zhang, T.; Zhu, Q.; Chen, H.; Zhao, P. Monitoring and estimating drought-induced impacts on forest structure, growth, function, and ecosystem services using remote-sensing data: Recent progress and future challenges. *Environ. Rev.* **2013**, *21*, 103–115. [CrossRef]

26.	AghaKouchak, A.; Farahmand, A.; Melton, F.S.; Teixeira, J.; Anderson, M.C.; Wardlow, B.D.; Hain, C.R. Remote sensing of drought: Progress, challenges and opportunities. *Rev. Geophys.* **2015**, *53*, 452–480. [CrossRef]

27.	Norman, S.P.; Koch, F.H.; Hargrove, W.W. Review of broad-scale drought monitoring of forests: Toward an integrated data mining approach. *For. Ecol. Manag.* **2016**, *380*, 346–358. [CrossRef]

28.	Svoboda, M.; LeComte, D.; Hayes, M.; Heim, R.; Gleason, K.; Angel, J.; Rippey, B.; Tinker, R.; Palecki, M.; Stooksbury, D.; et al. The drought monitor. *Bull. Am. Meteorol. Soc.* **2002**, *83*, 1181–1190. [CrossRef]

29.	NDMC. U.S. Drought Monitor. 2020. Available online: https://droughtmonitor.unl.edu/About.aspx (accessed on 3 September 2020).

30.	Iverson, L.R.; Prasad, A.M. Predicting abundance of 80 tree species following climate change in the eastern United States. *Ecol. Monogr.* **1998**, *68*, 465–485. [CrossRef]

31.	USDA Forest Service. National Forest Type Dataset. 2020. Available online: https://data.fs.usda.gov/geodata/rastergateway/forest_type/ (accessed on 3 September 2020).

32.	Ruefenacht, B.; Finco, M.V.; Nelson, M.D.; Czaplewski, R.; Helmer, E.H.; Blackard, J.A.; Holden, G.R.; Lister, A.J.; Salajanu, D.; Weyermann, D.; et al. Conterminous US and Alaska forest type mapping using forest inventory and analysis data. *Photogramm. Eng. Remote Sens.* **2008**, *74*, 1379–1388. [CrossRef]

33.	USGS. EROS Moderate Resolution Imaging Spectroradiometer (eMODIS) Digital Object Identifier (DOI) Number: /10.5066/F7H41PNT). 2020. Available online: https://www.usgs.gov/centers/eros/science/usgs-eros-archive-vegetation-monitoring-eros-moderate-resolution-imaging?qt-science_center_objects=0#qt-science_center_objects (accessed on 3 September 2020).

34.	Edwards, D.C.; McKee, T.B. *"Characteristics of 20th Century Drought in the United States at Multiple Time Scales," Climatology Report Number 97-2, Department of Atmospheric Science*; Colorado State University: Fort Collins, CO, USA, 1997.

35.	Vicente-Serrano, S.M.; Beguería, S.; López-Moreno, J.I. A multiscalar drought index sensitive to global warming: The standardized precipitation evapotranspiration index. *J. Clim.* **2010**, *23*, 1696–1718. [CrossRef]

36. Hobbins, M.T.; Wood, A.; McEvoy, D.J.; Huntington, J.L.; Morton, C.; Anderson, M.; Hain, C. The evaporative demand drought index. Part I: Linking drought evolution to variations in evaporative demand. *J. Hydrometeorol.* **2016**, *17*, 1745–1761. [CrossRef]

37. McEvoy, D.J.; Huntington, J.L.; Hobbins, M.T.; Wood, A.; Morton, C.; Anderson, M.; Hain, C. The evaporative demand drought index. Part II: CONUS-wide assessment against common drought indicators. *J. Hydrometeorol.* **2016**, *17*, 1763–1779. [CrossRef]

38. Bhanja, S.N.; Mukherjee, A.; Rodell, M. Groundwater storage change detection from in situ and GRACE-based estimates in major river basins across India. *Hydrol. Sci. J.* **2020**, *65*, 650–659. [CrossRef]

39. Li, B.; Rodell, M.; Kumar, S.; Beaudoing, H.K.; Getirana, A.; Zaitchik, B.F.; de Goncalves, L.G.; Cossetin, C.; Bhanja, S.; Mukherjee, A.; et al. Global GRACE data assimilation for groundwater and drought monitoring: Advances and challenges. *Water Resour. Res.* **2019**, *55*, 7564–7586. [CrossRef]

40. NASA GSFC Hydrological Sciences Laboratory—Nasa Gesdisc Data Archive, 2020. Available online: https://hydro1.gesdisc.eosdis.nasa.gov/data/GLDAS/GLDAS_CLSM025_DA1_D.2.2/ (accessed on 3 September 2020).

41. Keyantash, J.; Dracup, J.A. The quantification of drought: An evaluation of drought indices. *Bull. Am. Meteorol. Soc.* **2002**, *83*, 1167–1180. [CrossRef]

42. Nearing, G.S.; Mocko, D.M.; Peters-Lidard, C.D.; Kumar, S.V.; Xia, Y. Benchmarking NLDAS-2 soil moisture and evapotranspiration to separate uncertainty contributions. *J. Hydrometeorol.* **2016**, *17*, 745–759. [CrossRef]

43. Xia, Y.; Hao, Z.; Shi, C.; Li, Y.; Meng, J.; Xu, T.; Wu, X.; Zhang, B. Regional and global land data assimilation systems: Innovations, challenges, and prospects. *J. Meteorol. Res.* **2019**, *33*, 159–189. [CrossRef]

44. Kumar, S.V.; Peters-Lidard, C.D.; Mocko, D.; Reichle, R.; Liu, Y.; Arsenault, K.R.; Xia, Y.; Ek, M.; Riggs, G.; Livneh, B.; et al. Assimilation of remotely sensed soil moisture and snow depth retrievals for drought estimation. *J. Hydrometeorol.* **2014**, *15*, 2446–2469. [CrossRef]

45. Cai, X.; Yang, Z.L.; Xia, Y.; Huang, M.; Wei, H.; Leung, L.R.; Ek, M.B. Assessment of simulated water balance from Noah, Noah-MP, CLM, and VIC over CONUS using the NLDAS test bed. *J. Geophys. Res. Atmos.* **2014**, *119*, 13751–13770. [CrossRef]

46. Liu, Y.Y.; Parinussa, R.M.; Dorigo, W.A.; De Jeu, R.A.; Wagner, W.; Van Dijk, A.I.J.M.; McCabe, M.F.; Evans, J.P. Developing an improved soil moisture dataset by blending passive and active microwave satellite-based retrievals. *Hydrol. Earth Syst. Sci.* **2011**, *15*, 425–436. [CrossRef]

47. NOAA. NLDAS Drought Monitor Soil Moisture. 2020. Available online: https://www.emc.ncep.noaa.gov/mmb/nldas/drought/ (accessed on 3 September 2020).

48. Yuan, W.; Zheng, Y.; Piao, S.; Ciais, P.; Lombardozzi, D.; Wang, Y.; Ryu, Y.; Chen, G.; Dong, W.; Hu, Z.; et al. Increased atmospheric vapor pressure deficit reduces global vegetation growth. *Sci. Adv.* **2019**, *5*, eaax1396. [CrossRef]

49. Fletcher, A.L.; Sinclair, T.R.; Allen, L.H., Jr. Transpiration responses to vapor pressure deficit in well watered 'slow-wilting' and commercial soybean. *Environ. Exp. Bot.* **2007**, *61*, 145–151. [CrossRef]

50. Li, P.; Omani, N.; Chaubey, I.; Wei, X. Evaluation of Drought Implications on Ecosystem Services: Freshwater Provisioning and Food Provisioning in the Upper Mississippi River Basin. *Int. J. Environ. Res. Public Health* **2017**, *14*, 496. [CrossRef]

51. Daly, C.; Halbleib, M.; Smith, J.I.; Gibson, W.P.; Doggett, M.K.; Taylor, G.H.; Curtis, J.; Pasteris, P.A. Physiographically-sensitive mapping of temperature and precipitation across the conterminous United States. *Int. J. Climatol.* **2008**, *28*, 2031–2064. [CrossRef]

52. Daly, C.; Smith, J.I.; Olson, K.V. Mapping atmospheric moisture climatologies across the conterminous United States. *PLoS ONE* **2015**, *10*, e0141140. [CrossRef]

53. PRISM Climate Group; Oregon State University. Available online: http://prism.oregonstate.edu (accessed on 1 July 2020).

54. Philip, J.R. Plant water relations: Some physical aspects. *Annu. Rev. Plant Physiol.* **1966**, *17*, 245–268. [CrossRef]

55. Scholander, P.F.; Bradstreet, E.D.; Hemmingsen, E.A.; Hammel, H.T. Sap pressure in vascular plants: Negative hydrostatic pressure can be measured in plants. *Science* **1965**, *148*, 339–346. [CrossRef]

56. Baughn, J.W.; Tanner, C.B. Leaf Water Potential: Comparison of Pressure Chamber and in situ Hygrometer on Five Herbaceous Species 1. *Crop Sci.* **1976**, *16*, 181–184. [CrossRef]

57. Monteith, J.L. Evaporation and Environment. In *Symposia of the Society for Experimental Biology 19*; Cambridge University Press: Cambridge, UK, 1965; pp. 205–234.

58. Jarvis, P.G. The interpretation of the variations in leaf water potential and stomatal conductance found in canopies in the field. *Philos. Trans. R. Soc. Lond. B Biol. Sci.* **1976**, *273*, 593–610.

59. Cowan, I.R.; Farquhar, G.D. Stomatal function in relation to leaf metabolism and environment. *Symp. Soc. Exp. Biol.* **1977**, *31*, 471–505.

60. Ouimette, A.P.; Ollinger, S.V.; Richardson, A.D.; Hollinger, D.Y.; Keenan, T.F.; Lepine, L.C.; Vadeboncoeur, M.A. Carbon fluxes and interannual drivers in a temperate forest ecosystem assessed through comparison of top-down and bottom-up approaches. *Agric. For. Meteorol.* **2018**, *256*, 420–430. [CrossRef]

61. Hollinger, D.Y.; Aber, J.; Dail, B.; Davidson, E.A.; Goltz, S.M.; Hughes, H.; Leclerc, M.Y.; Lee, J.T.; Richardson, A.D.; Rodrigues, C.; et al. Spatial and temporal variability in forest–atmosphere CO2 exchange. *Glob. Chang. Biol.* **2004**, *10*, 1689–1706. [CrossRef]

62. Gu, L.; Pallardy, S.G.; Hosman, K.P.; Sun, Y. Drought-influenced mortality of tree species with different predawn leaf water dynamics in a decade-long study of a central US forest. *Biogeosciences* **2015**, *12*, 2831–2845. [CrossRef]

63. Noormets, A.; Gavazzi, M.J.; McNulty, S.G.; Domec, J.C.; Sun, G.E.; King, J.S.; Chen, J. Response of carbon fluxes to drought in a coastal plain loblolly pine forest. *Glob. Chang. Biol.* **2010**, *16*, 272–287. [CrossRef]

64. Clark, K.L.; Skowronski, N.; Gallagher, M.; Renninger, H.; Schäfer, K. Effects of invasive insects and fire on forest energy exchange and evapotranspiration in the New Jersey pinelands. *Agric. For. Meteorol.* **2012**, *166*, 50–61. [CrossRef]

65. Clark, K.L.; Renninger, H.J.; Skowronski, N.; Gallagher, M.; Schäfer, K.V. Decadal-scale reduction in forest net ecosystem production following insect defoliation contrasts with short-term impacts of prescribed fires. *Forests* **2018**, *9*, 145. [CrossRef]

66. Desai, A.R.; Bolstad, P.V.; Cook, B.D.; Davis, K.J.; Carey, E.V. Comparing net ecosystem exchange of carbon dioxide between an old-growth and mature forest in the upper Midwest, USA. *Agric. For. Meteorol.* **2005**, *128*, 33–55. [CrossRef]

67. Gough, C.M.; Vogel, C.S.; Schmid, H.P.; Su, H.B.; Curtis, P.S. Multi-year convergence of biometric and meteorological estimates of forest carbon storage. *Agric. For. Meteorol.* **2008**, *148*, 158–170. [CrossRef]

68. Cook, B.D.; Davis, K.J.; Wang, W.; Desai, A.; Berger, B.W.; Teclaw, R.M.; Martin, J.G.; Bolstad, P.V.; Bakwin, P.S.; Yi, C.; et al. Carbon exchange and venting anomalies in an upland deciduous forest in northern Wisconsin, USA. *Agric. For. Meteorol.* **2004**, *126*, 271–295. [CrossRef]

69. Elmore, A.J.; Nelson, D.; Guinn, S.M.; Paulman, R. *Landsat-based Phenology and Tree Ring Characterization, Eastern US Forests, 1984–2013*; ORNL DAAC: Oak Ridge, TN, USA, 2017. [CrossRef]

70. Elmore, A.J.; Nelson, D.M.; Craine, J.M. Earlier springs are causing reduced nitrogen availability in North American eastern deciduous forests. *Nat. Plants* **2016**. [CrossRef]

71. Kulkarni, S.S.; Wardlow, B.D.; Bayissa, Y.A.; Tadesse, T.; Svoboda, M.D.; Gedam, S.S. Developing a Remote Sensing-Based Combined Drought Indicator Approach for Agricultural Drought Monitoring over Marathwada, India. *Remote Sens.* **2020**, *12*, 2091. [CrossRef]

72. Bayissa, Y.A.; Tadesse, T.; Svoboda, M.; Wardlow, B.; Poulsen, C.; Swigart, J.; Van Andel, S.J. Developing a satellite-based combined drought indicator to monitor agricultural drought: A case study for Ethiopia. *GIscience Remote Sens.* **2019**, *56*, 718–748. [CrossRef]

73. Hanson, P.J.; Weltzin, J.F. Drought disturbance from climate change: Response of United States forests. *Sci. Total Environ.* **2000**, *262*, 205–220. [CrossRef]

74. Fritts, H. *Tree Rings and Climate*; Elsevier: Amsterdam, The Netherlands, 2012; p. 582.

75. Niinemets, Ü.; Valladares, F. Tolerance to shade, drought, and waterlogging of temperate Northern Hemisphere trees and shrubs. *Ecol. Monogr.* **2006**, *76*, 521–547. [CrossRef]

76. Abrams, M.D. Adaptations and responses to drought in Quercus species of North America. *Tree Physiol.* **1990**, *7*, 227–238. [CrossRef]

**Publisher's Note:** MDPI stays neutral with regard to jurisdictional claims in published maps and institutional affiliations.

 *remote sensing*

*Article*

# Large Uncertainty on Forest Area Change in the Early 21st Century among Widely Used Global Land Cover Datasets

He Chen [1,2], Zhenzhong Zeng [2,*], Jie Wu [2], Liqing Peng [3], Venkataraman Lakshmi [4], Hong Yang [5] and Junguo Liu [2]

[1] School of Environment, Harbin Institute of Technology, Harbin 150001, China; 11849584@mail.sustech.edu.cn
[2] School of Environmental Science and Engineering, Southern University of Science and Technology, Shenzhen 518055, China; wuj6@mail.sustech.edu.cn (J.W.); liujg@sustech.edu.cn (J.L.)
[3] World Resources Institute, Washington, DC 20002, USA; lpeng@princeton.edu
[4] Department of Engineering Systems and Environment, University of Virginia, Charlottesville, VA 22904, USA; vlakshmi@virginia.edu
[5] Department of Systems Analysis, Integrated Assessment and Modelling, Swiss Federal Institute for Aquatic Science and Technology (Eawag), 8600 Dübendorf, Switzerland; Hong.Yang@eawag.ch
* Correspondence: zengzz@sustech.edu.cn

Received: 25 September 2020; Accepted: 23 October 2020; Published: 25 October 2020

**Abstract:** Forests play an important role in the Earth's system. Understanding the states and changes in global forests is vital for ecological assessments and forest policy guidance. However, there is no consensus on how global forests have changed based on current datasets. In this study, five global land cover datasets and Global Forest Resources Assessments (FRA) were assessed to reveal uncertainties in the global forest changes in the early 21$^{st}$ century. These datasets displayed substantial divergences in total area, spatial distribution, latitudinal profile, and annual area change from 2001 to 2012. These datasets also display completely divergent conclusions on forest area changes for different countries. Among the datasets, total forest area changes range from an increase of $1.7 \times 10^6$ km$^2$ to a decrease of $1.6 \times 10^6$ km$^2$. All the datasets show deforestation in the tropics. The accuracies of the datasets in detecting forest cover changes were evaluated by a global land cover validation dataset. The spatial patterns of accuracies are inconsistent among the datasets. This study calls for the development of a more accurate database to support forest policies and to contribute to global actions against climate change.

**Keywords:** forest area change; data assessment; uncertainty evaluation; inconsistency

---

## 1. Introduction

As one of the most widely distributed land cover types, forests have a total area of $40.6 \times 10^6$ km$^2$, which accounts for 31% of the world's total land area [1]. Forests play an important role in maintaining the balance of the global ecosystem by acting as a carbon sink [2,3], prompting water conservation [4], providing a habitat for species [5], improving landscape functions [6], and regulating the climate [7]. Forests have experienced great changes, including deforestation and afforestation. Understanding states of global forest and its changes help to provide guidelines for forest conservation, protection, and management [8,9].

According to the Food and Agriculture Organization (FAO) of the United Nations (UN), more than $1.29 \times 10^6$ km$^2$ of forest have been lost globally since 1990 [1]. Fires, insects, severe weather events, and ecological evolution are important natural factors that cause changes in forest distribution [10,11]. Human activities have significant influence on forests. This includes sourcing wood for industrial and fuel usage [12,13]. Land expansions caused by human activities, e.g., farmland and urban expansion, are also major causes of deforestation [14]. Such changes can result in significant climate consequences by initiating considerable climate feedbacks in biochemical and biogeophysical processes [7]. For example, the Amazon suffers from the most severe deforestation. As a consequence of the biogeophysical feedback, it has become drier and warmer [15]. Forest changes also affect humans through their ecological functions, such as mitigating the heat island effect [16]. Given that forest changes can greatly affect the Earth systems and human society, it is critical to obtain accurate descriptions of the status and changes of the global forest. However, there is no consensus on how global forests have changed in the past several decades.

There are two main methods to monitor forests at large scales: i.e., remote sensing monitoring and forest censuses. Many global land cover datasets were produced to quantify the global land cover situations and its changes with respect to the advances in remote sensing theory and technology [17]. Large-scale forest monitoring by remote sensing has improved in terms of its availability, accuracy, and spatiotemporal resolution [18,19]. These datasets fostered the development of large-scale forest researches [20,21]. Forest censuses by governments are another commonly used method to assess forest status. The FAO's FRA reports were the most commonly used national scale forest datasets. It can provide a perspective on the impact of policies and other human factors on forests, e.g., how the Three-North Shelter Forest Program in China significantly improved forest coverage in northern China [22].

Based on existing datasets and studies, global forests have substantially changed, including changes in total area [23], spatial distribution [24], and tree biodiversity [25]. However, these studies came to inconsistent and even contradictory conclusions on global forest changes. Hansen et al. [26] detected global forest change with the Advanced Very High Resolution Radiometer (AVHRR) data, revealing a decrease in the tree cover percentage from 1982 to 1999. Among different regions, Latin America and Southeast Asia were the dominant regions for forest loss. This finding was contradictory to that from the FRA. According to Song et al. [27], global tree cover area increased by $2.2 \times 10^6$ km$^2$ from 1982 to 2016; the gross loss and gain of tree cover both increased, but the rate of tree cover gain was greater than that of tree cover loss, especially after the first decades of the 21st century. In contrast, Hansen et al. [28] reported that global forest area decreased by $1.6 \times 10^6$ km$^2$ from 2001 to 2012 [28]. According to FRA reports, the global forest area decreased by $1.3 \times 10^6$ km$^2$ between 1990 and 2015, including decreases of $7.3 \times 10^5$ km$^2$ from 1990 to 2001 and $5.6 \times 10^5$ km$^2$ from 2001 to 2015 [29–31]. Current studies reached a consensus: the tropics have undergone rapid forest loss over the past several decades. Keenan et al. [23] concluded the tropical forest area decreased by $1.95 \times 10^6$ km$^2$ from 1990 to 2015. Ordway et al. [32] found that agricultural expansion increased forest loss in South America and Southeast Asia; Qin et al. [33] also demonstrated a forest loss in the Brazilian Amazon from 2000 to 2017. Current studies also revealed that cropland expansion was an important driver of the forest loss in the tropics [24,34]. However, when it comes to the specific area of forest change, the conclusions are diverse. So far, these studies paid more attention to analyzing the results based on their own datasets, while comparisons between different datasets were often ignored. Comparisons of how forests changed based on the commonly used land cover datasets can help us to understand the states of forest and recognize the inconsistencies among the datasets.

The primary objective of this study was to investigate inconsistencies of global forest changes among five commonly used globally land cover datasets and FRA reports from 2001 to 2012. The selected land cover datasets including Vegetation Continuous Fields (VCF), Global Forest Change (Hansen), Terra and Aqua combined Moderate Resolution Imaging Spectroradiometer Land Cover Climate Modeling Grid (MCD12C1), Land Cover project of the Climate Change Initiative (CCI-LC), and the new generation of Land-Use Harmonization (LUH2). FRA reports also participated in the comparison at country and continental scales. To evaluate the inconsistencies of forest area changes among these datasets, total forest area change, latitudinal profile of forest change, forest area change in different climatic zones, annual area change, and spatial distribution of forest change for the selected datasets were compared. The global land cover validation dataset from the United States Geological Survey (USGS) was utilized to evaluate the accuracy of the land cover datasets.

## 2. Materials and Methods

### 2.1. Materials

#### 2.1.1. VCF

The Making Earth System Data Records for Use in Research Environments (the MEaSURES) VCF products use a non-parametric trend analysis in each pixel to measure land cover changes [27]. VCF provides the annual global fraction of tree canopy (TC) cover, short vegetation (SV) cover, and bare ground (BG) cover from 1982 to 2016. This dataset is based on the AVHRR and other supplementary data, including Landsat Enhanced Thematic Mapper Plus (ETM+), the Moderate Resolution Imaging Spectroradiometer (MODIS), and other very high-resolution satellite images [27]. This dataset can be obtained from Global Land Analysis and Discovery [35]. In this study, TC was selected to calculate forest area changes from 2001 to 2012 (Table 1).

#### 2.1.2. Hansen

Hansen is the high-resolution global maps of twenty-first-century forest cover, including tree canopy data for the year 2000, global forest cover gain between 2000 and 2012, year of gross forest cover loss event, and data mask at 30 m × 30 m resolution [28,36]. Forest cover gain is displayed as a binary value that provides information on whether forest gain occurred in each pixel from 2000 to 2012. Hansen also provides information of the years when net forest loss occurred in each pixel. Hansen was produced based on 654,178 growing season Landsat 7 ETM+ images. We used the forest cover gain for 2001–2012, forest cover loss for 2001–2017, and land mask (whether the specific pixel represents land or not) to calculate the global forest area change from 2001 to 2012 (Table 1) [28].

#### 2.1.3. MCD12C1

Version 6 of MCD12C1 product provides land cover data using three land cover classification schemes, i.e., International Geosphere-Biosphere Programme (IGBP), University of Maryland (UMD), and Leaf Area Index (LAI). For each land cover classification scheme, MCD12C1 provides global land cover fraction at a $0.05 \times 0.05°$ spatial resolution from 2001 to 2018. In this study, the IGBP classification scheme was utilized to calculate global forest area changes. Forest cover is defined by the pixels of evergreen broadleaf forest, deciduous needle leaf forest, or deciduous broadleaf forest layers [37,38].

#### 2.1.4. CCI-LC

CCI-LC product is produced by the Land Cover (LC) project of the Climate Change Initiative (CCI), which is led by the Europe Space Agency (ESA) [39]. Three original data types were used to develop CCI-LC, including AVHRR data from 1992 to 1999, Systeme Probatoire d'Observation de la Terre-Vegetation (SPOT-VGT) data from 1999 to 2013, and Project for On-Board Autonomy Vegetation (PROBA-V) data for 2013, 2014, and 2015. CCI-LC is at a 300 m × 300 m resolution, based on the Glob Cover unsupervised classification chain, which utilizes a machine learning algorithm [40]. CCI-LC uses the hierarchical classification system of the United Nations Land Cover Classification System (UN-LCCS), which was developed by the FAO.

#### 2.1.5. LUH2

LUH2 is the new generation of global land-use forcing datasets that provide a standard format of historical land use and future projections for climate models. The dataset provides historical reconstructions of land-use, which can be used with future projections via Earth system models.

LUH2 provides global land-use states and transitions at a 0.25 × 0.25° resolution from 850 to 2100. The classification of LUH2 is more detailed than that of the previous versions (Table 1) [41,42].

**Table 1.** Basic information of Vegetation Continuous Fields (VCF), Global Forest Change (Hansen), Terra and Aqua combined Moderate Resolution Imaging Spectroradiometer Land Cover Climate Modeling Grid (MCD12C1), Land Cover project of the Climate Change Initiative (CCI-LC), and Land-Use Harmonization (LUH2).

| Dataset | Temporal Coverage | Temporal Resolution | Spatial Resolution | Spatial Coverage | Reference |
|---|---|---|---|---|---|
| VCF | 1982–2016 | Yearly | 0.05° | Global | Song et al., 2018 [27] |
| Hansen | 2001–2017 | Yearly | 30 m | Global | Hansen et al., 2013 [28] |
| MCD12C1 | 2001–2017 | Yearly | 0.05° | Global | Sulla-Menashe and Friedl, 2018 [38] |
| CCI-LC | 1992–2015 | Yearly | 300 m | Global | Bontemps et al., 2013 [40] |
| LUH2 | 850–2100 | Yearly | 0.25° | Global | Hurtt et al., 2020 [41] |

### 2.1.6. FRA Reports

FRA reports are the most comprehensive forest assessment datasets. The dataset provides the global forest resources and their changes every five year. FRA reports are based on country level government inventories and remote sensing data. FRA reports provided every five-year assessment about the global forest resources and their changes. FRA reports were widely used for forest conditions popularizations, policy guidance, and land cover data accuracy validations. FRA report 2000, FRA report 2010, FRA report 2015, and FRA report 2020 were selected to compare in the forest area change with the global land cover datasets at country level.

### 2.1.7. USGS Global Land Cover Validation Data

Global land cover validation data were developed by USGS based on high resolution commercial satellite imagery. Over 1100 individual scenes of commercial data imagery were utilized to produce the validation data from several satellites, including QuickBird, WorldView-1, WorldView-2, IKONOS, OrbView-3, and GeoEye-1. The USGS validation dataset, includes 500 5 km × 5 km samples at a global scale, with samples locations that were randomly selected based on the modified Köppen climate zones and population [43,44]. The validation dataset provides the dominant land cover type in each pixel at a 2 m spatial resolution, spanning 2001–2014.

### 2.2. Data Processing

To ensure that the global land cover datasets at different temporal and spatial scales (VCF, Hansen, MCD12C1, CCI-LC, and LUH2) are comparable, all of the land cover datasets were aggregated to the same temporal period, spatial resolution, and spatial coverage. As Hansen exhibits the least latitude coverage (80°N–60°S) and shortest temporal period (2001–2012) among the five land cover datasets, all of the datasets were adjusted to the same spatiotemporal extent (latitude: 80°N–60°S; time period: 2001–2012). We also processed all the five datasets at a 0.05° resolution following the procedures outlined below.

Hansen provides information on when forest loss occurred from 2001 to 2017 and whether forest gain occurred between 2001 and 2012 in each 30 m × 30 m pixel. The number of pixels of forest loss, forest gain, and the land pixel in every 200 × 200 Hansen block (matched to a 0.05 × 0.05° pixel) were calculated. The forest change ratio in a 0.05 × 0.05° pixel was calculated by dividing the number of net change pixels (the number of forest gain pixels minus the number of forest loss pixels) by the total number of land pixels:

$$Change\ ratio_{[i,j]} = \frac{num_{gain} - num_{loss}}{num_{land}}, \tag{1}$$

where $i$ and $j$ refers to the row number and column number of the $0.05 \times 0.05°$ pixel; $num_{gain}$ refers to the number of forest gain in the pixel $[i,j]$; $num_{loss}$ refers to the number of forest loss event range from 1 to 12 in the pixel; $num_{land}$ refers to the number of land surface in the pixel.

CCI-LC provides the land cover types in each 250 m × 250 m pixel. In this study, regions with the class code 30, 40, 50, 60, 70, 80, 90, and 100 were defined as forest areas based on the legend system. We counted the pixel number of the above layers as forest cover in every 18 × 18 CCI-LC block (matched to a $0.05 \times 0.05°$ pixel), representing the forest cover ratio in each $0.05°$ pixel.

Forest cover ratio of VCF was obtained from the Percent Tree Cover layer. For the MCD12C1, evergreen needle leaf forest, evergreen broadleaf forest, deciduous needle leaf forest, and deciduous broadleaf forest land cover types were selected as forest cover under Land_Cover_Type_1_Percent for the IGBP scheme. For LUH2, the variables of primf (forested primary land) and secdf (potentially forested secondary land) in the states layer were selected to obtain the percentage of forest cover in each pixel. For these three datasets, forest area change in each pixel was calculated by multiplying the forest cover percentage by the area of the pixel.

To assess the accuracy of the land cover datasets, the forest cover percentage of each sample was calculated. First, we transformed the projection of the global land cover validation data from Universal Transverse Mercator (UTM) Projection to latitude and longitude. Second, the corresponding row number and column number for each sample were calculated based on the latitude and longitude. Then the numbers of forest pixel in each sample were calculated. The forest cover percentage in the sample is equal to the pixel numbers of forest divided by the total pixel numbers in the sample. We only found date information for 300 samples out of the 500 samples, such that the accuracy validation was based on the 300 samples with date information.

We also compared the forest area changes at country level. We calculated the forest area for the years 2001 and 2012 for each country; forest area change for each country was calculated by subtracting forest area of 2001 from that of 2012.

## 2.3. Statistical Indicators

To understand the spatial distribution of forest area changes for the five land cover datasets, we calculated the standard deviation and deviation among the five land cover datasets to describe the magnitude of the difference. For pixel $[i, j]$, the standard deviation and deviation were defined as follows:

$$Standard\ deviation_{[i,j]} = \sqrt{\frac{\sum_{k=1}^{5}\left(x_{i,j,k} - \overline{x}_{i,j}\right)^2}{5}}, \tag{2}$$

$$Deviation_{[i,j]} = \frac{\sum_{k=1}^{5}\left|x_{i,j,k} - \overline{x}_{i,j}\right|}{5}, \tag{3}$$

where $i$ and $j$ refers to the row number and column number, respectively; $x_{i,j,k}$ refers to the forest area change for pixel $[i, j]$ of dataset $k$; $\overline{x}_{i,j}$ refers to the average of forest area change in the five land cover datasets for pixel $[i, j]$.

## 3. Results

### 3.1. Total Forest Area Change from 2001 to 2012

There was substantial inconsistency in total global forest area change among the five global land cover datasets. Figure 1 displays the total forest area change from 2001 to 2012 in 80°N–60°S for the five land cover datasets and FRA reports. The VCF was the only dataset that displayed an increase in forest area ($1.7 \times 10^6$ km$^2$). Hansen exhibited the largest forest area decrease, up to $1.6 \times 10^6$ km$^2$, followed by the forest losses of $0.4 \times 10^6$ km$^2$ for MCD12C1, $0.2 \times 10^6$ km$^2$ for CCI-LC, and $0.1 \times 10^6$ km$^2$ for the LUH2. The gap in the forest area change between VCF and Hansen was $3.3 \times 10^6$ km$^2$, accounting for 8.8% of the total forest area worldwide ($4.1 \times 10^7$ km$^2$) based on FRA report 2015 [31].

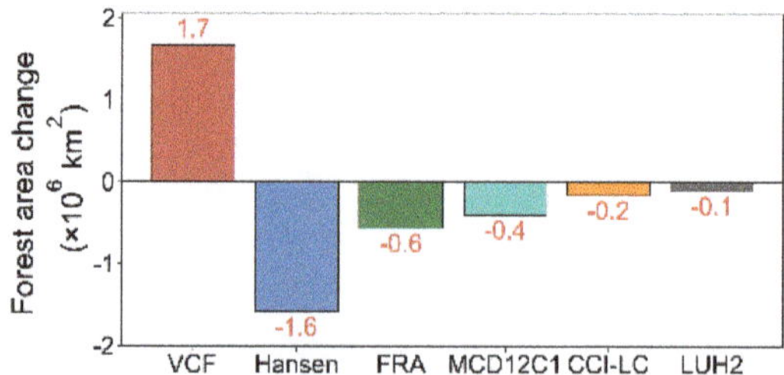

**Figure 1.** Total forest area changes in 60°S–80°N for VCF, Hansen, MCD12C1, CCI-LC, and LUH2 from 2001 to 2012.

Forest area changes across climatic zones are displayed in Figure 2. In the boreal zone (66.5°N–80.0°N in this study), VCF, FRA, and CCI-LC displayed increases in forest area, while the other datasets demonstrated decreases. VCF indicated the largest forest area increase ($2.2 \times 10^5$ km$^2$) (Figure 2a). In the temperate zone (35°N–66.5°N and 35°S–60°S) (Figure 2b), Hansen indicated forest loss while other datasets indicated forest gain. Among them, VCF indicated the largest forest gain, followed by FRA. In the sub tropics, VCF indicated the largest forest gain ($1.9 \times 10^5$ km$^2$), and Hansen indicated the largest forest loss ($1.5 \times 10^5$ km$^2$). As for the tropics (23.5°S–23.5°N), all the datasets indicated forest loss. FRA reported the largest forest loss area ($10.0 \times 10^5$ km$^2$), which approximately seven times that of VCF ($1.2 \times 10^5$ km$^2$). Hansen also indicated a large forest loss ($8.5 \times 10^5$ km$^2$). The tropics was the only climatic zone where all the datasets indicated a same trend (decrease) in forest change. This conclusion was consistent with previous studies: Hansen et al. [45] quantified the gross forest cover loss in the tropics as $47.6 \times 10^4$ km$^2$ from 2000 to 2005, and Zeng et al. [24] found that there was a rapid forest loss in Southeast Asia in the early 21st century (about $29.3 \times 10^4$ km$^2$). In general, VCF displayed increases in forest area in the boreal, temperate, and subtropical zones; it also indicated the least forest gain in the tropics. Hansen reported the largest forest loss among these datasets for all climatic zones except for the tropics.

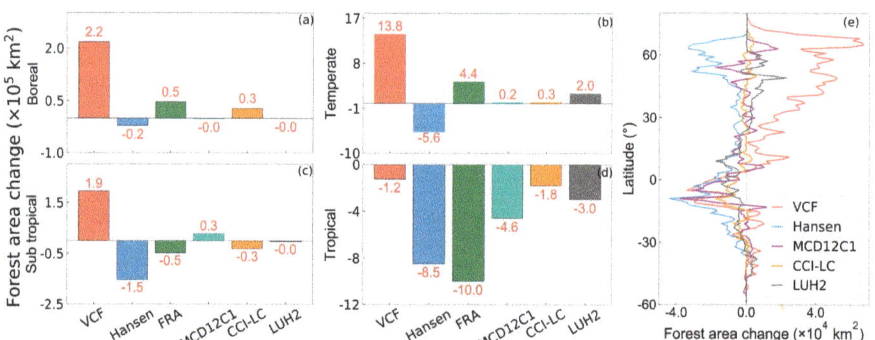

**Figure 2.** Forest area change in different climatic zones from 2001 to 2012: (**a**) boreal (66.5°N–80°N), (**b**) temperate (35°N–66.5°N and 35°S–60°S), (**c**) subtropical (23.5°N–35°N and 23.5°S–35°S), (**d**) tropical (23.5°S–23.5°N), and (**e**) the latitudinal profiles of forest area change in 60°S–80°N.

We also compared the latitudinal profiles of forest area changes based on the five datasets. In Figure 2e, VCF exhibited forest area gains in the northern hemisphere and the southern temperate zone but identified forest losses in the tropics and the southern hemisphere. VCF also had the largest latitudinal average in forest area change. Hansen showed gross forest loss along the latitudes. The latitudinal profiles of forest area changes for MCD12C1, CCI-LC, and LUH2 fluctuated with latitude, but all the three datasets indicated overall forest losses. Most of the inconsistencies occurred in the northern hemisphere, where all the datasets demonstrated decreases in the 10°S–10°N latitudinal range, which corresponds to the Amazon and Southeast Asia. The difference in the latitudinal profile of forest area change between VCF and Hansen was the largest, which was consistent with the results displayed in Figure 1.

Forest area changes in different continents were displayed in Figure 3. VCF demonstrated the largest forest gains in Asia, Northern America, Europe, and Oceania. Hansen demonstrated the largest forest losses in Asia, Northern America, Europe, and Oceania. In Africa and South America, regions with severe tropical rainforests, most datasets displayed large forest loss. For other continents, there were inconsistencies in the sign and magnitude of the forest area changes among the datasets.

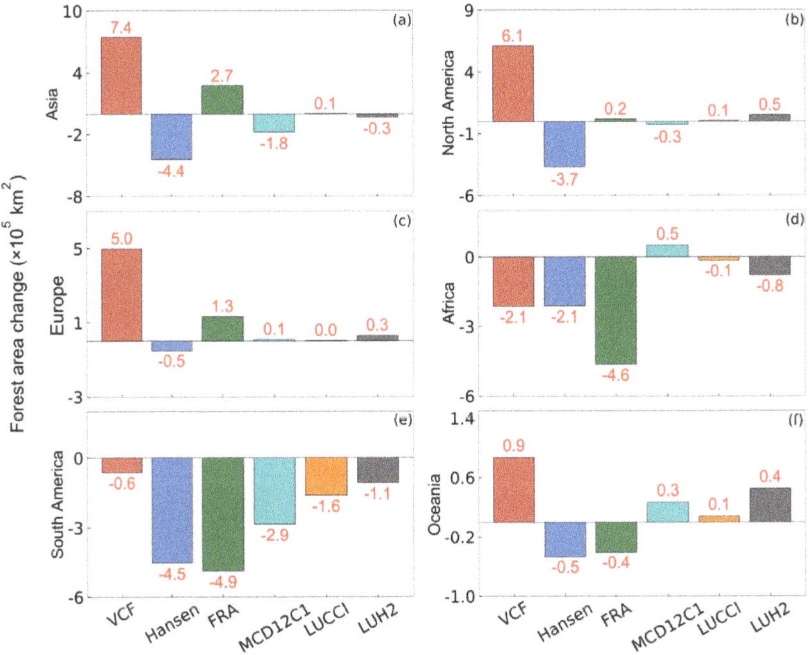

**Figure 3.** Forest area change in different continents from 2001 to 2012: (**a**) Asia, (**b**) North America, (**c**) Europe, (**d**) Africa, (**e**) South America, and (**f**) Oceania.

### 3.2. Spatial Distribution of Forest Change from 2001 to 2012

Figure 4 displays the spatial distribution of forest area changes in the five land cover datasets. For VCF, forest area increased in Europe, the USA, China, India, and Siberia (Figure 4a). The spatial distribution of Hansen was significantly different from that of VCF. Based on Figure 4b, the most apparent difference was that Hansen indicated forest losses of in more pixels than that in VCF, especially in Southeast Asia, Congo, and the Amazon. Combining with Figure 4c–e, the magnitudes of forest area changes in MCD12C1, CCI-LC, and LUH2 were much lower than those of VCF and Hansen. Figure 4f displays the ensemble mean of forest area change in the five land cover datasets. The Congo Rainforest, the Amazon Rainforest, and Southeast Asia displayed severe deforestation, while Eastern China, Siberia, and Eastern USA displayed forest gain.

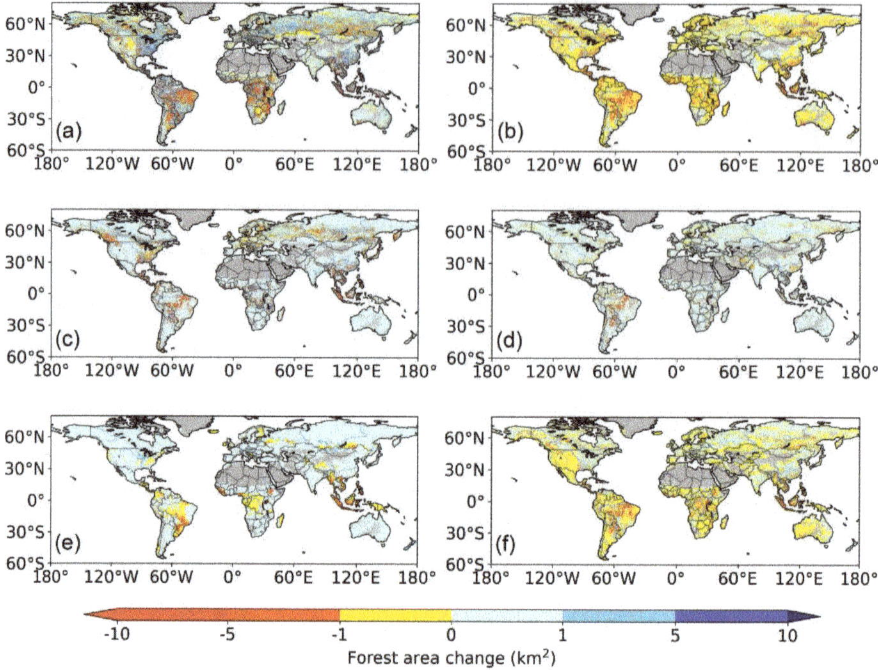

**Figure 4.** Distributions of forest area changes from 2001 to 2012: (**a**) VCF, (**b**) Hansen, (**c**) MCD12C1, (**d**) CCI-LC, (**e**) LUH2, and (**f**) ensemble mean.

The spatial patterns of standard deviation and deviation of forest area change between the five datasets are demonstrated in Figure 5. Standard deviation and deviation have very similar spatial patterns. The Congo Rainforest, Amazon Rainforest, and Southeast Asia have high standard deviation and deviations, which means discrete distributions of forest change area in these areas. Siberia, India, and Western USA have small standard deviations and deviations in forest area change.

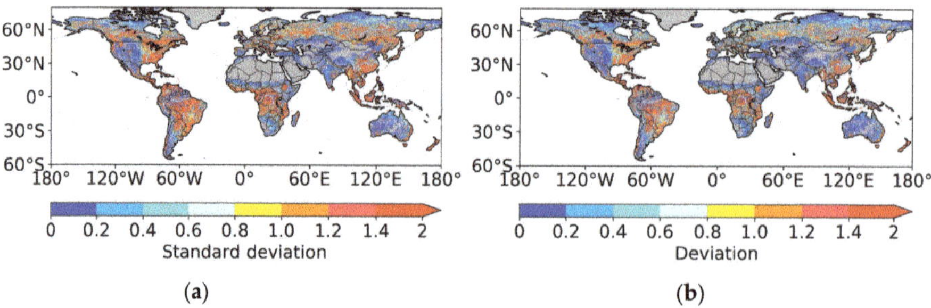

**Figure 5.** Distributions of standard deviation and deviation of forest area changes from 2001 to 2012: (**a**) standard deviation and (**b**) deviation.

To assess the inconsistencies of forest area change between VCF and Hansen, which displayed the largest inconsistency, the differences in forest change trends between the two datasets were displayed in Figure 6. Both VCF and Hansen demonstrated forest loss in the Amazon, South Africa, and the majority of Southeast Asia; most of these areas were located in the tropics. In contrast, both VCF and Hansen demonstrated forest gain in Europe, Eastern USA, and Western Siberia. VCF and Hansen demonstrated different change trends in forest area in other areas. Several hot spots, where heavy forest changes or large differences occurred in the base map, were also selected to compare differences among the five datasets. The hot spots include North America (37.0°N–50.0°N, 95.0°W–85.0°W), Europe (45.0°N–54.0°N, 0.0–15.0°E), Eastern Siberia (55.0°N–66.5°N, 90.0°E–150.0°E), the Amazon (10.0°S–10.0°N, 73.0°W–40.0°W), South Africa (5.0°S–8.0°N, 10.0°E–40.0°E), and Southeast Asia (10.0°S–28.0°N, 92.0–140.0°E). Apart from the Amazon, where all five datasets displayed forest losses, all other regions demonstrated inconsistent trends in forest change (Figure 6d). Forest losses were clear for regions in the southern hemisphere (e.g., the Amazon and the Congo rainforests), but the uncertainties were larger for regions in the northern hemisphere.

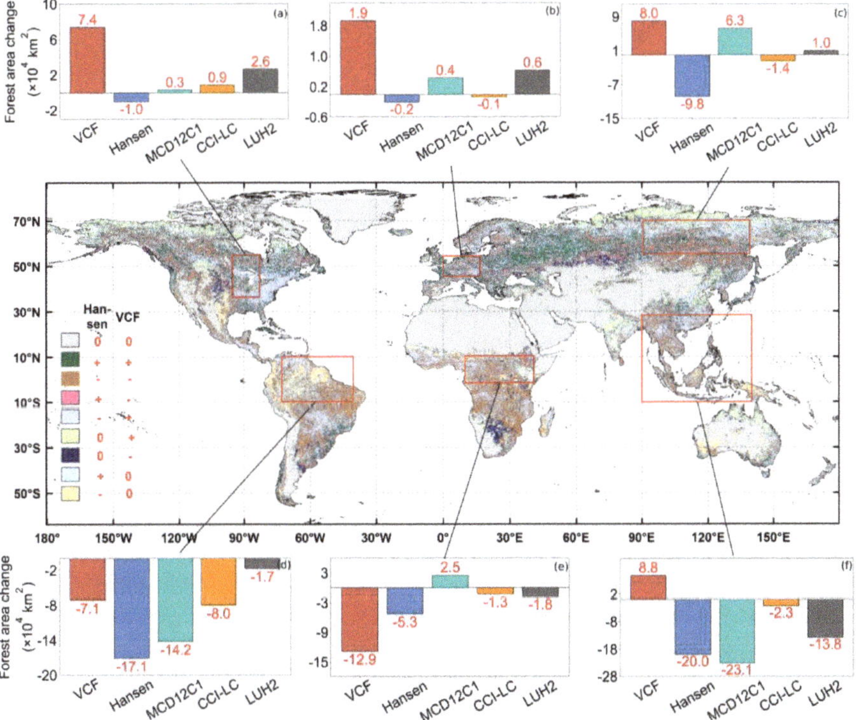

**Figure 6.** Difference in forest area change between VCF and Hansen, and comparisons of forest area changes among the five datasets in several hotspots. (**a**) North America (37.0°N–50.0°N, 95.0°W–85.0°W), (**b**) Europe (45.0°N–54.0°N, 0.0–15.0°E), (**c**) Eastern Siberia (55.0°N–66.5°N, 90.0°E–150.0°E), (**d**) Amazon (10.0°S–10.0°N, 73.0°W–40.0°W), (**e**) Southern Africa (5.0°S–8.0°N, 10.0°E–40.0°E), and (**f**) Southeast Asia (10.0°S–28.0°N, 92.0°E–140.0°E).

### 3.3. Annual Forest Area Change in Five Datasets from 2001 to 2012

Hansen displayed a relatively stable annual forest area change, while other datasets exhibited irregular changes (Figure 7). Hansen was the only dataset that displayed net forest loss for all years during the research period, and its annual forest loss areas were greater than $5.0 \times 10^4$ km$^2$. The subfigure displayed the magnitude of annual forest area change for these datasets, which represented the range of annual forest area change. VCF had the largest magnitude in forest area change. Forest gain exhibited by VCF was larger than $2.0 \times 10^6$ km$^2$ in 2004; forest loss in 2009 was larger than $1.5 \times 10^6$ km$^2$. The number of deforestation years was greater than the number of afforestation years for VCF. Annual forest area changes in MCD12C1, CCI-LC, and LUH2 fluctuated but exhibited a decreasing trends from 2001 to 2012. LUH2 demonstrated a minor total forest area changes with dramatic annual fluctuations in 2009 and 2010.

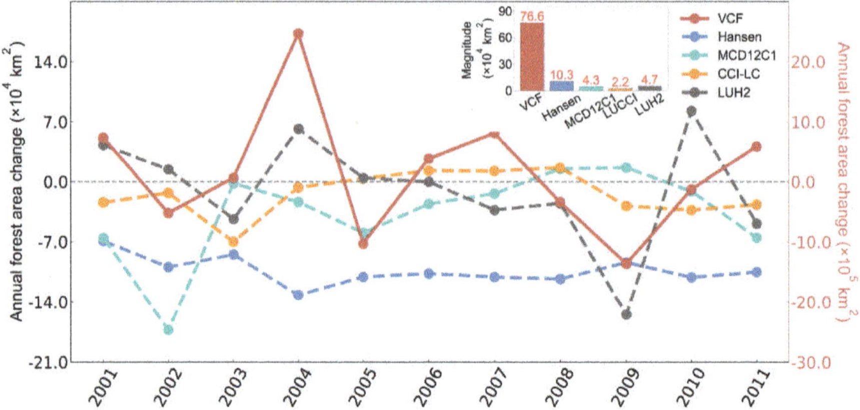

**Figure 7.** Annual forest area changes in VCF, Hansen, MCD12C1, CCI-LC, and LUH2 from 2001 to 2012.

*3.4. Comparison of Forest Area Change at Country Level*

Forest area changes for these datasets at the country level were calculated. Table 2 lists the forest area changes for countries with land areas over $10.0 \times 10^7$ km$^2$. According to Table 2, the datasets exhibited substantial inconsistencies at the country level. Hansen demonstrated decreases in forest area for all the selected countries. Hansen and FRA indicated very close forest area changes in Brazil, Australia, Kazakhstan, India, and Argentina, but there were huge inconsistencies in other countries. The increased forest area indicated by VCF was the largest among the five datasets. Forest loss was indicated in Brazil and Kazakhstan for VCF. From a country perspective, Hansen demonstrated reductions in Russia and India, while the other datasets exhibited increases in these two countries. All the datasets indicated forest loss in Brazil and Argentina, while they exhibited different patterns of forest changes in other countries.

**Table 2.** Comparison of forest area changes during 2001–2012 for VCF, Hansen, FRA, MCD12C1, CCI-LC, and LUH2 for nine major countries with total land areas over $10.0 \times 10^7$ km$^2$.

| | | VCF | Hansen | FRA | MCD12C1 | CCI-LC | LUH2 |
|---|---|---|---|---|---|---|---|
| **Country** | **Forest Area ($\times 10^4$ km$^2$)** | \multicolumn{6}{c}{**Forest Area Change ($\times 10^4$ km$^2$)**} | | | | | |
| Russia | 814.93 | 65.44 | −21.89 | 5.66 | 0.24 | 2.54 | 4.64 |
| Canada | 347.07 | 30.10 | −19.13 | −0.73 | 5.07 | 2.45 | 0.64 |
| USA | 310.10 | 26.88 | −13.24 | 6.56 | −5.19 | −0.97 | 4.67 |
| China | 208.32 | 24.03 | −4.14 | 31.32 | 7.44 | −0.49 | 2.30 |
| Brazil | 493.54 | −11.61 | −29.60 | −27.74 | −19.26 | −9.81 | −8.41 |
| Australia | 124.75 | 6.48 | −4.57 | −4.09 | 2.68 | 0.48 | 1.08 |
| Kazakhstan | 3.31 | −0.43 | −0.06 | −0.06 | −0.29 | 2.08 | −0.03 |
| India | 70.68 | 2.24 | −0.69 | 5.29 | 2.98 | 0.23 | 0.06 |
| Argentina | 27.11 | −4.42 | −4.28 | −4.75 | −0.63 | −3.12 | −0.14 |

## 4. Discussion

### *4.1. Accuracy Assessment Using Global Land Cover Validation*

In this study, five global land cover datasets and FRA reports were evaluated to reveal the inconsistencies in global forest changes from the perspectives of total area, spatial distribution, latitudinal profile, and annual area change from 2001 to 2012. There were huge inconsistencies in global forest changes for the above aspects among the selected datasets. Thus, assessments of data accuracy were essential to understanding the reliabilities of the selected datasets. However, the selected land cover datasets in this study have highly different spatial resolutions (Table 1). Certain very high resolution satellite images (e.g., Google Earth images and QuickBird images) were effective tools to validate several high-resolution datasets (e.g., Hansen data, 30 m spatial resolution; CCI-LC data, 250 m spatial resolution) based on visual interpretation [24] but not suitable to validate datasets with lower resolution (e.g., LUH2 data, with a spatial resolution of $0.25 \times 0.25°$).

Here, the global land cover validation data from USGS was applied to evaluate the accuracies of the selected land cover datasets. Forest cover percentage in each sample was calculated and compared with the forest cover percentages of the land cover datasets (Figure 8). Spatially, underestimation of forest cover percentage commonly occurred in the Amazon, Western USA, and Eastern Europe; on the contrary, overestimation occurred in Eastern Asia, Western Europe, and Northern USA. The correlations between the land cover datasets and the validation data were not very strong: the correlation coefficients were 0.63, 0.77, 0.59, 0.68, and 0.37 for VCF, Hansen, MCD12C1, CCI-LC, and LUH2, respectively. Hansen had the highest correlation with the validation data, while LUH2 had the lowest correlation. Zeng et al. [24] also proved that Hansen had a pretty good accuracy in Southeast Asia across Google Earth imagery, an overall accuracy of 98.4% can be achieved. Therefore, Hansen is recommended to be utilized in forest and forest changes estimates.

We note that Hansen provided the forest gain information as a binary value from 2000 to 2012, so that years when forest gain occurred were not available. To calculate forest cover percentage of Hansen, the pixel number of forest gain in the validation pixel ($0.05 \times 0.05°$) for each year was assumed to be equal in this study. The accuracy of Hansen forest cover percentage based on this hypothesis was lower than the actual accuracy.

### *4.2. Possible issues in VCF*

Previous studies have widely reported that the global forest area has decreased over the past several decades [23,28,31]. According to FRA reports, the total global forest area in the world was $41.3 \times 10^6$ km$^2$, $40.6 \times 10^6$ km$^2$, $40.3 \times 10^6$ km$^2$, $40.2 \times 10^6$ km$^2$, and $40.0 \times 10^6$ km$^2$ in 1990, 2000, 2005, 2010, and 2015, respectively [31]. It is also indicated that the gross forest cover losses in the boreal zone and the temperate zone were $3.5 \times 10^5$ km$^2$ and $1.7 \times 10^5$ km$^2$ from 2000 to 2005 [45]. However, the results based on VCF were contradictory to the general understanding of global forest changes, which indicated an increase in global forest area [27]. VCF revealed an increase of $1.7 \times 10^6$ km$^2$ in total forest area, whereas Hansen, FRA, MCD12C1, CCI_LC, and LUH2 demonstrated decreases of $1.6 \times 10^6$ km$^2$, $0.6 \times 10^6$ km$^2$, $0.4 \times 10^6$ km$^2$, $0.2 \times 10^6$ km$^2$, and $0.1 \times 10^6$ km$^2$, respectively (Figure 1). VCF also exhibited an order of magnitude inter annual forest area change than that in the other land cover datasets. According to these results, inconsistencies among the datasets primarily occurred between VCF and the other datasets. Another implausible behavior of VCF was that its forest area change fluctuated erratically in 2004, 2005, and 2009, indicating the occurrence of rapid forest gains or forest losses (Figure 7). For trees, defined as vegetation taller than 5 m in VCF, it is unreasonable to expect such a rapid change within a short period.

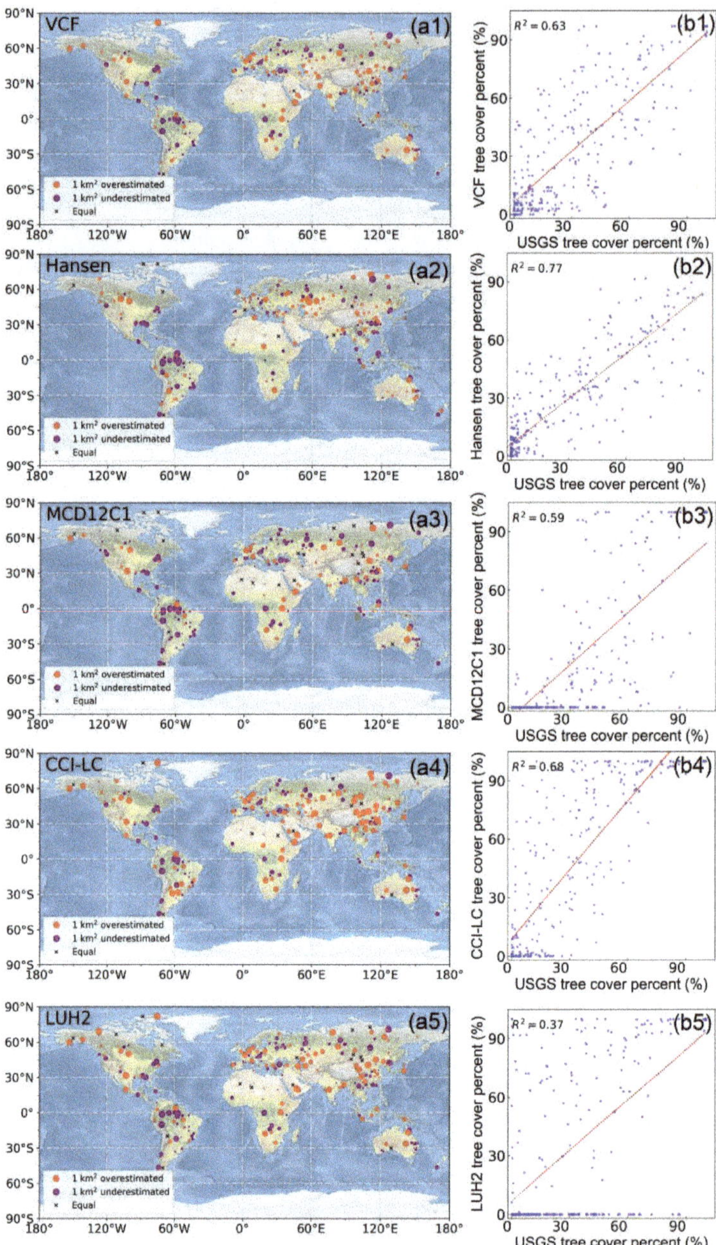

**Figure 8.** Accuracy assessment of the land cover datasets using USGS's global land cover validation data. Spatial pattern of differences in forest cover percentage between the land cover datasets and global land cover validation data (**a1–a5**) and correlation coefficient between the land cover datasets and global land cover validation data (**b1–b5**) for VCF (1), Hansen (2), MCD12C1 (3), CCI-LC (4), and LUH2 (5). Size of dots in (**a1–a5**) represent the difference of forest cover percentage between the land cover datasets and validation data.

To evaluate these inconsistencies between VCF and the other land cover datasets, we compared the spatial distributions of forest area changes from VCF with the land cover from MCD12C1 in 2004 and 2005, during which VCF reported drastic fluctuations in forest area. VCF indicated that the forest area increased by $2.4 \times 10^5$ km$^2$ in 2004 (Figure 9a) and decreased by $1.0 \times 10^5$ km$^2$ in 2005 (Figure 9b), but the forest distribution did not substantially change in 2004 and 2005 (Figure 9c or Figure 9d). MCD12C1 demonstrated that the main land cover types in Siberia were grassland and shrub land. Forest area did not change significantly in this region, while VCF displayed clear forest change in the corresponding period. Minor changes in the Amazon forest were indicated by MCD2C1 data, while VCF indicated a noticeable increase in 2004 and a decrease in 2005 for the same area.

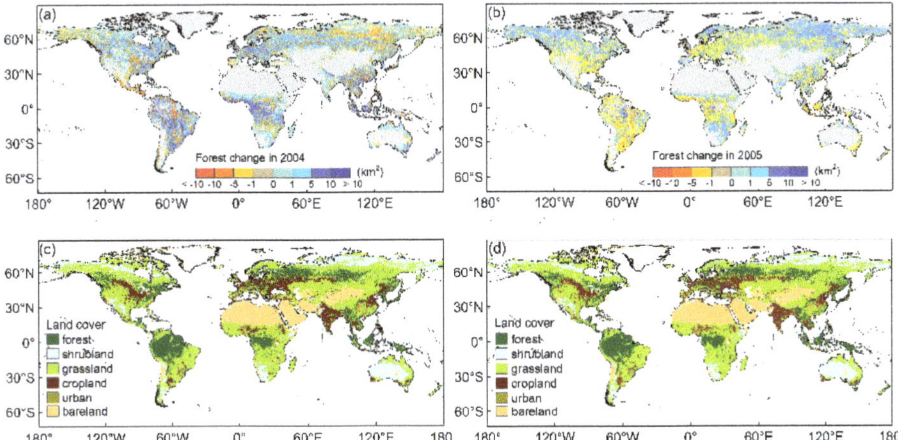

**Figure 9.** VCF forest area change and MCD12C1 land cover in 2004 and 2005, (**a**) forest area change in 2004, (**b**) forest area change in 2005, (**c**) MCD12C1 land cover in 2004, and (**d**) MCD12C1 land cover in 2005.

To investigate the rapid forest gain or loss in 2004, 2005, and 2009 indicated by VCF, we utilized the Google Earth images as reference data to validate the reliability of VCF. The fine resolution Google Earth images displayed land surface conditions for a visual inspection of land surface changes. We selected a target pixel in China (110.525°E, 21.775°N) as an example (the black dot in Figure 10a,b), where VCF reported a large forest increase (+19.2%) in 2004 (Figure 10a) and a substantial forest decrease (−16.6%) in 2005 (Figure 10b). According to the Google Earth images, short vegetation appeared in 2005, but the tree cover did not change significantly (far less than VCF indicated) (Figure 10c–e). We also validated pixels in Southeast Asia and the US with similar results. The validation indicated that VCF cannot distinguish short vegetation (e.g., crops) from forest cover sufficiently. The seasonal growth of short vegetation, such as crops may be one reason for the inconsistencies between VCF and the other datasets.

### 4.3. Impacts of Tree Cover Definition and Land Cover Classification System

The differences in concepts and definitions of "forest" between different land cover datasets lead to different results in forest detection, forest ranges, and forest transitions [46]. The definitions of "forest" in the selected land cover datasets were different from each other. VCF describes the "tree canopy" as consisting of tall vegetation (≥5 m in height) [27]. Hansen defines the tree as all vegetation taller than 5 m in height [28]. The IGBP classification system in MCD12C1 describes "forest" as land dominated by trees with a percent cover >60% and height exceeding 2 m [38]. CCI-LC defines forest as a parcel or unit of land of at least 0.5 hectares in size that is covered by 10% or more trees that are 5 m or taller [39]. FRA defines forest as an area over 0.5 ha with a minimum tree cover of 10%. However, "forest" in FRA reports represents a kind of land use rather than physical trees compared to the satellite-based land cover datasets [1]. An area will be classified as forest if it is registered as "forest" land use even if there is no tree. Thus, forest area measurement in different land cover datasets were not consistent. For example, some short shrubs, which are taller than 2 m, but shorter than 5 m, will be counted as forest in MCD12C1 but will not be counted in VCF and Hansen. Besides, different land cover classification systems in land cover datasets also lead to inconsistencies in forest area, and we cannot compare forest area change directly. In addition, for spatial resolution of the land cover datasets, algorithms to identify forest can also lead to inconsistencies in forest change.

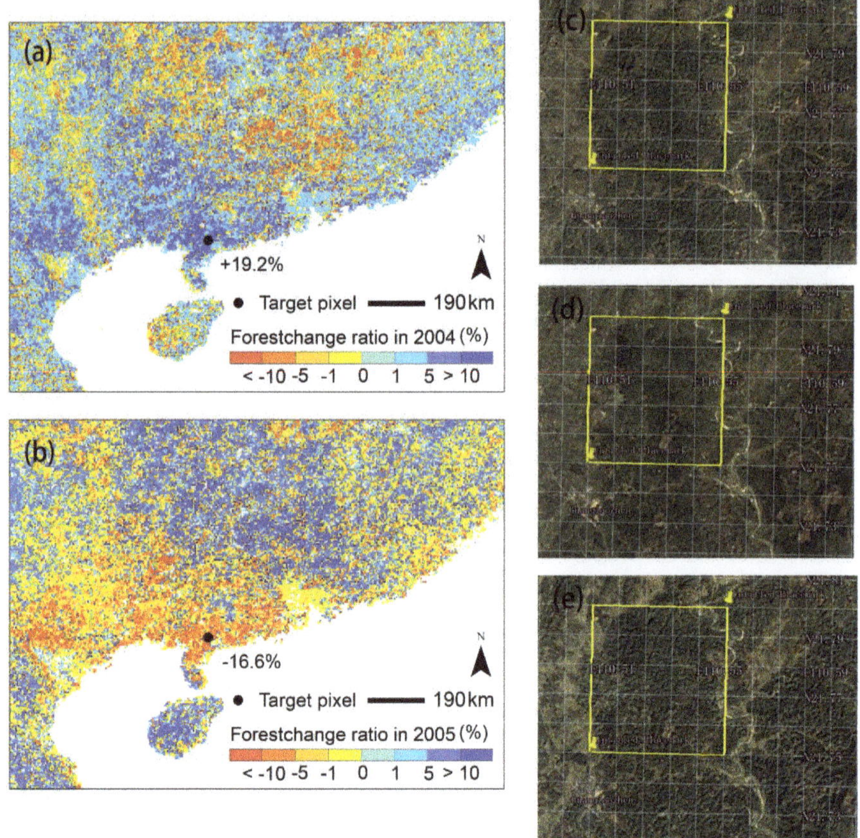

**Figure 10.** Contrast of forest change between VCF and Google Earth images. (**a**) VCF forest area change ratio in 2004, (**b**) VCF forest area change ratio in 2005, (**c**) land surface of target pixel from Google Earth in December 2004, (**d**) land surface of target pixel from Google Earth in December 2005, and (**e**) land surface of target pixel from Google Earth in December 2006.

The inconsistent definition of "forest" was not the major factor for the inconsistencies in the overall global forest area change. VCF and Hansen have the same definition of "forest", but the inconsistencies between them were the largest. VCF displayed the largest forest gain, while Hansen displayed the largest forest loss. So the differences in definition of "forest" were not the decisive factor for the inconsistencies. Another problem for the satellite-based land cover datasets is that they do not assess what happened to the land after deforestation. Regrowth after deforestation is a slow, incremental process compared to deforestation; the latter is easy to spot when comparing satellite images. In addition, it is more difficult to identify deforestation from year to year. Identifying regrowth and low shrubs is also challenging. Therefore, there is no clear answer on how global forest area has changed in the early 21st century. We recommend interpreting these datasets with caution and performing detailed local-scale spatial validation before use. We also highly advocate more accurate forest cover data with higher resolution and a better classification system.

## 5. Conclusions

This study analyzed global forest area changes from 2001 to 2012 using five global land cover datasets and FRA reports. To evaluate the inconsistencies in global forest area changes among these datasets, global forest total area change, spatial distribution, latitudinal profile, and annual area change were compared. We found large inconsistencies on forest changes among these datasets. The majority of the datasets indicated decreases in total global forest area, but VCF demonstrated a net increase trend, which was conflicted with the other datasets and previous studies. The inconsistencies in forest changes primarily occurred in the northern hemisphere, while the southern hemisphere showed more consistent results. These datasets demonstrated forest loss in the tropics, but inconsistent forest changes in other zones. Global land cover validation data from USGS was applied to evaluate the accuracies of the land cover datasets. The results indicated that all of the land cover datasets displayed mediocre accuracies in forest cover. Hansen had the highest correlation with validation data, while LUH2 had the lowest correlation. Therefore, Hansen is recommended for forest cover and forest cover change estimations.

The differences in definition of "forest", classification system, spatial resolution, and algorithm can affect the detection of forest and its changes. Due to these differences and the large uncertainties among these datasets, we need to interpret the results based on these land cover datasets with caution. Validations and assessments of forest and its changes using ground-level field data or very high-resolution imagery (e.g., Google Earth or USGS' validation dataset) are needed. In addition, some methods including data fusion and cross-dataset calibration are useful to reconcile the inconsistencies of the differences in forest definition [47,48].

**Author Contributions:** Conceptualization, Z.Z. and J.L.; methodology, Z.Z.; validation, H.C. and Z.Z.; formal analysis, H.C.; writing—original draft preparation, H.C.; writing—review and editing, Z.Z., J.W., L.P., V.L., H.Y., and J.L.; visualization, H.C. and Z.Z.; supervision, J.L. and Z.Z.; funding acquisition, J.L. and Z.Z. All authors have read and agreed to the published version of the manuscript.

**Funding:** This study was supported by the Strategic Priority Research Program of the Chinese Academy of Sciences (Grant No. XDA20060402), the National Natural Science Foundation of China (NSFC) (Grant No. 41625001; 42071022), and the start-up fund for Dr. Zeng provided by Southern University of Science and Technology (No. 29/Y01296122).

**Acknowledgments:** We thank Bruce Pengra from USGS for providing the global land cover dataset.



## References

1. FAO. *Global Forest Resources Assessments 2020: Main Report*; The Food and Agricultural Organization of the United Nations: Rome, Italy, 2020.
2. Kurz, W.A.; Dymond, C.C.; Stinson, G.; Rampley, G.J.; Neilson, E.T.; Carroll, A.L.; Ebata, T.; Safranyik, L. Mountain pine beetle and forest carbon feedback to climate change. *Nature* **2008**, *452*, 987–990. [CrossRef] [PubMed]
3. Seidl, R.; Thom, D.; Kautz, M.; Martin-Benito, D.; Peltoniemi, M.; Vacchiano, G.; Wild, J.; Ascoli, D.; Petr, M.; Honkaniemi, M.P.J.; et al. Forest disturbances under climate change. *Nat. Clim. Chang.* **2017**, *7*, 395–402. [CrossRef] [PubMed]
4. Keenan, T.F.; Hollinger, D.Y.; Bohrer, G.; Dragoni, D.; Munger, J.W.; Schmid, H.P.; Richardson, A.D. Increase in forest water-use efficiency as atmospheric carbon dioxide concentrations rise. *Nat. Cell Biol.* **2013**, *499*, 324–327. [CrossRef] [PubMed]
5. Williams, S.E.; Marsh, H.; Winter, J. Spatial scale, species diversity, and habitat structure: Small mammals in Australian tropical rain forest. *Ecology* **2002**, *83*, 1317–1329. [CrossRef]

6.    McDonnell, M.J.; Pickett, S.T.A. Ecosystem Structure and Function along Urban-Rural Gradients: An Unexploited Opportunity for Ecology. *Ecology* **1990**, *71*, 1232–1237. [CrossRef]

7.    Bonan, G.B. Forests and Climate Change: Forcings, Feedbacks, and the Climate Benefits of Forests. *Science* **2008**, *320*, 1444–1449. [CrossRef]

8.    Brack, C. Pollution mitigation and carbon sequestration by an urban forest. *Environ. Pollut.* **2002**, *116*, S195–S200. [CrossRef]

9.    Nowak, D.J.; Hirabayashi, S.; Bodine, A.; Greenfield, E. Tree and forest effects on air quality and human health in the United States. *Environ. Pollut.* **2014**, *193*, 119–129. [CrossRef]

10.   North, M.; Stephens, S.L.; Collins, B.M.; Agee, J.K.; Aplet, G.; Franklin, J.F.; Fule, P.Z. Reform forest fire management. *Science* **2015**, *349*, 1280–1281. [CrossRef]

11.   Pureswaran, D.S.; Roques, A.; Battisti, A. Forest Insects and Climate Change. *Curr. For. Rep.* **2018**, *4*, 35–50. [CrossRef]

12.   Bethel, J.S.; Schreuder, G.F. Forest Resources: An Overview. *Science* **1976**, *191*, 747–752.

13.   Zhu, Z.; Piao, S.; Myneni, R.B.; Huang, M.; Zeng, Z.; Canadell, J.G.; Ciais, P.; Sitch, S.; Friedlingstein, P.; Arneth, A.; et al. Greening of the Earth and its drivers. *Nat. Clim. Chang.* **2016**, *6*, 791–795. [CrossRef]

14.   Kovacic, Z.; Salazar, O.V. The lose-lose predicament of deforestation through subsistence farming: Unpacking agricultural expansion in the Ecuadorian Amazon. *J. Rural Stud.* **2017**, *51*, 105–114. [CrossRef]

15.   Feldpausch, T.R.; Phillips, O.L.; Brienen, R.J.W.; Gloor, E.; Lloyd, J.; Lopez-Gonzalez, G.; Monteagudo-Mendoza, A.; Malhi, Y.; Alarcón, A.; Dávila, E.Á.; et al. Amazon forest response to repeated droughts. *Glob. Biogeochem. Cycles* **2016**, *30*, 964–982. [CrossRef]

16.   McPherson, E.G.; Nowak, D.; Heisler, G.; Grimmond, C.; Souch, C.; Grant, R.; Rowntree, R. Quantifying urban forest structure, function, and value: The Chicago Urban Forest Climate Project. *Urban Ecosyst.* **1997**, *1*, 49–61. [CrossRef]

17.   Zak, M.; Cabido, M.; Cáceres, D.; Díaz, S. What Drives Accelerated Land Cover Change in Central Argentina? Synergistic Consequences of Climatic, Socioeconomic, and Technological Factors. *Environ. Manag.* **2008**, *42*, 181–189. [CrossRef]

18.   Yang, J.; Gong, P.; Fu, R.; Zhang, M.; Chen, J.; Liang, S.; Xu, B.; Shi, J.; Dickinson, R.E. The role of satellite remote sensing in climate change studies. *Nat. Clim. Chang.* **2013**, *3*, 875–883. [CrossRef]

19.   Pérez-Hoyos, A.; Rembold, F.; Kerdiles, H.; Gallego, J. Comparison of Global Land Cover Datasets for Cropland Monitoring. *Remote Sens.* **2017**, *9*, 1118. [CrossRef]

20.   Fermi, V.E. Identifying and quantifying uncertainty and spatial disagreement in the comparison of Global Land Cover for different applications. *Glob. Chang. Biol.* **2008**, *14*, 1057–1075. [CrossRef]

21.   Townshend, J.R.; Masek, J.G.; Huang, C.; Vermote, E.F.; Gao, F.; Channan, S.; Sexton, J.O.; Feng, M.; Narasimhan, R.; Kim, D.; et al. Global characterization and monitoring of forest cover using Landsat data: Opportunities and challenges. *Int. J. Digit. Earth* **2012**, *5*, 373–397. [CrossRef]

22.   Qiu, B.; Chen, G.; Tang, Z.; Lu, D.; Wang, Z.; Chen, C. Assessing the Three-North Shelter Forest Program in China by a novel framework for characterizing vegetation changes. *ISPRS J. Photogramm. Remote Sens.* **2017**, *133*, 75–88. [CrossRef]

23.   Keenan, R.J.; Reams, G.A.; Achard, F.; De Freitas, J.V.; Grainger, A.; Lindquist, E. Dynamics of global forest area: Results from the FAO Global Forest Resources Assessment 2015. *For. Ecol. Manag.* **2015**, *352*, 9–20. [CrossRef]

24.   Zeng, Z.; Estes, L.; Ziegler, A.D.; Chen, A.; Searchinger, T.; Hua, F.; Guan, K.; Jintrawet, A.; Wood, E.F. Highland cropland expansion and forest loss in Southeast Asia in the twenty-first century. *Nat. Geosci.* **2018**, *11*, 556–562. [CrossRef]

25.   Gardner, T.A.; Barlow, J.; Chazdon, R.; Ewers, R.M.; Harvey, C.A.; Peres, C.A.; Sodhi, N.S. Prospects for tropicalal forest biodiversity in a human-modified world. *Ecol. Lett.* **2009**, *12*, 561–582. [CrossRef]

26.   Hansen, M.; DeFries, R.S. Detecting Long-term Global Forest Change Using Continuous Fields of Tree-Cover Maps from 8-km Advanced Very High Resolution Radiometer (AVHRR) Data for the Years 1982–99. *Ecosystems* **2004**, *7*, 695–716. [CrossRef]

27.   Song, X.-P.; Hansen, M.C.; Stehman, S.V.; Potapov, P.V.; Tyukavina, A.; Vermote, E.F.; Townshend, J.R. Global land change from 1982 to 2016. *Nat. Cell Biol.* **2018**, *560*, 639–643. [CrossRef]

28. Hansen, M.C.; Potapov, P.V.; Moore, R.; Hancher, M.; Turubanova, S.A.; Tyukavina, A.; Thau, D.; Stehman, S.V.; Goetz, S.J.; Kommareddy, A.; et al. High-Resolution Global Maps of 21st-Century Forest Cover Change. *Science* **2013**, *342*, 850–853. [CrossRef]

29. FAO. *Global Forest Resources Assessments 2000*; FAO Forestry Paper 140; The Food and Agricultural Organization of the United Nations: Rome, Italy, 2001.

30. FAO. *Global Forest Resources Assessments 2010*; FAO Forestry Paper 163; The Food and Agricultural Organization of the United Nations: Rome, Italy, 2010.

31. FAO. *Global Forest Resources Assessments 2015*; FAO Forestry Paper 1; The Food and Agricultural Organization of the United Nations: Rome, Italy, 2015.

32. Ordway, E.M.; Asner, G.P.; Lambin, E.F. Deforestation risk due to commodity crop expansion in sub-Saharan Africa Deforestation risk due to commodity crop expansion in sub-Saharan Africa. *Environ. Res. Lett.* **2017**, *12*, 044015. [CrossRef]

33. Qin, Y.; Xiao, X.; Dong, J.; Zhang, Y.; Wu, X.; Shimabukuro, Y.; Arai, E.; Biradar, C.; Wang, J.; Zou, Z.; et al. Improved estimates of forest cover and loss in the Brazilian Amazon in 2000–2017. *Nat. Sustain.* **2019**, *2*, 764–772. [CrossRef]

34. Gibbs, H.K.; Ruesch, A.S.; Achard, F.; Clayton, M.K.; Holmgren, P.; Ramankutty, N.; Foley, J.A. Tropical forests were the primary sources of new agricultural land in the 1980s and 1990s. *Proc. Natl. Acad. Sci. USA* **2010**, *107*, 16732–16737. [CrossRef]

35. Song, X.; Hansen, M.C.; Stephen, V.; Peter, V.; Tyukavina, A.; Vermote, E.F.; Townshend, J.R. The Vegetation Continuous Fields. 2018. Available online: https://glad.umd.edu/dataset/long-term-global-land-change (accessed on 28 June 2019).

36. Hansen, M.C.; Potapov, P.V.; Moore, R.; Hancher, M.; Turubanova, S.A.; Tyukavina, A.; Thau, D.; Stehman, S.V.; Goetz, S.J.; Loveland, T.R.; et al. Global Forest Change 2000–2017 Data, Version 1.5. 2013. Available online: https://earthenginepartners.appspot.com/science-2013-global-forest/download_v1.5.html (accessed on 28 June 2019).

37. Terra and Aqua combined Moderate Resolution Imaging Spectroradiometer Land Cover Climate Modeling Grid (MCD12C1). Available online: https://lpdaac.usgs.gov/products/mcd12c1v006/ (accessed on 3 July 2019).

38. Sulla-menashe, D.; Friedl, M.A. *User Guide to Collection 6 MODIS Land Cover (MCD12Q1 and MCD12C1) Product*; USGS: Reston, VA, USA, 2018; pp. 1–18.

39. The Europe Space Agency (ESA). Land Cover Project of the Climate Change Initiative (CCI-LC) Data. Available online: https://www.esa-landcover-cci.org/?q=node/1 (accessed on 3 July 2019).

40. Bontemps, S.; Herold, M.; Kooistra, L.; Van Groenestijn, A.; Hartley, A.; Arino, O.; Moreau, I.; Defourny, P. Revisiting land cover observation to address the needs of the climate modeling community. *Biogeosciences* **2012**, *9*, 2145–2157. [CrossRef]

41. Hurtt, G.C.; Chini, L.; Sahajpal, R.; Frolking, S.; Bodirsky, B.L.; Calvin, K.; Doelman, J.C.; Fisk, J.; Fujimori, S.; Goldewijk, K.K.; et al. Harmonization of Global Land-Use Change and Management for the Period 850–2100 (LUH2) for CMIP6. *Geosci. Model Dev. Discuss.* **2020**, *2020*, 1–65. [CrossRef]

42. Global Ecology Laboratory, University of Maryland. The New Generation of Land-Use Harmonization (LUH2). Available online: https://www.wcrp-climate.org/wgcm-cmip/wgcm-LUH2 (accessed on 28 June 2019).

43. Olofsson, P.; Stehman, S.V.; Woodcock, C.E.; Sulla-Menashe, D.; Sibley, A.M.; Newell, J.D.; Friedl, M.A.; Herold, M. A global land-cover validation data set, part I: Fundamental design principles. *Int. J. Remote Sens.* **2012**, *33*, 5768–5788. [CrossRef]

44. Pengra, B.; Long, J.; Dahal, D.; Stehman, S.V.; Loveland, T.R. A global reference database from very high resolution commercial satellite data and methodology for application to Landsat derived 30 m continuous field tree cover data. *Remote Sens. Environ.* **2015**, *165*, 234–248. [CrossRef]

45. Hansen, M.C.; Stehman, S.V.; Potapov, P.V. Quantification of global gross forest cover loss. *Proc. Natl. Acad. Sci. USA* **2010**, *107*, 8650–8655. [CrossRef]

46. Chazdon, R.L.; Brancalion, P.H.S.; Laestadius, L.; Bennett-Curry, A.; Buckingham, K.; Kumar, C.; Moll-Rocek, J.; Vieira, I.C.G.; Wilson, S.J. When is a forest a forest? Forest concepts and definitions in the era of forest and landscape restoration. *Ambio* **2016**, *45*, 538–550. [CrossRef] [PubMed]

*Remote Sens.* **2020**, *12*, 3502

47. Lesiv, M.; Moltchanova, E.; Schepaschenko, D.; See, L.; Shvidenko, A.; Comber, A.; Fritz, S. Comparison of data fusion methods using crowd sourced data in creating a hybrid forest cover map. *Remote Sens.* **2016**, *8*, 261. [CrossRef]
48. Schepaschenko, D.; See, L.; Lesiv, M.; McCallum, I.; Fritz, S.; Salk, C.; Moltchanova, E.; Perger, C.; Shchepashchenko, M.; Shvidenko, A.; et al. Development of a global hybrid forest mask through the synergy of remote sensing, crowdsourcing and FAO statistics. *Remote Sens. Environ.* **2015**, *162*, 208–220. [CrossRef]

**Publisher's Note:** MDPI stays neutral with regard to jurisdictional claims in published maps and institutional affiliations.

*Article*

# A Model-Based Volume Estimator that Accounts for Both Land Cover Misclassification and Model Prediction Uncertainty

Jessica Esteban [1,2,*], Ronald E. McRoberts [3], Alfredo Fernández-Landa [2], José Luis Tomé [2] and Miguel Marchamalo [1]

[1] Departamento de Ingeniería y Morfología del Terreno, Laboratorio de Topografía y Geomática, Universidad Politécnica de Madrid, 28040 Madrid, Spain; miguel.marchamalo@upm.es
[2] AGRESTA Sociedad Cooperativa, 28012 Madrid, Spain; afernandez@agresta.org (A.F.-L.); jltome@agresta.org (J.L.T.)
[3] Department of Forest Resources, University of Minnesota, St. Paul, MN 55108, USA; mcrob001@umn.edu
* Correspondence: j.esteban@upm.es

Received: 12 August 2020; Accepted: 10 October 2020; Published: 15 October 2020

**Abstract:** Forest/non-forest and forest species maps are often used by forest inventory programs in the forest estimation process. For example, some inventory programs establish field plots only on lands corresponding to the forest portion of a forest/non-forest map and use species-specific area estimates obtained from those maps to support the estimation of species-specific volume (V) totals. Despite the general use of these maps, the effects of their uncertainties are commonly ignored with the result that estimates might be unreliable. The goal of this study is to estimate the effects of the uncertainty of forest species maps used in the sampling and estimation processes. Random forest (RF) per-pixel predictions were used with model-based inference to estimate V per unit area for the six main forest species of La Rioja, Spain. RF models for predicting V were constructed using field plot information from the Spanish National Forest Inventory and airborne laser scanning data. To limit the prediction of V to pixels classified as one of the main forest species assessed, a forest species map was constructed using Landsat and auxiliary information. Bootstrapping techniques were implemented to estimate the total uncertainty of the V estimates and accommodated both the effects of uncertainty in the *Landsat forest species map* and the effects of plot-to-plot sampling variability on training data used to construct the RF V models. Standard errors of species-specific total V estimates increased from 2–9% to 3–22% when the effects of map uncertainty were incorporated into the uncertainty assessment. The workflow achieved satisfactory results and revealed that the effects of map uncertainty are not negligible, especially for open-grown and less frequently occurring forest species for which greater variability was evident in the mapping and estimation process. The effects of forest map uncertainty are greater for species-specific area estimation than for the selection of field plots used to calibrate the RF model. Additional research to generalize the conclusions beyond Mediterranean to other forest environments is recommended.

**Keywords:** random forests; error propagation; bootstrapping; Landsat; LiDAR; La Rioja

---

## 1. Introduction

Forest and other wooded land environments provide numerous ecosystem services that contribute directly or indirectly to improving our well-being. They play an important role in mitigating climate change through carbon sequestration, combating desertification, maintaining biodiversity, protecting soil and water resources, and supplying wood and other forest products [1] (pp. 285–299). In light of the significance of these roles, accurate and updated information suitable for assessing forest resources is needed. National forest inventories (NFIs) were originally motivated by a need for information on forest area, volume, and increment of growing stock and the amount of timber [2]. Since then, many countries have established their NFI programs as the primary sources of forest information necessary for forest management and policy formulation [3,4].

NFI data include measurements of individual trees on plots selected using probabilistic sampling designs. Tree measurements include height and diameter, which are then used to predict individual tree attributes such as volume and biomass [5]. To increase the efficiency of inventory sampling designs, maps can be used in the design phase to select the locations of the field plots [6] and in the estimation phase to increase precision [7,8]. Of the 37 countries for which Tomppo et al. [9] include NFI reports, six countries use a forest/non-forest map to determine where to establish their field plots. Among them is Spain, where permanent sampling plots are established systematically at the intersections of a 1-km × 1-km grid in the portion of the Spanish National Forest Map (SNFM) classified as forest land. Therefore, the SNFM is the base mapping layer for the Spanish National Forest Inventory (SNFI), both of which are updated on a 10-year cycle [10]. The most recent versions of the SNFM are based on the photointerpretation of aerial photography and digitization of polygons that are classified with respect to the vegetation present in the area [11].

Traditional forest inventories have the disadvantage of being expensive, especially in areas with poor road infrastructure [12]. In the last decades, remote sensing (RS) technologies have emerged as an auxiliary data source that alleviates some of this limitation by increasing the precision of inventory estimates and reducing the cost of forest resources assessment [13]. These technologies typically entail constructing a model that represents the relationship between the RS data and field plot data. The model is applied to predict forest attributes where field plots are lacking, thereby producing spatially continuous forest attribute information [14]. These modeling applications have led to inferential approaches guided by the properties of the ground data [15]. In particular, model-based (MB) inferential approaches can be used when models are constructed using data collected from non-probability training samples and data external to the area of interest [16]. This feature makes MB inference an interesting option for areas where design-based inference is limited due to remoteness and inaccessibility [17]. However, the effectiveness of this approach relies on the correctness of the model, and consequently, population parameter estimators may be biased and imprecise if the model is incorrect [18,19]. Hence, the estimation of forest population parameters and a rigorous assessment of their uncertainties are necessary to produce reliable estimates.

In this context, reliable methods for propagating the main sources of uncertainty associated with forest predictions have been developed [12,20,21]. As noted previously, forest maps can be used as masks in the design phase to restrict the establishment of field plots and in the estimation phase to restrict the application of the models. Most current reports in the literature assume that these maps are without error [17,18], but very few have analyzed the effects of this source of uncertainty on the uncertainty of forest attribute estimates. Rodríguez-Veiga et al. [22] used multiple forest masks constructed using MODIS and ALOS PALSAR data to estimate total aboveground biomass for Mexico. Their study demonstrated that different forest masks had large impacts on the estimation of national carbon stocks due to the discrepancies in the forest extent estimated from each forest mask. Li et al. [23] found substantial differences in the regional climate modeling outputs when the uncertainty of the MODIS land cover products used was considered.

These studies showed how the uncertainty inherent in forest maps affects the reliability of estimates. Both national estimates and uncertainties of the estimates are required for NFI reporting and are specifically required for greenhouse gas inventories. In particular, the Intergovernmental Panel on Climate Change (IPCC) states as a guideline that uncertainties should be reduced as far as practicable [24]. However, the satisfaction of this guideline implies that before uncertainties can be reduced, they must first be properly estimated. Failure to estimate the uncertainty associated with forest maps used to restrict the establishment of ground plots and application of models could lead to imprecise estimates and under-estimated uncertainties, thereby leading to reporting estimates that would fail to comply with the IPCC guidelines. In addition, sampling completely within a forest mask facilitates the estimation of deforestation but not reforestation or afforestation outside the map unless the mask is updated frequently. In the last decades, Spain has reported an expansion of forest area because of the naturalization of marginal agricultural land due to rural abandonment and of afforestation policies through the Common Agricultural Policy [25].

The goal of this study is to estimate the effects of the uncertainty of forest species maps used in the sampling and estimation processes. To this end, we addressed the following objectives: (1) to provide pixel-level volume (V) predictions for a large region using three sources of information: SNFI ground plot data, multispectral data, and airborne laser scanning (ALS) data; (2) to estimate the total MB uncertainty of the large area V estimates taking into account both the uncertainty of a forest species map that guided the selection of plot locations and application of models, and sampling variability; and (3) to estimate the relative contributions to the total uncertainty in the large area estimates of each of the components. MB inference used random forests V models specific to each of the main forest species of La Rioja whose spatial distributions were initially determined from a forest species map constructed using Landsat imagery. Random forests (RF) predictive V models and RF classification models were constructed.

## 2. Materials and Methods

### 2.1. Study Area

The study area was La Rioja, Spain, covering an area of 5045 km$^2$ (Figure 1). This province in the north of Spain borders mostly Navarre, Castile, and Leon, and also the Basque Country and Aragon. La Rioja is surrounded by two large relief units with elevations ranging from 260 to approximately 2300 m. This large altitudinal gradient contributes to rich vegetative diversity ranging from coniferous forests, deciduous forests, mixed forests, and shrub lands to grasslands in a relatively small area. According to the SNFI, forest land (without considering shrublands and grasslands) covers 34.7% of La Rioja with the main forest species, in order of area: Pyrenean oak (*Quercus pyrenaica Willd*), Scots pine (*Pinus sylvestris* L.), beech (*Fagus sylvatica* L.), and holly oak (*Quercus ilex* L.) [26] (p. 11). The remaining part of the study area is lowlands composed of unirrigated and irrigated fields where the landscape becomes more homogenous. La Rioja includes important natural environmental aspects such as Sierra de Cebollera Natural Park and Urbion Lake, among others. In addition, 24% of La Rioja was declared a Biosphere Reserve by UNESCO.

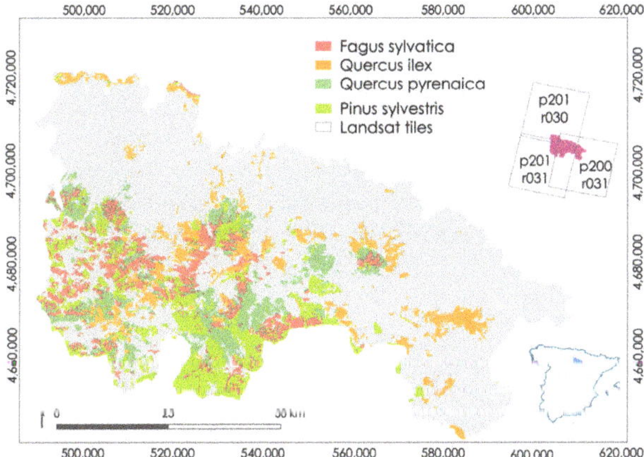


**Figure 1.** La Rioja study area and distribution of the main forest species based on the Spanish National Forest Map. ETRS89/UTM zone 30N (N-E) (EPSG: 3042). The blue polygon in the lower right corner depicts the location of the study area in north central Spain.

*2.2. Data*

Three types of data were used in this research: the SNFI field data, multispectral data, and ALS data.

### 2.2.1. Spanish National Forest Inventory (SNFI)

The 4th SNFI in La Rioja was conducted between 2011 and 2012 using permanent sample plots established systematically at the intersections of a 1 km × 1 km grid in areas identified as forest land by the SNFM (E: 1:25,000) (SNFM25) [10]. The methodology for producing SNFM25 includes three phases: (i) the manual digitalization of polygons with homogenous forest structure and forest species on screen by photo-interpretation, (ii) field monitoring, and (iii) quality control. Field visits are programmed for quality control over the polygons whose digitalization are problematic [10]. Discrepancies between SFM25 and SNFI are analyzed, and modifications are made if needed.

The SNFI sample units consist of four circular concentric plots of radius 5, 10, 15, and 25 m. Trees with diameter at breast height (DBH) of at least 75 mm are measured in a 5 m radius plot; trees with DBH > 125 mm are measured in a 10 m radius plot; trees with DBH > 225 mm are measured in a 15 m radius plot; and only trees with a DBH > 425 mm are measured in the 25 m radius plot [10,27]. For each tree, DBH, total height, forest species, and the tree's position relative to the plot center (direction and distance) are recorded. Species-specific allometric models were used with measured DBH and height to predict individual tree volumes which were weighted to predict the total plot volume. The main difficulty in combining SNFI plot data with ALS measurements is discrepancies between the attribute measured in field and the ALS data assigned due to center plot coordinate errors [28]. Therefore, only plots for which the maximum height of a measured tree, hmax, was within ±4 m of the 99th ALS height percentile were considered. The deletion of 153 of these plots altered the systematic nature of the grid-based sample with the result that the remaining sample consisting of 1095 SNFI plots was considered a purposive sample, albeit with systematic components. This sample was used to construct the volume models (Section 2.4.3).

### 2.2.2. Multispectral and Auxiliary Data

The study area is covered by three Landsat scenes with paths (p) and rows (r): p201 r031, p200 r031, and p200 r030 (Figure 1). For each scene, predominantly cloud-free Landsat 5 Thematic Mapper (TM) images from 1 June to 31 August for 2010 were used to construct an updated forest species map (Section 2.4.2). The reason behind the decision to use Landsat 5 data is their availability for free and absence of sensor malfunction problems as occurred in Landsat 7. Annual summer composites based on the greenest pixels available defined by Normalized Difference Vegetation Index (NDVI) were constructed using the Google Earth Engine Python and Javascript Application Programming Interfaces [29] and resampled to the corresponding 25 m × 25 m ALS cell size (Section 2.2.3). For this study, we used the following bands: blue (0.45–0.52 μm), green (0.52–0.60 μm), red (0.63–0.69 μm), near infrared (NIR; 0.76–0.90 μm), and two shortwave IR bands (SWIR1, 1.55–1.75 μm; and SWIR2, 2.08–2.35 μm). Three vegetation indexes were derived from the annual summer composites: NDVI, the Normalized Difference Moisture Index (NDMI) and the Normalized Burn Ratio (NBR). Finally, elevation information was derived from a 25 m × 25 m digital elevation model downloaded from the Spanish National Center of Geographic Information.

### 2.2.3. Airborne Laser Scanning (ALS) Data

ALS data were acquired between August and October 2010, during leaf-on conditions by the Spanish National Programme of Aerial Orthophotography with a mean pulse density of 0.5 points per $m^2$ and vertical root mean square error (RMSE) ≤ 0.20 m. ALS tiles were processed using FUSION software [30] to construct a 2-m digital elevation model from the ground points, thereby facilitating the estimation of height above the ground surface for each ALS vegetation point. Following the removal of ground points with heights less than 2 m, 15 forest structure metrics were calculated for both the

SNFI plots and for the 25 m × 25 m cells that tessellated the study area and served as population units. ALS metrics included mean, variance (varia), standard deviation (stdev), coefficient of variation (cv), interquartile range (iq), kurtosis (kurto), percentiles (ranging from the 1st to 99th percentile: p1, p5, p25, p50, p75, p95, and p99), canopy relief ratio (crr), and forest canopy cover (lfcc).

The methodological framework is illustrated in Figure 2.

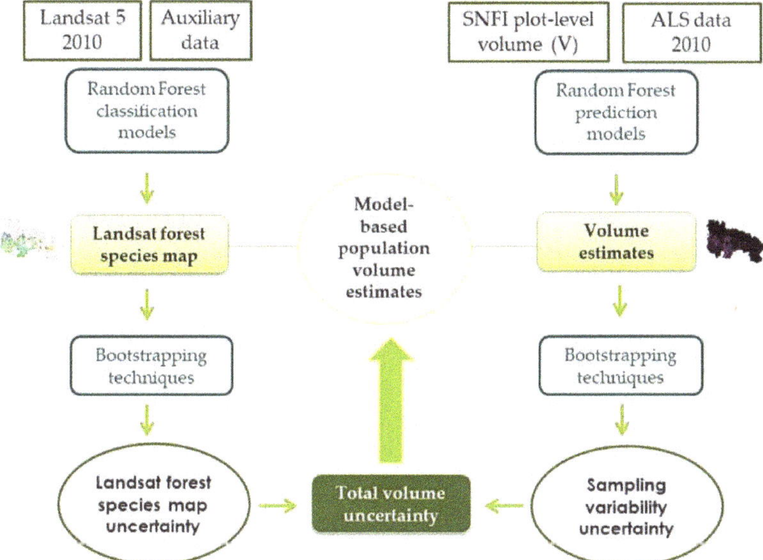


**Figure 2.** Workflow diagram of the processing steps conducted to account for the sources of uncertainty involved in the volume estimation.

## 2.3. Statistical Techniques

### 2.3.1. Overview

Three primary statistical techniques were used in the study. First, the RF method (Section 2.3.2) was used for classification to construct a forest species map (Section 2.4.2) and to predict volume (V) for individual population units (Sections 2.4.3 and 2.4.4). Second, the sampling design used to acquire the model calibration data had both systematic and purposive features (Section 2.2.1). As a result of the purposive features, model-based (MB) population parameter estimation methods were used (Section 2.3.3). Third, because the V estimation procedure had multiple estimation and uncertainty components and because analytical procedures for estimating uncertainties associated with the RF approach are generally not available, bootstrapping procedures were used. For the estimation of uncertainties associated with the forest species map, *pairs bootstrapping* (*p*) was used (Section 2.3.4). However, for the estimation of uncertainties associated with volume estimation, for which the model calibration data were acquired using a design with substantial systematic components, *wild bootstrapping* (*w*) was used (Section 2.3.4). These three statistical techniques are described in greater detail in the sections that follow immediately.

### 2.3.2. Random Forests (RF)

Forest population parameter estimation using models that represent relationships between inventory variables and RS observations has become relatively common in recent decades [31–33]. For this study, RF [34] was selected as the model or prediction technique due to its utility and its popularity [35,36]. Multiple studies support the reliability of RF as a robust classifier and prediction model for forest attributes when used with RS auxiliary data [37–41]. Although the term "regression" has often been used in combination with RF, in this study, the term "prediction model" was used to avoid confusion with linear or nonlinear regression models.

RF consists of a combination of decision tree predictors, each of which is preceded by a bootstrap resample usually consisting of 2/3 of the original sample data. The remaining 1/3 of the original sample is retained as another subset called out-of-bag (oob) and is used for internal error estimation [42]. For this study, RF models were calibrated using the R package *RandomForest* [43] with the default settings of number of trees (500) and of predictor variables (mtry) tested at each node (mtry = p/3 for prediction or mtry = sqrt (p) for classification, where p is the total number of predictor variables).

### 2.3.3. Model-Based (MB) Estimation

For the estimation of species-specific V mean, MB inferential methods that do not require probability samples were used. As previously noted (Section 2.2.1), the available SNFI sample was considered a non-probability, purposive sample. In addition, new forest species maps constructed as part of the uncertainty estimation procedure (Sections 2.4.2 and 2.4.4) may extend into areas originally classified as non-forest in which case no sample plots will be available, thereby further altering the systematic nature of the sample and deviating from a probability sample. The MB estimator of the species-specific V mean is

$$\hat{\mu}_{MB} = \frac{1}{N} \sum_{i=1}^{N} \hat{v}_i, \tag{1}$$

where $\hat{v}_i$ is the RF prediction of $v$ for the $i$th population unit (pixel, map unit) and $N$ is the total number of population units for each forest species for which V prediction models were applied.

### 2.3.4. Bootstrapping

Four primary sources of uncertainty are associated with all MB predictions: (i) model misspecification, (ii) uncertainty in the values of the independent variables, (iii) residual variability around model predictions, and (iv) uncertainty in the model predictions resulting from the effects of sampling variability as they affect the model calibration dataset [16]. For this study, the RF prediction technique was assumed to be sufficiently accurate that model misspecification was not considered problematic (Section 3.1.2). Furthermore, uncertainty in the values of the independent variables were considered negligible. Finally, for large forest areas, the effects of model prediction uncertainty resulting from sampling variability dominate the effects of residual uncertainty [44]. Therefore, for this study, all the sources were ignored except the effects of sampling variability on the V model calibration dataset. When a regression model is used for prediction purposes, the effects of this sampling variability can be expressed in terms of the covariances for the model parameter estimates. However, when non-parametric prediction techniques such as RF are used, more complex techniques merit consideration [44]. For these situations, bootstrapping techniques [33,45,46] have emerged as robust alternatives to analytical variance estimators and uncertainty propagation [47,48].

Two bootstrapping techniques were considered for this study to estimate the uncertainty of the MB population parameter estimates: pairs bootstrapping [49] and wild bootstrapping [50]. Pairs bootstrapping, also characterized as non-parametric bootstrapping [51], entails randomly selecting with replacement a resample of the original data of the same size as the original sample. With pairs bootstrapping, the resample will omit some of the original sample units but include some of the original sample units multiple times. A basic bootstrap principle is that the resampling procedure should mimic the original sampling scheme [50–53] (p. 2). Thus, unless the original sample was acquired using simple random sampling, conventional RF that uses pairs bootstrapping does not preserve the underlying spatial structure of systematic sample, or samples with substantial systematic components such as the SNFI field plot sample.

As an alternative to pairs bootstrapping, Liu [50] proposed the *wild bootstrap*, which retains the full set of original sample units and, therefore, retains any original spatial structure in the sample data. Wild bootstrapping entails two steps. First, predictions and residuals for the original sample are calculated using an arbitrary prediction technique. Second, the wild bootstrap resample is constructed as the predictions plus products of the residuals and a randomly selected variable from a distribution with mean 0 and standard deviation of 1 [54–56]. For this study, a standard normal distribution was used but with truncation of values less than −2.0 and greater than 2.0. The advantage of wild bootstrapping is that it preserves the original spatial structure of the data, but it requires the calculation of predictions and residuals for the original sample units using another technique for which pairs bootstrapping is a viable candidate and is easily implemented for continuous response variables.

*2.4. Analyses*

2.4.1. Overview

The analyses focused on estimating species-specific mean and total V. A key feature of the study was that RF models for predicting V were constructed using data for only those field plots that were located in the forest portion of a forest species map. Thus, uncertainty in the estimates of mean and total V must incorporate the uncertainty from two sources: uncertainty in the forest species map and the sampling variability in the model calibration data. The analyses focused on four tasks. First, a forest species map was constructed using training and Landsat data. The map was first used to estimate species-specific areas using MB methods and their standard errors using pairs bootstrapping (Section 2.4.2). Second, the effects of sampling variability associated with the portion of the SNFI field dataset used to calibrate the models were estimated using wild bootstrapping (Section 2.4.3). Third, the map was used to limit the plots whose data were used to construct the RF V models and to limit the population units to which the models were applied. MB methods were used to estimate mean and total V and their standard errors (SE) (Section 2.4.4). Finally, total uncertainty was estimated by combining the effects of uncertainty in the forest species map and sampling variability in the model calibration data (Section 2.4.5). Methods for accomplishing these tasks are described in the sections that follow immediately.

2.4.2. The Forest Species Map and the Effects of Its Uncertainty on Area Estimates

A forest species map, hereafter called the *Landsat forest species map*, was constructed for the study area (Figure 1) to depict the six dominant forest species of La Rioja: *Fagus sylvatica* (FS), *Pinus halepensis* (PH), *Pinus nigra* (PN), *Pinus sylvestris* (PS), *Quercus pyrenaica/faginea* (Q), and *Quercus ilex* (QI). The remaining forest species occurring in the study area were classified into two general groups designated as "Other coniferous" (OC) and "Other broadleaves" (OB). Non-forest areas such as water bodies, agricultural areas, bare soil, urban fabric, shrublands and grasslands were merged into a single non-forest (NF) class. Training areas were digitalized using the combination of fine resolution imagery and information on forest species from the SNFM25 map which, for purposes of training area delimitation, were considered as "ground truth" and not subject to uncertainty. Once the training areas were delineated, a stratified sample of 100 points was selected for which the nine species classes served as strata. The number of points per training area could be one, several, or no points. Spectral bands, vegetation indexes, and auxiliary information (Section 2.2.2) were used as predictor variables for the calibration of the RF classification models. To assess the accuracy of the *Landsat forest species map*, the RF oob error was analyzed. This oob error estimation is considered a reliable source that can replace a test dataset independent of the training dataset for the algorithm [42].

Uncertainty in the *Landsat forest species map* induces uncertainty in the species-specific area estimates. Hence, to estimate the effects of this source of uncertainty, a 4-step bootstrap procedure was used:

(1)  A pairs bootstrap resample was selected from the training data used to calibrate the RF classification model,

(2)  A new *Landsat forest species map* was constructed by applying a new RF classification model based on the resample from step (1),

(3)  The area for each of the dominant forest species, k, for each bootstrap iteration, b, was estimated as the product of the number of pixels classified as the species and the pixel area and was denoted $\hat{A}^k_{p\,b}$ where the subscript "$p$" indicates that pairs bootstrapping was used,

(4)  Steps (1)–(3) were replicated 2000 times,

(5)  The MB estimates of species-specific areas and their SEs were estimated as,

$$\hat{A}^k_{p\,map} = \frac{1}{n_{boot}} \sum_{b=1}^{nboot} \hat{A}^k_{p\,b} \tag{2}$$

and

$$SE\left(\hat{A}^k_{p\,map}\right) = \sqrt{\frac{1}{n_{boot}-1} \sum_{b=1}^{nboot} \left(\hat{A}^k_{p\,b} - \hat{A}^k_{p\,map}\right)^2} \tag{3}$$

where the subscript "*map*" indicates that only the uncertainty in the *Landsat forest species map* was incorporated into Equation (3), and the subscript "*b*" indexes the bootstrap resamples.

2.4.3. The Effects of Sampling Variability in the Volume Model Calibration Data on Volume Estimates

Species-specific RF models of the relationship between mean V per unit area (m$^3$/ha) as the response variable and the set of ALS metrics as the predictor variables (Section 2.2.3) were constructed for each of the six dominant forest species in La Rioja (FS, PH, PN, PS, Q, QI) (see Table A1 in Appendix A). Although OB and OC were discriminated on the *Landsat forest species map*, V models for OB and OC were not constructed and, therefore, mean and total V for these forest species were not estimated for this study.

RF models were calibrated using the original SNFI field plot dataset (Table 1) subject to two constraints: (i) only data for plots that were in forest land portion of the original *Landsat forest species map*, and (ii) only data for plots whose ground measurements were of that forest species without regard to the forest species predicted by the *Landsat forest species map*.

Table 1. Statistics for mean volume (m$^3$/ha) obtained from the Spanish National Forest Inventory (SNFI) field plots classified as forest according to the *Landsat forest species map* and used for the construction of the V models.

| Forest Species * | Number of SNFI Plots | Mean | Standard Deviation | Minimum | Maximum |
|---|---|---|---|---|---|
| FS | 182 | 192.62 | 93.07 | 5.89 | 530.81 |
| PH | 35 | 80.58 | 42.44 | 9.06 | 164.24 |
| PN | 82 | 137.69 | 94.96 | 3.71 | 415.36 |
| PS | 199 | 225.12 | 136.10 | 2.53 | 756.76 |
| Q | 272 | 90.31 | 55.36 | 1.44 | 305.75 |
| QI | 136 | 52.91 | 34.99 | 0.00 | 215.51 |

* FS: *Fagus sylvatica*; PH: *Pinus halepensis*; PN: *Pinus nigra*; PS: *Pinus sylvestris*; Q: *Quercus pyrenaica/faginea*; QI: *Quercus ilex*.

For each forest species, the species-specific V model was used to predict V for each population unit classified as that species in the original *Landsat forest species map*. Corresponding SEs for the estimates of mean and total V were estimated using a 7-step wild bootstrapping procedure (Figure 3):

(1)  Select a wild bootstrap resample from the SNFI field plot dataset subject to the two previously noted constraints,

(2)  Calibrate the species-specific RF V models,

(3) For each species, k, predict V for all population units classified as that species in the original *Landsat forest species map*,

(4) Estimate mean species-specific V as $\hat{V}^{k}_{w\,b}$ using Equation (1) and total V, $\widehat{VT}^{k}_{w\,b}$, as the product of the estimates of mean V and the area, $\hat{A}^{k}$, from the original *Landsat forest species map*,

$$\widehat{VT}^{k}_{w\,b} = \hat{V}^{k}_{w\,b} \cdot \hat{A}^{k} \tag{4}$$

where the subscript "$w$" indicates that wild bootstrapping was used.

(5) Repeat steps (1)–(4) 2000 times,

(6) Estimate species-specific mean V and its SE as,

$$\hat{V}^{k}_{w\,plot} = \frac{1}{n_{boot}} \sum_{b=1}^{nboot} \hat{V}^{k}_{w\,b} \tag{5}$$

with

$$SE\left(\hat{V}^{k}_{w\,plot}\right) = \sqrt{\frac{1}{n_{boot}-1} \sum_{b=1}^{nboot} \left(\hat{V}^{k}_{w\,b} - \hat{V}^{k}_{w\,plot}\right)^{2}}, \tag{6}$$

(7) Estimate species-specific total V and its *SE* as,

$$\widehat{VT}^{k}_{w\,plot} = \frac{1}{n_{boot}} \sum_{b=1}^{nboot} \widehat{VT}^{k}_{w\,b} \tag{7}$$

with

$$SE\left(\widehat{VT}^{k}_{w\,plot}\right) = \sqrt{\frac{1}{n_{boot}-1} \sum_{b=1}^{nboot} \left(\widehat{VT}^{k}_{w\,b} - \widehat{VT}^{k}_{w\,plot}\right)^{2}} \tag{8}$$

where the subscript "*plot*" indicates that only the effects of sampling variability in the RF model calibration dataset are incorporated into Equations (6) and (8).

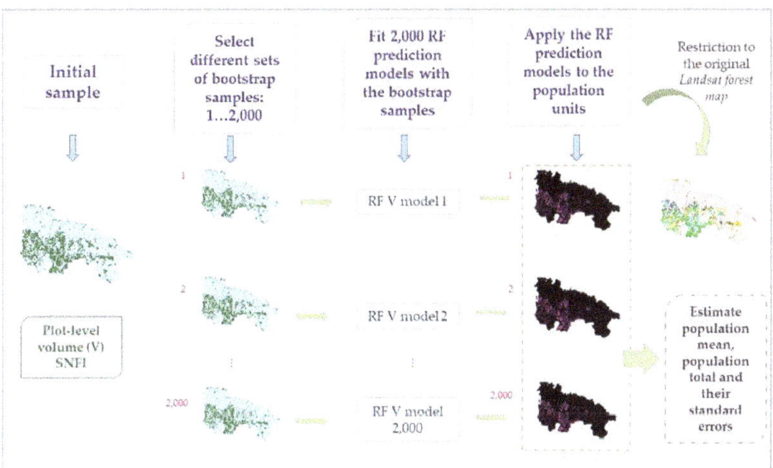


**Figure 3.** Wild bootstrapping scheme conducted in this study to account for the effects of plot-to-plot sampling variability in the model training data, ignoring uncertainty in the *Landsat forest species map*.

### 2.4.4. The Effects of Uncertainty in the Forest Species Map on Volume Estimates

Little is known about the cumulative effects of land cover product classification errors when they are used to limit the model calibration sample and application of the prediction models. The effects of uncertainty in the *Landsat forest species map* (map-to-map variability) on the V model calibration and application data were estimated using a 9-step pairs bootstrap procedure (Figure 4):

(1) Select a pairs bootstrap resample of the training areas used to calibrate the RF classification model,

(2) Construct a new *Landsat forest species map* and for each species, k, estimate the area, $\hat{A}_{p\,b}^{k}$. The oob error estimation for each RF classification model, recalibrated in each bootstrap iteration with the resample from step (1), was recorded to estimate the average user's and producer's accuracy for each of the classified forest species and to estimate the standard error of the user's and producer's accuracy,

(3) Select the subset of the SNFI field plot dataset located in the forest portion of the new *Landsat forest species map*,

(4) Construct new species-specific RF V prediction models using data for that species determined from the plot data, not the map species classification for plot,

(5) For each species, apply the model constructed in (4) to each pixel classified as that species in the map constructed in step (2),

(6) For each species, k, estimate mean V for each bootstrap iteration, b, as $\hat{V}_{p\,b}^{k}$ using Equation (1) and total V, $\widehat{VT}_{p\,b}^{k}$, as the product of the estimates of mean V and the area from step (2):

$$\widehat{VT}_{p\,b}^{k} = \hat{V}_{p\,b}^{k} * \hat{A}_{p\,b}^{k} \tag{9}$$

(7) Replicate steps (1)–(6) 2000 times,

(8) Estimate species-specific mean V and its SE as,

$$\hat{V}_{p\,map}^{k} = \frac{1}{n_{boot}} \sum_{b=1}^{nboot} \hat{V}_{p\,b}^{k} \tag{10}$$

with

$$SE\!\left(\hat{V}_{p\,map}^{k}\right) = \sqrt{\frac{1}{n_{boot}-1} \sum_{b=1}^{nboot} \left(\hat{V}_{p\,b}^{k} - \hat{V}_{p\,map}^{k}\right)^{2}}, \tag{11}$$

(9) Estimate species-specific total V and its SE as,

$$\widehat{VT}_{p\,map}^{k} = \frac{1}{n_{boot}} \sum_{b=1}^{nboot} \widehat{VT}_{p\,b}^{k} \tag{12}$$

with

$$SE\!\left(\widehat{VT}_{p\,map}^{k}\right) = \sqrt{\frac{1}{n_{boot}-1} \sum_{b=1}^{nboot} \left(\widehat{VT}_{p\,b}^{k} - \widehat{VT}_{p\,map}^{k}\right)^{2}} \tag{13}$$

where the subscript "*map*" indicates that only the effects of uncertainty in the *Landsat forest species map* were incorporated into Equations (11) and (13).

As the above procedure indicates, uncertainty in the *Landsat forest species map* affects V estimates in two ways: (1) induces uncertainty into the set of SNFI field plots falling within the forest land portion of the map which, in turn, affects the set of SNFI plots used to calibrate the RF V models and the ensuing pixel-level V predictions, and (2) induces uncertainty in the portion of the *Landsat forest species map* for which V is predicted. To assist in distinguishing the relative magnitudes of these effects of map uncertainty, two additional analyses were conducted.

First, each new *Landsat forest species map*, (one for each bootstrap iteration), was compared with the original *Landsat forest species map* to determine the population units for which the predicted forest species changed and those that retained the original classification. The percentages of population units whose predicted forest species did not change over the 2000 bootstrap iterations were calculated and hereafter designated as the *percentage of stable pixels* or *pixel stability*. If predicted forest species changed for only a small percentage of population units, then the effects of area changes on mean and total V estimates would be expected to be small.

Second, the percentages of SNFI field plots that were within the forest portions of the 2000 new *Landsat forest species maps* and, therefore, were used to calibrate the RF models were calculated and hereafter designated as *percentage of stable plots or plot stability*. If only a few plots were affected by uncertainty in the *Landsat forest species map*, then the effects of map uncertainty on the RF model predictions would be expected to be small. In addition, these analyses facilitate distinguishing among forest species with respect to how map uncertainty affects respective area estimates and V predictions for individual forest species.

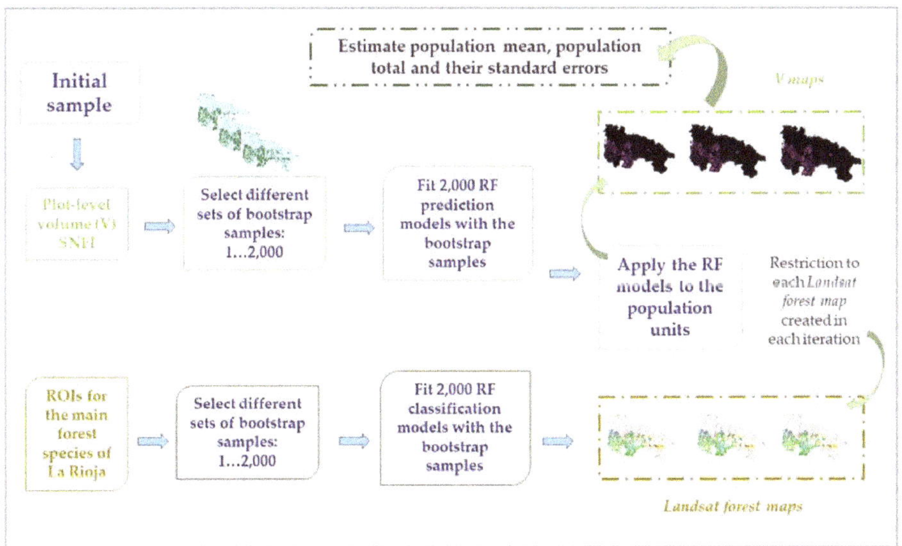


**Figure 4.** Wild and pairs bootstrapping schemes used to account for uncertainty in the *Landsat forest species map*.

### 2.4.5. Total Uncertainty

When accounting for both the uncertainty in the *Landsat forest species map* and sampling variability in the RF model calibration data, the effects of sampling variability change with each iteration of the *Landsat forest species map*, technically making it necessary to separately estimate the sampling variability effects for each map. This approach would entail on the order of 2000 × 2000 overall bootstrap iterations and require considerable computational intensity. However, the differences in plots selected for different forest species maps will be relatively small, and the effects of this sampling variability for the different maps are expected to be relatively constant. Therefore, instead of estimating the effects of this sampling variability for each new *Landsat forest species map*, we assume that the average SE over all map iterations would be about the same as the SE obtained based on the sampling variability effects for the original *Landsat forest species map* as estimated in Section 2.4.3. Thus, for species k, the overall SE, $SE\left(\widehat{VT^{k}}_{total}\right)$, which incorporates the effects of both sources of uncertainty was estimated as,

$$SE\left(\widehat{VT^k}_{total}\right) = \sqrt{SE^2\left(\widehat{VT}^k_{p\ map}\right) + SE^2\left(\widehat{VT}^k_{w\ plot}\right)},$$ (14)

where $SE\left(\widehat{VT}^k_{p\ map}\right)$ is obtained using Equation (13) and $SE\left(\widehat{VT}^k_{w\ plot}\right)$ is obtained using Equation (6).

## 3. Results

### 3.1. Accuracy Assessment

#### 3.1.1. Forest Species Map Accuracy

The accuracy assessment of the original *Landsat forest species map* used the oob RF error (Table 2) and produced an overall accuracy of 88.77%. User's and producer's accuracies for most individual forest species were greater than 80% and even greater than 90% for some. User's and producer's accuracies were less for the Q, OB, and OC classes. The former showed a commission error of 23% due to confusion of this species with the OB class (see Table A2 in Appendix A), which also explains the OB omission error (23%). The OB class's commission error (29%) is due to some of the points of this class erroneously classified as Q and FS. As for the OC class, it tends to be classified as some of the other *Pinus spp.* species and vice versa because of the similar spectral response among them. Nevertheless, the *Landsat forest species map* achieved an excellent overall accuracy of 98.8% for distinguishing between NF and the aggregation of all the forest species with producer's accuracies of 92.9% and 99.7% and user's accuracies of 98.1% and 98.8% for NF and the aggregation of forest classes, respectively.

**Table 2.** Out-of-bag confusion matrix estimated from the random forest (RF) classification model fitted to construct a forest species map for the study area.

| Forest Species * | User's Accuracy (%) | Commission Error (%) | Producer's Accuracy (%) | Omission Error (%) |
|---|---|---|---|---|
| NF | 98 | 2 | 93 | 7 |
| FS | 95 | 5 | 88 | 13 |
| PH | 97 | 3 | 94 | 6 |
| PN | 89 | 11 | 93 | 7 |
| PS | 93 | 7 | 96 | 4 |
| Q | 77 | 23 | 84 | 16 |
| QI | 91 | 9 | 84 | 16 |
| OB | 71 | 29 | 77 | 23 |
| OC | 81 | 19 | 88 | 12 |

* NF: *Non-forest*; FS: *Fagus sylvatica*; PH: *Pinus halepensis*; PN: *Pinus nigra*; PS: *Pinus sylvestris*; Q: *Quercus pyrenaica/faginea*; QI: *Quercus ilex*; OB: "*Other broadleaves*" (OB) and OC: "*Other coniferous*".

#### 3.1.2. RF Volume Models

A graph of V predictions versus observations for SNFI plots showed that in general, most observations were located close to the 1:1 line, although a few observations exhibited a different tendency in the form of greater distances from this line (Figure 5). The lack of systematic error supports the previous MB inferential assumption that the RF model is essentially correct.

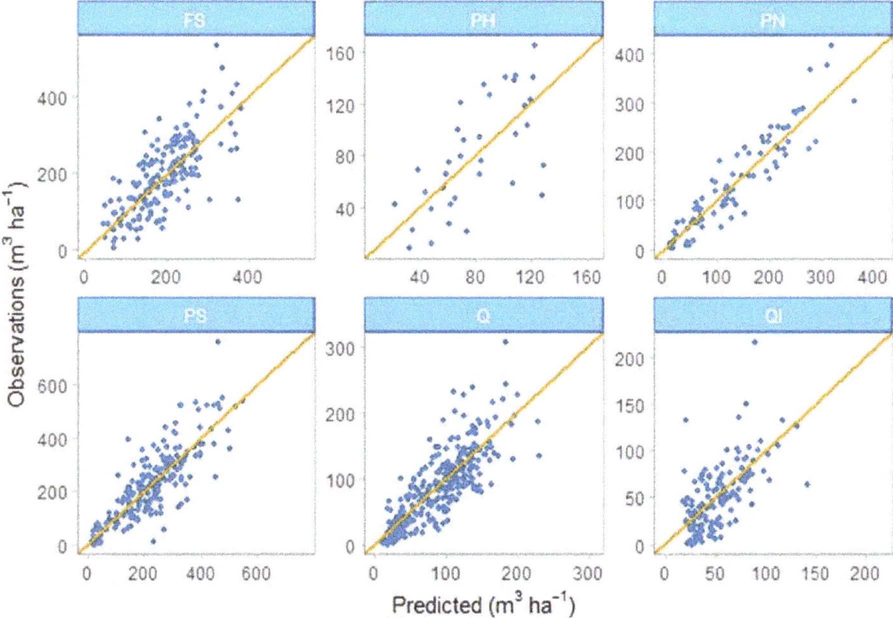

**Figure 5.** Volume reference values for each of the field plots, used to construct the different RF models, versus their predictions. Note: FS: *Fagus sylvatica*; PH: *Pinus halepensis*; PN: *Pinus nigra*; PS: *Pinus sylvestris*; Q: *Quercus pyrenaica/faginea*; QI: *Quercus ilex*. Orange line represents the 1:1 line.

*3.2. Uncertainty Assessment*

3.2.1. The Effects of Uncertainty in the *Landsat Forest Species Map* on Area Estimates

SEs for the area estimates were generally less than 10% with the exception of PH and PN for which $SE(\hat{A}^{k}_{p\ map})$ as percentages of estimates of the means reached 22.07% and 12.27%, respectively (Table 3). Greater SE estimates for areas for the latter two species are at least partially attributed to their less frequent occurrence among the six main forest species analyzed. These results indicate the overall variability in the *Landsat forest species maps* among the bootstrap iterations. The percentages of stable pixels were strongly positively correlated with the $SE\!\left(\hat{A}^{k}_{p\ map}\right)$ estimates. Among all species, more than 80% of the pixels were always classified as the same forest species, although for PH and PN, just 67% and 79% of the pixels remained stable. Regarding plot stability, among all species, nearly 80% of the field plots were in the forest portions of all 2000 *Landsat forest species maps*, one for each of the bootstrap iterations, and were therefore used for the calibration of V models. For individual species, PH exhibited the least plot stability with just 73% of the field plots selected for all 2000 *Landsat forest species maps* and with some plots selected for fewer than 50% of the maps. Excellent results were achieved for PS and Q for which the plot stabilities were nearly 100%. Although overall plot stabilities for these species were large, some field plots used to fit V models for these species were selected only a few times, particularly two plots that were selected for only about 40% of the maps. Generally, PH exhibited greater variability for area estimates and for plots selected to construct the RF V models, which was likely because this is an open grown forest species whose locations are more likely to be misclassified as non-forest.

**Table 3.** Area estimates (ha) and pixel and plot stabilities for the main forest species in La Rioja.

| Forest Species * | Area Estimates | | | | Stability | |
|---|---|---|---|---|---|---|
| | $\hat{A}^{k}$ (ha) | $\hat{A}^{k}_{p\,map}$ (ha) | $SE(\hat{A}^{k}_{p\,map})$ (ha) | $SE(\hat{A}^{k}_{p\,map})$ (%) | % of Stable Pixels | % of Stable Plots |
| | See Footnote (**) | Equation (2) | Equation (3) | Equations (2) and (3) | | |
| FS | $2.10 \times 10^4$ | $2.24 \times 10^4$ | 552.66 | 2.47 | 94.20 | 100.00 |
| PH | $1.09 \times 10^4$ | $0.97 \times 10^4$ | 2132.26 | 22.07 | 66.85 | 72.97 |
| PN | $0.67 \times 10^4$ | $0.63 \times 10^4$ | 779.12 | 12.27 | 79.52 | 86.59 |
| PS | $1.79 \times 10^4$ | $1.86 \times 10^4$ | 1243.83 | 6.67 | 92.01 | 96.48 |
| Q | $5.51 \times 10^4$ | $5.06 \times 10^4$ | 2664.66 | 5.27 | 83.54 | 95.96 |
| QI | $3.53 \times 10^4$ | $3.61 \times 10^4$ | 3202.95 | 8.84 | 84.51 | 86.76 |

* FS: *Fagus sylvatica*; PH: *Pinus halepensis*; PN: *Pinus nigra*; PS: *Pinus sylvestris*; Q: *Quercus pyrenaica/faginea*; QI: *Quercus ilex*. ** Calculated as the product of pixel size and the number of pixels classified as species k in the original *Landsat forest species map*.

A confusion matrix analysis for the *Landsat forest species maps*, constructed for the bootstrap iterations, was conducted using the RF oob error obtained with the pairs bootstrap resamples. The mean and the standard deviation of the accuracies of these maps over all iterations were estimated (Figure 6) with results similar to those for the original confusion matrix obtained for the original forest species map. Smaller user's and producer's accuracies were obtained for Q.

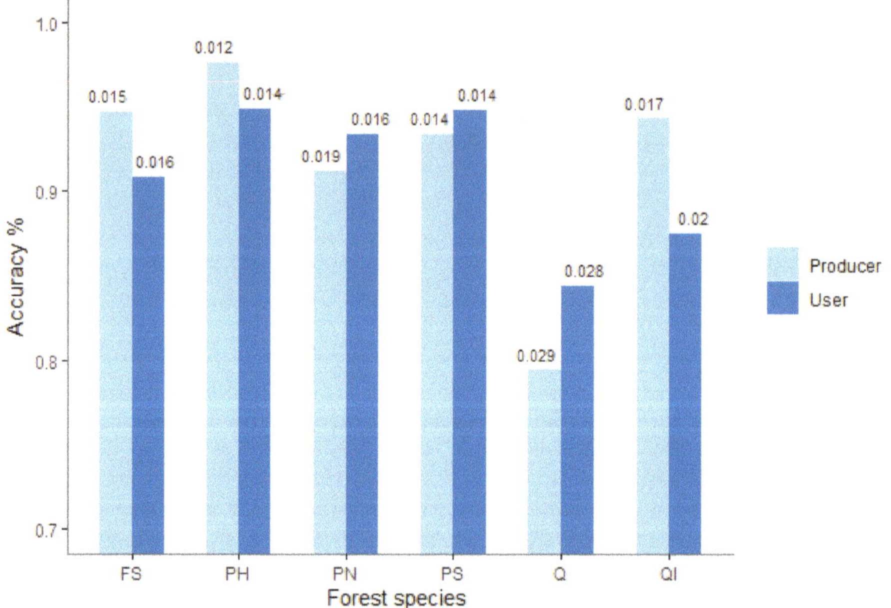


**Figure 6.** Mean and the standard error estimates (number at the top of each bar) of the accuracies (from 0 to 1) of the *Landsat forest species maps* over all the iterations for the main forest species of La Rioja. Note: FS: *Fagus sylvatica*; PH: *Pinus halepensis*; PN: *Pinus nigra*; PS: *Pinus sylvestris*; Q: *Quercus pyrenaica/faginea*; QI: *Quercus ilex*.

### 3.2.2. The Effects of Uncertainty in the *Landsat Forest Species Map* on Volume Estimates

Bootstrapping procedures were applied to assess the effects of map-to-map variability on the uncertainty of the V estimates for each of the most important forest species in La Rioja (Table 4). The results showed that the effects of map-to-map uncertainty were not negligible for any of the main forest species with $SE\left(\widehat{VT}^{k}_{p\,map}\right)$% ranging from 3% to 22%. The uncertainties in the total V estimates resulting from the effects of uncertainty in the *Landsat forest species map* were greatest for PH (21.95%) and for QI (12.05%). The least $SE\left(\widehat{VT}^{k}_{p\,map}\right)\%$ was achieved, from greatest to smallest, for FS, PS, and Q, as indicated by estimates of 3.17, 4.65, and 4.66%, respectively.

**Table 4.** Volume estimates and standard errors based on uncertainty in the *Landsat forest species map*.

| Forest Species * | Mean Volume (m³/ha) | | Total Volume (m³) | | |
| --- | --- | --- | --- | --- | --- |
| | $\widehat{V}^{k}_{p\,map}$ | $SE(\hat{V}^{k}_{p\,map})$ | $\widehat{VT}^{k}_{p\,map}$ | $SE(\widehat{VT}^{k}_{p\,map})$ | $SE(\widehat{VT}^{k}_{p\,map})$ (%) |
| | Equation (10) | Equation (11) | Equation (12) | Equation (13) | Equations (12) and (13) |
| FS | 204.05 | 5.46 | $4.57 \times 10^6$ | $1.45 \times 10^5$ | 3.17 |
| PH | 67.38 | 7.91 | $0.65 \times 10^6$ | $1.42 \times 10^5$ | 21.95 |
| PN | 144.05 | 8.75 | $0.91 \times 10^6$ | $0.79 \times 10^5$ | 8.71 |
| PS | 216.28 | 8.62 | $4.02 \times 10^6$ | $1.87 \times 10^5$ | 4.65 |
| Q | 70.50 | 2.80 | $3.56 \times 10^6$ | $1.66 \times 10^5$ | 4.66 |
| QI | 44.63 | 4.10 | $1.62 \times 10^6$ | $1.95 \times 10^5$ | 12.05 |

* FS: *Fagus sylvatica*; PH: *Pinus halepensis*; PN: *Pinus nigra*; PS: *Pinus sylvestris*; Q: *Quercus pyrenaica/faginea*; QI: *Quercus ilex* .

### 3.2.3. The Effects of Sampling Variability in the Model Calibration Dataset on Volume Estimates

Bootstrapping procedures were applied to assess the effects of plot-to-plot sampling variability for each of the most important forest species in La Rioja without consideration of the effects of map uncertainty. A total of 2000 iterations was sufficient for estimates of both means and SEs to stabilize (Figure 7). Overall, the $\hat{V}^{k}_{w\,plot}$ estimates of the means were very similar to the $\hat{\mu}_{MB}$ estimates (Table 5) with the greatest estimates for PS and FS and the least for PH and QI. The effects of plot-to-plot sampling variability on $SE\left(\widehat{VT}^{k}_{w\,plot}\right)\%$ (as percentages of estimates of the total V) varied among the forest species, ranging from 2% to 11%. PH and QI were the forest species with more dispersion in the total V estimates as suggested by their larger $SE\left(\widehat{VT}^{k}_{w\,plot}\right)\%$ values, 11.23% for the former and 9.07% for the latter. The smallest $SE\left(\widehat{VT}^{k}_{w\,plot}\right)\%$ values, from smallest to greatest, were for PS, FS, and Q.

**Table 5.** Volume estimates and standard errors based on sampling variability in model calibration dataset.

| Forest Species * | Mean Volume (m³/ha) | | | Total Volume (m³) | | | |
| --- | --- | --- | --- | --- | --- | --- | --- |
| | $\hat{\mu}_{MB}$ | $\hat{V}^{k}_{w\,plot}$ | $SE(\hat{V}^{k}_{w\,plot})$ (%) | $\widehat{VT}$ | $\widehat{VT}^{k}_{w\,plot}$ | $SE(\widehat{VT}^{k}_{w\,plot})$ | $SE(\widehat{VT}^{k}_{w\,plot})$ (%) |
| | Equation (1) | Equation (5) | Equation (6) | See Footnote (**) | Equation (7) | Equation (8) | Equations (7) and (8) |
| FS | 203.70 | 204.29 | 5.02 | $4.28 \times 10^6$ | $4.29 \times 10^6$ | $1.06 \times 10^5$ | 2.46 |
| PH | 61.52 | 66.49 | 7.47 | $0.67 \times 10^6$ | $0.73 \times 10^6$ | $0.82 \times 10^5$ | 11.23 |
| PN | 146.96 | 144.22 | 4.25 | $0.99 \times 10^6$ | $0.97 \times 10^6$ | $0.28 \times 10^5$ | 2.94 |
| PS | 223.08 | 220.43 | 5.22 | $3.99 \times 10^6$ | $3.95 \times 10^6$ | $0.94 \times 10^5$ | 2.37 |
| Q | 70.84 | 72.43 | 2.06 | $3.90 \times 10^6$ | $3.99 \times 10^6$ | $1.13 \times 10^5$ | 2.84 |
| QI | 43.27 | 44.46 | 4.03 | $1.53 \times 10^6$ | $1.57 \times 10^6$ | $1.43 \times 10^5$ | 9.07 |

* FS: *Fagus sylvatica*; PH: *Pinus halepensis*; PN: *Pinus nigra*; PS: *Pinus sylvestris*; Q: *Quercus pyrenaica/faginea*; QI: *Quercus ilex*. ** Calculated as the product of species-specific area, from original Landsat forest species map, and mean V based on plots from SNFI field dataset located within area classified as the species.

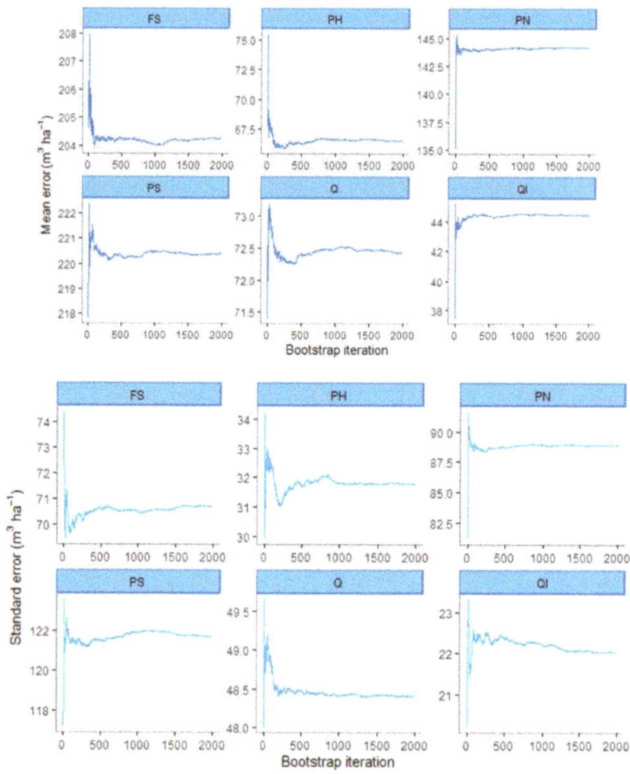

**Figure 7.** Estimates for means and standard errors for each bootstrap iteration and forest species analyzed. Note: FS: *Fagus sylvatica*; PH: *Pinus halepensis*; PN: *Pinus nigra*; PS: *Pinus sylvestris*; Q: *Quercus pyrenaica/faginea*; QI: *Quercus ilex*.

### 3.3. Total Uncertainty

The results (Table 6) revealed that the uncertainties in the V estimates were dominated by the effects of uncertainties in the *Landsat forest species map* as suggested by a greater $SE\left(\widehat{VT}^{k}_{p\,map}\right)$ relative to $SE\left(\widehat{VT}^{k}_{w\,plot}\right)$ obtained when considering only sampling variability associated with the model calibration data. When both sources of uncertainty were considered together, the SEs increased for all the forest species with increases greater than 5% for most species. Greater uncertainties were estimated for PH, with $SE\left(\widehat{VT}^{k}_{total}\right)$ increasing from 11% to 25%, and for QI with increases from 9% to 15%.

**Table 6.** Uncertainties in total volume estimates for the main forest species of La Rioja.

| Forest Species * | Uncertainty in the *Landsat Forest Species Map* | Sampling Variability in Model Calibration Data | Total Uncertainty |
| --- | --- | --- | --- |
| | $SE(\widehat{VT}^{k}_{p\,map})$ (%) | $SE(\widehat{VT}^{k}_{w\,plot})$ (%) | $SE(\widehat{VT}^{k}_{total})$ (%) |
| | Equations (12) and (13) | Equations (7) and (8) | Equation (14) |
| FS | 3.17 | 2.46 | 4.01 |
| PH | 21.95 | 11.23 | 24.66 |
| PN | 8.71 | 2.94 | 9.19 |
| PS | 4.65 | 2.37 | 5.22 |
| Q | 4.66 | 2.84 | 5.46 |
| QI | 12.05 | 9.07 | 15.08 |

* FS: *Fagus sylvatica*; PH: *Pinus halepensis*; PN: *Pinus nigra*; PS: *Pinus sylvestris*; Q: *Quercus pyrenaica/faginea*; QI: *Quercus ilex*.

## 4. Discussion

### 4.1. The Statistical Techniques

In this study, mean and total V estimates were inferred using an MB estimator based on RF predictions for the six main forest species of La Rioja (Spain). RF V models were constructed using SNFI field plot information and ALS data. To limit the prediction of V to pixels classified as one of the main forest species assessed, a forest species map was constructed using Landsat and auxiliary information and RF classification models. RF performed well with respect to both classification and prediction accuracy. Both RF models were calibrated with the default settings included in the R package *RandomForest* [43]. Balanced training samples were constructed to calibrate the RF classification, although RF is not known to be sensitive to the characteristics of the training sample [39]. An assessment of the effects of correlation among the ALS metrics on RF performance for volume prediction and therefore, procedures for selecting the most suitable variables would be appropriate [40,57].

Bootstrapping techniques were the lynchpin of this study and facilitated accounting for two sources of uncertainty: the uncertainty in the *Landsat forest species map* and the RF model prediction uncertainty resulting from the effects of sampling variability in the model training data. The similarity in the bootstrap estimates and the MB estimates indicates the lack of any substantial bias in the bootstrap procedure. Estimates of bootstrap means and standard errors stabilized by 2000 bootstrap iterations. However, this result should be generalized with caution and would better be considered a parameter that must be tuned for each forest species. As per Figure 6, the number of iterations necessary for estimates of the bootstrap means and standard errors to stabilize varied by forest species. Computer intensity will depend not only on the number of iterations but also on the size of the study area, with smaller study areas requiring less computation time. If a large number of iterations are necessary for a large study area, the development of the methodology could be conditional on access to a great deal of computationally capability. The workflow developed in this study was performed on a multi-processor Core i-7 6800 K box, with 12 cores and 64 GB memory because of the computational intensity involved. The cumulative effects of both sources of uncertainty were estimated with the effects of plot-to-plot sampling variability for the different maps assumed to be relatively constant. Apart from this assumption, the computational intensity associated with the methodology could become prohibitive, particularly if a large number of iterations were necessary to achieve stability in the estimates of means and standard errors. Nevertheless, the use of cloud-based platforms could overcome these difficulties providing the users with computing power at affordable prices.

### 4.2. Effects of Uncertainty in the Landsat Forest Species Map

The *Landsat forest species map* constructed for this study exhibited satisfactory accuracies considering the large number of forest species and their level of heterogeneity. Our results are in line with those reported by Fernández-Landa et al. [28] who used RF classification models to construct a map for PS and FS in a subset area of La Rioja of 16,000 ha. They reported user's accuracies of 0.80 and 0.97 and producer's accuracies of 0.97 and 0.89 for FS and PS, respectively. In their study, they claimed confusion between FS and OB as the main reason for the FS classification errors. Among the other forest species mapped, Q yielded the greatest commission errors coinciding with the results of other studies that mapped forest species using remotely sensed data in Mediterranean environments [58,59]. Despite the difficulty involved in mapping open forest species such as PH, the accuracies obtained for this study are in accordance with other studies conducted for Mediterranean species using multispectral information [60].

Uncertainty in the *Landsat forest species map* affects the V estimates in two manners: (1) it affects which plots are used to construct the species-specific V models, and (2) it affects the estimate of the area for each forest species that is multiplied by the estimate of the V mean. For the sake of simplicity, we assumed that the effects of plot-to-plot sampling variability in the model training data on V predictions for the different *Landsat forest species maps* to be relatively constant. Overall, the results justify this decision, because the differences in the field plots selected for the different *Landsat forest species maps* were relatively small. Hence, the results obtained suggest that the *Landsat forest species map* effects on area estimates are more important than the effects on field plot selection on RF model predictions. However, caution must be exercised with forest species in open woodlands such as PH, because the *Landsat forest species map* has a similar effect on both factors. On one hand, V models for PH were calibrated using a smaller calibration dataset (35 field plots) relative to the other forest species for which V models were calibrated. On the other hand, even though we refined the training dataset by removing the SNFI field plots that showed greater discrepancies, this dataset is characterized by a lack of location precision for its plot centers [28,61]. For species growing in open areas such as PH, this effect is exacerbated, resulting in a greater sampling variability (Figure 8) and increasing variance estimates [32].

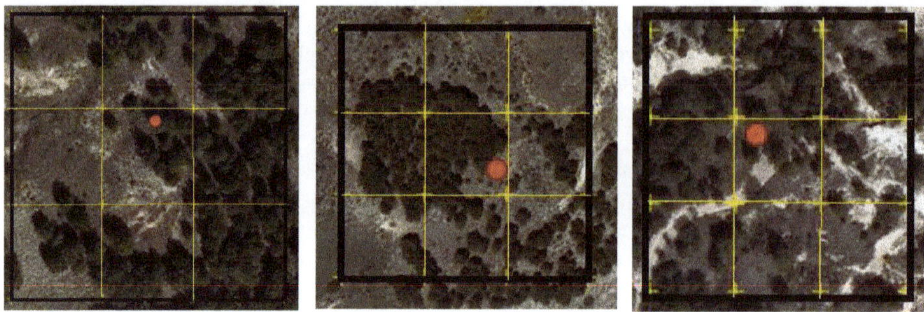

**Figure 8.** Examples of SNFI field plots used for V prediction purposes for PH not included in 100% of the total bootstrap iterations conducted to assess plot-to-plot variability. Yellow squares represent population units of cell size of 25 × 25 m.

An interesting pattern observed in this study is the strong positive relationship between pixel stability and SE for estimates of V totals. The results indicate that greater stability in the forest species classification produces less uncertainty in V estimates. In our study, we did not consider other sources of uncertainty such as tree measurement errors and ALS errors. However, both are expected to contribute very little to total uncertainty [62]. Nevertheless, if uncertainty in V estimates for smaller areas is desired, an approach accounting for residual uncertainty should be developed [63].

There was no positive relationship between species-specific pixel stabilities and accuracies obtained for the different *Landsat forest species maps*. Even though PH exhibited the largest variability in the pixels classified as such among the bootstrap iterations, it did not produce smaller accuracies. This is likely due to the location of the points used to train the RF classification models and therefore used to calculate the oob error, most of which were in dense forest areas. These points are less likely to be misclassified, although PH is a Mediterranean forest species occurring in open forest. These areas represent mixed pixels that were likely to be classified both as forest or non-forest, which could explain the larger PH classification variability. Uncertainty results at pixel scale facilitate analysis of the accuracy of the spatial distribution of V estimates and contribute to the identification of special patterns [12,56]. It is important to bear in mind that the reference data used to fit the RF classification models did not represent a probability sample but just an approximation. Although probability samples are not required for training data, future research could assess the effects on classification accuracy of probability-based samples of training data. Even though the RF oob error has been reported as a reliable accuracy measurement that can replace the use of an independent dataset [42], further research using a variety of datasets in different application scenarios would be appropriate [39].

### 4.3. Effects of Sampling Variability for the Model Calibration Datas

In this study, we only considered the uncertainty in the model predictions resulting from the effects of sampling variability. We did not include tree measurement errors for which the effects are generally regarded as negligible, thereby producing no meaningful adverse consequences [5]. Spatial correlation among pixel predictions was not included because for large contiguous area, distances between the vast majority of pairs of pixels are greater than the range of spatial correlation, thus, when averaging over all pairs of pixels, the overall effect is negligible [44]. Spatial correlation needs to be assessed in future research especially for smaller forest areas for which forest attributes are calculated. In addition, there were other sources of uncertainty involved in the V estimation framework such as the uncertainty associated with the V allometric model used in the SNFI and with the model linking plot-level V with ALS metrics. When these two sources of uncertainty are considered, uncertainty estimates are larger [12]. Sampling variability in the model calibration data produced SE% estimates in the range of 2–11% with greater values for PH and QI. Chirici et al. [64] estimated growing stock volume using NFI field plots and remotely sensed data (Landsat and SAR data) for a large area in Italy. Their SE estimates varied from 2% to 4%, although SE estimates were not reported for specific forest species. Irulappa-Pillai-Vijayakumar et al. [65] reported V estimates for a French region using NFI field plot data and remotely sensed data. They reported SE estimates of approximately 3% for oak V estimates and 5.76% for PS. For oak volume estimates, their SEs were slightly less than the SE of 5.46% for our study, although they did not consider the uncertainty of the forest species map.

The greater SEs for PH and QI are in accordance with the map uncertainty results as suggested by greater variability for PH and QI mapping. PH and QI are Mediterranean forest species that intermix with other vegetation types in complex patterns or in open woodlands. This complexity poses a challenge when using remote sensing-based classification methods in these kinds of environments as opposed to those in boreal forests [11,64]. A relationship between SEs for V totals and plot stabilities was not clearly observed as it was for pixel stability. That being said, it seems that there is a relationship between plot stability and the number of field plots used to calibrate the RF V models. Plot stabilities were less for PH and PN for which field plot sample sizes were smaller.

### 4.4. Total Uncertainty

Many studies have reported large area V estimates, even at a country level, using models and remotely sensed data [21,64,66]. These studies have demonstrated the potential of assessing forest resources using different sources of information already acquired and therefore posing an opportunity to replicate the methodology in other countries with well-established NFI programs. However, the results of our study demonstrated that if similar approaches are to be replicated, it is necessary to include the uncertainty of the forest species map used. Uncertainty results can be underestimated when the uncertainty of the forest species map is ignored in the modeling approach. This is particularly relevant for the V estimates for forest species: (1) for less representative species such as PN, incorporating the effects of map uncertainty increased SE% from 2.94% to 9.19%; and (2) for Mediterranean forest species occurring in open areas such as PH, incorporation of the effects of map uncertainty increase SE% from 11.23% to 24.66%. Further work is recommended to construct uncertainty maps at a pixel scale that represents the spatial distributions of the accumulated sources of uncertainty considered. Such maps would be key to distinguishing uncertainties between site conditions and estimated volume levels [12].

### 4.5. Operational Consequences

This study is unique because of our approach to propagating uncertainty to account for the effects of uncertainty in the map of the spatial distribution of the main forest species estimated from Landsat and from the effects of sampling variability on RF V model predictions. Even though we used a different approach for constructing the forest map than the SNFM, the results are relevant for the SNFI. On one hand, although the SNFI used a forest map constructed by photo-interpretation, it could also have potential errors that might affect the V estimates. On the other hand, until now, national V estimates have been sample-based, but because of the economic crisis, the number of field plots has been reduced [10]. Nevertheless, an appropriate use of remotely sensed data could guarantee that accuracy is not lost in forest attribute predictions, even if the number of field plots was reduced [67], although it is crucial to account for the time interval between field data and remotely sensed data used. However, if NFI plots are to be established only in the forest portion of a remote sensing-based forest map, then the uncertainty associated with the map should be considered. Otherwise, countries will underestimate uncertainty and fail to comply with the IPCC good practice guidelines [24].

## 5. Conclusions

Our study assessed the effects of uncertainty in a forest species map involved in the selection of the field plots used to calibrate the volume models and in the estimation process on the uncertainty of large area volume estimates. Five conclusions were drawn from the study: (1) the effects of uncertainty in the forest species map on the uncertainty of large area volume estimates are not negligible, and ignoring the effects could jeopardize the reliability of the forest volume estimates; (2) overall, the effects of uncertainty in the forest species map on area estimates were greater than the effects of uncertainty in the map on the selection of field plots used to calibrate the RF volume prediction model; (3) the effects of the forest species map uncertainty increased for open forest species or less representative forest species; (4) bootstrapping estimates demonstrated the suitability of this technique to accommodate the effects of uncertainty from more than one source; and (5) the results are relevant for countries that use a remote sensing-based forest/non-forest map to guide the establishment of field plots. Further work in a variety of forest environments to assess whether the conclusions can be generalized beyond Mediterranean environments is recommended.

**Author Contributions:** Conceptualization, J.E., R.E.M. and A.F.-L.; methodology, J.E., R.E.M. and A.F.-L.; software, J.E.; formal analysis, J.E., R.E.M., A.F.-L., J.L.T. and M.M.; resources, J.E., A.F.-L. and J.L.T.; data curation, J.E., A.F.-L. and J.L.T.; writing—original draft preparation, J.E., R.E.M., A.F.-L., J.L.T. and M.M.; writing—review and editing, J.E., R.E.M., A.F.-L., J.L.T. and M.M.; visualization, J.E.; funding acquisition, M.M. All authors have read and agreed to the published version of the manuscript.

**Funding:** First author is supported by a pre-doctoral contract co-funded by Universidad Politécnica de Madrid and Agresta S.Coop.

**Acknowledgments:** We are grateful to the National Geographic Institute from Spain and thank to the Spanish National Forest Inventory for providing free access to ALS data and forest inventory information.

**Conflicts of Interest:** The authors declare no conflict of interest.

## Abbreviations

| Abbreviation | Full Description |
| --- | --- |
| ALS | Airborne laser scanning |
| crr | ALS canopy relief ratio |
| cv | ALS height coefficient of variation |
| DBH | Diameter at breast height |
| FS | *Fagus sylvatica* |
| IPCC | Intergovernmental Panel on Climate Change |
| iq | ALS height interquartile range |
| kurto | ALS height kurtosis |
| lfcc | Forest canopy cover |
| MB | Model-based |
| MSE | Mean square error |
| NBR | Normalized Burn Ratio |
| NDMI | Normalized Difference Moisture Index |
| NDVI | Normalized Difference Vegetation Index |
| NFI | National Forest Inventory |
| NIR | Near infrared |
| OB | Other broadleaves |
| OC | Other coniferous |
| oob | Out-of-bag |
| p1-p99 | ALS percentiles (ranging from the 1st to 99th percentile) |

| | |
|---|---|
| PH | *Pinus halepensis* |
| PN | *Pinus nigra* |
| PS | *Pinus sylvestris* |
| Q | *Quercus faginea or Quercus pyrenaica* |
| QI | *Quercus ilex* |
| RF | Random forest |
| RMSE | Root mean square error |
| rRMSE | Relative root mean square error |
| RS | Remote sensing |
| SE | Standard error |
| SNFI | Spanish National Forest Inventory |
| SNFM | Spanish National Forest Map |
| stdev | ALS height standard deviation |
| TM | Thematic Mapper |
| V | Mean volume per hectare |
| varia | ALS height variance |

## Appendix A

**Table A1.** Summary of the V models using 2010 airborne laser scanning (ALS) data and SNFI field plots for the six main forest species of La Rioja (Spain). Measures of predictive accuracy were obtained from the out-of-bag (oob) internal error derived from the RF models.

| Forest Species * | % Variance Explained | MSE (m³/ha) | RMSE (m³/ha) | rRMSE (%) |
|---|---|---|---|---|
| FS | 52.62 | 4081.84 | 63.89 | 33.27 |
| PH | 44.37 | 973.24 | 31.20 | 38.72 |
| PN | 85.64 | 1279.01 | 35.76 | 25.97 |
| PS | 69.26 | 5665.13 | 75.27 | 33.43 |
| Q | 64.47 | 1084.85 | 32.94 | 36.47 |
| QI | 36.38 | 769.52 | 27.74 | 52.43 |

* FS: *Fagus sylvatica*; PH: *Pinus halepensis*; PN: *Pinus nigra*; PS: *Pinus sylvestris*; Q: *Quercus pyrenaica/faginea*; QI: *Quercus ilex*. Note: MSE: mean square error; RMSE: root mean square error, and rRMSE: relative root mean square error.

**Table A2.** Out-of-bag confusion matrix derived from the RF classification model fitted to construct a forest species map for the study area. Columns represent true classes, while rows represent the classifier's predictions.

| Forest Species * | NF | FS | PH | PN | PS | Q | QI | OB | OC |
|---|---|---|---|---|---|---|---|---|---|
| NF | 158 | 0 | 1 | 1 | 0 | 0 | 0 | 0 | 1 |
| FS | 0 | 146 | 1 | 3 | 1 | 0 | 1 | 2 | 3 |
| PH | 1 | 0 | 104 | 0 | 0 | 1 | 0 | 0 | 1 |
| PN | 0 | 2 | 3 | 100 | 0 | 6 | 1 | 0 | 0 |
| PS | 0 | 1 | 0 | 0 | 84 | 5 | 13 | 6 | 0 |
| Q | 3 | 0 | 1 | 1 | 4 | 128 | 0 | 1 | 3 |
| QI | 2 | 1 | 0 | 1 | 9 | 6 | 89 | 12 | 5 |
| OB | 0 | 0 | 0 | 0 | 2 | 0 | 7 | 161 | 0 |
| OC | 6 | 2 | 1 | 1 | 0 | 7 | 4 | 2 | 97 |

* NF: *Non-forest*; FS: *Fagus sylvatica*; PH: *Pinus halepensis*; PN: *Pinus nigra*; PS: *Pinus sylvestris*; Q: *Quercus pyrenaica/faginea*; QI: *Quercus ilex*; OB: "*Other broadleaves*" (OB) and OC: "*Other coniferous*".

## References


1. Food and Agriculture Organizations of the United Nations. State of Europe's Forests 2015 Report. 2015. Available online: https://foresteurope.org/state-europes-forests-2015-report/ (accessed on 12 October 2020).
2. McRoberts, R.E.; Tomppo, E.O.; Næsset, E. Advances and emerging issues in national forest inventories. *Scand. J. For. Res.* **2010**, *25*, 368–381. [CrossRef]
3. Vidal, C.; Alberdi, I.; Redmond, J.; Vestman, M.; Lanz, A.; Schadauer, K. The role of European National Forest Inventories for international forestry reporting a Legally Binding Agreement. *Ann. For. Sci.* **2016**, *73*, 793–806. [CrossRef]
4. Alberdi, I.; Vallejo, R.; Álvarez-González, J.G.; Condés, S.; González-Ferreiro, E.; Guerrero, S.; Hernández, L.; Martínez-Jauregui, M.; Montes, F.; Oliveira, N.; et al. The multi-objective Spanish National Forest Inventory. *For. Syst.* **2017**, *26*, 1–17. [CrossRef]

5.   McRoberts, R.E.; Westfall, J.A. Propagating uncertainty through individual tree volume model predictions to large-area volume estimates. *Ann. For. Sci.* **2016**, *73*, 625–633. [CrossRef]

6.   Grafström, A.; Saarela, S.; Ene, L. Efficient sampling strategies for forest inventories by spreading the sample in auxiliary space. *Can. J. For. Res.* **2014**, *44*, 1156–1164. [CrossRef]

7.   McRoberts, R.E.; Cohen, W.B.; Erik, N.; Stehman, S.V.; Tomppo, E.O. Using remotely sensed data to construct and assess forest attribute maps and related spatial products. *Scand. J. For. Res.* **2010**, *25*, 340–367. [CrossRef]

8.   McRoberts, R.E.; Næsset, E.; Sannier, C.; Stehman, S.V.; Tomppo, E.O. Remote sensing support for the gain-loss approach for greenhouse gas inventories. *Remote Sens.* **2020**, *12*, 1891. [CrossRef]

9.   Tomppo, E.; Gschwantner, T.; Lawrence, M.; McRoberts, R.E. *National Forest Inventories: Pathways for Common Reporting*; Springer: Dordrecht, The Netherlands, 2010; ISBN 978-90-481-3233-1.

10.  Alberdi, I.; Cañellas, I.; Vallejo Bombín, R. The Spanish National Forest Inventory: History, development, challenges and perspectives. *Pesqui. Florest. Bras.* **2017**, *37*, 361. [CrossRef]

11.  Gomez, C.; Alejandro, P.; Hermosilla, T.; Montes, F.; Pascual, C.; Ruiz, L.Á.; Alvarez-taboada, F.; Tanase, M.A.; Valbuena, R. Remote sensing for the Spanish forests in the 21 st century: A review of advances, needs, and opportunities. *For. Syst.* **2019**, *28*, eR001.

12.  Saarela, S.; Wästlund, A.; Holmström, E.; Mensah, A.A.; Holm, S.; Nilsson, M.; Fridman, J.; Ståhl, G. Mapping aboveground biomass and its prediction uncertainty using LiDAR and field data, accounting for tree-level allometric and LiDAR model errors. *For. Ecosyst.* **2020**, *7*, 43. [CrossRef]

13.  McRoberts, R.E.; Tomppo, E.O. Remote sensing support for national forest inventories. *Remote Sens. Environ.* **2007**, *110*, 412–419. [CrossRef]

14.  White, J.C.; Coops, N.C.; Wulder, M.A.; Vastaranta, M.; Hilker, T.; Tompalski, P. Remote Sensing Technologies for Enhancing Forest Inventories: A Review. *Can. J. Remote Sens.* **2016**, *42*, 619–641. [CrossRef]

15.  Ståhl, G.; Saarela, S.; Schnell, S.; Holm, S.; Breidenbach, J.; Healey, S.P.; Patterson, P.L.; Magnussen, S.; Næsset, E.; McRoberts, R.E.; et al. Use of models in large-area forest surveys: Comparing model-assisted, model-based and hybrid estimation. *For. Ecosyst.* **2016**, *3*, 5. [CrossRef]

16.  McRoberts, R.E.; Moser, P.; Zimermann Oliveira, L.; Vibrans, A.C. A general method for assessing the effects of uncertainty in individual-tree volume model predictions on large-area volume estimates with a subtropical forest illustration. *Can. J. For. Res.* **2014**, *45*, 44–51. [CrossRef]

17.  McRoberts, R.E.; Næsset, E.; Gobakken, T. Estimation for inaccessible and non-sampled forest areas using model-based inference and remotely sensed auxiliary information. *Remote Sens. Environ.* **2014**, *154*, 226–233. [CrossRef]

18.  Hansen, M.H.; Madow, W.G.; Tepping, B.J. An evaluation of model-dependent and probability-sampling inferences in sample surveys. *J. Am. Stat. Assoc.* **1983**, *78*, 776–793. [CrossRef]

19.  Royall, R.M.; Herson, J. Robust Estimation in Finite Populations I. *J. Am. Stat. Assoc.* **1973**, *68*, 880–889. [CrossRef]

20.  Shettles, M.; Temesgen, H.; Gray, A.N.; Hilker, T. Comparison of uncertainty in per unit area estimates of aboveground biomass for two selected model sets. *For. Ecol. Manag.* **2015**, *354*, 18–25. [CrossRef]

21.  Urbazaev, M.; Thiel, C.; Cremer, F.; Dubayah, R.; Migliavacca, M.; Reichstein, M.; Schmullius, C. Estimation of forest aboveground biomass and uncertainties by integration of field measurements, airborne LiDAR, and SAR and optical satellite data in Mexico. *Carbon Balance Manag.* **2018**, *13*, 1–12. [CrossRef]

22.  Rodríguez-Veiga, P.; Saatchi, S.; Tansey, K.; Balzter, H. Magnitude, spatial distribution and uncertainty of forest biomass stocks in Mexico. *Remote Sens. Environ.* **2016**, *183*, 265–281. [CrossRef]

23.  Li, X.; Messina, J.P.; Moore, N.J.; Fan, P.; Shortridge, A.M. MODIS land cover uncertainty in regional climate simulations. *Clim. Dyn.* **2017**, *49*, 4047–4059. [CrossRef]

24.  IPCC. Volume 4: Agriculture, Forestry and other land use. In *Refinement to the 2006 IPCC Guidelines for National Greenhouse Gas Inventories*; Calvo Buendia, E., Tanabe, K., Kranjc, A., Baasansuren, J., Fukuda, M., Ngarize, S., Osako, A., Pyrozhenko, Y., Shermanau, P., Federici, S., Eds.; IPCC: Geneva, Switzerland, 2019; Available online: https://www.ipcc-nggip.iges.or.jp/public/2019rf/index.html (accessed on 12 October 2020).

25.  Bravo, F.; Guijarro, M.; Cámara, A.; Balteiro, L.D.; Fernández-Rebollo, P.; Pajares, J.A.; Pemán, J.; Ruiz-Peinado, R. Informe de Situación de los bosques y sector forestal en España—ISFE 2017. *Sociedad Española de Ciencias Forestales*. 2017. Available online: http://secforestales.org/sites/default/files/archivos/7cfe_avance_isfe_final.pdf (accessed on 12 October 2020).

26.  Cuarto Inventario Forestal Nacional La Rioja. Edited by Ministerio de Agricultura, Alimentación y Medio Ambiente. Available online: https://docplayer.es/4459716-Cuarto-inventario-forestal-nacional-la-rioja.html (accessed on 12 October 2020).

27. Álvarez-González, J.G.; Cañellas, I.; Alberdi, I.; Gadow, K.V.; Ruiz-González, A.D. National Forest Inventory and forest observational studies in Spain: Applications to forest modeling. *For. Ecol. Manag.* **2014**, *316*, 54–64. [CrossRef]

28. Fernández-Landa, A.; Fernández-Moya, J.; Tomé, J.L.; Algeet-Abarquero, N.; Guillén-Climent, M.L.; Vallejo, R.; Sandoval, V.; Marchamalo, M. High resolution forest inventory of pure and mixed stands at regional level combining National Forest Inventory field plots, Landsat, and low density lidar. *Int. J. Remote Sens.* **2018**, *39*, 4830–4844. [CrossRef]

29. Gorelick, N.; Hancher, M.; Dixon, M.; Ilyushchenko, S.; Thau, D.; Moore, R. Google Earth Engine: Planetary-scale geospatial analysis for everyone. *Remote Sens. Environ.* **2017**, *202*, 18–27. [CrossRef]

30. McGaughey, R.; Forester, R.; Carson, W. Fusing LIDAR data, photographs, and other data using 2D and 3D visualization techniques. *Proc. Terrain Data Appl. Vis. Connect.* **2003**, *28–30*, 16–24.

31. Næsset, E.; Bollandsås, O.M.; Gobakken, T.; Solberg, S.; McRoberts, R.E. The effects of field plot size on model-assisted estimation of aboveground biomass change using multitemporal interferometric SAR and airborne laser scanning data. *Remote Sens. Environ.* **2015**, *168*, 252–264. [CrossRef]

32. Saarela, S.; Grafström, A.; Ståhl, G.; Kangas, A.; Holopainen, M.; Tuominen, S.; Nordkvist, K.; Hyyppä, J. Model-assisted estimation of growing stock volume using different combinations of LiDAR and Landsat data as auxiliary information. *Remote Sens. Environ.* **2015**, *158*, 431–440. [CrossRef]

33. McRoberts, R.E.; Chen, Q.; Domke, G.M.; Ståhl, G.; Saarela, S.; Westfall, J.A. Hybrid estimators for mean aboveground carbon per unit area. *For. Ecol. Manag.* **2016**, *378*, 44–56. [CrossRef]

34. Breiman, L. Random Forests. *Mach. Learn.* **2001**, *45*, 5–32. [CrossRef]

35. Shataee, S.; Kalbi, S.; Fallah, A.; Pelz, D. Forest attribute imputation using machine-learning methods and ASTER data: Comparison of k-NN, SVR and random forest regression algorithms. *Int. J. Remote Sens.* **2012**, *33*, 6254–6280. [CrossRef]

36. Penner, M.; Pitt, D.G.; Woods, M.E. Parametric vs. nonparametric LiDAR models for operational forest inventory in boreal Ontario. *Can. J. Remote Sens.* **2013**, *39*, 426–443.

37. Rodriguez-Galiano, V.F.; Chica-Rivas, M. Evaluation of different machine learning methods for land cover mapping of a Mediterranean area using multi-seasonal Landsat images and Digital Terrain Models. *Int. J. Digit. Earth* **2012**, *7*, 492–509. [CrossRef]

38. Grinand, C.; Rakotomalala, F.; Gond, V.; Vaudry, R.; Bernoux, M.; Vieilledent, G. Estimating deforestation in tropical humid and dry forests in Madagascar from 2000 to 2010 using multi-date Landsat satellite images and the random forests classifier. *Remote Sens. Environ.* **2013**, *139*, 68–80. [CrossRef]

39. Belgiu, M.; Drăguţ, L. Random forest in remote sensing: A review of applications and future directions. *ISPRS J. Photogramm. Remote Sens.* **2016**, *114*, 24–31. [CrossRef]

40. Moser, P.; Vibrans, A.C.; McRoberts, R.E.; Næsset, E.; Gobakken, T.; Chirici, G.; Mura, M.; Marchetti, M. Methods for variable selection in LiDAR-assisted forest inventories. *Forestry* **2016**, *90*, 112–124. [CrossRef]

41. Navarro, J.A.; Algeet, N.; Fernández-Landa, A.; Esteban, J.; Rodríguez-Noriega, P.; Guillén-Climent, M.L. Integration of UAV, Sentinel-1, and Sentinel-2 data for mangrove plantation aboveground biomass monitoring in Senegal. *Remote Sens.* **2019**, *11*, 77. [CrossRef]

42. Rodriguez-Galiano, V.F.; Ghimire, B.; Rogan, J.; Chica-Olmo, M.; Rigol-Sanchez, J.P. An assessment of the effectiveness of a random forest classifier for land-cover classification. *ISPRS J. Photogramm. Remote Sens.* **2012**, *67*, 93–104. [CrossRef]

43. Liaw, A.; Wiener, M. Classification and Regression by randomForest. *R News* **2002**, *2*, 18–22.

44. McRoberts, R.E.; Næsset, E.; Gobakken, T.; Chirici, G.; Condés, S.; Hou, Z.; Saarela, S.; Chen, Q.; Ståhl, G.; Walters, B.F. Assessing components of the model-based mean square error estimator for remote sensing assisted forest applications. *Can. J. For. Res.* **2018**, *48*, 642–649. [CrossRef]

45. Condés, S.; McRoberts, R.E. Updating national forest inventory estimates of growing stock volume using hybrid inference. *For. Ecol. Manag.* **2017**, *400*, 48–57. [CrossRef]

46. Hou, Z.; Xu, Q.; McRoberts, R.E.; Greenberg, J.A.; Liu, J.; Heiskanen, J.; Pitkänen, S.; Packalen, P. Effects of temporally external auxiliary data on model-based inference. *Remote Sens. Environ.* **2017**, *198*, 150–159. [CrossRef]

47. McRoberts, R.E.; Magnussen, S.; Tomppo, E.O.; Chirici, G. Parametric, bootstrap, and jackknife variance estimators for the k-Nearest Neighbors technique with illustrations using forest inventory and satellite image data. *Remote Sens. Environ.* **2011**, *115*, 3165–3174. [CrossRef]

48. Hou, Z.; McRoberts, R.E.; Ståhl, G.; Packalen, P.; Greenberg, J.A.; Xu, Q. How much can natural resource inventory benefit from finer resolution auxiliary data? *Remote Sens. Environ.* **2018**, *209*, 31–40. [CrossRef]

49. Efron, B.; Tibshirani, R.J. *An Introduction to the Bootstrap*; Chapman and Hall: New York, NY, USA, 1994.

50. Liu, R. Bootstrap Procedures under some Non-I.I.D. Models. *Ann.Stat.* **1988**, *16*, 1696–1708. [CrossRef]

51. Carpenter, J.; Bithell, J. Bootstrap confidence intervals: When, which, what? A practical guide for medical statisticians. *Stat. Med.* **2000**, *19*, 1141–1164. [CrossRef]

52. Diaconis, P.; Efron, B. Computer-Intensive Methods in Statistics. *Sci. Am.* **1983**, *248*, 116–130. [CrossRef]

53. Ranalli, M.G.; Mecatti, F. Comparing Recent Approaches For Bootstrapping Sample Survey Data: A First Step Towards A Unified Approach. *Proc. Surv. Res. Methods Sect. Am. Stat. Assoc.* **2012**, 4088–4099. Available online: http://boa.unimib.it/retrieve/handle/10281/41947/62652/Ranalli_Mecatti_Proc2013.pdf (accessed on 12 October 2020).

54. Flachaire, E. Bootstrapping heteroskedastic regression models: Wild bootstrap vs. pairs bootstrap. *Comput. Stat. Data Anal.* **2005**, *49*, 361–376. [CrossRef]

55. Freedman, D.A. Bootstrapping Regression Models. *Ann. Stat.* **1981**, *6*, 1218–1228. [CrossRef]

56. Esteban, J.; McRoberts, R.E.; Fernández-Landa, A.; Tomé, J.L.; Næsset, E. Estimating forest volume and biomass and their changes using random forests and remotely sensed data. *Remote Sens.* **2019**, *11*, 1944. [CrossRef]

57. Domingo, D.; Alonso, R.; de la Riva, J.; Lamelas, M.T.; Rodríguez, F.; Montealegre, A.L. Temporal Transferability of Pine Forest Attributes Modeling Using Low-Density Airborne Laser Scanning Data. *Remote Sens.* **2019**, *11*, 261. [CrossRef]

58. Rodriguez-Galiano, V.; Chica-Olmo, M. Land cover change analysis of a Mediterranean area in Spain using different sources of data: Multi-seasonal Landsat images, land surface temperature, digital terrain models and texture. *Appl. Geogr.* **2012**, *35*, 208–218. [CrossRef]

59. Stavrakoudis, D.G.; Dragozi, E.; Gitas, I.Z.; Karydas, C.G. Decision Fusion Based on Hyperspectral and Multispectral Satellite Imagery for Accurate Forest Species Mapping. *Remote Sens.* **2014**, *6*, 6897–6928. [CrossRef]

60. Pesaresi, S.; Mancini, A.; Quattrini, G.; Casavecchia, S. Mapping Mediterranean Forest Plant Associations and Habitats with Functional Principal Component Analysis Using Landsat 8 NDVI Time Series. *Remote Sens.* **2020**, *12*, 1132. [CrossRef]

61. Valbuena, R.; Mauro, F.; Rodriguez-Solano, R.; Manzanera, J.A. Accuracy and precision of GPS receivers under forest canopies in a mountainous environment. *Span. J. Agric. Res.* **2013**, *8*, 1047. [CrossRef]

62. Chen, Q.; Vaglio Laurin, G.; Valentini, R. Uncertainty of remotely sensed aboveground biomass over an African tropical forest: Propagating errors from trees to plots to pixels. *Remote Sens. Environ.* **2015**, *160*, 134–143. [CrossRef]

63. Breidenbach, J.; Astrup, R. Small area estimation of forest attributes in the Norwegian National Forest Inventory. *Eur. J. For. Res.* **2012**, *131*, 1255–1267. [CrossRef]

64. Chirici, G.; Giannetti, F.; McRoberts, R.E.; Travaglini, D.; Pecchi, M.; Maselli, F.; Chiesi, M.; Corona, P. Wall-to-wall spatial prediction of growing stock volume based on Italian National Forest Inventory plots and remotely sensed data. *Int. J. Appl. Earth Obs. Geoinf.* **2020**, *84*, 101959. [CrossRef]

65. Irulappa-Pillai-Vijayakumar, D.B.; Renaud, J.P.; Morneau, F.; McRoberts, R.E.; Vega, C. Increasing precision for French forest inventory estimates using the k-NN technique with optical and photogrammetric data and model-assisted estimators. *Remote Sens.* **2019**, *11*, 991. [CrossRef]

66. Maselli, F.; Chiesi, M.; Mura, M.; Marchetti, M.; Corona, P.; Chirici, G. Combination of optical and LiDAR satellite imagery with forest inventory data to improve wall-to-wall assessment of growing stock in Italy. *Int. J. Appl. Earth Obs. Geoinf.* **2014**, *26*, 377–386. [CrossRef]

67. McRoberts, R.E.; Næsset, E.; Gobakken, T. Inference for lidar-assisted estimation of forest growing stock volume. *Remote Sens. Environ.* **2013**, *128*, 268–275. [CrossRef]

**Publisher's Note:** MDPI stays neutral with regard to jurisdictional claims in published maps and institutional affiliations.

*Article*

# An End-to-End Deep Fusion Model for Mapping Forests at Tree Species Levels with High Spatial Resolution Satellite Imagery

Ying Guo [1,2], Zengyuan Li [1,2,*], Erxue Chen [1,2], Xu Zhang [1,2], Lei Zhao [1,2], Enen Xu [3], Yanan Hou [1,2] and Rui Sun [4]

[1] Research Institute of Forest Resource Information Techniques, Beijing 100091, China; guoying@ifrit.ac.cn (Y.G.); chenerx@ifrit.ac.cn (E.C.); zhangxu@ifrit.ac.cn (X.Z.); zhaolei@ifrit.ac.cn (L.Z.); houyanan@ifrit.ac.cn (Y.H.)
[2] Key Laboratory of Forestry Remote Sensing and Information System, NFGA, Chinese Academy of Forestry, Beijing 100091, China
[3] College of Geomatics, Xi'an University of Science and Technology, Xi'an 710054, China; xuee@ifrit.ac.cn
[4] School of Information Mechanical & Electrical Engineering, Jiangsu Open University, Nanjing 210017, China; sunrui@jsou.cn
* Correspondence: lizengyuan@ifrit.ac.cn; Tel.: +86-13611067375

Received: 7 September 2020; Accepted: 8 October 2020; Published: 13 October 2020

**Abstract:** Mapping the distribution of forest resources at tree species levels is important due to their strong association with many quantitative and qualitative indicators. With the ongoing development of artificial intelligence technologies, the effectiveness of deep-learning classification models for high spatial resolution (HSR) remote sensing images has been proved. However, due to the poor statistical separability and complex scenarios, it is still challenging to realize fully automated and highly accurate forest types at tree species level mapping. To solve the problem, a novel end-to-end deep learning fusion method for HSR remote sensing images was developed by combining the advantageous properties of multi-modality representations and the powerful features of post-processing step to optimize the forest classification performance refined to the dominant tree species level in an automated way. The structure of the proposed model consisted of a two-branch fully convolutional network (dual-FCN8s) and a conditional random field as recurrent neural network (CRFasRNN), which named dual-FCN8s-CRFasRNN in the paper. By constructing a dual-FCN8s network, the dual-FCN8s-CRFasRNN extracted and fused multi-modality features to recover a high-resolution and strong semantic feature representation. By imbedding the CRFasRNN module into the network as post-processing step, the dual-FCN8s-CRFasRNN optimized the classification result in an automatic manner and generated the result with explicit category information. Quantitative evaluations on China's Gaofen-2 (GF-2) HSR satellite data showed that the dual-FCN8s-CRFasRNN provided a competitive performance with an overall classification accuracy (OA) of 90.10%, a Kappa coefficient of 0.8872 in the Wangyedian forest farm, and an OA of 74.39%, a Kappa coefficient of 0.6973 in the GaoFeng forest farm, respectively. Experiment results also showed that the proposed model got higher OA and Kappa coefficient metrics than other four recently developed deep learning methods and achieved a better trade-off between automaticity and accuracy, which further confirmed the applicability and superiority of the dual-FCN8s-CRFasRNN in forest types at tree species level mapping tasks.

**Keywords:** forest type; deep learning; FCN8s; CRFasRNN; GF2; dual-FCN8s

---

## 1. Introduction

Forest classification at tree species' levels is important for the management and sustainable development of forest resources [1]. Mapping the distribution of forest resources is important due to their strong association with qualitative monitoring indicators such as spatial locations, as well as with many quantitative indicators like forest stocks, forest carbon storage, and biodiversity [2]. Satellite images have been widely used to map forest resources due to their efficiency and increasing availability [3].

With more accurate and richer spatial and textural information, high spatial resolution (HSR) remote sensing images can be used to extract more specific information of forest types [4]. In the recent past, a growing number of studies have been conducted on this topic [5,6]. Important milestones have been achieved, but there remain limitations [7,8]. One of the key limitations is that there is poor statistical separability of the images spectral range as there are a limited number of wavebands in such images [9]. As a result, in the case of forests with complex structures and more tree species, the phenomenon of "same objects with different spectra" and "different objects with the same spectra" can lead to serious difficulties in extracting relevant information. Thus, it raises the requirements for advanced forest information extraction methods.

Deep learning models are a kind of deep artificial neural network methods that have attracted substantial attention in recent years [10]. They have been successfully applied in land cover classification as they can adaptively learn discriminative characteristics from images through supervised learning, in addition to extracting and integrating multi-scale and multi-level remote sensing characteristics [11–14]. Compared with traditional machine learning methods, these models are capable of significantly improving the classification accuracy of land cover types, especially in areas with more complicated land cover types [15–17].

With the ongoing development of artificial intelligence technologies, several efficient deep-learning-based optimization models for the classification of land cover types have been proposed [18]. According to several recent studies, fusion individual deep-learning classifiers such as a convolutional neural network (CNN) into a multi-classifier can further improve the classification capacity of each classifier [19–21]. At the same time, other studies have shown that a CNN designed with a two-branched structure can extract panchromatic and multispectral information in remote sensing images individually, ensuring better classification quality compared to single structures [22]. In addition, some related research also indicated that the combination of a CNN and traditional image analysis technology such as conditional random fields (CRF) [23] is conducive to further improve the classification accuracy [24].

In recent years, to explore the classification effectiveness of deep-learning models for mapping forest types at tree species level, some scholars have attempted to apply advanced deep-learning classification methods to HSR satellite images. Guo (2020) proposed a two-branched fully convolutional network (FCN8s) method [25] and successfully extracted forest type distribution information at the dominant tree species level in the Wangyedian forest farm of Chifeng City [26] with China's GF2 data. The study revealed that the deep characteristics extracted by the two-branch FCN8s method showed a certain diversity and can enrich the input data sources of the model to some degree. Thus, it is a simple and effective optimization method. At the same time, compared with the traditional machine learning algorithms such as support vector machine and so on, there is a significant improvement in the resultant overall classification accuracy (OA). However, the classification accuracy of some forest types or dominate tree species, such as White birch (*Betula platyphylla*) and Aspen (*Populus davidiana*) in the study area, needs to be improved further.

Relevant studies have shown that when classifying forest types using FCN8s, the use of CRF as an independent post-processing step can further improve the classification accuracy [27,28]. However, even though the CRF method can improve the results of classification to some extent as a post-processing step independent of the FCN8s training, the free structures are unable to fully utilize CRF's inferential capability. This is because the operation of the model is separated from the training of the deep neural network model. Consequently, the model parameters cannot be updated with the iterative update of weights in the training phase. To address this limitation, Zheng (2015) has proposed a conditional random field as recurrent neural networks (CRFasRNN) model in which the CNN and CRF are constructed into a recurrent neural network (RNN) structure. Then, training the deep neural network model and operation of the CRF post-processing model can be implemented in an end-to-end manner. The advantages of the CNN and CRF models are thus fully combined. At the same time, the parameters of the CRF model can also be synchronously optimized during the whole network training, resulting in significant improvements in the classification accuracy [29].

In the paper, we proposed a novel end-to-end deep learning fusion method for mapping the forest types at tree species level based on HSR remote sensing data. The proposed model based on the previous published two-branch FCN8s method and imbedded a CRFasRNN layer into the network as the post-processing step, which is named dual-FCN8s-CRFasRNN in the paper.

The main contributions of this paper are listed as follows:

1. An end-to-end deep fusion dual-FCN8s-CRFasRNN network was constructed to optimize the forest classification performance refined to the tree species level in an automated way by combining the advantageous properties of multi-modality representations and the powerful features of post-processing step to recover a high-resolution feature representation and to improve the pixel-wise mapping accuracy in an automatic way.
2. A CRFasRNN module was designed to insert into the network to comprehensively consider the powerful features of post-processing step to optimize the forest classification performance refined to the dominant tree species level in an end-to-end automated way.

The remainder of the paper was structured as follows. The Materials and Method are presented in detail in Section 2. The Results are given in Section 3. Section 4 then discusses the feasibility of the optimized model. Finally, Section 5 concludes the paper.

## 2. Materials and Method

### 2.1. Study Areas

In this research, two study areas in China were selected, namely the Wangyedian forest farm and the GaoFeng forest farm, which are located in the North and South of China respectively (Figure 1). The reason why the Wangyedian forest farm and the GaoFeng forest farm were chosen as study areas was that both of them represented typical forest plantations in North and South China, respectively. Among them, the coniferous and broad-leaved tree species in the Wangyedian forest farm had clear spectral differences; however, the tree species in the GaoFeng forest farm belonged to the evergreen species, which made the spectral characteristics less affected by the seasonal changes and had more challenging for the classifier. Thus, the validation of the two above test areas could better illustrate the effectiveness and limitations of the proposed method.

The Wangyedian forest farm was founded in 1956. The geographical location is 118°09′E~118°30′E, 41°35′N~41°50′N, which lies to the southwest of Harqin Banner, Chifeng City, Inner Mongolia Autonomous Region, China, at the juncture of the Inner Mongolia, Hebei, and Liaoning provinces. The altitude is 800–1890 m, and the slope is 15–35°. The annual average temperature is 7.4 °C and the annual average precipitation is 400 mm, the climate zone belongs to moderate-temperate continental monsoon climate. The area of the forest farm is 2.47 ha with the forest area covers 2.33 ha, of which the plantation is 1.16 ha and the forest coverage rate 92.10%. The tree species of the plantation are mainly Chinese pine (*Pinus tabuliformis* Carrière) and Larix principis (*Larix principis-rupprechtii* Mayr); the dominant tree species in natural forests includes White birch (*Betula platyphylla*) and Aspen (*Populus davidiana*).

The GaoFeng forest farm was established in 1953. Its geographical location is 108°20′E~108°32′E, 22°56′~23°4′N (Figure 1), which is in Nanning City, Guangxi Zhuang Autonomous Region, China. The relative height of the mountain is generally 150–400 m and the slopes are 20–30°. It is in the humid subtropical monsoon climate area, with an average annual temperature of 21.6 °C and an average annual rainfall of 1300 mm. The total area of the forest land under management is 8.70 ha and forest coverage is 87.50%. Eucalyptus (*Eucalyptus robusta* Smith) and Chinese fir (*Cunninghamia lanceolate*) are the main plantations. It should be noted that the GaoFeng forest farm test area in our study is part of the GaoFeng forest farm. This is because after we searched all China's Gaofen-2 (GF-2) images within the GaoFeng Forest Farm in recent three years, only the DongSheng and JieBei sub-forest farms were cloud-free during this period. Therefore, these two sub-forest farms were selected in the research.

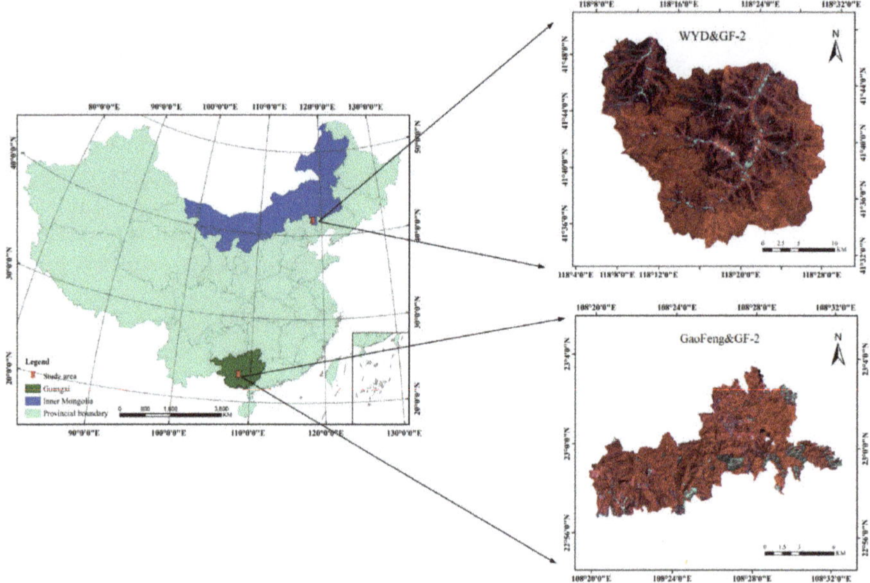

**Figure 1.** Schematic diagram of study area location, data source. WYD—Wangyedian Forest Farm, GaoFeng—Gaofeng Forest Farm, Gaofen-2 satellite (bands 4, 3, 2 false-color combinations).

*2.2. Test Data*

2.2.1. Land Cover Types, Forest and Tree Species Definition

The land cover, forest, and tree species classification used in this study mainly refer to the regulation of forest resources planning, design, and measurement [30], which are the technical standards of national forest resources planning and design survey. Based on the analysis of potential land classification results through pre-classifications from China's GF-2 images, the classification system of this study was determined as shown in Tables 1 and 2. The classes were divided into 11 categories for the Wangyedian forest farm (Table 1), including Chinese pine (*Pinus tabulaeformis*), Larix principis (*Larix principis-rupprechtii*), Korean pine (*Pinus koraiensis*), White birch and Aspen (*Betula platyphylla* and *Populus davidiana*, respectively), Mongolian oak (*Quercus mongolica*), Shrub land, Other forest land, Cultivated land, Grassland, Construction land, and Other non-forest lands. For simplicity, the above categories were abbreviated as CP, LP, KP, WA, MO, SL, OFL, CUL, GL, COL, ONFL as shown in Table 1.

For the GaoFeng forest farm test area (Table 2), it was divided into 7 categories, including Eucalyptus (*Eucalyptus spp.*), Chinese fir (*Cunninghamia lanceolata Hook.*), Masson pine (*Masson pine Lamb.*), Star anise (*Illicium verum Hook.f.*), Miscellaneous wood, Logging site, and Other non-forest lands. Here, Eucalyptus, Chinese fir, Masson pine, Star anise, and Miscellaneous wood belong to the subdivision category of forest land; Other non-forest lands mainly include construction land, water, and so on. For simplicity, the above types were abbreviated as EP, CF, MP, SA, MW, LS, ONFL as shown in Table 2.

**Table 1.** Classification system of the Wangyedian study areas.

| Test Area | Level One | Level Two | Level Three |
|---|---|---|---|
| The Wangyedian forest farm | Forest land | Woodland | Chinese pine (CP)<br>Larix principis (LP)<br>Korean pine (KP)<br>White birch and aspen (WA)<br>Mongolian oak (MO) |
| | | Shrub land(SL) | / |
| | | Other forest land | Include part of Mixed Broadleaf-conifer Forest and part of Mixed Broadleaf Forest (OFL) |
| | Non-forest land | Cultivated land (CUL) | / |
| | | Grassland (GL) | / |
| | | Construction land (COL) | / |
| | | Other non-forest land (ONFL) | / |

**Table 2.** Classification system of the GaoFeng study areas.

| Test Area | Level One | Level Two | Level Three |
|---|---|---|---|
| The GaoFeng forest farm test area | Forest land | Woodland | Eucalyptus (EP)<br>Chinese fir (CF)<br>Masson pine (MP)<br>Star anise (SA)<br>Miscellaneous wood (MW) |
| | | Logging site (LS) | / |
| | Non-forest land | Other non-forest land (ONFL) | Include part of Construction land and part of Water |

## 2.2.2. Remote Sensing Data

Launched on 19 August 2014, China's GF-2 was the first sub-meter HSR satellite successfully launched by the China High-resolution Earth Observation System (CHEOS). The GF-2 satellite is equipment with two panchromatic and multispectral (PMS) cameras which can provide pan and multispectral data with nadir resolutions of 1 m and 4 m, respectively, across an imaging swath of 45 km. Radiometric resolution of GF-2 is 10 bit. The GF-2 multispectral remote sensing images include the blue band(B) (0.45 μm–0.52 μm), green band(G)(0.52 μm–0.59 μm), red band (R) (0.63 μm–0.69 μm), and near infrared band(NIR)(0.77 μm–0.89 μm).The study used China's GF-2 panchromatic and multispectral (PMS) remote sensing imagery, which comprised a panchromatic band (1-m resolution) and four multispectral bands (4-m resolution). The WYD study area was covered by four scenes and the GaoFeng study area by one scene (specific image information is shown in Table 3).

**Table 3.** Parameter information for Gaofen-2 (GF-2) remote sensing images in the two study areas.

| Research Area | Scenery Serial Number | Image Time | Solar Elevation Angle (°) | Solar Azimuth (°) | Cloud Cover (%) |
|---|---|---|---|---|---|
| Wangyedian (WYD) | 4074551 | 5 September 2017 | 36.139 | 163.305 | 2% |
| | 4074552 | 5 September 2017 | 35.978 | 163.166 | 2% |
| | 4082058 | 5 September 2017 | 36.039 | 163.724 | 0% |
| | 4082059 | 5 September 2017 | 35.878 | 163.586 | 0% |
| GaoFeng (GF) | 2835829 | 21 September 2016 | 26.133 | 147.107 | 0% |

The imaging time was September 5, 2017 for the Wangyedian forest farm and September 21, 2016 for the GaoFeng forest farm test area. The preprocessing procedure of the satellite images comprised four steps, which were radiometric calibration, atmospheric correction, ortho-rectification, and image fusion. The radiometric calibration was the first step. The pixel brightness values of satellite observations were converted to apparent radiance by using the absolute radiometric calibration coefficients released by China Resources Satellite Data and Application Center [31]. Then, the fast line-of-sight atmospheric analysis of hypercube method [32] was used to perform atmospheric correction of multi-spectral and panchromatic data. In the next step, the parameters of the multi-spectral and panchromatic images and one digital elevation model (DEM) of 5 m resolution were used to perform ortho-rectification aided by the ground control points automatically extracted by image to image registration using a scale invariant feature transformation algorithm [33] with ZY-3 digital ortho-photo map (DOM) in 2 m spatial resolution [34] as reference. Finally, by using the nearest-neighbor diffusion-based pan-sharpening algorithm [35], the multi-spectral and panchromatic images were fused to obtain a 0.8 m high spatial resolution multi-spectral remote sensing image.

### 2.2.3. Ground Reference Data

Being aided by multi-temporal high-resolution remote sensing imagery and forest sub-compartment map and field survey data, 154 samples for the Wangyedian forest farm (Figure 2) and 136 samples for the GaoFeng forest farm test area (Figure 3) were constructed by visual interpretation. Each sample was composed of a pre-processed remote sensing image block and a matching image interpretation block at pixel levels. The size of each sample was 310 × 310 pixels.

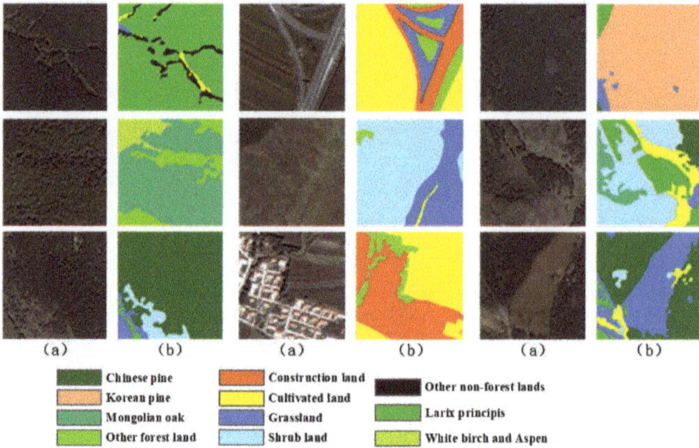


**Figure 2.** Examples in details for some of the training samples in the Wangyedian forest farm (**a**) Original image blocks; (**b**) Ground truth (GT) blocks showing the labels corresponding to the image blocks in (**a**).

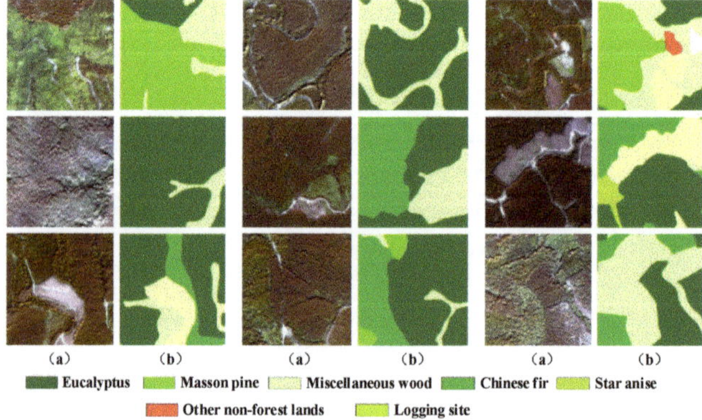


**Figure 3.** Examples in details for some of the training samples in the GaoFeng forest farm test area (**a**) Original image blocks; (**b**) Ground truth (GT) blocks showing the labels corresponding to the image blocks in (**a**).

At the same time, to verify the classification accuracy of the deep learning model, field surveys were carried out in September 2017 in the Wangyedian forest farm and January 2018 in the GaoFeng forest farm test area. A total of 404 samples in the Wangyedian forest farm and 289 samples in the GaoFeng forest farm test area were collected, as shown in Figure 4. It should be noted that the selected time of remote sensing image used in GaoFeng Forest Farm test area was inconsistent with that of field survey. This was because the cloud cover of China's GF-2 data in 2018 is large, which affected the classification accuracy. After the search of the data in experimental area during recent three years, we chose the cloud-free data in 2016. By the comparison among the multi-period data, the land cover types during that period had little changes, which was classified into logging sites.

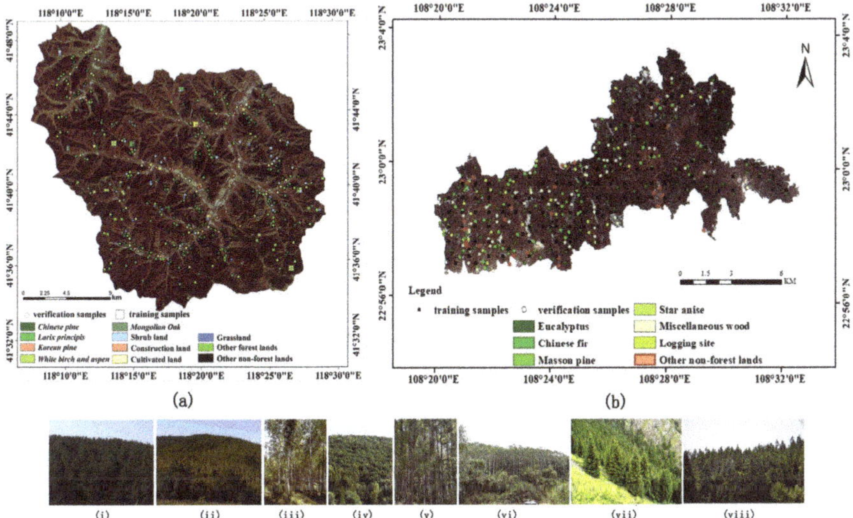

**Figure 4.** Spatial distribution map of the field survey sample points and some of the training samples in (**a**) the Wangyedian forest farm; (**b**) the GaoFeng forest farm test area, (**i**) Chinese pine, (**ii**) Larix principis, (**iii**) White birch and aspen, (**iv**) Mongolian oak, (**v**) Korean pine, (**vi**) Eucalyptus, (**vii**) Chinese fir, and (**viii**) Masson pine.

## 2.3. Workflow Description

In this study, a novel dual-fcn8s-crfasrnn method was developed to classify forest type at tree species level for HSR images. The network consisted with a two-branch FCN8s model with the ResNet50 network [36] as the backbone and a CRFasRNN model as a post-processing module. The classification process was shown in Figure 5: 310 × 310 image blocks were cut from the entire image and the real feature categories were labeled as training samples. During the training process, it was divided into the training samples and the validation samples according to the proportion of 80% and 20%. Then the dual-fcn8s-crfasrnn method was built. The test image contained eleven and seven feature types for Wangyedian forest farm and the GaoFeng forest farm, respectively. The same method was used to label the true feature types. The image block instead of the pixel unit was sent to the network for training, and the model loss was obtained after training. The model parameters were updated by using the back-propagation algorithm [37] until the optimal parameters were obtained. In the classification stage, the test image was sent to the trained network to obtain the final classification map.

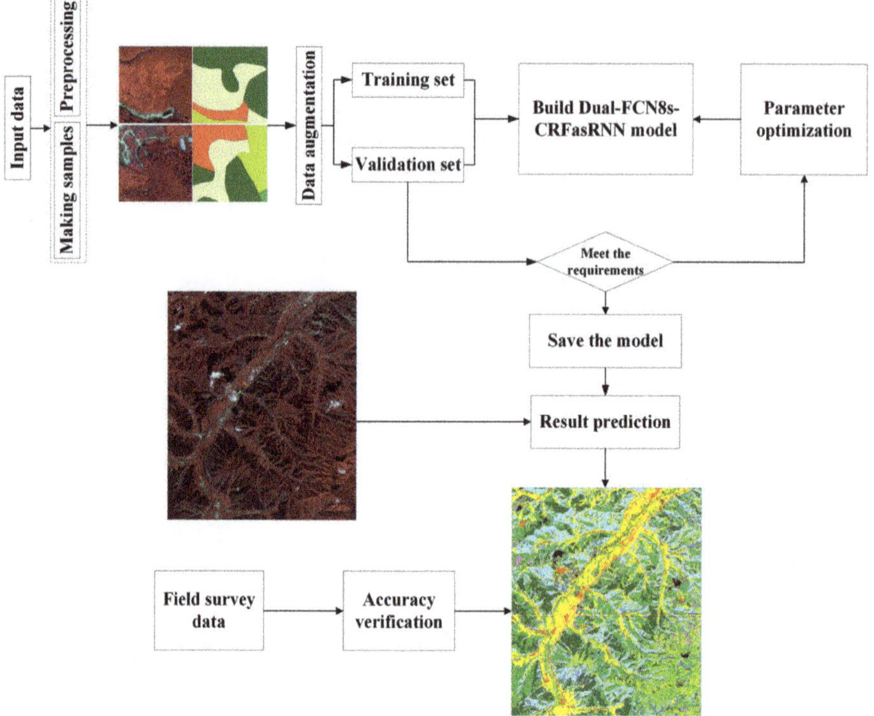

**Figure 5.** Workflow for the proposed model for forest type classification at tree species level based on high spatial resolution (HSR) image.

## 2.4. Network Structure

The proposed dual-FCN8s-CRFasRNN model is a kind of deep learning fusion model. It bases on a two-branched fully convolutional network, which predicts pixel-level labels without considering the smoothness and the consistency of the label assignments, followed by a CRFasRNN stage, which performs CRF-based probabilistic graphical modelling for structured prediction. The general workflow of the model is shown in Figure 6.

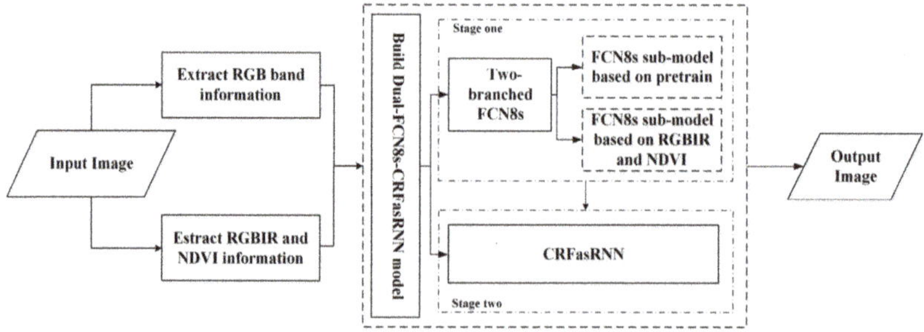

**Figure 6.** The general workflow of the Dual-FCN8s-CRFasRNN classification method.

### 2.4.1. Two-Branched Fully Convolutional Network

The two-branched fully convolutional network (FCN8s) was proposed recently for the forest type classification at a dominant tree species level using HSR remotely sensed imagery [26], and the

FCN8s [25] is the basic structure of the network which has showed impressive performance in terms of accuracy and computation time with many benchmark datasets. The architecture of FCN8s consists of down-sampling and up-sampling parts. The down-sampling part has convolutional and max-pooling layers to extract high-level abstract information, which is widely used in the classification related tasks in CNN. The convolutional and deconvolutional layers are contained in the up-sampling part which up-samples feature maps to output the score masks.

The structure of the two-branched FCN8s method contains two FCN8s sub-models which used Resnet50 [36] as its base classifier. Among them, one of the FCN8s sub-model used the RGB band information of image with fine-tuning strategy to construct the network by using the pre-trained weights of ImageNet dataset [38] and another sub-model made full use of the original 4-band information and the Normalized Difference Vegetation Index (NDVI) to build the model.

In the study, the two-branched FCN8s model was used to extract multi-modality features, with each FCN8s sub-model associated to one specific modality. For the model's output, each FCN8s sub-model was separated into five blocks according to the resolution of the feature maps, and the features with the identical resolution from different branches were combined using convolutional blocks. Then, the combined features from different branches gradually up sampled to the original resolution of the input image. The whole model can be defined as the minimum total loss between the prediction results of the training data and the ground truth value during the training process. Meanwhile, the parameters of the network can be iteratively updated by using a stochastic gradient descent (SGD) algorithm [39].

### 2.4.2. Conditional Random Field as Recurrent Neural Networks

CRFasRNN is an end-to-end deep learning model to solve the problem of pixel level semantic image segmentation. This approach combines the advantages of the CNN and CRF based graphics model in a unified framework. To be more specific, the model formulates mean-field approximate inferences for the dense CRF with Gaussian pairwise potentials as an RNN model. Since the parameters can be learnt in the RNN setting using the standard back-propagation, the CRFasRNN model can refine coarse outputs from a traditional CNN in the forward pass, while passing error differentials back to the CNN during training. Thus, the whole network can be trained end-to-end by utilizing the usual back-propagation algorithm [37].

### 2.4.3. Implementation Details

The implementation details of the proposed dual-FCN8s-CRFasRNN model were as follows. During the training procedure, we first trained a two-branched FCN8s architecture for semantic segmentation and the error at each pixel can be computed using the standard SoftMax cross-entropy loss [40] with respect to the ground truth segmentation of the image. Then, a CRFasRNN layer was inserted into the network and continued to train with the network. The detailed structure of the model is shown in Figure 7.

For more detail, once the computation enters the CRFasRNN model in the forward pass, it takes five iterations for the data to leave the loop created by the RNN. Neither the two-branched FCN8s that provides unary values nor the layers after the CRFasRNN need to perform any computations during this time since the refinement happens only inside the RNN's loop. Once the output Y leaves the loop, the next stages of the deep network after the CRFasRNN can continue the forward pass. During the backward pass, once the error differentials reach the CRFasRNN's output Y, they similarly spend five iterations within the loop before reaching the RNN input. In each iteration inside the loop, error differentials are computed inside each component of the mean-field iteration. After the CRFasRNN block, the output probability graph was obtained by using a softmax layer. Then, the probability output was thresholded to generate a classification result for each pixel.

Based on this, the complete system unifies strengths of both two-branched FCN8s and CRFs and is trainable end-to-end using the back-propagation algorithm [40] and the SGD procedure [41]. It is important to mention that in all of the convolutional block, we use a batch normalization layer [42] followed by a rectifier linear unit activation function [43].

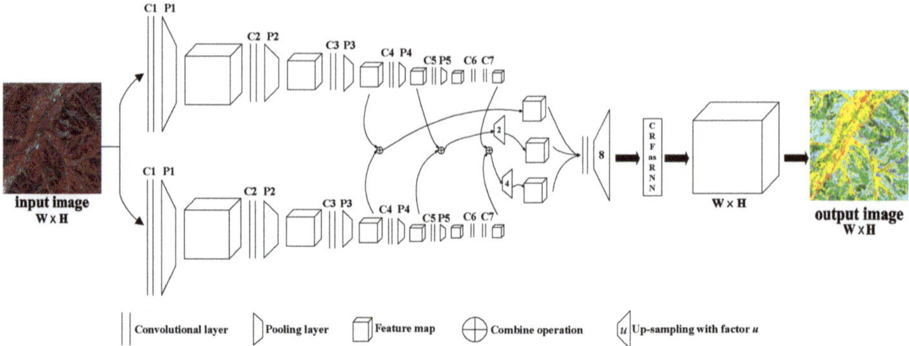


**Figure 7.** The structure of the proposed Dual-FCN8s-CRFasRNN.

In our all experiments, we initialized the first part of the network using the publicly available weights of the two-branched FCN8s network [26]. The compatibility transform parameters of the CRFasRNN were initialized using the Potts model [29]; the kernel width and weight parameters were obtained from a cross-validation process. During the training phase, the parameters of the whole network were optimized end-to-end using the back-propagation algorithm. We used full image training with the learning rate fixed at $10^{-9}$ and momentum set to 0.99. The loss function used was the standard SoftMax loss function, that is, the log-likelihood error function described in [40].

The proposed model was implemented in the Python language using Keras [44] with a TensorFlow [45] backend. All of the experiments were performed on a Nvidia Tesla K40C GPU. To optimize the network weights and early stopping criterion, the training set was divided into subsets (training and validation). We trained the proposed model using the Adam optimizer [46] with minibatches of size 16. The maximum number of training epochs was fixed to 10,000 for all experiments, and the training computation time was approximately 36 h.

*2.5. Accuracy Evaluation Index*

The evaluation index includes the OA, Kappa coefficient, as well as the user's (UA) and producer's accuracy (PA) values. OA is calculated as a percentage of correctly classified samples relative to all verified samples. The Kappa coefficient measures the consistency between classification results. The UA is expressed as the proportion of the total number of correctly classified samples for a specific class with respect to the total number of reference sample. The PA is calculated by the proportion of a specific class of accurate classification reference samples to the whole reference samples of that class and is a supplement of the omission error. The confusion matrix represents the number of pixels which are classified into a specific class.

## 3. Results

*3.1. Classification Results of the Dual-FCN8s-CRFasRNN*

### 3.1.1. The Wangyedian Forest Farm

The results of the dual-FCN8s-CRFasRNN approach showed a high level of agreement with the forest status on the ground in the Wangyedian forest farm. The OA was 90.10% and the Kappa coefficient was 0.8872 (Table 4). Among the individual forest type or tree species, it showed better results for the coniferous tree species compared to the broad-leaved tree species. For the three coniferous tree species, Larix principis and Korean pine performed well reaching PA and UA values above 90%, respectively. In contrast, the classification of the broad-leaved tree species (White birch and Aspen, and Mongolian Oak), performed worse than the coniferous trees. The accuracy obtained with PA and UA was only above 80%. For White birch and aspen, the UA was 90.91%, and the PA was only 80.65% with ten samples being assigned to the coniferous forest and two samples to the other forest lands category. Mongolian Oak had better accuracies (UA:86.67%; PA:92.86%). However, the mixed class called the other forest lands had a UA of 58.33% and a PA of 48.75% with some misclassification of the Cultivated land class, Mongolian Oak, and Chinese pine class. This was probably due to the similarity of the spectrum in the reference dataset.

**Table 4.** Confusion Matrix of Classification Result of Dual-FCN8s-CRFasRNN for the Wangyedian forest farm.

| | CP | LP | KP | WA | MO | OFL | CUL | COL | SL | GL | ONFL | Total | UA (%) |
|---|---|---|---|---|---|---|---|---|---|---|---|---|---|
| CP | 70 | 3 | 0 | 10 | 0 | 0 | 0 | 2 | 0 | 0 | 0 | 85 | 82.35 |
| LP | 1 | 63 | 0 | 0 | 0 | 0 | 0 | 1 | 0 | 0 | 0 | 65 | 96.92 |
| KP | 1 | 0 | 15 | 0 | 0 | 0 | 0 | 0 | 0 | 0 | 0 | 16 | 93.75 |
| WA | 1 | 2 | 0 | 50 | 2 | 0 | 0 | 0 | 0 | 0 | 0 | 55 | 90.91 |
| MO | 0 | 0 | 0 | 0 | 26 | 1 | 0 | 2 | 1 | 0 | 0 | 30 | 86.67 |
| OFL | 0 | 0 | 0 | 0 | 0 | 32 | 0 | 3 | 1 | 1 | 0 | 37 | 86.49 |
| CUL | 0 | 0 | 0 | 0 | 0 | 0 | 40 | 0 | 0 | 0 | 0 | 40 | 100.00 |
| COL | 1 | 1 | 0 | 2 | 0 | 1 | 0 | 7 | 0 | 0 | 0 | 12 | 58.33 |
| SL | 0 | 0 | 0 | 0 | 0 | 2 | 0 | 1 | 21 | 0 | 0 | 24 | 87.50 |
| GL | 0 | 0 | 0 | 0 | 0 | 0 | 0 | 0 | 0 | 11 | 0 | 11 | 100.00 |
| ONFL | 0 | 0 | 0 | 0 | 0 | 0 | 0 | 0 | 0 | 0 | 29 | 29 | 100.00 |
| Total | 74 | 69 | 15 | 62 | 28 | 36 | 40 | 16 | 23 | 12 | 29 | 404 | |
| PA (%) | 94.59 | 91.30 | 100.00 | 80.65 | 92.86 | 88.89 | 100.00 | 43.75 | 91.30 | 91.67 | 100.00 | | |
| Overall accuracy: 90.10%; Kappa coefficient: 0.8872 | | | | | | | | | | | | | |

### 3.1.2. The GaoFeng Forest Farm Test Area

The quantitative evaluation of this test site is shown in Table 5. It showed that the OA was 74.39% and the Kappa coefficient was 0.6973. Compared to the non-forest lands, the classification effect of the dual-FCN8s-CRFasRNN model on single tree species or forest land was better. The Eucalyptus type had the highest classification accuracy (PA:93.33%; UA: 70.89%). In addition to the good extraction effect on Eucalyptuses, the classification accuracy on other four types of tree species or forests was also about 70%. In these four tree species or forest types, the classification effect of the Chinese fir tree species was good, which may be related to the wide distribution of this tree species in the test site. The Other non-forest lands type performed poor and a large amount of Other non-forest lands was misclassified as Eucalyptus, which may be due to the sample size involved in modeling.

**Table 5.** Confusion Matrix of Classification Result of Dual-FCN8s-CRFasRNN for the GaoFeng forest farm.

| | EP | CF | MP | SA | MW | LS | ONFL | Total | UA (%) |
|---|---|---|---|---|---|---|---|---|---|
| EP | 56 | 1 | 5 | 6 | 3 | 3 | 5 | 79 | 70.89 |
| CF | 0 | 35 | 3 | 0 | 4 | 0 | 0 | 42 | 83.33 |
| MP | 1 | 5 | 26 | 2 | 2 | 3 | 1 | 40 | 65.00 |
| SA | 1 | 1 | 0 | 25 | 2 | 0 | 1 | 30 | 83.33 |
| MW | 1 | 5 | 4 | 3 | 25 | 3 | 2 | 43 | 58.14 |
| LS | 1 | 0 | 0 | 0 | 1 | 27 | 3 | 32 | 84.38 |
| ONFL | 0 | 0 | 0 | 0 | 0 | 2 | 21 | 23 | 91.30 |
| Total | 60 | 47 | 38 | 36 | 37 | 38 | 33 | 289 | |
| PA (%) | 93.33 | 74.47 | 68.42 | 69.44 | 67.57 | 71.05 | 63.64 | | |
| Overall accuracy: 74.39%; Kappa coefficient: 0.6973 | | | | | | | | | |

### 3.1.3. The Complementarity of the Case Study

It could be seen from the result of Table 4, the OA of the dual-FCN8s-CRFasRNN model in the Wangyedian forest farm was 90.10%; however, it was only 43.75% for the less distributed broad-leaved tree species or mixed forest (such as White birch and Aspen). To verify the effect of the model for broad-leaved tree species or forest, this study further carried out the experiment in the GaoFeng forest farm. The main forest type in this area was broad-leaved mixed forest represented by Eucalyptus. The results showed that (Table 5), the OA of the dual-FCN8s-CRFasRNN model was 74.39%; however, it was 93.33% for the Eucalyptus species with wide distribution. This indicated that the classification effect of the model was not restricted by tree species, but rather by the distribution of tree species. It also could be seen from the results of the two test areas that (Tables 4 and 5) the dual-FCN8s-CRFasRNN model performed well to extract the plantation with the accuracy all above 90%.

### 3.2. Impact of Adding NDVI and Using Fine-Tuning Strategy on Performance

According to the conclusion presented in [27], the two-branched FCN8s model can optimize the classification results after including NDVI features and using the fine-tuning strategy. Based on their conclusion, this paper further analyzed the influences of the use of NDVI features and fine-tuning strategies on the classification effect of the dual-FCN8s-CRFasRNN model.

In the follows, we will firstly analyze the classification effects of the dual-FCN8s-CRFasRNN model on the two test sites using the fine-tuning strategy without NDVI features, of which the results will be denoted as dual-FCN8s-noNDVI-CRFasRNN. Then, we will analyze the classification performance of the dual-FCN8s-CRFasRNN model with NDVI features but not using the fine-tuning strategy, whose results will be denoted as dual-FCN8s-noFinetune-CRFasRNN.

### 3.2.1. Impact of Adding NDVI on Performance

The classification results of the dual-FCN8s-noNDVI-CRFasRNN model on the two test sites were shown in Tables 6 and 7. It can be seen from Table 6 that the OA of this model on the Wangyedian forest farm was 88.37%, and the Kappa coefficient was 0.8678. Its classification accuracy decreased by 1.73% compared with the dual-FCN8s-CRFasRNN model. The classification accuracy of this model on the GaoFeng forest farm test area is shown in Table 7. It can be found that the OA of the model in this test site was 71.63%, and the Kappa coefficient was 0.6645. Its classification accuracy deceased by 2.76% compared with the dual-FCN8s-CRFasRNN model. From the further analysis of the results, it can be known that the dual-FCN8s-CRFasRNN model could optimize the effect on broad-leaved trees after adding NDVI. From the comparative analysis of Tables 6 and 7, it can be found that after including NDVI, the classification accuracies of the two broad-leaved mixed tree species of White birch and Aspen and Mongolian Oak were both improved. Especially for White birch and Aspen, the classification accuracy was improved by approximately 10%. The comparison results of the GaoFeng forest farm test site showed that, after the including of NDVI, the classification performance of broad-leaved forest was also improved.

### 3.2.2. Impact of Using a Fine-Tuning Strategy on Performance

The classification results of the dual-FCN8s-noFinetune-CRFasRNN model are shown in Table 6 Table 7. It can be seen from the Table 4 that the classification accuracy decreased by 4.95% and 12.8% on the Wangyedian forest farm and the GaoFeng forest farm test area, respectively, compared with the dual-FCN8s-CRFasRNN model. Through the further analysis of the results, it can be known that after applying the fine-tuning strategy, the dual-FCN8s-CRFasRNN model can optimize the classification effect of most types in the classification systems of these two test sites. From the comparative analysis of Table 6, after the fine-tuning strategy was used, the optimization effect of broad-leaved forests was significant in the Wangyedian forest farm. From the comparative analysis of the GaoFeng forest farm shown in Table 7, the classification accuracy of Chinese fir improved significantly.

**Table 6.** Impact of NDVI Features and Fine-tuning Strategy on Classification Accuracy of Dual-FCN8s-CRFasRNN for the Wangyedian forest farm.

| | dual-FCN8s-noNDVI-CRFasRNN | | dual-FCN8s-noFinetune-CRFasRNN | |
| --- | --- | --- | --- | --- |
| | PA (%) | UA (%) | PA (%) | UA (%) |
| CP | 94.59 | 81.4 | 93.24 | 77.53 |
| LP | 91.30 | 96.92 | 89.86 | 88.57 |
| KP | 100.00 | 93.75 | 100.00 | 100.00 |
| WA | 70.97 | 89.80 | 62.90 | 92.86 |
| MO | 89.29 | 100.00 | 85.71 | 85.71 |
| OFL | 81.25 | 72.22 | 68.75 | 47.83 |
| CUL | 88.89 | 84.21 | 94.44 | 79.07 |
| COL | 100.00 | 100.00 | 100.00 | 100.00 |
| SL | 95.65 | 75.86 | 82.61 | 100.00 |
| GL | 83.33 | 76.92 | 83.33 | 76.92 |
| ONFL | 79.31 | 92.00 | 72.41 | 95.45 |
| OA (%) | 88.37 | | 85.15 | |
| Kappa coefficient | 0.8678 | | 0.831 | |

**Table 7.** Impact of NDVI Features and Fine-tuning Strategy on Classification Accuracy of Dual-FCN8s-CRFasRNN for the Gaofeng forest farm test area.

| | dual-FCN8s-noNDVI-CRFasRNN | | dual-FCN8s-noFinetune-CRFasRNN | |
| --- | --- | --- | --- | --- |
| | PA (%) | UA (%) | PA (%) | UA (%) |
| EP | 91.67 | 68.75 | 91.67 | 76.39 |
| CF | 61.7 | 76.32 | 46.81 | 78.57 |
| MP | 60.53 | 48.94 | 68.42 | 55.32 |
| SA | 69.44 | 89.29 | 72.22 | 81.25 |
| MW | 59.46 | 56.41 | 64.86 | 48.98 |
| LS | 84.21 | 91.43 | 78.95 | 85.71 |
| ONFL | 63.64 | 95.45 | 63.64 | 80.77 |
| OA(%) | 71.63% | | 70.59% | |
| Kappa coefficient | 0.6645 | | 0.6538 | |

*3.3. Impact of Using a CRFasRNN Post-Procedureon Performance*

For clearly observing the difference before and after the CRFasRNN post-processing, we compared the results of the proposed dual-FCN8s-CRFasRNN and the two-branched FCN8s, as well as the results obtained from comparing the FCN8s model with the CRFasRNN module called FCN8s-CRFasRNN with the FCN8s.

3.3.1. The Wangyedian Forest Farm

The comparison of the results of the dual-FCN8s-CRFasRNN model and the two-branched FCN8s model on the Wangyedian forest farm are shown in Table 8. It can be found that after the embedding of the CRFasRNN post-processing module, the OA of the two-branched FCN8s model increased from 87.38% to 90.1%, and the Kappa coefficient increased from 0.8567 to 0.8872. The classification result of single category showed that the Grassland had the best improvement effect, followed by White birch and Aspen.

In addition, the results in Table 8 show the OA of the FCN8s model increased from 86.63% to 88.12%, and the Kappa coefficient increased from 0.8482 to 0.8646 after the inserting the CRFasRNNpost-processing module. The classification result of single category showed that White birch and aspen had the best improvement effect.

**Table 8.** Classification accuracies of the FCN8s, FCN8s-CRFasRNN, Dual-FCN8s and Dual-FCN8s-CRF for the Wangyedian forest farm.

| | FCN8s | | FCN8s-CRFasRNN | | Dual-FCN8s | | Dual-FCN8s-CRF | |
|---|---|---|---|---|---|---|---|---|
| | PA (%) | UA (%) | PA (%) | UA (%) | PA (%) | UA (%) | PA (%) | UA (%) |
| CP | 90.54 | 81.71 | 86.49 | 81.01 | 95.95 | 86.59 | 95.71 | 84.81 |
| LP | 94.20 | 87.84 | 89.86 | 96.88 | 86.96 | 92.31 | 94.20 | 94.20 |
| KP | 100.00 | 93.75 | 100.00 | 100.00 | 100.00 | 93.75 | 100.00 | 100.00 |
| WA | 56.45 | 92.11 | 79.03 | 77.78 | 67.74 | 93.33 | 60.53 | 88.46 |
| MO | 100.00 | 82.35 | 92.86 | 81.25 | 92.86 | 92.86 | 92.86 | 96.30 |
| OFL | 50.00 | 40.00 | 31.25 | 83.33 | 62.50 | 43.48 | 62.50 | 55.56 |
| CUL | 97.22 | 87.50 | 97.22 | 83.33 | 94.44 | 75.56 | 90.44 | 80.95 |
| COL | 100.00 | 100.00 | 100.00 | 100.00 | 100.00 | 100.00 | 100.00 | 100.00 |
| SL | 86.96 | 95.24 | 95.65 | 91.67 | 91.30 | 87.5 | 95.65 | 95.65 |
| GL | 83.33 | 100.00 | 83.33 | 90.91 | 50.00 | 85.71 | 83.33 | 83.33 |
| ONFL | 93.10 | 93.10 | 96.55 | 100.00 | 96.55 | 96.55 | 86.21 | 100.00 |
| OA (%) | 86.63 | | 88.12 | | 87.38 | | 89.63 | |
| Kappa coefficient | 0.8482 | | 0.8646 | | 0.8567 | | 0.882 | |

Furthermore, the classification performance of the dual-FCN8s-CRFasRNN and the FCN8s-CRFasRNN models were better than the two-branched FCN8s model and FCN8s models. These results indicate that the learned features obtained by the CRFasRNN post-procedure achieve a level of performance that is complementary to the deep features extracted by the original model.

3.3.2. The GaoFeng Forest Farm Test Area

The comparison of the classification results of the dual-FCN8s-CRFasRNN model and the two-branched FCN8s model on the GaoFeng forest farm is shown in Table 9. After the embedding of the CRFasRNN post-processing module, the OA of the two-branched FCN8s model increased from 72.32% to 74.39%, and the Kappa coefficient increased from 0.6735 to 0.6973. The classification result of single category showed that the classification effect of Chinese fir, Masson pine, and Logging site were all improved.

**Table 9.** Classification accuracies of the FCN8s, FCN8s-CRFasRNN, Dual-FCN8s and Dual-FCN8s-CRF for the GaoFeng forest farm test area.

| | FCN8s | | FCN8s-CRFasRNN | | Dual-FCN8s | | Dual-FCN8s-CRF | |
|---|---|---|---|---|---|---|---|---|
| | PA (%) | UA (%) | PA (%) | UA (%) | PA (%) | UA (%) | PA (%) | UA (%) |
| EP | 95.00 | 66.28 | 93.33 | 78.87 | 93.33 | 76.71 | 86.67 | 89.66 |
| CF | 59.57 | 66.67 | 55.32 | 81.25 | 63.83 | 73.17 | 70.21 | 68.75 |
| MP | 50.00 | 70.37 | 78.95 | 55.56 | 63.16 | 61.54 | 63.16 | 61.54 |
| SA | 72.22 | 76.47 | 58.33 | 84.00 | 69.44 | 75.76 | 72.22 | 86.67 |
| MW | 67.57 | 40.32 | 70.27 | 54.17 | 72.97 | 54 | 62.16 | 43.40 |
| LS | 63.16 | 88.89 | 73.68 | 90.32 | 65.79 | 89.29 | 78.95 | 85.71 |
| ONFL | 27.27 | 81.82 | 75.76 | 89.29 | 66.67 | 88.00 | 78.79 | 100.00 |
| OA (%) | 65.05% | | 73.36% | | 72.32% | | 74.05% | |
| Kappa coefficient | 0.5857 | | 0.6863 | | 0.6735 | | 0.695 | |

In addition, the results in Table 9 reveal the OA of the FCN8s model increased from 65.05% to 73.36%, and the Kappa coefficient increased from 0.5857 to 0.6863 after embedding the CRFasRNN post-processing module. The classification result of single category showed that the classification effect of Masson pine has the most obvious improvement.

From the OA results of the two test sites, the embedding of the CRFasRNN post-processing module can optimize the classification effect of the two-branched FCN8s and FCN8s models. At the same time, the results showed that the FCN8s model with the CRFasRNN post-processing module had better optimization effects on the classification results compared with the optimization of the dual structure on the FCN8s model.

### 3.4. Benchmark Comparison for Classification

In order to further investigate the validity of the proposed method, this study compared the classification accuracy of the dual-FCN8s-CRFasRNN method with the performances of other four models: the two-branched FCN8s model using CRF as post processing method (named dual-FCN8s-CRF); the two-branched FCN8s model without any post-processing procedure (named dual-FCN8s); the traditional FCN8s method (named FCN8s); and the FCN8s inserted with a CRFasRNN layer (named FCN8s-CRFasRNN). Both study areas were tested with the above models (Tables 8 and 9).

The backbone network of the FCN8s model mentioned in the four above approaches was Resnet50 [36]. The input features for the four above approaches were four-band spectral information and the NDVI feature. It can be seen from the classification results of the two test sites that among the above models that the dual-FCN8s-CRFasRNN model had the best classification performance; the FCN8s model had the poorest accuracy. Among the remaining three models, the classification accuracy of the dual-FCN8s-CRF model was better than that of the FCN8s-CRFasRNN model. The classification performance of the FCN8s-CRFasRNN model was better than that of the dual-FCN8s model. From the perspective of a single category classification, the accuracy of the dual-FCN8s-CRFasRNN model on the White birch and Aspen in the Wangyedian forest farm test site and the Chinese fir in the GaoFeng forest farm test site were better than those of other models.

Figures 8 and 9 present the land cover type derived from the above five models at a more detailed scale. After post processing by the CRF and CRFasRNN algorithms, the classification result images are refined. The algorithms both reduce the degree of roughness of the image; the boundaries of Larix principis and White birch and Aspen become smoother, and the land cover shapes become clearer during the segmentation process. However, due to the excessive expansion of certain land covers during the CRF post-processing, some small land covers are incorrectly classified by the surrounding land covers, which is inferior to the CRFasRNN as the post-processing method. From the visual comparison of the model with and without the CRFasRNN post-procedure, we observed classes that tend to represent large homogeneous areas benefit substantially from the post-procedure. For the dual-FCN8s method, some small objects are easily misclassified as the surrounding categories because they are merged with surrounding pixels into the same objects during the segmentation phase. In addition, the local inaccurate boundaries generated by the segmentation method also caused deviation from the real edges. The result also shows that some details were missing from the land cover border. Small land cover areas tend to be round, and some incorrect classifications are exaggerated by directly applying the common-structure FCN8s to the image classification problem.

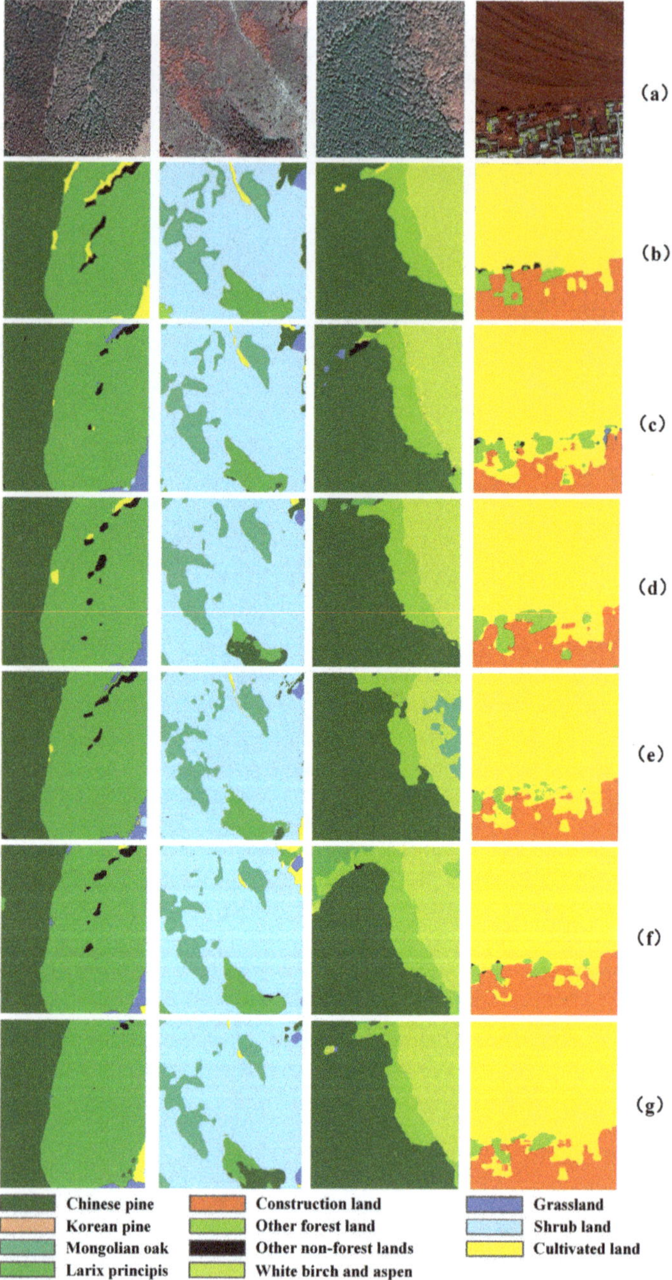



**Figure 8.** The detailed classification results of the Wangyedian forest farm (**a**) GF2-PMS (**b**) Label (**c**) Dual-FCN8s-CRFasRNN (**d**) Dual-FCN8s-CRF (**e**) Dual-FCN8s (**f**) FCN8s (**g**) FCN8s-CRFasRNN.

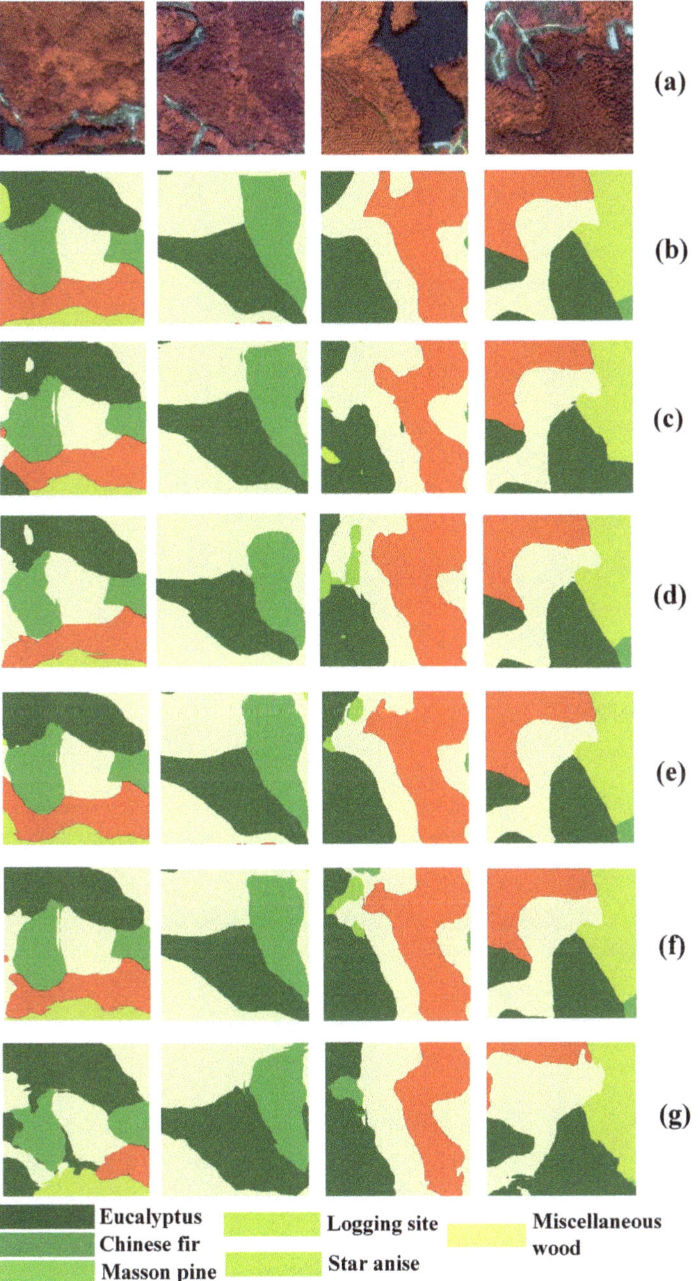



**Figure 9.** The detailed classification results of the GaoFeng forest farm test area (**a**) GF2-PMS (**b**) Label (**c**) Dual-FCN8s-CRFasRNN (**d**) Dual-FCN8s-CRF (**e**) FCN8s-CRFasRNN (**f**) Dual-FCN8s (**g**) FCN8-CRFasRNN.

## 4. Discussion

The study showed that the deep learning fusion model has great potential in the classification of forest types and tree species. The proposed dual-FCN8s-CRFasRNN method indicated its applicability for forest type classification at a tree species level from HSR remote sensing imagery. The experimental results showed that, the dual-FCN8s-CRFasRNN model could extract the dominant tree species or forest types which were widely distributed in two study areas, especially for the extraction of the plantation species such as Chinese pine, Larix principis, and Eucalyptus which all had an OA above 90%. However, for the other forest land types in the Wangyedian forest farm and the other category in the GaoFeng forest farm, the classification accuracy was poor. That may be because the above two categories included many kinds of surface features, presenting spectral characteristics that are complex and difficult to distinguish.

Through the comparative analysis of the research results of the two experimental areas, it could be seen that the classification accuracy of the GaoFeng forest farm in Guangxi was relatively low compared with the classification results of the Wangyedian forest farm in North China. The PA of Masson pine, Star anise, and Miscellaneous wood were less than 70%. This may have been caused by two reasons. Firstly, the spectral is similar among these forest land types, which inevitably increased the difficulty of distinguishing them. Secondly, the number of collected samples of these forest land types compared to Eucalyptus was relatively small, which leads to an under-fitting phenomenon of the model.

To verify the effectiveness of the proposed method and clarify the optimization effect of embedding CRFasRNN layer in the model published earlier, the paper not only compared the dual-FCN8s-CRFasRNN model with the deep learning model published earlier, but also compared the results of post-processing with CRF. The results showed that the optimal classification results were obtained after adding CRFasRNN layer in both experimental areas compared with the above two methods. The classification effect was improved obviously especially for the category with less distribution range.

To further prove that embedding CRFasRNN layer into deep learning model is a general and effective optimization method. This study further compared and analyzed the optimization effect of embedding CRFasRNN layer in the classic FCN8s model. The results showed that the model with CRFasRNN layer got better classification accuracy than the classic FCN8s model and the previously published two branch fcn8s model. At the same time, in terms of processing efficiency, it also reduced the processing time compared with using CRF post-processing method.

For including NDVI indices and using a fine-tuning strategy, the proposed method outperformed those without NDVI feature and fine-tuning strategy in terms of the OA and the Kappa coefficient achieved. To further clarify the effectiveness of the use of NDVI index, this study further replaced the NDVI index with green NDVI (GNDVI) [47] index in two experimental areas as showed in Tables 10 and 11. The results showed that the accuracy of the dual-FCN8s-CRFasRNN model with GNDVI index was with OA of 89.47% and 73.70% and Kappa coefficient of 0.8804 and 0.6898 for Wangyedian and GaoFeng forest farm, respectively, which was very similar with the model using NDVI index and the difference between them was less than 1%. From the classification results of a single category, the use of GNDVI index improved the effect of broad-leaved mixed forest, such as the other forest types in Wangyedian forest farm and the Miscellaneous in GaoFeng forest farm.

Compared with the previous research results of the fine classification of forest types using HSR data, the method proposed in the study got better performance. Immitzer et al. (2012) carried out tree species classification with random forest using WorldView-2 satellite data and the overall accuracy for classifying 10 tree species was around 82% [48]. Adelabu et al. (2015) employed ground and satellite based QuickBird data and random forest to classify five tree species in a Southern African Woodland with OA of 86.25% [49]. Waser et al. (2014) evaluated the potential of WorldView-2 data to classify tree species using object-based supervised classification methods with OA of 83% [50]. Cho et al. (2015) based on WorldView-2 data performed tree species and canopy gaps mapping in South Africa with

OA of 89.3% [51]. Sun et al. (2019) optimized three different deep learning methods to classify the tree species, the results showed that VGG16 had the best performance, with an overall accuracy of 73.25% [52]. Cao (2020) based on the airborne charge coupled device (CCD) orthophoto proposed an improved Res-UNet model for tree species classification in GaoFeng forest farm. The result showed that the proposed method got an OA of 87% which was higher than this study; however, the experimental area of this study was smaller, and the accuracy of Eucalyptus classification is 88.37%, which is lower than this paper [53]. Xie (2019) based on multi-temporal ZY-3 data, carried out the classification of tree species, forest type, and land cover type in the Wangyedian forest farm [54], which is same test area utilized in our research. The OA (84.9%) was much low than our result, but the accuracy of broad-leaved tree species such as White birch and Aspen (approximately 85%) was a little bit higher than ours. This may be due to the use of multi-temporal data to optimize the performance. The effect of using multi-temporal data for the fine classification of forest types had also been proven by other studies. For example, Ren et al. based on multi-temporal SPOT-5 and China GF-1 data achieved the fine classification of forest types with an accuracy of up to 92% [55]. Agata (2019) classified tree species over a large area based on multi-temporal Sentinel-2 and DEM with a classification accuracy of 94.8% [56]. Based on these results, studies are planned to combine multi-temporal satellite data with the deep learning method for forest type fine classification as the next step in the research. In addition, the proposed method also needs to be assessed in other forest areas to evaluate the effect of different forest structures and other tree species.

**Table 10.** Classification accuracy of dual-FCN8s-CRFasRNN with the green NDVI index in the Wangyedian forest farm.

| | PA (%) | UA (%) |
|---|---|---|
| CP | 95.89 | 86.42 |
| LP | 91.30 | 96.92 |
| KP | 100.00 | 88.24 |
| WA | 68.42 | 86.67 |
| MO | 89.29 | 100.00 |
| OFL | 81.25 | 72.22 |
| CUL | 88.89 | 84.21 |
| COL | 100.00 | 100.00 |
| SL | 95.65 | 75.86 |
| GL | 83.33 | 83.33 |
| ONFL | 80.00 | 96.00 |
| OA(%) | 89.4737 | |
| Kappa coefficient | 0.8804 | |

**Table 11.** Classification accuracy of dual-FCN8s-CRFasRNN with GNDVI index in the GaoFeng forest farm.

| | PA (%) | UA (%) |
|---|---|---|
| EP | 93.33 | 75.68 |
| CF | 61.70 | 76.32 |
| MP | 57.89 | 52.38 |
| SA | 69.44 | 83.33 |
| MW | 70.27 | 56.52 |
| LS | 81.58 | 91.18 |
| ONFL | 72.73 | 96.00 |
| OA(%) | 73.7024% | |
| Kappa coefficient | 0.6898 | |

## 5. Conclusions

As an effort to optimize the forest mapping performance refined to the tree species level in an automated way, this study developed a novel end-to-end deep learning fusion method (dual-FCN8s-CRFasRNN) for HSR remote sensing images by combining the advantageous properties of multi-modality representations and the powerful features of post-processing step, and verified its applicability for the two areas which are located in the North and South of China respectively. With an overall accuracy of 90.1% and 74.39% for two test areas, respectively, we could demonstrate the high potential of the model for forest mapping at tree species level. The results also showed that it could get a remarkable result for some plantation tree species, such as Chinese pine and Larix principis in the northern test area, and Eucalyptus in the southern test area. The embedding of the CRFasRNN post-processing module could effectively optimize the classification result. Especially for the tree species with small distribution range, the improvement effect is obvious. Through comprehensive comparison of classification accuracy and processing time, embedding CRFasRNN layer in deep learning model not only automatically completed post-processing operation in an end-to-end manner, but also improved classification effect and reduced processing time.

Given the importance of mapping forest resources, the proposed dual-FCN8s-CRFasRNN model provided a feasible optimized idea for mapping the forest type at tree species levels for HSR image, and will substantially contribute to the improvement for the management and sustainable development of forest resources in the country.

In the future, we will further exploit the potentials of deep learning based on multi-temporal data, as well as investigating the means to build the model with limited number of training samples for the forest type classification at tree species level of high-spatial-resolution images.

**Author Contributions:** Y.G. conceptualized the manuscript topic and was in charge of overall direction and planning. Z.L. and E.C. reviewed and edited the first draft of the manuscript. X.Z. provided valuable information for field survey site selection. L.Z. was involved in field data collection. E.X., Y.H., and R.S. carried out the preprocessing of the images and were involved in field data collection. All co-authors contributed to the final writing. All authors have read and agreed to the published version of the manuscript.

**Funding:** This research was funded by the Special Funds for Fundamental Research Business Expenses of Central Public Welfare Research Institutions "Study on Dual-branch uNet Optimal Combination Model for Forest Type Remote Sensing Classification (CAFYBB2019SY027)" project and National Key R&D Program of China "Research of Key Technologies for Monitoring Forest Plantation Resources (2017YFD0600900)" project.

**Acknowledgments:** We express our sincere thanks to Yuancai Lei from Institute of Forest Resource Information Techniques, Chinese Academy of Forestry, for reviewing and editing the draft of the manuscript. We also would like to thank the Wangyedian Forest Farm and the GaoFeng Forest Farm for providing the wonderful datasets. The authors are also grateful to the editors and referees for their constructive criticism on this paper.

**Conflicts of Interest:** The authors declare no conflict of interest.

## References

1. Liu, X.; Zhang, X. Research Advances and Countermeasures of Remote Sensing Classification of Forest Vegetation. *For. Resour. Manag.* **2004**, *1*, 61–64. (In Chinese)
2. Xiao, X.; Boles, S.; Liu, J.; Zhuang, D.; Liu, M. Characterization of forest types in Northeastern China, using multi-temporal SPOT-4 VEGETATION sensor data. *Remote Sens. Environ.* **2002**, *82*, 335–348. [CrossRef]
3. Xie, Y.; Sha, Z.; Yu, M. Remote sensing imagery in vegetation mapping: A review. *J. Plant Ecol.* **2008**, *1*, 9–23. [CrossRef]
4. Sun, J. Study on Automatic Cultivated Land Extraction from High Resolution Satellite Imagery Based on Knowledge. Ph.D. Thesis, China Agricultural University, Beijing, China, 2014. (In Chinese).
5. Eisfelder, C.; Kraus, T.; Bock, M.; Werner, M.; Buchroithner, M.F.; Strunz, G. Towards automated forest-type mapping—A service within GSE Forest Monitoring based on SPOT-5 and IKONOS data. *Int. J. Remote Sens.* **2009**, *30*, 5015–5038. [CrossRef]
6. Li, X.; Shao, G. Object-Based Land-Cover Mapping with High Resolution Aerial Photography at a County Scale in Midwestern USA. *Remote Sens.* **2014**, *6*, 11372–11390. [CrossRef]
7. Duro, D.C.; Franklin, S.E.; Dubé, M.G. A comparison of pixel-based and object-based image analysis with selected machine learning algorithms for the classification of agricultural landscapes using SPOT-5 HRG imagery. *Remote Sens. Environ.* **2012**, *118*, 259–272. [CrossRef]

8. Zhang, P.; Lv, Z.; Shi, W. Object-Based Spatial Feature for Classification of Very High Resolution Remote Sensing Images. *IEEE Geosci. Remote Sens. Lett.* **2013**, *10*, 1572–1576. [CrossRef]
9. Yu, Q.; Gong, P.; Clinton, N.; Biging, G.; Kelly, M.; Schirokauer, D. Object-based detailed vegetation classification with airborne high spatial resolution remote sensing imagery. *Photogramm. Eng. Remote Sens.* **2006**, *7*, 799–811. [CrossRef]
10. Reichstein, M.; Camps-Valls, G.; Stevens, B.; Jung, M.; Denzler, J.; Carvalhais, N.; Prabhat, N. Deep learning and process understanding for data-driven Earth system science. *Nature* **2019**, *566*, 195–204. [CrossRef]
11. Zeiler, M.D.; Fergus, R. Visualizing and understanding convolutional networks. In Proceedings of the ECCV 2014 European Conference on Computer Vision, Zurich, Switzerland, 6–12 September 2014; pp. 818–833.
12. Huang, B.; Zhao, B.; Song, Y. Urban land-use mapping using a deep convolutional neural network with high spatial resolution multispectral remote sensing imagery. *Remote Sens. Environ.* **2018**, *214*, 73–86. [CrossRef]
13. Yuan, Q.; Shen, H.; Li, T.; Li, Z.; Li, S.; Jiang, Y.; Xu, H.; Tan, W.; Yang, Q.; Wang, J.; et al. Deep learning in environmental remote sensing: Achievements and challenges. *Remote Sens. Environ.* **2020**, *241*, 111716. [CrossRef]
14. Zhang, C.; Sargent, I.; Pan, X.; Li, H.; Gardiner, A.; Hare, J.; Atkinson, P.M. Joint deep learning for land cover and land use classification. *Remote Sens. Environ.* **2019**, *221*, 173–187. [CrossRef]
15. Zhang, C.; Pan, X.; Li, H.; Gardiner, A.; Sargent, I.; Hare, J.; Atkinson, P.M. A hybrid MLP-CNN classifier for very fine resolution remotely sensed image classification. *ISPRS J. Photogramm. Remote Sens.* **2018**, *140*, 133–144. [CrossRef]
16. Zhao, W.; Du, S. Learning multiscale and deep representations for classifying remotely sensed imagery. *ISPRS J. Photogramm. Remote Sens.* **2016**, *113*, 155–165. [CrossRef]
17. Zhao, W.; Guo, Z.; Yue, J.; Zhang, X.; Luo, L. On combining multiscale deep learning features for the classification of hyperspectral remote sensing imagery. *Int. J. Remote Sens.* **2015**, *36*, 3368–3379. [CrossRef]
18. Mahdianpari, M.; Salehi, B.; Rezaee, M.; Mohammadimanesh, F.; Zhang, Y. Very deep convolutional neural networks for complex land cover mapping using multispectral remote sensing imagery. *Remote Sens.* **2018**, *10*. [CrossRef]
19. Ienco, D.; Interdonato, R.; Gaetano, R.; Minh, H.D.T. Combining Sentinel-1 and Sentinel-2 satellite image time series for land cover mapping via a multi-source deep learning architecture. *ISPRS J. Photogramm. Remote Sens.* **2019**, *158*, 11–22. [CrossRef]
20. Scott, G.J.; Hagan, K.C.; Marcum, R.A.; Hurt, J.A.; Anderson, D.T.; Davis, C.H. Enhanced fusion of deep neural networks for classification of benchmark high-resolution image data sets. *IEEE Geosci. Remote Sens. Lett.* **2018**, *15*, 1451–1455. [CrossRef]
21. Scott, G.J.; Marcum, R.A.; Davis, C.H.; Nivin, T.W. Fusion of deep convolutional neural networks for land cover classification of high-resolution imagery. *IEEE Geosci. Remote Sens. Lett.* **2017**, *14*, 1638–1642. [CrossRef]
22. Gaetano, R.; Ienco, D.; Ose, K.; Cresson, R. A two-branch CNN architecture for land cover classification of PAN and MS imagery. *Remote Sens.* **2018**, *10*, 1746. [CrossRef]
23. Lafferty, J.; McCallum, A.; Pereira, F.C.N. Conditional Random Fields: Probabilistic Models for Segmenting and Labeling Sequence Data. In Proceedings of the Eighteenth International Conference on Machine Learning (ICML-2001), Williams College, Williamstown, MA, USA, 1–28 June 2001.
24. Chen, L.C.; Papandreou, G.; Kokkinos, I.; Murphy, K.; Yuille, A.L. Semantic Image Segmentation with Deep Convolutional Nets and Fully Connected CRFs. *Comput. Sci.* **2016**, *26*, 357–361.
25. Shelhamer, E.; Long, J.; Darrell, T. Fully convolutional networks for semantic segmentation. *IEEE Trans. Pattern Anal. Mach. Intell.* **2017**, *39*, 640–651. [CrossRef] [PubMed]
26. Guo, Y.; Li, Z.; Chen, E.; Zhang, X.; Zhao, L.; Chen, Y.; Wang, Y. A Deep Learning Method for Forest Fine Classification Based on High Resolution Remote Sensing Images: Two-Branch FCN-8s. *Sci. Silvae Sin.* **2020**, *3*, 48–60. (In Chinese)
27. Guo, Y.; Li, Z.; Chen, E.; Zhang, X.; Zhao, L.; Chen, Y.; Wang, Y. A Deep Learning Forest Types Classification Method for Resolution Multispectral Remote Sensing Images: Dual-FCN8s-CRF. In Proceedings of the 2019 IEEE International Geoscience and Remote Sensing Symposium, Yokohama, Japan, July 28–August 2019.
28. Wang, Y.; Chen, E.; Guo, Y.; Li, Z.; Jin, Y.; Zhao, J.; Zhou, Y. Deep U-net Optimization Method for Forest Type Classification with High Resolution Multispectral Remote Sensing Images. *Forest Res.* **2020**, *33*, 11–18. (In Chinese)

29. Zheng, S.; Jayasumana, S.; Romera-Paredes, B.; Vineet, V.; Su, Z.; Du, D.; Huang, C.; Torr, P. Conditional Random Fields as Recurrent Neural Networks. In Proceedings of the 2015 IEEE International Conference on Computer Vision, Santiago, Chile, 7–13 December 2015; pp. 1529–1537.

30. GB/T 26424-2010 Technical Regulations for Inventory for Forest Management Planning and Design [S]. Available online: https://www.antpedia.com/standard/6160978.html/ (accessed on 10 October 2020).

31. China Centre for Resources Satellite Data and Application. The Introduction of GF-2 Satellite. Available online: http://218.247.138.119/CN/Satellite/3128.shtml (accessed on 24 September 2020).

32. Cooley, T.; Anderson, G.P.; Felde, G.W.; Hoke, M.L.; Ratkowski, A.J.; Chetwynd, J.H.; Gardner, J.A.; Adler-Golden, S.M.; Matthew, M.W.; Berk, A.; et al. FLAASH, a MODTRAN4-based atmospheric correction algorithm, its application and validation. In Proceedings of the 2002 IEEE International Geoscience and Remote Sensing Symposium, IGARSS 2002, Toronto, ON, Canada, 24–28 June 2002; pp. 1414–1418.

33. Gao, Y.; Zhang, W. Comparison test and research progress of topographic correction on remotely sensed data. *Geogr. Res.* **2008**, *27*, 467–477. (In Chinese)

34. Cao, H.; Gao, W.; Zhang, X.; Liu, X.; Fan, B.; Li, S. Overview of ZY-3 satellite research and application. In Proceedings of the 63rd IAC (International Astronautical Congress), Naples, Italy, 1–5 October 2012.

35. Sun, W.; Chen, B.; Messinger, D.W. Nearest-neighbor diffusion-based pansharpening algorithm for spectral images. *Opt. Eng.* **2014**, *53*, 013107. [CrossRef]

36. He, K.; Zhang, X.; Ren, S.; Sun, J. Identity Mappings in Deep Residual Networks. In Proceedings of the European Conference on Computer Vision, Amsterdam, The Netherlands, 11–14 October 2016.

37. Hirose, Y.; Yamashita, K.; Hijiya, S. Back-propagation algorithm which varies the number of hidden units. *IEEE Trans. Neural Netw.* **1991**, *4*, 61–66. [CrossRef]

38. Deng, J.; Dong, W.; Socher, R.; Li, L.-J.; Li, K.; Li, F.-F. ImageNet: A large-scale hierarchical image database. In Proceedings of the IEEE Conference on Computer Vision and Pattern Recognition, Miami, FL, USA, 20–25 June 2009; pp. 248–255.

39. Bottou, L. Large-Scale Machine Learning with Stochastic Gradient Descent. In Proceedings of the COMPSTAT'2010, Paris, France, 22–27 August 2010; pp. 177–186.

40. Krähenbühl, P.; Koltun, V. Parameter learning and convergent inference for dense random fields. In Proceedings of the 30th International Conference on Machine Learning, Atlanta, GA, USA, 16–21 June 2013.

41. LeCun, Y.; Bottou, L.; Bengio, Y.; Haffner, P. Gradient-based learning applied to document recognition. *Proc. IEEE* **1998**, *86*, 2278–2324. [CrossRef]

42. Ioffe, S.; Szegedy, C. Batch Normalization: Accelerating Deep Network Training by Reducing Internal Covariate Shift. In Proceedings of the 32nd International Conference on Machine Learning, ICML, Lille, France, 6–11 July 2015; pp. 448–456.

43. Nair, V.; Hinton, G. Rectified Linear Units Improve Restricted Boltzmann Machines. In Proceedings of the 27th International Conference on Machine Learning (ICML-10), Haifa, Israel, 21–24 June 2010; pp. 807–814.

44. Moolayil, J. *An Introduction to Deep Learning and Keras: A Fast-Track Approach to Modern Deep Learning with Python*; Apress: Berkeley, CA, USA, 2018; pp. 1–16.

45. Drakopoulos, G.; Liapakis, X.; Spyrou, E.; Tzimas, G.; Sioutas, S. Computing long sequences of consecutive fibonacci integers with tensorflow. In Proceedings of the International Conference on Artificial Intelligence Applications and Innovations, Dubai, UAE, 30 November 2019; pp. 150–160.

46. Kingma, D.; Ba, J. Adam: A Method for Stochastic Optimization. In Proceedings of the 3rd International Conference for Learning Representations, San Diego, CA, USA, 7–9 May 2015.

47. Gitelson, A.A.; Kaufman, Y.J.; Merzlyak, M.N. Use of a green channel in remote sensing of global vegetation from EOS-MODIS. *Remote Sens. Environ.* **1996**, *58*, 289–298. [CrossRef]

48. Immitzer, M.; Atzberger, C.; Koukal, T. Tree Species Classification with Random Forest Using Very High Spatial Resolution 8-Band WorldView-2 Satellite Data. *Remote Sens.* **2012**, *4*, 2661–2693. [CrossRef]

49. Adelabu, S.; Dube, T. Employing ground and satellite-based QuickBird data and random forest to discriminate five tree species in a Southern African Woodland. *Geocarto Int.* **2014**, *30*, 457–471. [CrossRef]

50. Waser, L.; Küchler, M.; Jütte, K.; Stampfer, T. Evaluating the Potential of WorldView-2 Data to Classify Tree Species and Different Levels of Ash Mortality. *Remote Sens.* **2014**, *6*, 4515–4545. [CrossRef]

51. Cho, M.A.; Malahlela, O.; Ramoelo, A. Assessing the utility WorldView-2 imagery for tree species mapping in South African subtropical humid forest and the conservation implications: Dukuduku forest patch as case study. *Int. J. Appl. Earth Obs.* **2015**, *38*, 349–357. [CrossRef]

52. Sun, Y.; Huang, J.; Ao, Z.; Lao, D.; Xin, Q. Deep Learning Approaches for the Mapping of Tree Species Diversity in a Tropical Wetland Using Airborne LiDAR and High-Spatial-Resolution Remote Sensing Images. *Forests* **2019**, *10*, 1047. [CrossRef]
53. Cao, K.; Zhang, X. An Improved Res-UNet Model for Tree Species Classification Using Airborne High-Resolution Images. *Remote Sens.* **2020**, *12*, 1128. [CrossRef]
54. Xie, Z.; Chen, Y.; Lu, D.; Li, G.; Chen, E. Classification of Land Cover, Forest, and Tree Species Classes with ZiYuan-3 Multispectral and Stereo Data. *Remote Sens.* **2019**, *11*, 164. [CrossRef]
55. Ren, C.; Ju, H.; Zhang, H.; Huang, J. Forest land type precise classification based on SPOT5 and GF-1 images. In Proceedings of the 2016 IEEE International Geoscience and Remote Sensing Symposium, Beijing, China, 10–15 July 2016; pp. 894–897.
56. Hościło, A.; Lewandowska, A. Mapping Forest Type and Tree Species on a Regional Scale Using Multi-Temporal Sentinel-2 Data. *Remote Sens.* **2019**, *11*, 929. [CrossRef]

*Article*

# Wide-Area Near-Real-Time Monitoring of Tropical Forest Degradation and Deforestation Using Sentinel-1

Dirk Hoekman [1,*], Boris Kooij [2], Marcela Quiñones [2], Sam Vellekoop [2], Ita Carolita [3], Syarif Budhiman [3], Rahmat Arief [4] and Orbita Roswintiarti [3,4]

1   Water Systems and Global Change Group, Wageningen University, Droevendaalsesteeg 3, 6708 PB Wageningen, The Netherlands

2   SarVision, Agro Business Park 10, 6708 PW Wageningen, The Netherlands; kooij@sarvision.nl (B.K.); quinones@sarvision.nl (M.Q.); vellekoop@sarvision.nl (S.V.)

3   Remote Sensing Applications Center, Jalan Kalisari No.8, Pekayon, Jakarta 13710, Indonesia; ita.carolita@lapan.go.id (I.C.); syarif.budhiman@lapan.go.id (S.B.); orbita@lapan.go.id (O.R.)

4   Remote Sensing Technology and Data Center, Jalan Lapan No.70, Pekayon, Jakarta 13710, Indonesia; rahmat.arief@lapan.go.id

*   Correspondence: dirk.hoekman@wur.nl; Tel.: +31-317482894; Fax: +31-317419000

Received: 26 August 2020; Accepted: 4 October 2020; Published: 8 October 2020

**Abstract:** The use of Sentinel-1 (S1) radar for wide-area, near-real-time (NRT) tropical-forest-change monitoring is discussed, with particular attention to forest degradation and deforestation. Since forest change can relate to processes ranging from high-impact, large-scale conversion to low-impact, selective logging, and can occur in sites having variable topographic and environmental properties such as mountain slopes and wetlands, a single approach is insufficient. The system introduced here combines time-series analysis of small objects identified in S1 data, i.e., segments containing linear features and apparent small-scale disturbances. A physical model is introduced for quantifying the size of small (upper-) canopy gaps. Deforestation detection was evaluated for several forest landscapes in the Amazon and Borneo. Using the default system settings, the false alarm rate (FAR) is very low (less than 1%), and the missed detection rate (MDR) varies between 1.9% ± 1.1% and 18.6% ± 1.0% (90% confidence level). For peatland landscapes, short radar detection delays up to several weeks due to high levels of soil moisture may occur, while, in comparison, for optical systems, detection delays up to 10 months were found due to cloud cover. In peat swamp forests, narrow linear canopy gaps (road and canal systems) could be detected with an overall accuracy of 85.5%, including many gaps barely visible on hi-res SPOT-6/7 images, which were used for validation. Compared to optical data, subtle degradation signals are easier to detect and are not quickly lost over time due to fast re-vegetation. Although it is possible to estimate an effective forest-cover loss, for example, due to selective logging, and results are spatiotemporally consistent with Sentinel-2 and TerraSAR-X reference data, quantitative validation without extensive field data and/or large hi-res radar datasets, such as TerraSAR-X, remains a challenge.

**Keywords:** Sentinel-1; NRT monitoring; deforestation; degradation; tropical forest; tropical peat

---

## 1. Introduction

Worldwide, forests disappear at alarming rates. In the last decade, the average annual net forest/non-forest conversion loss was estimated at 4.74 million ha [1]. Degradation of remaining natural forests is another major concern. A recent study of the Amazon region showed that losses in carbon were almost evenly split between cases attributable to forest conversion (e.g., biomass removals associated with commodity-driven deforestation) and cases due to forest degradation and disturbance (e.g., biomass reductions attributable to selective logging, drought, wildfire, etc.) [2]. Less well-known, but of equal concern, are losses of peat underneath tropical peat swamp forests. The largest tropical peat deposits are found in Indonesia, the Peruvian Amazon, and the Congo Basin, accounting for a total of approximately100 gigatons carbon (GtC), equal to 25% of the carbon stock stored globally in biomass [3–5]. For example, in degraded peat swamp forests in Indonesia, on average, approximately 0.4 GtC is lost annually because of oxidization and fires [6]. Forest loss, forest degradation, and peat degradation are important components of carbon accounting. The Intergovernmental Panel on Climate Change (IPCC) recently released the 2019 Refinement to the 2006 IPCC Guidelines for National Greenhouse Gas Inventories, where the use of Earth observation data plays a prominent role [7]. In addition to its role in carbon accounting, timely information on forest change is needed in support of other applications, such as forest management and peatland restoration. Optical satellite data have been used for decades for forest-change monitoring. Cloud cover, which can be very persistent in tropical rainforests, may pose problems when continuous timelines and timeliness of information is important. Radar data, such as Sentinel-1 data, offers an alternative. Radar's independence of cloud cover is a clear advantage. However, this may not be the only advantage, as is explored later in this paper.

Deforestation is commonly defined as land-use change from forest land to any other non-forest land-use category and forest degradation as long-term loss of forest carbon stocks, as well as forest values without land-use, change [8–11]. However, quantitative criteria to describe forest degradation are still under discussion. Specification of thresholds for carbon stock loss and minimum area and time affected are not given, but are mandatory to apply such a definition [9–11]. Deforestation detection (with optical data) is based on the easy differentiation between forest and non-forest classes, such as open areas, bare soil, agriculture, and settlements. Commonly used methods are based on sub-pixel approaches like spectral mixture analysis, which are also used to assess proxies of forest degradation [12–15]. The abovementioned recent study of the Amazon was based on MODIS data at ~500 m resolution [2]. It states that applying higher-resolution satellite data (e.g., 30 m Landsat imagery) would reduce uncertainty in carbon loss estimates, in particular, from degradation and disturbance. Another study, based on Landsat, mentions that subtle degradation signals are not easy to detect and are quickly lost over time due to fast re-vegetation [16]. This would mean that significant loss of carbon could remain undetected. Obviously, higher spatial resolution provides more detail; however, regrowth or remaining understory can limit disturbance detection. Radar imaging is fundamentally different in several ways and can also be used to detect subtle patterns of forest disturbance, such as patterns caused by selective logging. Good results have been obtained with airborne radar [17], and high-resolution satellite radar data, such as COSMO-SkyMed spotlight data [18] and TerraSAR-X 3 m stripmap data [19,20]. However, such data types are not practical for wide-area monitoring applications. Coarser resolution radar data, such as the Sentinel-1 IW data, may offer a good alternative.

Forest disturbance, either through forest loss or drainage, can result in tropical peat disturbance. Peat swamp forests are among the world's most threatened and least known ecosystems. In Southeast Asia, large areas of peat swamp forest have been drained, deforested (for timber), converted for agricultural projects (even though the soil is too acid), or are converted into plantations (such as oil palm, acacia, and Borneo rubber), even though peat systems are fragile and sensitive to hydrological disturbance (e.g., see Reference [21]). Drainage through canalization has frequently and severely disrupted groundwater-level dynamics. Besides resulting in $CO_2$ emissions due to oxidization [22,23], this process makes them particularly vulnerable to fire, especially during 'El Niño' years [24]. Emissions from the fires in Indonesia during 1997–1998, for example, have been estimated to be 0.8–2.5 GtC [25,26]. Water management is essential in addressing these disturbances. Indonesia currently makes efforts to restore degraded peatlands by "re-wetting", blocking canals and promoting paludiculture. For these vast areas, near-real-time (NRT) information is needed on the construction of new drainage canals in the forest, which are often illegal and a precursor to further forest and peat disturbance. This is currently achieved by using SPOT-6/7 and Pleiades data, even though costs and cloud cover pose severe limitations.

Wide-area and spatially detailed NRT data are needed not only for tropical peatland management and restoration. Other sectors needing such data include law enforcement, national forest monitoring systems, and indigenous communities; MRV systems, carbon accounting, and REDD+ projects; sustainable development of timber trade, forest plantations, and other commodities; protection of conservation areas, biodiversity, and ecological corridors; and early warning and disaster management [1,10,27–30].

Several wide-area NRT systems already exist. Since 2004, near-real-time deforestation monitoring over the Brazilian Amazon has been carried out by INPE based on the Real Time Deforestation Detection System (DETER) program. The system currently uses the optical AWiFS data with 56 m spatial resolution and five-day temporal resolution [31]. Because cloud cover poses a problem, the JJ-FAST system is considered as an additional NRT data source [32]. The JJ-FAST system is based on ALOS PALSAR-2 ScanSAR data and is the first SAR-based global early warning system for tropical forests (covering 77 countries) [33,34]. It currently offers deforestation data every 1.5 months, at a minimum mapping unit (MMU) of 2 ha. Validation based on Landsat data provided by the GLAD system [15] shows an overall user accuracy of 66.7% [35]. The ALOS PALSAR observation strategy was designed to provide consistent wall-to-wall observations, at fine resolution, of almost all land areas on Earth, on a repetitive basis [36]. In addition to providing the data for the JJ-FAST system, it is used to make annual global forest/non-forest maps [37] and provide insight in the spatiotemporal radar backscatter variation of tropical forests [38]. The L-band variability is relatively low for (dryland) tropical rainforest, higher for tropical moist deciduous forest and highest for tropical dry forest (for HH-polarization up to 3.5 dB standard deviation). The variability is also high for wetlands, such as peat swamp and floodplain forests [38]. Moreover, for C-band data, spatiotemporal backscatter variation of tropical rainforest is low [39] and can be substantial for tropical dry forest [40].

Though L-band radar, in general, provides better contrast between forest and non-forest classes [41–45]; C-band radar also seems well suitable for forest-change monitoring. From an operational point-of-view, this is particularly true for the Sentinel-1 radar, which reliably provides free medium-resolution data with a global coverage at a 6- or 12-day repeat cycle. From a technical point-of-view, there are a few challenges. (1) The first is the contrast between forest and non-forest, which, compared to L-band, is often low and of short duration (because of regrowth). (2) The spatiotemporal variability of forest can be substantial, notably for wetland forests; however, it seems lower than for L-band. (3) Spatial co-registration, with optical data and other radar, is difficult because of forest height, and limitations of available DEMs and radar parallaxes. This is especially true for the finer details, such as forest edges and disturbances. (4) Radar imaging is fundamentally different from optical imaging. Together with the previous issue, this complicates validation based on other satellite data. (5) Moreover, as stated in Reference [32], "radar image analysis must be conducted carefully because there is no single pattern of deforestation in the Brazilian Amazon". This is also the case for other tropical rainforest areas. Results depend on forest and terrain type and other environmental conditions, as discussed in this paper.

Several recent studies discuss the appropriateness of Sentinel-1 for forest/non-forest discrimination and deforestation detection. For a dryland forest site in the Peruvian Amazon, deforestation could be detected successfully (detection rate 95%) based on time-series analysis of radar shadows and the combined use of ascending and descending observations [43]. In Reference [44], several methods for forest/non-forest classification for a wide range of forest types are discussed. Classification accuracies for the tropical rainforest sites in this study are up to 81.6% (Sumatra) and up to 88.6% (Colombia). In Reference [45] the potential of time-series and recurrence metrics for deforestation mapping at test sites in Mexico is discussed. In Reference [46], for a site in the Amazon, it is shown that (time-series of) interferometric coherence has potential for deforestation detection.

The objective of this paper is to discuss the suitability of Sentinel-1 radar for wide-area, NRT monitoring of tropical forest change in terms of deforestation and degradation, for a variety of landscapes in Indonesia and the Amazon, including tropical peat swamp forest landscapes. This requires clear definitions of forest, deforestation, degradation, and disturbance. However, users in different countries and disciplines use different definitions. It is therefore practical to have a flexible system that can be adapted to the needs of the user. It is also necessary to indicate the limitations of such a system. In this paper, certain default settings and definitions are used. The forest class includes undisturbed and disturbed natural forest and excludes forest plantations, such as acacia, rubber, and eucalypt. Deforestation is defined as clear-cut areas exceeding a certain size. The default threshold is arbitrarily set at 1.0 ha, except for peat swamp forest, where 0.3 ha is used. Any detected forest loss smaller than the default size is labelled as degradation. Degradation in degraded forests is mapped as the additional degradation since the start of the monitoring. The defaults values for the change detection algorithms used in this study were established during earlier work in Malaysia and Sumatra and seem generally applicable in all areas studied so far, including the Guianas, Gabon, and the areas validated in this paper. Special characteristics of tropical peat swamp forests, viz, the flat terrain and occurrence of long and narrow straight gaps caused by drainage canal construction, form ideal conditions for theoretical studies relevant for development of models to quantify degradation. Section 2.1 introduces the study sites and supporting data. Sections 2.2–2.4 describe physical background, system design considerations, and types of errors and summarize system components. Section 2.5 provides a theoretical background for the radar imaging of linear canopy gaps (roads and canals) and small canopy gaps (tree logging gap disturbances) and introduces a physical model for radar imaging at high-resolution. Section 3 provides results for NRT canal gap detection, NRT deforestation monitoring, and NRT degradation monitoring. It also reflects on validation challenges. Sections 4 and 5 provide a synthesis and the main conclusions. The system is tested for a range of tropical forest landscapes and seems to function well, even in challenging environments, such as mountain slopes and wetlands. In this paper, the term "landscape" is used in the context of an ecosystem approach and stands for a vast area with a mosaic of forest types.

## 2. Materials and Methods

### 2.1. Study Sites and Supporting Data

The system was tested in several tropical forest areas, comprising wall-to-wall coverage of the island Borneo, Suriname, Guyana, and selected sites in Brazil, Colombia, Gabon, and the island Sumatra, for the entire period of Sentinel-1 observation, since launch (see Tables 6 and 7). System validation mainly focused on the Indonesian Province Central Kalimantan (154,000 km$^2$) on Borneo and the municipality Almeirim (73,000 km$^2$) in the State of Pará, Brazil (see Figure 1). These two validation areas have a wide range of forest types, as well as a wide range of deforestation and degradation characteristics. Central Kalimantan was almost entirely covered by tropical evergreen forest until the 1980s. Intensive logging of predominant commercial Dipterocarp species and conversion to cropland, oil palm, and timber plantations has reduced forest cover significantly. Other major natural vegetation types include peat swamp forests, which are found in the coastal and sub-coastal lowlands, freshwater swamps along rivers inland, and mangrove forests in the coastal plains. Large fractions of the peat swamp forests are drained, causing frequent forest fires, notably in El Niño periods. Almeirim, located between the Amazon River and Suriname, has moist evergreen forest, varied topography, and blackwater nutrient-poor rivers. Selective logging and conversion to pasture, agriculture and plantations occur in this area, however, the rates of deforestation and degradation are still relatively low.

Monitoring forest change requires knowledge on the location and characteristics of forest at the start of the monitoring period. A simple approach would be to use the globally available forest/non-forest maps derived from PALSAR-2 [37]. For this study, accurate regional baseline maps with more detailed thematic information were made based on the first available Sentinel-1 images, and concurrent PALSAR-2, Landsat-8, and, when available, Sentinel-2 images [47–49], using the systematics of the FAO Land Cover Classification System (LCCS) [50]. The combination of these three sensors allows accurate distinction of the different forest classes, such as peat forest, heath forest, high dipterocarp forest, mangrove forest, riparian forest, and forest plantations. The LCCS classification system describes vegetation in terms of vegetation structure and soil wetness conditions, such as flooding under the canopy. L-band radar is able to observe wetness under the canopy and is uniquely suitable to distinguish dryland and wetland forest classes. The baseline typically comprises 30 classes, depending on the eco-region. For Borneo, using 9000 reference areas, an accuracy of 85% was achieved [51]. In combination with additional historical data (mainly PALSAR), additional information on degradation, regrowth, and flood frequency (also under the canopy) is obtained.

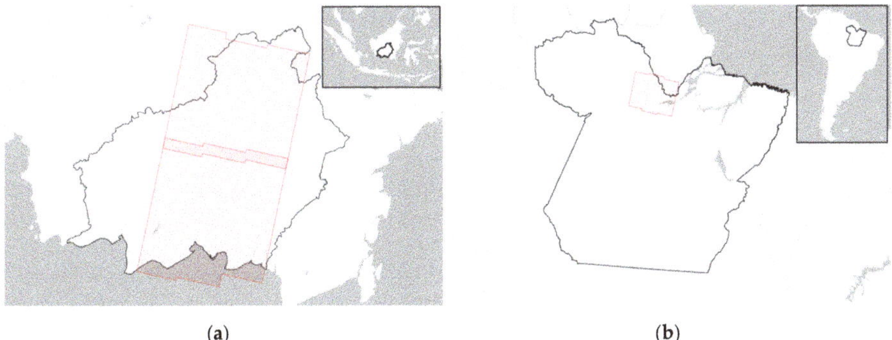

(a)        (b)

**Figure 1.** Overview test site locations. Two Sentinel-1 footprints in the Indonesian Province Central Kalimantan (**a**). One Sentinel-1 footprint in Brazilian State of Pará (**b**).

Sentinel-2, SPOT-6/7, Google Earth, and TerraSAR-X single polarization 3 m resolution stripmap data were used to validate the detected forest changes (see Tables 6 and 7). For Sentinel-1 data simulation use was made of field and LiDAR data collected in earlier experiments (see Section 2.5).

### 2.2. Physical Background of Methodology

In general, undisturbed forests have a relatively high and stable backscatter level; therefore, a significant decrease in backscatter level would indicate deforestation. Though true in general, this assumption, for several reasons, is too simple to allow for accurate change detection, notably for small-scale disturbances and wetland forests. Three reasons are discussed next.

(1) In the first place, good knowledge on the natural temporal backscatter variation of undisturbed forest is required. For tropical rainforests, the causes of natural variation include seasonality, foliage wetness, soil wetness, and flooding. The strengths and spatial scales of these phenomena also depend on the forest type, notably, because of flooding, on the distinction between dryland and wetland forest types. Moreover, rain affects backscatter level, though this strongly depends on rain intensity and radar wavelength. For the C-band of Sentinel-1, for example, heavy rain can result in a slight decrease in backscatter over large areas, while excessive rain can result in backscatter decrease exceeding 3 dB over smaller areas. A typical example is given in Figure 2. Approaches to address temporal backscatter variation are discussed later.

(2) The opposite can also be true. Radar images may not show any clearly perceivable change even though strong forest disturbance has taken place (such as in Figure 20). This phenomenon is partly explained in Figure 3. Logging a single large tree causes a small depression in the upper canopy. A small area of backscatter decrease (caused by radar shadow) and a small adjacent area of backscatter increase (caused by radar overlay) results, even when the lower canopy still covers the soil completely (Figure 3B). The mean backscatter is hardly affected, while the patterns which constitute the texture

change significantly. When the lower canopy is also removed, the radar textural change signal is somewhat stronger (Figure 3C). Optical systems, in the latter case, may detect spectral change because of a bare soil contribution. However, there are two other important differences between radar and optical imaging to consider. The first difference is the duration of change. Since it usually takes years for the canopy gap depression to fill up again, while it only takes a few weeks or months for the bare soil to be covered with regrowth again, radar change is well observable for a long period, while, for optical systems, this is only a short time and only in the absence of cloud cover. The second difference is the incidence angle. Theoretical models (to be discussed in Section 2.5) show that Sentinel-1 incidence angle has little effect, while, for optical systems, this may not be true. Since canopy gaps can be small and deep, incidence angles for optical systems may be too large to observe the soil surface in high forest. For Sentinel-2, the local incidence angle can be as large as 23.86° [52], while, for SPOT-6/7, using the oblique viewing capability, as is often done for areas with persistent cloud cover, this can be as large as 45°. Though the Sentinel-1 signal of a new small gap is stable for a long period, it is also weak. It relates to a few pixels only and the backscatter decrease and increase is roughly at the same level as the standard deviation of the radar speckle. Approaches to estimate this weak signal in the presence of speckle are discussed later.

(3) The backscatter level of clear-cut areas, in both polarizations, is usually significantly lower (>2.0 dB) than the original forest cover. Under certain circumstances, however, factors such as remaining debris and undergrowth, terrain slopes [53], soil roughness, and soil moisture can cause much smaller decreases. This loss of contrast can be temporarily or persistent. It is evident that good knowledge is needed on the causes, levels, and probabilities of contrast loss, for the landscapes to be monitored.

The monitoring system is designed to be capable of accommodating all the above mentioned issues as good as possible; however, at a certain point, compromises have to be made. These compromises are discussed in Section 2.4, after the introduction of the system components in Section 2.3.

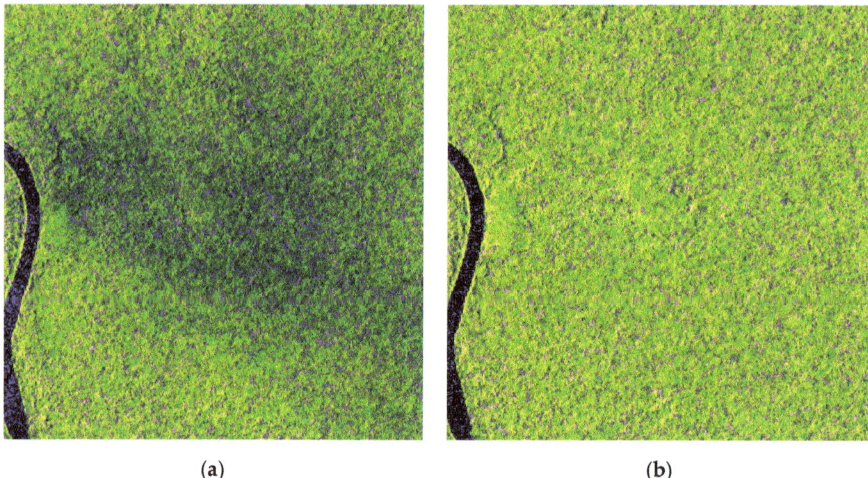

(a)            (b)

**Figure 2.** Sentinel-1 radar image of 20161229 (**a**) and next observation at 20170122 (**b**). Excessive rain at the first date reduces backscatter in VV- and VH-polarization with more than 3 dB. Location: Sebangau, Central Kalimantan. Size: 6 km × 6 km. Standard color scale: (red) VV with range −15.0 to −6.0 dB; (green) VH with range −24.0 to −13.0 dB; (blue) VV–VH with range 4.0 to 12.0 dB.

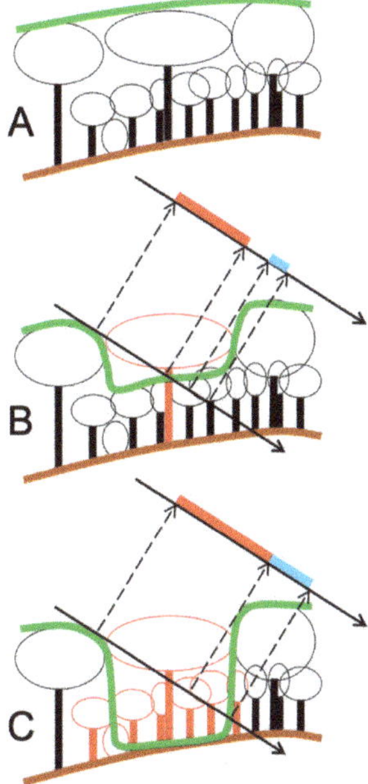


**Figure 3.** Radar response to forest degradation. Pre-disturbance situation (case **A**). Small gaps create small areas of (dark) "radar shadow" and (bright) "radar overlay", shown in red and cyan, respectively. Therefore, extraction of large canopy trees can be detected, even when the soil is still covered with lower vegetation (case **B**). When a soil fraction is visible (case **C**), both radar and optical detect extraction. The radar signal persists over long periods when canopy gaps are steep and deep.

## 2.3. System Description

*End-to-end system:* As soon as a new Sentinel-1 radar image is available [54], the system automatically downloads this image, updates time-series analyses, and produces new maps, including a deforestation map, a forest access-road map, and a forest-degradation map. Since the baseline reference is in the past, and updates have to be made chronologically, the system initially produces a series of historical maps, before commencing near-real-time map production. In this paper, these historical sequences are studied and validated. The end-to-end system consists of many steps. It is out of scope to discuss all of these technical steps in detail. Therefore, only a high-level description and some relevant details are presented next.

*Preprocessing:* Interferometric preprocessing is performed for radiometric calibration, geometric correction, and precise co-registration. This is done over tiles, each containing 30–50 bursts, covering the landscape (Figure 4). Preprocessing includes slope correction [53] and slant-range multi-temporal speckle reduction. The speckle reduction step combines a number of approaches [55–57] in order to preserve edges and texture well. This is relevant, in particular, for the quality of the access road and forest degradation assessment. Speckle filtering increases the equivalent number of looks (ENL) from 4 to 19 and reduces speckle in homogeneous regions from 2.3 to 1.0 dB. The result is an (updated) time-series of dual-polarization (VV- and VH-) intensity images at a 15 m pixel size and interferometric coherence data. It is noted that, throughout this paper, backscatter intensity is expressed by the backscattering coefficient $\gamma^0$ ($\gamma^0 = \sigma^0 / \cos(\theta_{inc})$ , where $\sigma^0$ is the normalized radar cross-section).

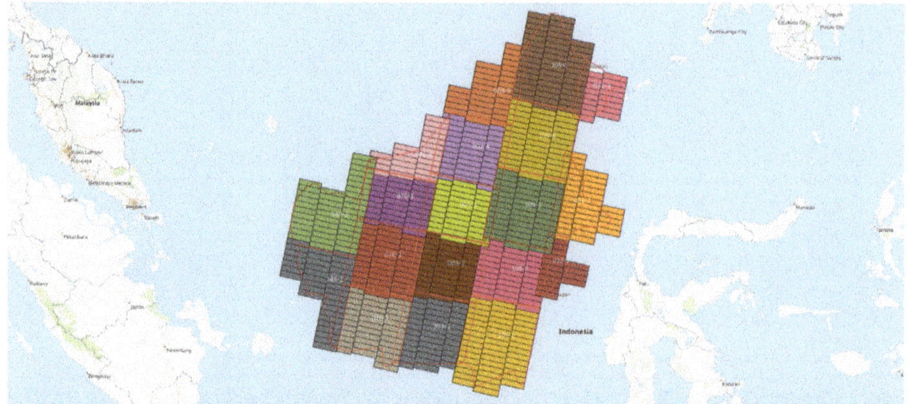

**Figure 4.** Area division is done in tiles typically comprising 30–50 Sentinel-1 bursts. Borneo Island.

There are two important reasons for using the Single Look Complex (SLC) products instead of the more generally used Ground Range Detected (GRD) products as input data. SLC data are in the original radar geometry and can be converted by multi-looking and orthorectification in the required ground range geometry. However, SLC data also contain phase information which allows for a very precise co-registration (typically 6 mm accuracy) of multi-looked data in the slant range. Another advantage is that all slant-range multi-look pixels have the same ENL, which allows for superior multi-temporal speckle reduction. Co-registration of GRD data is far less precise and typically has an accuracy of several meters in flat areas but can exceed the pixel dimension in mountainous areas. The increased sharpness and spatial co-registration as a result of interferometric preprocessing (i.e., using phase information) is essential to allow the applications as discussed in this paper, in particular, the quantification of degradation and the detection of subtle linear features.

*Baselines:* The baseline consists of a land-cover baseline map (described in Section 2.1) and a Sentinel-1 reference image from the same date. The land-cover map is used to stratify the landscape, allowing forest-type-dependent algorithm settings for change detection, as well as forest-type-dependent assessment of spatiotemporal backscatter variation in undisturbed forest. The latter assessment is done at a much coarser scales; however, it still captures changes caused by flooding of wetland forest. The backscatter decrease in cells of extreme rainfall cannot be corrected well and has to be dealt with in the thematic mapping steps.

*Change detection:* When the natural temporal backscatter variation is taken into account, change follows by detecting a substantial backscatter increase or decrease in one or both of the polarizations. Such a change can be flagged (i.e., provisionally detected) by using relatively simple time-series analysis techniques, such as Bayesian techniques [58]. A similar technique is adopted here; however, several important interconnected modifications have been made. (1) Segmentation allows time-series analysis at segment level, not only pixel level. (2) The threshold for change detection is not fixed. It not only depends on forest class and terrain slope angles but can also change in time and in relation to the position in the segment. For example, the threshold can be lowered after a confirmed change detection or for pixels at the edge of a segment. (3) To avoid error propagation and to allow a high sensitivity (i.e., low threshold values), the results of the pixel-based change detection are used in a feedback loop, to regulate the segment-based results. (4) The change is not a scalar value but is defined as the combined vectorial change of the two polarizations. (5) Several minimum mapping unit (MMU) sizes are used, depending on forest type and product type. For example, for deforestation in high dryland dipterocarp forest, the MMU is 1.0 ha, and for peat swamp forest, the MMU is 0.3 ha. Intermediate products with still lower MMUs are made to guide the production of forest-degradation maps. The latter information can be added to the deforestation maps as a qualitative indication where degradation is ongoing.

*Products:* The NRT system currently produces, next to the deforestation and degradation maps, two other types of thematic maps. Quantitative forest-degradation maps are based on a theoretical model for radar imaging of canopy gaps. This is discussed in Section 2.5. The third map type is based on change of linear features. It is used to update access road maps and canals, such as in peat swamp forest, and also gives qualitative indications of degradation. This is also discussed in Section 2.5. Consequently, three fundamentally different types of forest-change maps are generated, each focused on different aspects of forest change. A flowchart is shown in Figure 5.

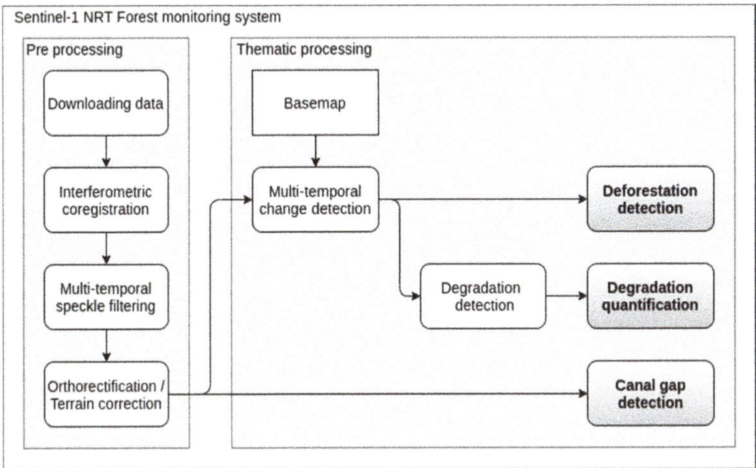


**Figure 5.** Flowchart: preprocessing and thematic processing steps of the Sentinel-1 near-real-time (NRT) forest-monitoring system.

## 2.4. Types of Errors and Compromises

For a near-real-time monitoring system, two types of compromises exist. To discuss these, the error types should be introduced first. Suppose Sentinel-1 NRT deforestation maps are validated using available Sentinel-2 images. Two factors would complicate this validation exercise. The first is the asynchrony of the radar and optical time-series and the second is cloud cover. These complications are addressed by defining six rule-based classes as shown in Table 1. Missing pixels in the optical reference dataset, because of clouds and cloud shadows, are classified as forest when the pixel has the forest class in future images and are classified as non-forest when the pixel has the non-forest class in a previous image. The remaining missing pixels are classified as unknown. Only pixels labelled in the baseline as one of the forest classes are evaluated.

**Table 1.** Quality control classes and color coding used for validation of deforestation.

| | | |
|---|---|---|
| | CD1 | Correct deforestation detection |
| | CD2 | Correct deforestation detection, prior to next optical reference date |
| | MD | Missed non-forest detection |
| | UN | Unknown |
| | CF | Correct forest classification |
| | FA | False alarm |

A pixel or segment is labelled as correct deforestation (CD1) when an optical image of the same or earlier date is deforested. It is also labelled as correct deforestation (CD2) in case the previous optical image shows the forest class and the next optical image the non-forest class. The label "missed non-forest detection" (MD) is assigned when the pixel is classified as forest in the radar map and non-forest in a previous optical image. The label false alarm (FA) is assigned when the pixel is classified as non-forest in the radar map and as forest in the previous optical image. The label correct forest classification (CF) is assigned when the pixel is classified as forest in the radar map and the previous optical image. Using these rules, not all the radar map pixels can be labelled with one of these five classes because of the presence of unknowns in the optical reference maps. In these cases, the label unknown (UN) is assigned. The false alarm rate (FAR) and missed detection rate (MDR) are calculated by using the following equations:

$$FAR = FA/(FA + CF) \tag{1}$$

$$MDR = MD/(MD + CD1 + CD2) \tag{2}$$

The performance of the system can be tuned to specific needs of the user. Basically, the user has to make two important compromises. The first relates to the interchangeability of the two types of detection error, i.e., the false alarm (FA) rate and the missed detection (MD) rate, also known as false positive and false negative. When algorithm settings are selected to decrease the FA rate, then the MD rate increases, and vice versa. Of course, multiple maps using different settings can be made. The second compromise relates to the interchangeability of overall accuracy and timeliness. The timeliness of NRT maps is defined on the basis of the dates of the available radar image time-series. When the first radar image has the time stamp $t_0$ (t zero), the second $t_1$, etc., and the last $t_p$ (t present), then the second to last image has time stamp $t_{p-1}$. An NRT map can be based on a $t_p$ radar image, a $t_{p-1}$ radar image or, in general, a $t_{p-n}$ radar image. Larger values for $n$ cause larger delays in map availability; however, in general, they result in larger overall accuracy. Of course, multiple maps using different values of $n$ can be made simultaneously and can be combined. To make a distinction between different types of NRT maps, these are denoted as NRT(N = 0), NRT(N = 1), etc. Within an NRT(N = 1) system the most recent radar image is only used as confirmation, which, for example, can be used to avoid false alarms caused by heavy rain cells. The default NRT system studied in this paper is an NRT(N = 1) system with a low FA rate. Nevertheless, evaluation of the performance of this relatively simple system can yield important insights. These insights, to be presented in this paper, support the design of more accurate and complicated systems which combine multiple maps made with different settings, tuned for the landscapes of interest.

*2.5. Theoretical Background of Canal-Gap Mapping and Forest-Degradation Quantification*

Models of the physical interaction, the forest structure, and the canal gap geometry can be used to simulate radar imaging of canal gaps. The canal gap geometry was derived from SPOT-6/7 data and is expressed as canal width and orientation. The description of forest structure is based on field observations from previous studies [17,19,59,60]. Relevant parameters include forest height and canopy roughness. The physical interaction is modelled at high resolution, accounting for the three-dimensional structure of canopy roughness and incidence angle, as described in Reference [61] and canal gap geometry.

Radar profiles of canal gaps were extracted from the radar images in the east–west direction by re-sampling and averaging over straight canal sections of approximately 45 radar image rows, which strongly reduces the variation caused by speckle. Figure 6 shows a comparison between an observed profile and a simulated profile. The observation differs from the simulation because of remaining speckle and texture effects. However, across the canal profile, the fit is very good, with a standard error of estimate of only 0.5 dB. Results over several canal sections are summarized in Table 2 and Figure 7. Since realistic simulations can be made, the radar backscatter model can be used as a theoretical tool to support further quantitative analysis. In Section 3.1, this is done to study limiting factors related to canal gap detection and in Section 3.3 to study possibilities to quantify small forest-gap dimensions in relation to forest degradation.

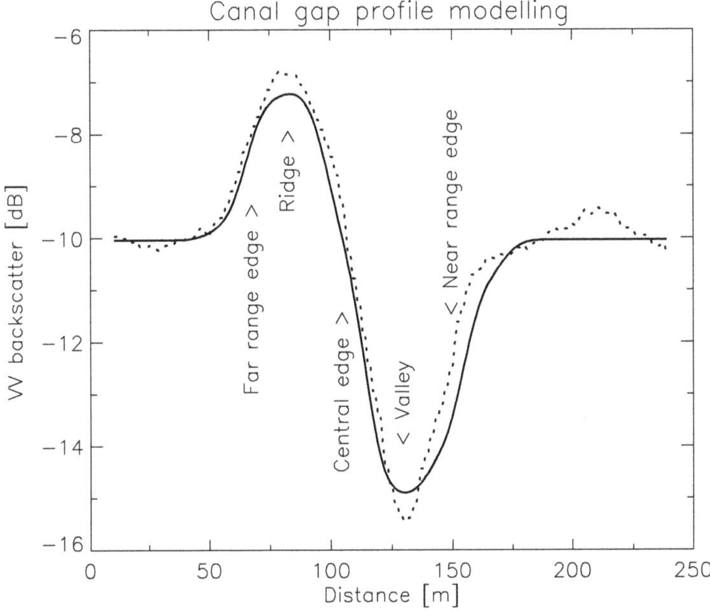


**Figure 6.** Comparison between simulated radar backscatter profile across a canal gap (solid curve) and an observed profile (dotted curve). The positions of the ridge, valley, and edges are indicated.

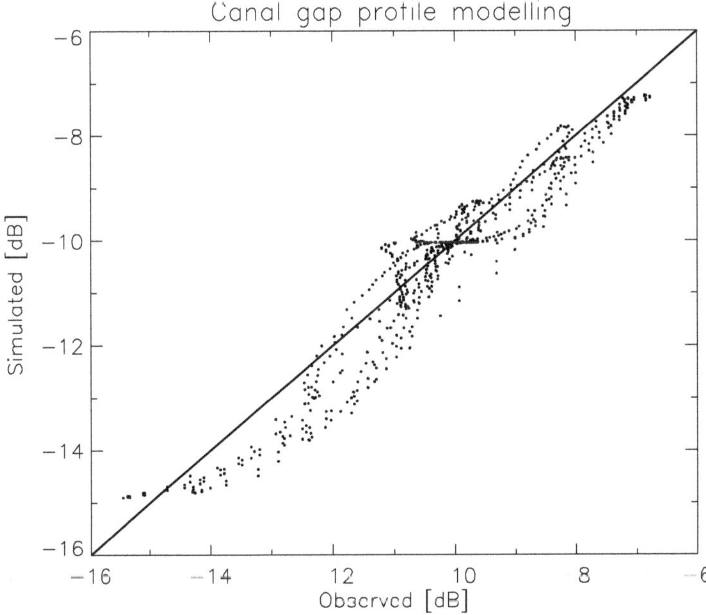


**Figure 7.** Comparison of observed and predicted VV-backscatter across five canal gaps.

**Table 2.** Canal gap simulation results. Width of gap (m), orientation with respect to radar azimuth direction (degrees), RMSE (dB), and Pearson correlation between observed and predicted values. Note that the RMSE for canal B is relatively large. This is caused by its large width, which causes visibility of the gap floor, and strong regrowth on the gap floor, which causes an increase in backscatter.

| Canal | Width | Orientation | RMSE | $r^2$ |
|---|---|---|---|---|
| A | 56.2 | 32.6 | 0.501 | 0.987 |
| B | 50.0 | 32.6 | 0.813 | 0.959 |
| C | 32.8 | 32.4 | 0.556 | 0.964 |
| D | 25.0 | 31.3 | 0.386 | 0.954 |
| E | 18.7 | 63.4 | 0.285 | 0.870 |
| Average | | | 0.539 | 0.959 |

Since the radar data are acquired near the equator in descending orbit, the azimuth direction is $-168.0°$ with respect to north and the radar look direction, which is towards the right, is $-78.0°$, i.e., almost west. For descending data, as shown in Figure 6, the radar profile of the canal gap, shows a ridge positioned left of a valley. The valley results from radar shadowing and the ridge from radar overlay. The widths and heights of the ridges and valleys vary as function of canal gap width and orientation. The characteristic shape of the radar gap profile suggests several alternative approaches for linear feature detection. For descending data in the direction from east to west (or right to left) the profile shows a negative edge (or sharp decrease) followed by a valley, a sharp increase, a ridge and a second negative edge. It suggests that several classes of operators are suitable to detect the canal gaps, such as edge detectors, ridge-valley (or line) detectors [62,63], and matching filters (for the characteristic valley-ridge pattern in descending data). The application of these operators is the first step in the process of generating canal-gap maps. Subsequent steps include thresholding of the detections, applying spatial shifts (because the operators act on different parts of the canals gaps), linking small segments into larger segments (by evaluating canal gap directions), and time-series analysis (to reduce false alarms). The operators used for detection are briefly described first.

The Sobel operator was used for edge detection. It uses two $3 \times 3$ kernels which are convolved with the original image to calculate approximate edge gradients in the horizontal and vertical direction. In subsequent steps, for computational efficiency, the edge gradients in only eight discrete directions (at 45-degree intervals) are used. Therefore, in the initial step, eight $3 \times 3$ kernels are applied as shown in Figure 8 (top). The same approach was used for the ridge and valley detection (see Figure 8, middle) and the matched filter detection (Figure 8, bottom). Therefore, in this approach, in total, 24 types of detection per pixel can be made. Since these detections are not independent, a selection of a subset of these detections would be sufficient. A careful evaluation showed that 10 types suffice without decreasing performance and that the main value of the matched filter is the improvement of the detection of small canals. The latter also explains the shape of the matched filter, which works well on narrow canals and is less efficient for wider canals.

Before discussing experimental results of canal gap mapping, the utility of the theoretical model introduced above should be discussed in more detail. Canal gaps in peat swamp forest show up more prominently in radar images when they are oriented more closely in azimuth direction and when they are wider. The theoretical model can be used to quantify these relationships; moreover, it can be used to predict the effect of forest structural parameters and incidence angle on these relationships.

This can be done by introducing the parameter "contrast", which simply is the sum of the absolute radar backscatter change (in dB) of the disturbance in the forest canopy caused by the canal gap, as shown in Figure 6. This sum is taken over pixels of a single row (i.e., east–west direction) matching the canal disturbance section. Higher contrast values can be related to higher visibility of canals gaps in the radar image. Higher contrast values are found for canals gaps wider than 10 m in combination with a canal orientation smaller than 75 degrees from azimuth direction (see Figure 9).

| | | |
|---|---|---|
| -1 | 0 | 1 |
| -2 | 0 | 2 |
| -1 | 0 | 1 |

| | | |
|---|---|---|
| 0 | 1 | 2 |
| -1 | 0 | 1 |
| -2 | -1 | 0 |

Sobel grad, 0 and 45 deg

| | | |
|---|---|---|
| -1 | -2 | -1 |
| 2 | 4 | 2 |
| -1 | -2 | -1 |

| | | |
|---|---|---|
| -2 | -1 | 2 |
| -1 | 4 | -1 |
| 2 | -1 | -2 |

Ridge, 0 and 45 deg

| | | | | |
|---|---|---|---|---|
| 0 | 0 | 0 | 0 | 0 |
| 0 | 0 | 0 | 0 | 0 |
| 0 | 1 | 2 | -2 | -1 |
| 0 | 0 | 0 | 0 | 0 |
| 0 | 0 | 0 | 0 | 0 |

| | | | | |
|---|---|---|---|---|
| 0 | 0 | 0 | 0 | -1 |
| 0 | 0 | 0 | -2 | 0 |
| 0 | 0 | 2 | 0 | 0 |
| 0 | 1 | 0 | 0 | 0 |
| 0 | 0 | 0 | 0 | 0 |

Matched filter, 0 and 45 deg

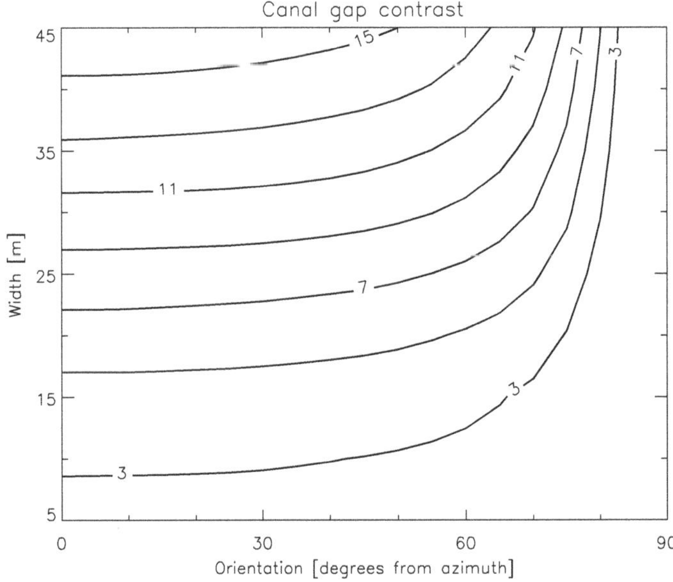

**Figure 8.** Sobel gradient, ridge and matched filters used for canal gap detection. Only the 0° (east–west) and 45° orientations are shown.

**Figure 9.** Contour plot of modelled contrast (dB) as function of orientation (degrees from radar azimuth) and canal gap width (m).

Since the contrast parameter is independent from canal length, it also applies for gaps of very short canals, which resemble gaps caused by selective logging. These small canopy gaps, or forest degradation gaps, are usually not elongated. Therefore, it may be assumed that contrast values for small orientation angles apply. Furthermore, it can be noted that, for small angles, the ratio between contrast and gap width is almost constant when the gap width is above 20 m. The latter relation can be computed by using the same model and depends on incidence angle and forest structure. In Figure 10, the relation between contrast and degradation gap width for a peat swamp forest at three incidence angles is shown. This example shows that lower incidence angles give higher contrast. Simulations also show that higher forest, in general, gives higher contrast. Therefore, when the right model is applied and contrast is not computed over a single gap section but over a certain fixed area (e.g., 10 × 10 pixels), then the averaged contrast can be related to the fraction of the forest canopy lost because of degradation. Examples for quantification of degradation are discussed in Section 3.3.

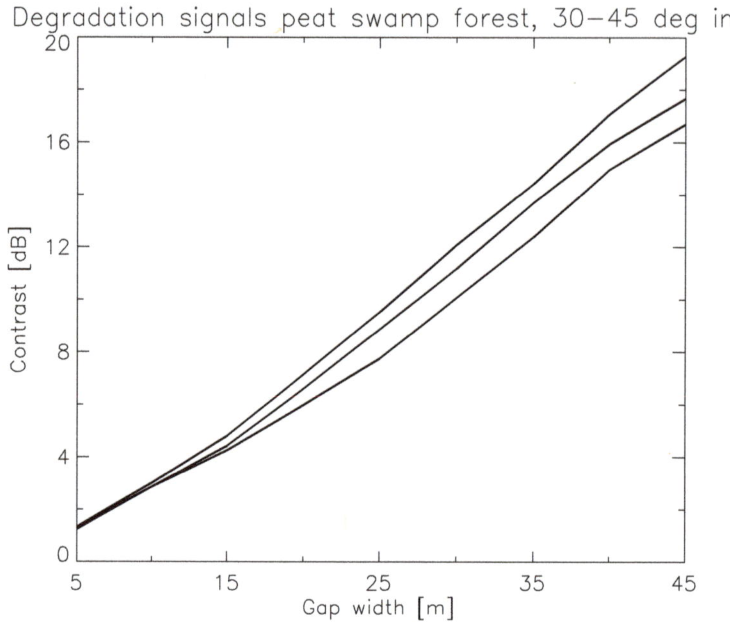


**Figure 10.** Modelled contrast (dB) as a function of canal gap width (m) for a peat swamp forest in Central Kalimantan for three radar incidence angles. Steeper incidence angles yield higher contrast: 30.0 degrees (top), 37.6 degrees (middle), and 45.0 degrees (bottom).

## 3. Results

### 3.1. Results for Canal Gap Detection

The Sentinel-1 NRT canal maps were validated by using results of visual interpretation of SPOT-6/7 images as reference. For each canal visible, the length, width, and orientation were determined. The detection rate was studied by comparing the lengths of these canals with the corresponding lengths in the Sentinel-1 map. This was done as a function of canal width and orientation. The false-alarm rate was studied by evaluating Sentinel-1 canal detections not present in the initial reference dataset. A large fraction of the initial reference map for 8 August 2017 is shown in Figure 11a, while, in Figure 11b, the corresponding Sentinel-1 NRT canal map for 7 August 2017 is shown.

The overall detection rate is 85.5%, i.e., 9.3 km of canal length is missed out of a total 64.2 km. Tables 3–5 divide this result over several width and orientation classes. Only for the smallest width class (5–10 m range, Table 3) and the orientation two classes closest to the range direction (more than 80 degrees from azimuth direction, Table 4), the accuracy drops below 50%. Table 5 combines these 2 classes showing that for small canals (smaller than 20 m) in radar look direction (within ±15 degrees from range direction) the accuracy drops to 27.3%. In all other cases, the accuracy is much higher, which is in agreement with the simulated result presented in Figure 9.

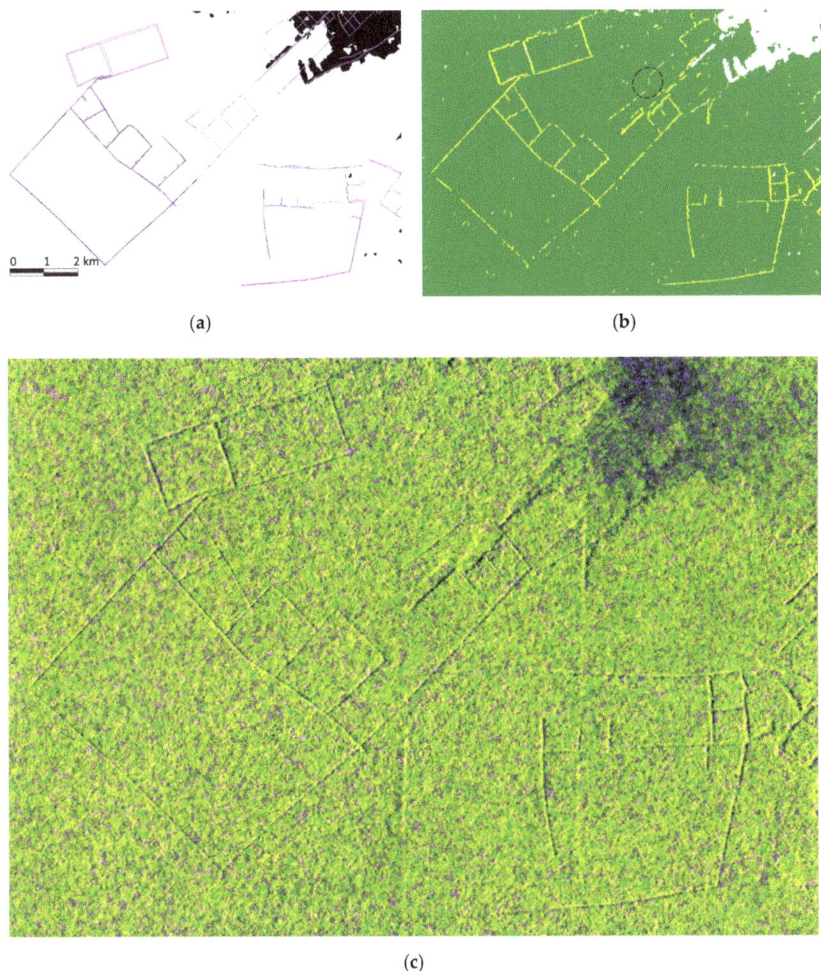



(a)

(b)

(c)

**Figure 11.** (**a**) Reference data from SPOT-6/7 20170908; (**b**) detail Sentinel-1 (S1) NRT canal map 20170907, 11.6 km × 7.9 km; (**c**) corresponding S1 radar image 20170907. (Standard color scale: see caption for Figure 2.)

**Table 3.** Detection rate based on canal width.

| Width (m) | Length (m) | Correct (%) |
|:---:|:---:|:---:|
| 5–10 | 1559 | 47.4 |
| 10–15 | 15,634 | 77.8 |
| 15–20 | 13,462 | 91.3 |
| 20–25 | 7938 | 84.1 |
| 25–30 | 9351 | 99.9 |
| 30–35 | 3859 | 70.7 |
| 35–40 | 5838 | 84.4 |
| 40–45 | 2511 | 92.4 |
| 45–50 | 2861 | 87.8 |
| >50 | 1186 | 100.0 |
| Total | 64,197 | 85.5 |

**Table 4.** Detection rate based on look direction.

| Azimuth | Length (m) | Correct (%) |
| --- | --- | --- |
| 0–5 | 2283 | 97.8 |
| 5–10 | 152 | 100.0 |
| 10–15 | 720 | 100.0 |
| 15–20 | 3555 | 99.7 |
| 20–25 | 1511 | 99.3 |
| 25–30 | 1119 | 98.7 |
| 30–35 | 12,283 | 95.0 |
| 35–40 | 4470 | 83.2 |
| 40–45 | 3255 | 94.8 |
| 45–50 | 2562 | 83.2 |
| 50–55 | 2827 | 95.6 |
| 55–60 | 6883 | 79.3 |
| 60–65 | 7317 | 81.4 |
| 65–70 | 903 | 84.5 |
| 70–75 | 6048 | 81.1 |
| 75–80 | 6610 | 68.4 |
| 80–85 | 1422 | 45.2 |
| 85–90 | 275 | 27.3 |
| Total | 64,197 | 85.5 |

**Table 5.** Detection rate for wide and narrow canals, oriented in look direction or other direction. Here, wide means >20 m; in look direction means within ±15 degrees from range direction.

| Combination Classes | Length (m) | Correct (%) |
| --- | --- | --- |
| Wide, not in look direction | 25,510 | 96.2 |
| Wide, in look direction | 8032 | 64.3 |
| Narrow, not in look direction | 30,379 | 82.7 |
| Narrow, in look direction | 275 | 27.3 |
| Total | 64,197 | 85.5 |

Sentinel-1 canal detections not present in the initial reference set could be divided in two different categories. The first category consists of true canal gap segments very poorly visible in the SPOT images. These canal gaps are often narrow and often show regrowth. An example is given in Figure 12. Once these canals are recognized in SPOT images, aided by the Sentinel-1 maps, additional visual interpretation is possible. In this study, 7.3 km of additional canal gaps could be found in the SPOT data, of which 4.6 km (or 62.7%) was actually already mapped by Sentinel-1. This includes three canals smaller than 10 m, all oriented at 55 degrees from azimuth direction: (1) 9.7 m width, 114 m length, and 100.0% detected; (2) 8.1 m width, 340 m length, and 100.0% detected; and (3) 6.5 m width, 348 m length, and 48.3% detected. The second category consists of small canal gap segments in the NRT map which are not visible in the SPOT images, even after careful re-evaluation. While a part of these false-alarm detections may constitute true false alarms, another part may be true detections (or "false false" alarms) not visible in the SPOT image, for example, related to small canopy gaps caused by illegal selective logging. This notion is based on an evaluation of a time-series of canal gap maps. For example, in the 7 August 2017 NRT canal gap [NRT(N = 0)] map, the false-alarm rate is 9.5%. However, in subsequent maps, an increasing number of these false alarms disappear. Therefore, these false alarms may be related to noise effects and could be regarded as true false alarms. After approximately two months of persistent false alarm detections remain. These persistent false alarm detections, unlike the non-persistent false alarm detections, are not located at random, but are located near canals and rivers or forest edges. These places are much more accessible and prone to illegal logging activities. Thus, the false alarm rate of 9.5%, after approximately two months, may be divided in a non-persistent false alarm rate of 3.9% and a non-verifiable false alarm rate of 5.6%, which may relate, to a large extent, to true disturbances such as illegal logging.

**Figure 12.** Location of canals hardly visible in SPOT-6/7 corresponding with the detections by Sentinel-1 in the black ellipse of Figure 11b. The inset located in the upper-left shows the corresponding area of the Sentinel-1 NRT canal gap map.

## 3.2. Results NRT Deforestation Monitoring

Sentinel-1 NRT deforestation maps were validated in Borneo and Brazil, using a careful visual interpretation of all available Sentinel-2 images and Google Earth. The result is an optical reference set of image segments labelled either as completely forested or completely deforested. These optical segments are compared with individual pixels in the radar maps. Before discussing the results of this validation, it may be insightful to compare an S1 radar NRT deforestation map with an S2 deforestation map, where all pixels of the S2 map which have at least a small bare soil fraction in at least one of the images used for validation are marked as deforested (Figure 13). This map clearly shows large agreement between optical and radar data for deforested areas. However, for forest areas recovering from fire, newly degraded areas, and narrow linear gaps (new roads and canals), large fractions are missing in the S2 map.

**Table 6.** Overview Sentinel-1 NRT map series and Sentinel-2 reference data series. For Brazil, two Sentinel-2 series were used.

| Series | Number | Period |
| --- | --- | --- |
| Indonesia S1 | 91 | 20150930–20190804 |
| Indonesia S2 | 157 | 20170119–20191120 |
| Brazil S1 | 81 | 20170108–20191001 |
| Brazil S2 | 63 | 20170616–20191029 |
| Brazil S2 | 56 | 20180524–20191130 |

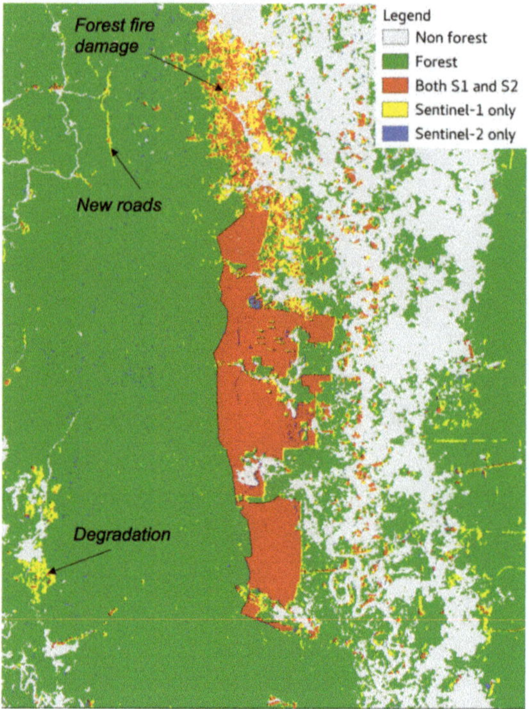



**Figure 13.** A 30 km × 40 km section of an S1 deforestation map of August 2019 is compared with a Sentinel-2 (S2) reference map for an oil palm development area on shallow peat in Central Kalimantan. The S2 reference map shows the cumulative detection (for all available S2 images, see Table 6) of pixels with a bare soil fraction. For large-scale deforestation, there is a large agreement (red). In other areas, such as forests recovering from fire damage and degraded forest, the S2 reference misses large areas (yellow). The same is true for new road and canal gaps.

Results of (quantitative validation at radar pixel level) can be shown in charts, as in Figure 14a. This example is the aggregated result for three representative landscapes in Central Kalimantan with a total area of 194,235 ha, and with major deforestation events in the Sentinel-1 observation period. Within these landscapes, all areas interpreted as deforested in the Sentinel-2 images while being classified as forest in the 2015 baseline map were evaluated. An overview of the S1 NRT maps and the S2 images used for validation are given in Table 6. In the chart of Figure 14, the transition in time from forest to non-forest is visible in terms of the classes used for validation. Only the classes MD (orange) and FA (red) represent errors. In the vertical direction the relative strength of the errors is visible and in the horizontal direction the duration of the errors.

The FAR (Equation (1)), in this example (Figure 14a), is very low, and is discussed later. The MDR (Equation (2)) is sometimes substantial and varies over time. For example, on 30 April 2019, at a 90% confidence level, the MDR has a value of 18.6% ± 1.0%, while, on 17 June 2019, the MDR is 1.9% ± 1.1%. This variation can be explained partly in methodological terms and partly in physical terms. The presented result relates to the default NRT(N = 1) system (see Section 2.4). This means detected deforestation is only (or mostly, depending on system settings) mapped when it can be confirmed by the next radar image. This is often not the case, as is illustrated by the time-series of radar images in Figure 15. This series of eight consecutive radar images, covering an oil palm plantation development area on shallow peat, clearly demonstrates the backscatter contrast between forest and new clear-cut can go up and down. This may be explained, physically, by the relatively large soil roughness in combination with changes in soil moisture. The same phenomenon is illustrated in segment-averaged temporal backscatter signals for VH, VV, and VH–VV ratio for the same area in Figure 16a. The VH and VV signals jointly go up and down, and deforestation is detected at the first moment it stays down. However, the VH–VV ratio stays low from the moment the VH and VV signals go down for the first time. The lowered VH–VV ratio is a sign of vegetation loss and the fluctuation of the VV and VH are signs of soil moisture fluctuations. High levels of soil moisture and large soil roughness in combination with the NRT(N = 1) methodological rules explain the delay in deforestation detection in this shallow peat landscape. It could also be noted that, on average, the delays are larger in the deep peat landscape and absent in the dry forest landscape. This may be related to other soil roughness and/or soil drainage conditions.

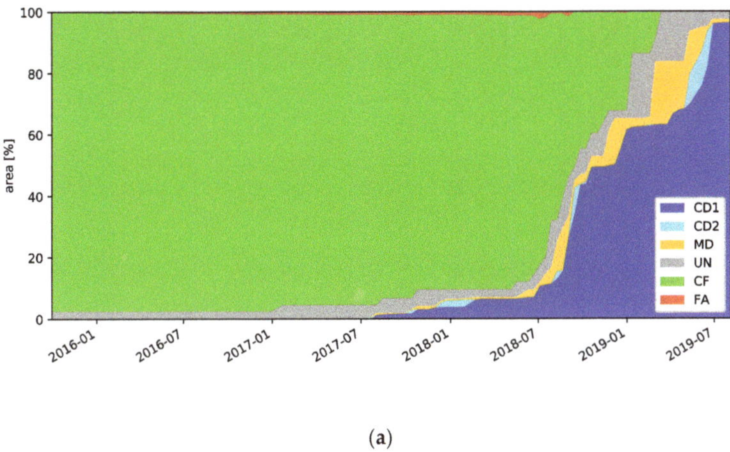

(**a**)

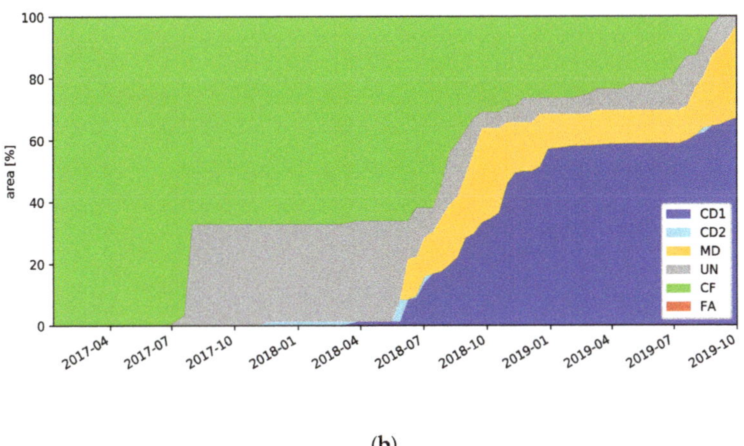

(**b**)

**Figure 14.** (**a**) Aggregated validation results for three representative landscapes in Central Kalimantan for the period September 2015 until August 2019. (**b**) Aggregated validation results for representative landscapes in Pará, Brazil, for the period January 2017 until September 2019.

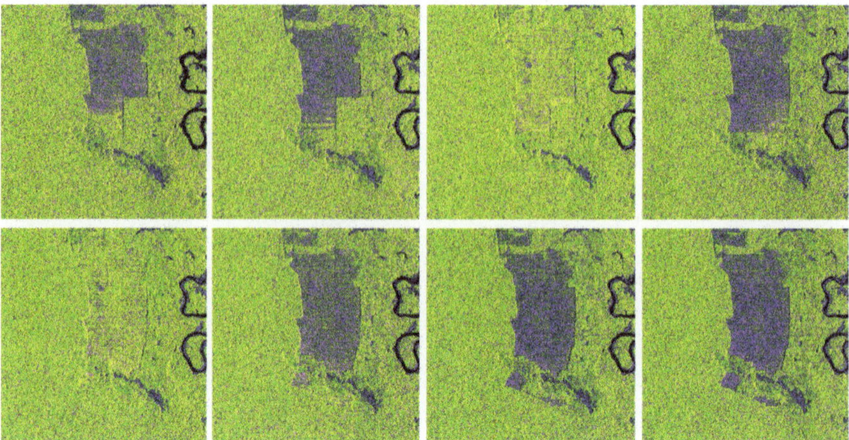

**Figure 15.** Eight successive radar observations at a 12-day interval for the period 20180704–20180926 (from top left to bottom right). Area: 10 km × 10 km. Location: Central Kalimantan. (Standard color scale: see caption Figure 2.)

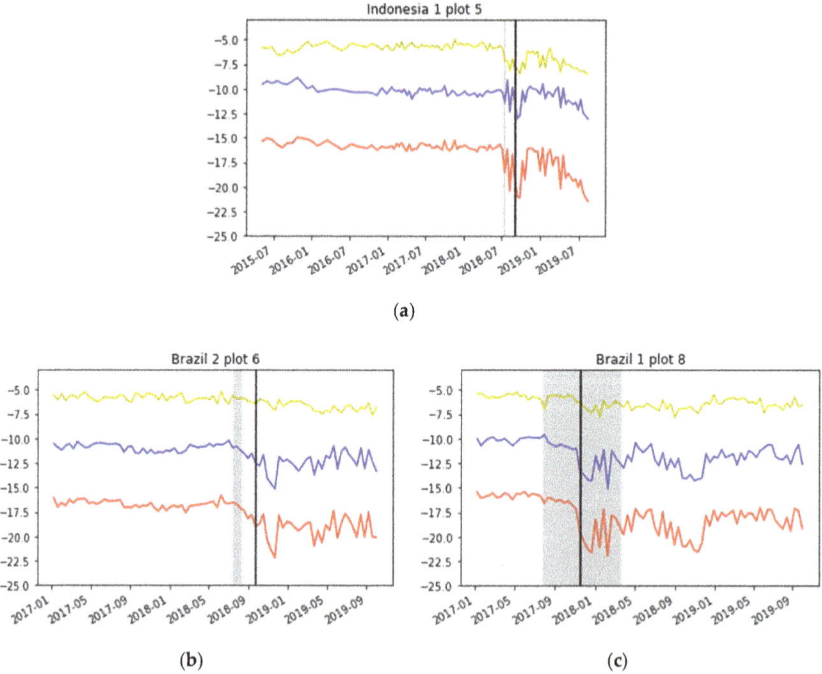

**Figure 16.** (**a**) Temporal radar signature for an area located in Figure 15. VH-polarization (red), VV-(blue), VH/VV-polarization (red). The black line shows the time of detection for the NRT (N = 1) system, and the gray line is the reference time following from visual interpretation of Sentinel-2. (**b**) Temporal radar signature for an area in Brazil where, after the first indication of deforestation on S2, radar backscatter decreases slowly. This causes a delayed detection by an NRT (N = 1) system. (**c**) Temporal radar signature for an area in Brazil with much more cloud cover. In this case, Sentinel-2 data can only show that deforestation occurred within the gray interval.

These results can be compared with a representative landscape north of the Amazon River in the State of Pará, Brazil, with a total area of 57,792 ha, as shown in Figure 14b. This landscape is characterized with small-scale deforestation, often of the slash-and-burn type. Deforestation is often preceded by severe degradation and sometimes changed in low secondary vegetation without ever passing through a bare soil state. In the latter case, the change in averaged radar backscatter is low and deforestation is only detected well along the near range forest edges as radar shadow and partly as degradation (see Section 3.3). This explains the aggregated validation results for the Brazil landscape shown in Figure 14b. Like in Indonesia, the FAR is very low and the MDR is sometimes substantial and varies over time. The MDR may be split for MD which is followed by correct detections within a short period and persistent MD related to areas that never experience a stage without substantial vegetation cover. For example, at 10 January 2019, at a 90% confidence level, the MDR is 17.0% ± 1.6%, which can be split into a temporary part of 3.3% ± 1.5% and a persistent part of 13.8% ± 1.4%. These numbers do not change much until 21 July 2019, when significant temporary MD cases start to occur. Figure 16b illustrates the gradual deforestation process typical for this landscape. The backscatter decreases over a period of several months before deforestation is detected. Gradual deforestation may partly explain the delay in deforestation detection, which is unlike the situation in Indonesia where delayed detection may be better explained by high levels of soil moisture (mainly on peat).

In summary, it can be concluded that the FA rate is very low (because of selected user settings) and the MD rate can be significant and varies because of delays in detection; however, the MD error is not permanent. The detection delay is a typical feature of the NRT(N = 1) system. Such delays are absent in the NRT(N = 0) system at the expense of a higher FA rate (for example caused by heavy rain cells). In an NRT(N ≥ 2) system, the delays are much shorter at the expense of having less timely maps. This may illustrate the importance of proper user settings or adopting a more complex system with multiple sets of user settings.

Opposed to detection delays in radar data, there are also detection delays in optical data. The class CD2 shows radar detection prior to the first available next optical image. However, there are many more cases where radar detection precedes optical detection. In the validation procedure, these cases are present in the class unknown (UN), but these cases cannot be validated, by definition, by optical data. An example is given in Figure 16c. Because of cloud cover, the optical data can only be used to show the deforestation occurs in the period August 2017 until April 2018. The radar detection is in the middle of this period, where a significant drop in the radar backscatter occurs. In cases where validation could be done, a drop of such a magnitude leads to a correct deforestation detection. An evaluation of the radar signatures of all test areas reveals that, for the Brazilian landscape, in more than 20% of the cases, the radar detection precedes the optical detection by at least two months and up to 11 months. For the Indonesian landscapes, all relatively close to the coast, where there is less cloud cover, the radar detection precedes the optical detection in approximately 10% of the cases by at least two months and up to eight months.

In the presented examples (Figure 14a,b), the FAR is always very low. This result should be interpreted as follows. FAR and MDR are dependent. The selected settings of the algorithm used for this example favor a low FAR at the expense of a relatively high MDR. The results shown are at pixel level, while the optical reference data, based on visual interpretation, are at segment level. Within the segment a very small fraction of pixels may relate to small scale deforestation such as road development prior to deforestation, or small remnant patches of trees after deforestation. These fractions cause small FAR and MDR errors.

### 3.3. Results NRT Degradation Monitoring

#### 3.3.1. Brazilian Study Site

Since the theoretical model introduced in Section 2.5 applies equally well for gaps of very short canals, which resemble gaps caused by selective logging, it can be used to quantify degradation.

When the right model parameters are applied and contrast is computed over a certain fixed area (e.g., 10 × 10 pixels), then the averaged contrast can be related to the fraction of the forest canopy lost resulting from degradation (e.g., Figure 10). A mapping example for an active section in a Brazilian timber concession area is shown in Figure 17.

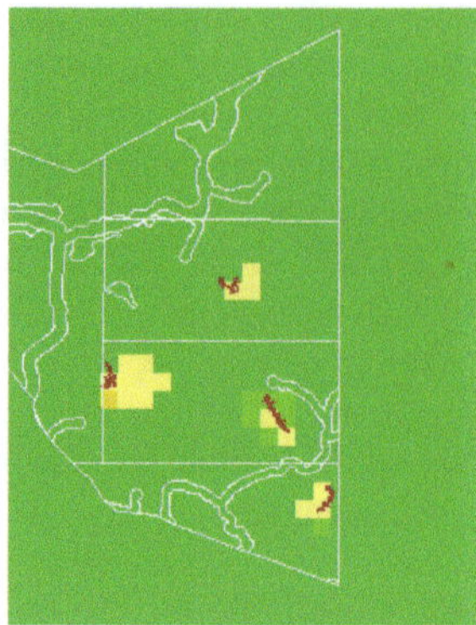

**Figure 17.** A section of a Sentinel-1 radar derived map showing deforestation at pixel level (red) and quantitative degradation for 150 m × 150 m blocks in a timber concession area (demarcated with white lines). Area: 4 × 5 km. Degradation classes expressed in canopy loss fraction: <20% (light green), 20–30% (yellow), 30–40% (tan), and >40% (pink). Almeirim, Brazil, February 2018.

Radar is a suitable instrument to quantify degradation. Unlike optical data, which detect degradation mainly by the signal fraction from the bare soil, the radar detects degradation by signals from gaps in the canopy, even when the understory still covers the soil. Therefore, the radar signal is very persistent (gaps in the upper canopy do not fill up fast), while the optical signal is visible for a short time window only (secondary regrowth on bare soil appears fast). The latter is even more troublesome when cloud cover is frequent. This is illustrated well by the example given in Figure 18. Here, for a selective logging concession area in Brazil a comparison is made between the Sentinel-1 radar and Sentinel-2 optical results. The solid line shows the total forest canopy fraction loss for each radar observation as a function of time. For optical data, such a result is not feasible because of cloud cover; instead, the accumulated detections can be shown (dotted line with diamonds). This accumulated result sums all detections, even when they are not visible anymore because of regrowth or cloud cover. It can also be noted that in this period where 81 radar observations were made only 12 partly cloud-free optical images (diamonds) are available (see Table 7). From the comparison, it is clear that most degradation in the wet season (December–May) is not detected by the optical system. Obviously, optical data have severe limitations to detect degradation and, thus, are less suitable for the validation of radar degradation maps. An alternative is the use of high-resolution radar data, such as TerraSAR-X, which are used to map selective logging at the level of individual canopy trees [20]. Results for the wet season, in January–February 2018, show a clear correspondence in time and location of degradation. The 85 trees logged in this period (mapped by TerraSAR-X) compare with an effective forest canopy fraction loss of 4.5 ha (mapped by Sentinel-1). This would relate to an average loss of ±500 m$^2$ per logged canopy tree.

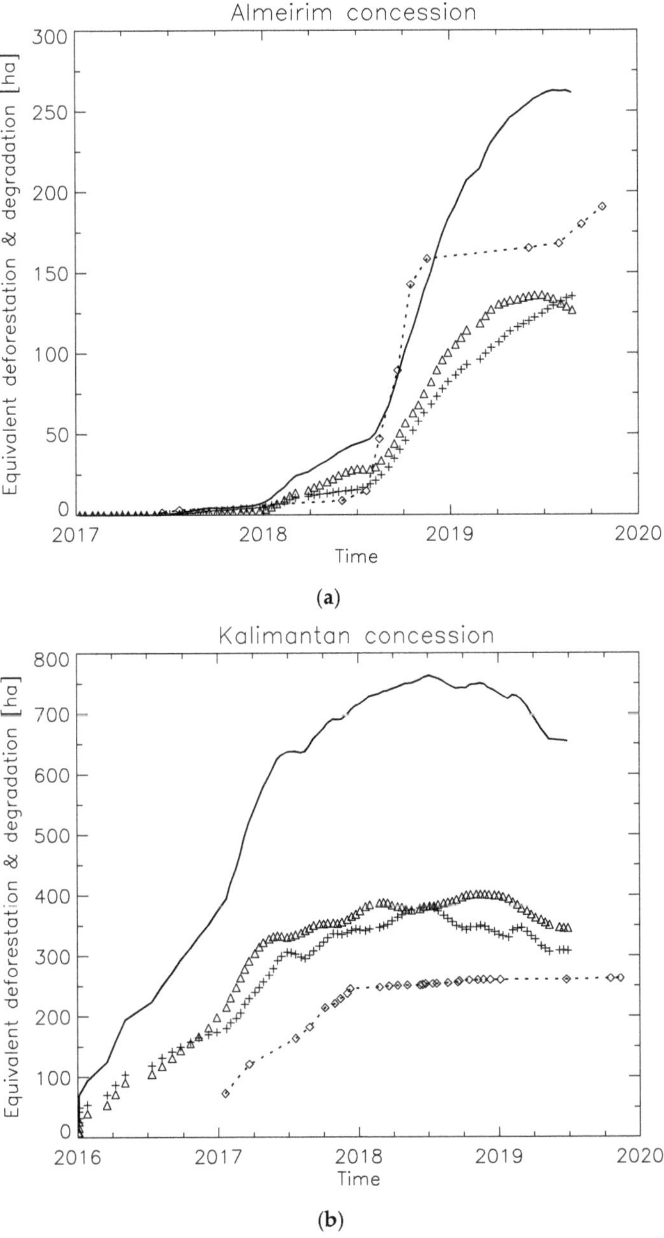



(a)

(b)

**Figure 18.** (a) Results of degradation mapping for a selective logging timber concession in Almeirim, Brazil. Total forest canopy fraction loss as a function of time for radar (solid line), which is the sum of the fraction related to canopy gaps (triangles) and the fraction related to timber trails (plusses). For optical data, the accumulated detections are shown (dotted line with diamonds). The symbols indicate the time of observation (scaled between January 2017 and December 2020). (b) As Figure 18a, for a timber concession in Central Kalimantan.

**Table 7.** Overview Sentinel-1 NRT map series and Sentinel-2 and TerraSAR-X (3 m single-pol stripmap) reference data series.

| Series | Number | Period |
|---|---|---|
| Indonesia S1 | 88 | 20160104–20190804 |
| Indonesia S2 | 29 | 20170119–20191110 |
| Brazil S1 | 81 | 20170108–20191001 |
| Brazil S2 | 12 | 20170616–20191024 |
| Brazil TerraSAR-X | 3 | 20171203–20180207 |

### 3.3.2. Indonesian Study Site

A second example is given for a timber concession located in Central Kalimantan on undulating terrain (250–1100 m altitude). For this area Sentinel-1 degradation maps have been made from January 2016 onward. For Sentinel-2 29 partly clouded images are available since January 2017. These images were used to delineate all disturbance through careful visual interpretation. Results (Figure 18b) are shown in a similar way, as was done for the Brazilian site. The study area (shown in Figures 19 and 20) is 8000 ha large. The radar shows that an equivalent area of approximately 400 ha is lost in the year 2016 and another 300 ha in 2017. The first Sentinel-2 image shows a loss of only ±80 ha. This is partly explained by partial cloud cover; however, almost all of the 2016 losses detected by radar were not detected, either, in the following 28 Sentinel-2 images. The accumulated losses visible in the optical data increase until ±280 ha, while the instantaneous losses mapped by radar peaks at ±750 ha in mid-2018, followed by a slight decrease, which may be attributed to regrowth.

The evaluation of the radar and optical maps shown in Figure 19a–e and the corresponding radar images (Figure 20) of this area allows for a more detailed discussion. This is done for three moments in time: one of the first radar degradation maps (August 2016: Figure 19a); the first optical reference map and the associated radar degradation map (January 2017; Figure 19b,c); and the optical reference and radar map around the end of the logging operation (December 2017; Figure 19d,e). The radar images coincide with the August 2016 radar map (Figure 20a) and the December 2017 radar map (Figure 20b). A visual comparison of these two radar images clearly shows a change in the road network but no clear signs (i.e., backscatter level or textural change) of degradation. Nevertheless, degradation can be mapped based on model-based calculations using the changes in the patterns that constitute the texture (Figure 19a,c,e). In the first Sentinel-2 reference map (Figure 19b) the main road network is well visible as well as some small canopy gaps. These gaps correspond almost completely with the area of current activity, while most of the gaps visible in the August 2016 radar map, five months earlier, are not visible. The December 2017 Sentinel-2 reference image shows the accumulated visibility of roads and gaps. New roads and gaps show up at the same location as in the radar map. However, in general, there are two qualitative differences. The first, as discussed above, radar detects more forest loss related to gaps. Secondly, the radar maps show less roads than the optical reference image, for which there are two causes. The first cause is that the radar maps only show change with respect to the baseline. Therefore, existing and stable road gaps do not show up while in the optical reference image they are still mapped. One clear example is the gap caused by a river visible in the upper left corner of Figures 19d and 20. The second cause is the use of an MMU of 1 ha for mapping of clear-cut in this forest type. Since narrow roads may break in multiple smaller segments, some smaller than the MMU, these missed sections may be mapped as degradation instead. The same is true for wider roads oriented in the radar range direction, as discussed in Section 3.1, which is visible, for example, in the top-right of Figure 19d,e. It is noted that, for the calculation of the total equivalent area of forest loss (as shown in Figure 18a,b), it does not matter whether a small narrow road section is classified as deforestation or as degradation.

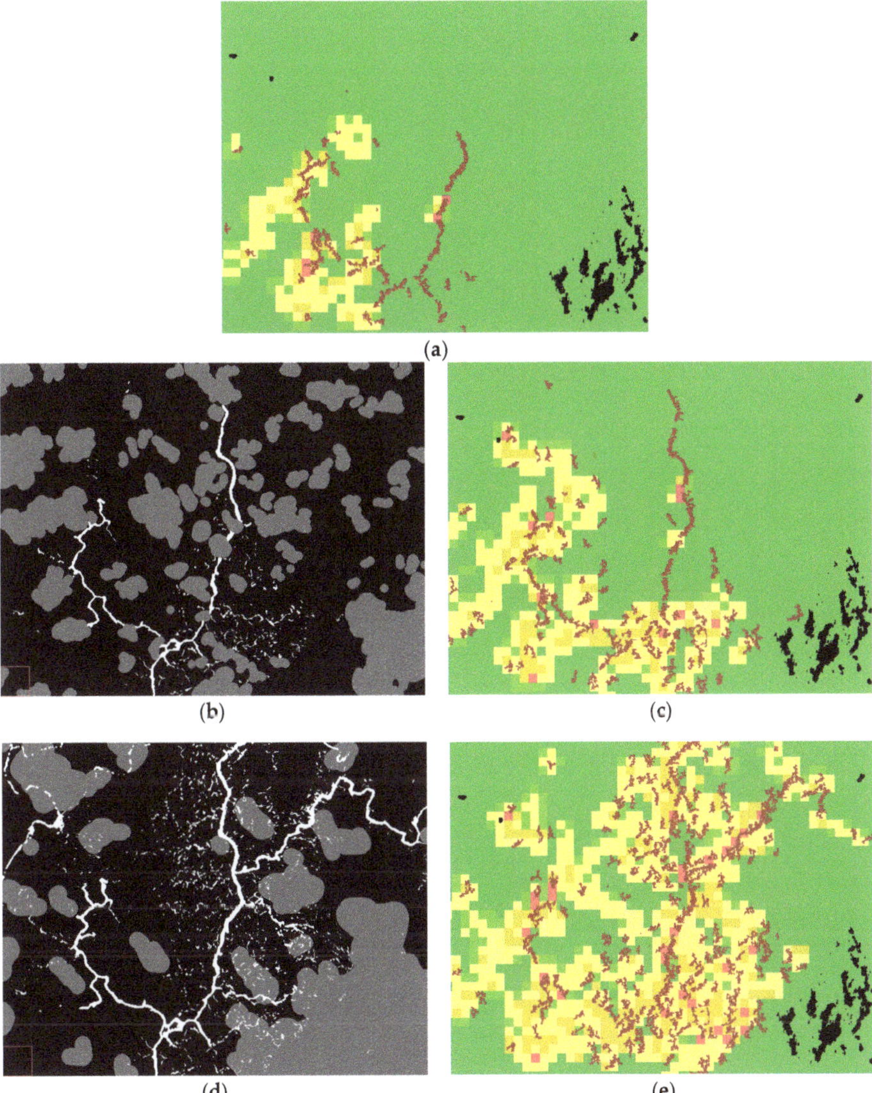

**Figure 19.** (**a**) S1 degradation map 20160807. Detail of a timber concession area in Central Kalimantan. Area: 7.3 × 5.6 km. (Legend: see caption Figure 17). (**b**) S2 reference map 20170119 and (**c**) corresponding S1 degradation map 20170122. In the S2 reference map, cloud cover in the last available image is shown in gray, and the accumulated visibility of roads and gaps is shown in white. (**d**) S2 reference map 20171210 and (**e**) corresponding S1 degradation map 20171212.

(a)         (b)

**Figure 20.** (**a**) S1 radar image 20160807 and (**b**) S1 radar image 20171212. Same area as in Figure 19. (Standard color scale: see caption Figure 2.)

## 4. Discussion

### 4.1. System and Disturbance Model

The Sentinel-1 NRT radar monitoring system is an automated system based on interferometric preprocessing and time-series analysis of small image segments, linear features, and small-scale disturbances. This results in a system that can accurately map different phenomena simultaneously, such as deforestation (clear-cut and fire scars), degradation, selective logging impact, and new narrow canals in peat swamp forest. Radar imaging of canal gaps and the canal-gap-detection mapping results were shown to be in agreement with a physical-interaction model. Small gaps caused by selective logging are too small to be detected individually; however, the same theoretical model (that describes canal gaps) can be used to quantify the canopy disturbance in a statistical sense. Therefore, change mapping is done in three fundamentally different ways. Deforestation is detected by using segment-based time-series analysis and uses a decrease of backscatter as an indicator of deforestation. Canal-gap detection is based on time-series analysis of linear features, using edge, line, and matched filters. Degradation is quantified by using a time-series analysis of textural change based on a physical model. These three approaches are not completely independent, not from a data processing point-of-view, nor from a forest-change-interpretation point-of-view.

Several examples of interdependency can be given. For deforestation mapping in peat swamp forests an MMU of 15 pixels (0.3 ha) applies. This means that wide canal gaps are often mapped as (a row of individual) segments. The same canal gaps show up in the canal gap maps as linear features. Of course, the dedicated canal gap product shows more canals, including some very narrow ones which are hardly visible in SPOT-6/7 data. Very small deforestation segments or very short canal gap detections are often part of degradation areas. At the Brazil test sites some areas of deforestation, the ones that gradually change from forest to low secondary forest without going through a bare soil stage, are not detected with the deforestation mapping approach. However, the near range edges of such areas are still visible as elongated segments, and parts of these areas are detected as degraded. Using such interdependencies explicitly may contribute to a better interpretation of ongoing forest-change processes.

### 4.2. Deforestation

Deforestation detection success is evaluated by using results of a careful visual interpretation of Sentinel-2 time-series as a reference. These results are independent of any issues related to baseline class definitions and timing. In summary, for the Central Kalimantan landscapes, it is shown that the false alarm rate (FAR) is very low (less than 1%) and the missed detection rate (MDR) varies between

18.6% ± 1.0% and 1.9% ± 1.1% (90% confidence level). However, results also depend on user settings. FAR and MDR are interchangeable. Settings were selected to favor a low FAR at the expense of a slightly higher MDR. Another compromise to be made is between overall accuracy and timeliness. In other words, the faster the maps should be made available after radar observation, the lower the accuracy. Settings were selected to favor a relatively fast system, which results in significant detection delays in the map time-series. It was found that peatlands are a typical case where detection delays up to two months occur which are caused by the combination of rough soil surface and high soil moisture. This is causing the high MDR, but these missed detections are only temporary, not permanent. Other settings could decrease such delays to a few weeks. These delays were not found outside the peat areas or in the Amazon. Because of cloud cover, radar can be much faster than optical systems, but this cannot be validated by optical systems. It was found that radar very often detects deforestation two months and up to 10 months faster than optical systems.

### 4.3. Canal Gap Detection

Results of visual interpretation of SPOT-6/7 images were used as reference. The overall detection rate is 85.5%; however, results strongly depend on canal gap orientation and, to a lesser extent, on canal gap width. Only for the smallest width class (5–10 m range) or for orientations of more than 80 degrees from azimuth direction, the accuracy drops below 50%. In total, 9.3 km of canal length was missed out of a total 64.2 km. Sentinel-1 canal gap detections which were not present in the initial reference set could be divided in two different categories. The first category consists of true canal gap segments very poorly visible in the SPOT images. These canal gaps are often narrow and often show regrowth. Once these canals are recognized in SPOT images, aided by the Sentinel-1 maps, additional visual interpretation is possible. In this study, 7.3 km of additional canal gaps could be found in the SPOT data. The second category consists of small canal gap segments in the NRT map which are not visible in the SPOT images, even after careful re-evaluation. A part of these false alarms is persistent while others disappear within two months. These persistent false-alarm detections, unlike the non-persistent false-alarm detections, are not located at random, but are located near canals and rivers or forest edges. These places are much more accessible and prone to illegal logging activities. Therefore, the false alarm rate of 9.5%, after approximately two months, could be divided into a non-persistent false alarm rate of 3.9% and a non-verifiable false alarm rate of 5.6%, which may relate to a large extent to true disturbances, such as illegal logging.

### 4.4. Degradation

Like for deforestation, degradation detection success is evaluated using results of a careful visual interpretation of Sentinel-2 time-series as a reference. Radar is a suitable instrument to quantify degradation. Unlike optical data, which detect degradation mainly by the signal fraction from the bare soil, the radar detects degradation by signals from gaps in the canopy, even when the understory still covers the soil. Therefore, the radar signal is very persistent (gaps in the upper canopy do not fill up fast), while the optical signal is visible for a short time window only (secondary regrowth on bare soil appears fast). The latter is even more troublesome when cloud cover is frequent. Validation is difficult using optical data since degradation is detected in a fundamentally different way and a lot of degradation is missed. Nevertheless, results are spatiotemporally consistent. It may be much better to use TerraSAR-X for validation of degradation, notably for quantitative validation. The result presented here, for Brazil, is based on limited data only but provides high spatiotemporal, as well as quantitative, agreement.

### 4.5. Comparison with Other Approaches Based on C-Band

Other existing methods based on C-band [43–46], which focus either on deforestation monitoring or forest/non-forest mapping, can be compared with the deforestation results presented here. The method for deforestation detection presented in Reference [43] provides a similar accuracy level, however, requires both ascending and descending data. These are not available for most tropical forests, nor will they become available in the near future. The method for deforestation detection based on interferometric coherency [46] is less accurate; however, it may provide useful additional information. The results of forest/non-forest classification presented in Reference [44] also seem less accurate when used for the purpose of deforestation monitoring. The comparison with other systems is difficult because the system presented in this paper uses C-band for the monitoring part and a combination of L-band, C-band, and optical data for the (baseline) classification part.

## 5. Conclusions

Like JJ-FAST, the automated Sentinel-1 system presented here can be used for wide-area NRT forest-change monitoring. Results are available two days after satellite overpass, with higher spatial and temporal resolution and a high accuracy for deforestation detection. Unlike JJ-FAST, it utilizes multiple approaches for change detection, to allow monitoring of finer scaled features, such as narrow linear elements (roads and canals) and low-intensity degradation (selective logging). However, these refinements require the availability of a good forest baseline and system tuning to optimize it for local conditions, as well as local user requirements. The Sentinel-1 system is not meant primarily as a single system for pan-tropical coverage, but as a system to be operated and customized by local operators at the national level.

Though C-band, in general, has a lower forest/non-forest contrast than L-band, the Sentinel-1 radar in IW mode provides a higher temporal and spatial resolution than the PALSAR-2 ScanSAR mode, which makes Sentinel-1 equally suitable for the purpose of forest-change monitoring. One of the novelties in this paper is the use of the Sentinel-1 phase information for very precise co-registration (Section 2.3), which allows for the detection of subtle changes.

Though tested for large areas in Indonesia and the Amazon, the validation requires more efforts, which should result in more refined local tuning. For example, a preliminary test for entire Borneo over the entire period revealed that there are still some deforested areas which were not detected but would have been detected with slightly different system settings. The quantitative estimation of degradation is difficult to validate with optical data. More work is needed, using extensive field data as reference, to calibrate the radar proxy for degradation and compare it with optical proxies.

Even though degradation information is still not yet properly calibrated and validated, several interesting applications emerge. For example, the result in Figure 17 suggests that selective logging is not too intensive and is absent in the protected buffer zones along the river. This information is already helpful for planning field inspections in remote areas, certification, and transparency. Another example is early warning. Low-impact changes, such as new narrow roads, degradation, and small clear-cut areas in remote places are the first indications of potential future threats of forest-cover change. This information is already successfully used as input for predictive modelling based on machine learning by the World Wide Fund for Nature for the development of their Early Warning System (EWS) to predict deforestation [29]. The data also support and could modernize conventional approaches based on hi-res optical data, to get more information out of these data, such as discussed above for the detection of narrow canal gaps in SPOT-6/7 data. Alternatively, hi-res optical data acquisition, for example, after a long period of cloud cover, could be focused on areas where change actually has occurred. The availability of free Sentinel-1 radar data with a systematic and complete coverage is a great asset for future development of efficient wide-area forest-monitoring systems.

**Author Contributions:** Conceptualization, D.H., B.K., I.C., S.B., R.A. and O.R.; Formal analysis, D.H. and B.K.; Investigation, D.H., B.K. and R.A.; Methodology, D.H., B.K. and M.Q.; Resources, O.R.; Software, D.H. and B.K.; Validation, S.V. and R.A.; Visualization, B.K.; Writing—original draft, D.H.; Writing—review and editing, B.K., S.V., I.C., R.A. and O.R. All authors have read and agreed to the published version of the manuscript.

**Funding:** Work on tropical peatland monitoring is supported and partly funded through the ESA EO Science for Society project "Tropical Peat View" (ESA AO/1-9101/17/I-NB; September 2018–March 2020).

**Acknowledgments:** The development of NRT Sentinel-1 monitoring technology is partly developed for the international EWS system of WWF Netherlands, for which WWF Netherlands provided useful feedback.

**Conflicts of Interest:** The authors declare no conflict of interest.

## References

1. The State of the World's Forests 2020. Available online: http://www.fao.org/state-of-forests/en/ (accessed on 6 October 2020).
2. Walker, W.S.; Gorelik, S.R.; Baccini, A.; Aragon-Osejo, J.L.; Josse, C.; Meyer, C.; Macedo, M.N.; Augusto, C.; Rios, S.; Katan, T.; et al. The role of forest conversion, degradation, and disturbance in the carbon dynamics of Amazon indigenous territories and protected areas. *Proc. Natl. Acad. Sci. USA* **2020**, *117*, 3015–3025. [CrossRef] [PubMed]
3. Joosten, H.; Couwenberg, J. Peatlands and carbon. In *Assessment on Peatlands, Biodiversity and Climate Change*; Global Environment Centre: Kuala Lumpur, Malaysia & Wetlands International: Wageningen, The Netherlands, 2007; pp. 99–117.
4. Dargie, G.C.; Lewis, S.L.; Lawson, I.; Mitchard, E.T.A.; Page, S.E.; Bocko, Y.E.; Ifo, S.A. Age, extent and carbon storage of the central Congo Basin peatland complex. *Nat. Cell Biol.* **2017**, *542*, 86–90. [CrossRef] [PubMed]
5. Lähteenoja, O.; Page, S.E. High diversity of tropical peatland ecosystem types in the Pastaza-Marañón basin, Peruvian Amazonia. *J. Geophys. Res. Space Phys.* **2011**, *116*, 02025. [CrossRef]
6. Crump, J.; Avagyan, A.; Baker, E.; Barthelmes, A.; Velarde, C.; Dargie, G.; Guth, M.; Hergoualc'h, K.; Johnson, L.; Joosten, H.; et al. *Smoke on Water—Countering Global Threats from Peatland Loss and Degradation. A UNEP Rapid Response Assessment*; United Nations Environment Programme: Nairobi, Kenya and GRID-Arendal: Arendal, Norway, 2017; ISBN 978-82-7701-168-4.
7. IPCC. *2019 Refinement to the 2006 IPCC Guidelines for National Greenhouse Gas Inventories*; IPCC: Geneva, Switzerland, 2019.
8. Marziliano, P.A.; Veltri, A.; Menguzzato, G.; Pellicone, G.; Coletta, V. A comparative study between "default method" and "stock change method" of Good Practice Guidance for Land Use, Land-Use Change and Forestry (IPCC, 2003) to evaluate carbon stock changes in forest. *Secondo Congresso Internazionale di Selvicoltura Second Int. Congr. Silvic.* **2015**, 551–557. [CrossRef]
9. IPCC. *Definitions and Methodological Options to Inventory Emissions from Direct Human-Induced Degradation of Forests and Devegetation of Other Vegetation Types*; Institute for Global Environmental Strategies (IGES): Kanagawa, Japan, 2003; ISBN 4-88788-004-9.
10. GFOI. *Integrating Remote-Sensing and Ground-Based Observations for Estimation of Emissions and Removals of Greenhouse Gases in Forests: Methods and Guidance from the Global Forest Observations Initiative*; Group on Earth Observations: Geneva, Switzerland, 2014; ISBN 978-92-990047-4-6.
11. GOFC-GOLD. *A Sourcebook of Methods and Procedures for Monitoring and Reporting Anthropogenic Greenhouse Gas Emissions and Removals Associated with Deforestation, Gains and Losses of Carbon Stocks in Forests Remaining Forests, and Forestation*; GOFC-GOLD Report Version COP22-1; GOFC-GOLD Land Cover Project Office, Wageningen University: Wageningen, The Netherlands, 2016.
12. Souza, J.C.M.; Roberts, D.A.; Cochrane, M.A. Combining spectral and spatial information to map canopy damage from selective logging and forest fires. *Remote. Sens. Environ.* **2005**, *98*, 329–343. [CrossRef]
13. Asner, G.P.; Knapp, D.E.; Balaji, A.; Paez-Acosta, G. Automated mapping of tropical deforestation and forest degradation: CLASlite. *J. Appl. Remote. Sens.* **2009**, *3*, 033543. [CrossRef]
14. Franke, J.; Navratil, P.; Keuck, V.; Peterson, K.; Siegert, F. Monitoring Fire and Selective Logging Activities in Tropical Peat Swamp Forests. *IEEE J. Sel. Top. Appl. Earth Obs. Remote. Sens.* **2012**, *5*, 1811–1820. [CrossRef]
15. Potapov, P.; Hansen, M.; Kommareddy, I.; Kommareddy, A.; Turubanova, S.; Pickens, A.; Adusei, B.; Tyukavina, A.; Ying, Q. Kommareddy Landsat Analysis Ready Data for Global Land Cover and Land Cover Change Mapping. *Remote. Sens.* **2020**, *12*, 426. [CrossRef]
16. Hirschmugl, M.; Steinegger, M.; Gallaun, H.; Schardt, M. Mapping Forest Degradation due to Selective Logging by Means of Time Series Analysis: Case Studies in Central Africa. *Remote. Sens.* **2014**, *6*, 756–775. [CrossRef]
17. Hoekman, D.; Varekamp, C. Observation of tropical rain forest trees by airborne high-resolution interferometric radar. *IEEE Trans. Geosci. Remote. Sens.* **2001**, *39*, 584–594. [CrossRef]

18. Deutscher, J.; Perko, R.; Gutjahr, K.-H.; Hirschmugl, M.; Schardt, M. Mapping Tropical Rainforest Canopy Disturbances in 3D by COSMO-SkyMed Spotlight InSAR-Stereo Data to Detect Areas of Forest Degradation. *Remote. Sens.* **2013**, *5*, 648–663. [CrossRef]

19. De Grandi, E.C.; Mitchard, E.; Hoekman, D. Wavelet Based Analysis of TanDEM-X and LiDAR DEMs across a Tropical Vegetation Heterogeneity Gradient Driven by Fire Disturbance in Indonesia. *Remote. Sens.* **2016**, *8*, 641. [CrossRef]

20. Hoekman, D.H.; Kahwage, C. Monitoramento por radar do desmatamento na area central da TIARG. In *Gestão Ambiental e Territorial da Terra Indígena Alto Rio Guamá*; IDEFLOR-Bio: Belém, Brazil, 2017; pp. 327–346, ISBN 978-85-92612-04-7.

21. Hoekman, D. Satellite radar observation of tropical peat swamp forest as a tool for hydrological modelling and environmental protection. *Aquat. Conserv. Mar. Freshw. Ecosyst.* **2007**, *17*, 265–275. [CrossRef]

22. Harris, N.L.; Brown, S.; Hagen, S.C.; Saatchi, S.S.; Petrova, S.; Salas, W.; Hansen, M.C.; Potapov, P.V.; Lotsch, A. Baseline Map of Carbon Emissions from Deforestation in Tropical Regions. *Science* **2012**, *336*, 1573–1576. [CrossRef]

23. Zarin, D.J.; Behie, S.W.; Zelisko, P.M.; Bidochka, M.J. Carbon from Tropical Deforestation. *Science* **2012**, *336*, 1518–1519. [CrossRef]

24. Van Der Werf, G.R.; Morton, D.C.; DeFries, R.S.; Olivier, J.G.J.; Kasibhatla, P.S.; Jackson, R.B.; Collatz, G.J.; Randerson, J.T. CO2 emissions from forest loss. *Nat. Geosci.* **2009**, *2*, 737–738. [CrossRef]

25. Page, S.E.; Siegert, F.; Rieley, J.O.; Boehm, H.-D.V.; Jaya, A.; Limin, S. The amount of carbon released from peat and forest fires in Indonesia during 1997. *Nat. Cell Biol.* **2002**, *420*, 61–65. [CrossRef]

26. Kool, D.M.; Buurman, P.; Hoekman, D.H. Oxidation and compaction of a collapsed peat dome in Central Kalimantan. *Geoderma* **2006**, *137*, 217–225. [CrossRef]

27. ITTO & IUCN. *ITTO/IUCN Guidelines for the Conservation and Sustainable Use of Biodiversity in Tropical Timber Production Forests*; ITTO Policy Development Series No. 17; ITTO: Yokohama, Japan, 2009; ISBN 4-902045-41-9.

28. Vasconcelos, A.; Bernasconi, P.; Guidotti, V.; Silgueiro, V.; Valdiones, A.; Carvalho, T.; Bellfield, H.; Guedes Pinto, L.F. Illegal Deforestation and Brazilian Soy Exports: The Case of Mato Grosso. Available online: http://resources.trase.earth/documents/issuebriefs/TraseIssueBrief4_EN.pdf (accessed on 6 October 2020).

29. van Stokkom, A.; Dallinga, J.; Debuyser, M.; Hoekman, D.; Kooij, B.; Pacheco, P.; Thau, D.; Valkman, S.; Beukeboom, H. An Innovative Early Warning System To Tackle Illegal Deforestation (10667). In Proceedings of the FIG Working Week 2020, Smart surveyors for land and water management, Amsterdam, The Netherlands, 10–14 May 2020.

30. UNFCCC. Reducing Emissions from Deforestation in Developing Countries: Approaches to Stimulate Action. UNFCCC/COP-13 Draft Decision. 2007. Available online: http://unfccc.int/files/meetings/cop_13/application/pdf/cp_redd.pdf (accessed on 2 October 2020).

31. Diniz, C.G.; Souza, A.A.D.A.; Santos, D.C.; Dias, M.C.; Da Luz, N.C.; De Moraes, D.R.V.; Maia, J.S.A.; Gomes, A.R.; Narvaes, I.D.S.; Valeriano, D.M.; et al. DETER-B: The New Amazon Near Real-Time Deforestation Detection System. *IEEE J. Sel. Top. Appl. Earth Obs. Remote. Sens.* **2015**, *8*, 3619–3628. [CrossRef]

32. Sano, E.E.; Freitas, D.M.; Souza, R.A.; Matos, F.L.; Ferreira, G.P. Detecting new deforested areas in the Brazilian Amazon using ALOS-2 PALSAR-2 imageries. In *ALOS Kyoto & Carbon Initiative Science Team Reports Phase 4 (2015–2019)*; JAXA EORC NDX-2019009; Japan Aerospace Exploration Agency, Earth Observation Research Center: Ibaraki, Japan, March 2020; pp. 269–277.

33. Watanabe, M.; Koyama, C.N.; Hayashi, M.; Nagatani, I.; Shimada, M. Early-Stage Deforestation Detection in the Tropics With L-band SAR. *IEEE J. Sel. Top. Appl. Earth Obs. Remote. Sens.* **2018**, *11*, 2127–2133. [CrossRef]

34. Watanabe, M.; Koyama, C.; Hayashi, M.; Nagatani, I.; Tadono, T.; Shimada, M. Improvement of Deforestation Detection Algorithms Used In JJ-FAST. In Proceedings of the IGARSS 2019—2019 IEEE International Geoscience and Remote Sensing Symposium, Yokohama, Japan, 28 July–2 August 2019; pp. 5328–5331.

35. Watanabe, M.; Koyama, C.; Hayashi, M.; Nagatani, I.; Tadono, T.; Shimada, M. JJ-FAST Update. JAXA Kyoto and Carbon Initiative Meeting 26. 2020. Available online: https://www.eorc.jaxa.jp/ALOS/kyoto/jan2020_kc26/pdf/1-07_PKC1_Watanabe.pdf (accessed on 2 October 2020).

36. Rosenqvist, A.; Shimada, M.; Ito, N.; Watanabe, M. ALOS PALSAR: A Pathfinder Mission for Global-Scale Monitoring of the Environment. *IEEE Trans. Geosci. Remote. Sens.* **2007**, *45*, 3307–3316. [CrossRef]

37. Shimada, M.; Itoh, T.; Motooka, T.; Watanabe, M.; Shiraishi, T.; Thapa, R.; Lucas, R. New global forest/non-forest maps from ALOS PALSAR data (2007–2010). *Remote. Sens. Environ.* **2014**, *155*, 13–31. [CrossRef]

38. Koyama, C.N.; Watanabe, M.; Hayashi, M.; Ogawa, T.; Shimada, M. Mapping the spatial-temporal variability of tropical forests by ALOS-2 L-band SAR big data analysis. *Remote. Sens. Environ.* **2019**, *233*, 111372. [CrossRef]

39. Woodhouse, I.H.; Van Der Sanden, J.J.; Hoekman, D.H. Scatterometer observations of seasonal backscatter variation over tropical rain forest. *IEEE Trans. Geosci. Remote. Sens.* **1999**, *37*, 859–861. [CrossRef]

40. Reiche, J.; Hamunyela, E.; Verbesselt, J.; Hoekman, D.; Herold, M. Improving near-real time deforestation monitoring in tropical dry forests by combining dense Sentinel-1 time series with Landsat and ALOS-2 PALSAR-2. *Remote. Sens. Environ.* **2018**, *204*, 147–161. [CrossRef]

41. Lucas, R.; Rosenqvist, A.; Kellndorfer, J.; Hoekman, D.; Shimada, M.; Clewley, D.; Walker, W.; Navarro de Mesquita Junior, H. *Chapter 4: Global forest monitoring with Synthetic Aperture Radar (SAR) data, In Global Forest Monitoring from Earth Observation*; CRC Press, Taylor & Francis Group: Boca Raton, FL, USA, 2013; ISBN 9781466552012.

42. Armston, J.; Brown, S.; Calders, K.; Cutler, M.; Disney, M.; Endo, T.; Falkowski, M.; Goetz, S.; Herold, M.; Hirata, Y.; et al. Status of Evolving Technologies, GOFC-GOLD Sourcebook Section 2.10, Release: December 2016. GOFC-GOLD Land Cover Project Office, Wageningen University, The Netherlands. Available online: http://www.gofcgold.wur.nl/redd/sourcebook/GOFC-GOLD_Sourcebook.pdf (accessed on 6 October 2020).

43. Bouvet, A.; Mermoz, S.; Ballère, M.; Koleck, T.; Le Toan, T. Use of the SAR Shadowing Effect for Deforestation Detection with Sentinel-1 Time Series. *Remote. Sens.* **2018**, *10*, 1250. [CrossRef]

44. Hansen, J.N.; Mitchard, E.T.A.; King, S. Assessing Forest/Non-Forest Separability Using Sentinel-1 C-Band Synthetic Aperture Radar. *Remote. Sens.* **2020**, *12*, 1899. [CrossRef]

45. Cremer, F.K.A.; Urbazaev, M.; Cortes, J.; Truckenbrodt, J.; Schmullius, C.; Thiel, C.J. Potential of Recurrence Metrics from Sentinel-1 Time Series for Deforestation Mapping. *IEEE J. Sel. Top. Appl. Earth Obs. Remote. Sens.* **2020**, 1. [CrossRef]

46. Sica, F.; Pulella, A.; Rizzoli, P. Forest Classification and Deforestation Mapping by Means of Sentinel-1 InSAR Stacks. In Proceedings of the IGARSS 2019—2019 IEEE International Geoscience and Remote Sensing Symposium, Yokohama, Japan, 28 July–2 August 2019; pp. 2635–2638.

47. Quiñones, M.J.; Vissers, M.; Palacios, S.; Hettler, B.; Mancera, J.R. *Mapa de tipos estructurales de Amazonia occidental, frecuencias de inundación y cambios de cobertura de vegetación: 10 años de línea base para el estudio del bioma Amazónico*; Report Amazon Conservation Team and SarVision: Brasília, Brasil, 2019.

48. Quiñones, M.; Vissers, M.; Pacheco-Pascaza, A.M.; Flórez, C.; Estupiñán-Suárez, L.M.; Aponte, C.; Jaramillo, Ú.; Huertas, C.; Hoekman, D. Un enfoque ecosistémico para el análisis de una serie densa de tiempo de imágenes de radar Alos PALSAR, para el mapeo de zonas inundadas en el territorio continental colombiano. *Biota Colomb.* **2016**, *16*, 63–84. [CrossRef]

49. Flórez, C.; Estupiñán-Suárez, L.M.; Rojas, S.; Aponte, C.; Quiñones, M.; Acevedo, Ó.; Vilardy, S.; Jaramillo, Ú. Identificación espacial de los sistemas de humedales continentales de Colombia. *Biota Colomb.* **2016**, *16*, 44–62. [CrossRef]

50. Di Gregorio, A.; Jansen, L.J.M. *Land Cover Classification System (LCCS): Classification Concepts and User Manual Environment and Natural Resources Service, GCP/RAF/287/ITA Africover—East Africa Project and Soil Resources, Management and Conservation Service*; FAO: Rome, Italy, 2000.

51. Quinones, M.; Sartika, L.; Kooij, B. Brief Technical Report WWF-EWS Project Phase 2. Baseline Map 2015–2019 & Oil Palm Map 2018, Borneo. SarVision Report, WWF Contract, SarVision, Wageningen, The Netherlands. 2020. Available online: https://www.sarvision.nl/contact/ (accessed on 6 October 2020).

52. Roy, D.P.; Li, J.; Zhang, H.K.; Yan, L.; Huang, H.; Li, Z. Examination of Sentinel-2A multi-spectral instrument (MSI) reflectance anisotropy and the suitability of a general method to normalize MSI reflectance to nadir BRDF adjusted reflectance. *Remote. Sens. Environ.* **2017**, *199*, 25–38. [CrossRef]

53. Hoekman, D.; Reiche, J. Multi-model radiometric slope correction of SAR images of complex terrain using a two-stage semi-empirical approach. *Remote. Sens. Environ.* **2015**, *156*, 1–10. [CrossRef]

54. Torres, R.; Snoeij, P.; Geudtner, D.; Bibby, D.; Davidson, M.; Attema, E.; Potin, P.; Rommen, B.; Floury, N.; Brown, M.; et al. GMES Sentinel-1 mission. *Remote. Sens. Environ.* **2012**, *120*, 9–24. [CrossRef]

55. Quegan, S.; Yu, J.J. Filtering of multichannel SAR images. *IEEE Trans. Geosci. Remote. Sens.* **2001**, *39*, 2373–2379. [CrossRef]
56. Lopes, A.; Touzi, R.; Nezry, E. Adaptive speckle filters and scene heterogeneity. *IEEE Trans. Geosci. Remote. Sens.* **1990**, *28*, 992–1000. [CrossRef]
57. Hoekman, D.H. Speckle ensemble statistics of logarithmically scaled data (radar). *IEEE Trans. Geosci. Remote. Sens.* **1991**, *29*, 180–182. [CrossRef]
58. Reiche, J.; De Bruin, S.; Hoekman, D.; Verbesselt, J.; Herold, M. A Bayesian Approach to Combine Landsat and ALOS PALSAR Time Series for Near Real-Time Deforestation Detection. *Remote. Sens.* **2015**, *7*, 4973–4996. [CrossRef]
59. Schlund, M.; Von Poncet, F.; Kuntz, S.; Schmullius, C.; Hoekman, D.H. TanDEM-X data for aboveground biomass retrieval in a tropical peat swamp forest. *Remote. Sens. Environ.* **2015**, *158*, 255–266. [CrossRef]
60. Ferraz, A.; Saatchi, S.S.; Xu, L.; Hagen, S.C.; Chave, J.; Yu, Y.; Meyer, V.; Garcia, M.; Silva, C.A.; Roswintiart, O.; et al. Carbon storage potential in degraded forests of Kalimantan, Indonesia. *Environ. Res. Lett.* **2018**, *13*, 095001. [CrossRef]
61. Varekamp, C.; Hoekman, D.H. High-resolution InSAR image simulation for forest canopies. *IEEE Trans. Geosci. Remote. Sens.* **2002**, *40*, 1648–1655. [CrossRef]
62. Haralick, R. Ridges and Valleys on Digital Images. *Comput. Vis. Graph. Image Process.* **1983**, *22*, 28–38. [CrossRef]
63. López, A.M.; Lumbreras, F.; Serrat, J.; Villanueva, J. Evaluation of methods for ridge and valley detection. *IEEE Trans. Pattern Anal. Mach. Intell.* **1999**, *21*, 327–335. [CrossRef]

 *remote sensing*

*Article*

# A Near Real-Time Method for Forest Change Detection Based on a Structural Time Series Model and the Kalman Filter

Martin Puhm *, Janik Deutscher, Manuela Hirschmugl, Andreas Wimmer, Ursula Schmitt and Mathias Schardt

Joanneum Research, Institute for Information and Communication Technologies, Steyrergasse 17, 8010 Graz, Austria; janik.deutscher@joanneum.at (J.D.); manuela.hirschmugl@joanneum.at (M.H.); andreas.wimmer@joanneum.at (A.W.); ursula.schmitt@joanneum.at (U.S.); mathias.schardt@joanneum.at (M.S.)
* Correspondence: martin.puhm@joanneum.at; Tel.: +43-316-876-5107

Received: 16 July 2020; Accepted: 22 September 2020; Published: 24 September 2020

**Abstract:** The increasing availability of dense time series of earth observation data has incited a growing interest in time series analysis for vegetation monitoring and change detection. Vegetation monitoring algorithms need to deal with several time series characteristics such as seasonality, irregular sampling intervals, and signal artefacts. While common algorithms based on deterministic harmonic regression models account for intra-annual seasonality, inter-annual variations of the seasonal pattern related to shifts in vegetation phenology due to different temperature and rainfall are usually not accounted for. We propose a transition to stochastic modelling and present a near real-time change detection method that combines a structural time series model with the Kalman filter. The model continuously adapts to new observations and allows to better separate phenology-related deviations from vegetation anomalies or land cover changes. The method is tested in a forest change detection application aiming at the assessment of damages caused by storm events and insect calamities. Forest changes are detected based on the cumulative sum control chart (CUSUM) which is used to decide if new observations deviate from model-based forecasts. The performance is evaluated in two test sites, one in Malawi (dry tropical forest) and one in Austria (temperate deciduous, coniferous and mixed forests) based on Sentinel-2 time series. Both forest areas are characterized by a distinct, but temporally varying leaf-off season. The presented change detection method shows overall accuracies above 99%, users' accuracies of 76.8% to 88.6%, and producers' accuracies of 68.2% to 80.4% for the forest change stratum (minimum mapping unit: 0.1 ha). Results are based on visually interpreted points derived by stratified random sampling. A further analysis revealed that increasing the time series density by merging data from two Sentinel-2 orbits yields better forest change detection accuracies in comparison to using data from one orbit only. The resulting increase in users' accuracy amounts to 7.6%. The presented method is capable of near real-time processing and could be used for a variety of automated forest monitoring applications.

**Keywords:** state space models; forest disturbance mapping; near real-time monitoring; Sentinel-2; CUSUM

---

## 1. Introduction

Current Earth observation (EO) missions employing optical sensors such as Sentinel-2 acquire a vast volume of data: a new image every 5 days of almost every place on earth. By taking orbit overlaps into account, the time between consecutive images of the same region is reduced even further and the chance of acquiring cloud-free observations is further increased. Through high-quality georeferencing and atmospheric correction of the satellite images, it is possible to create consistent time series of measured reflectance values for any given spectral band. The vast availability of high-resolution optical data allows—for the first time—to also map small changes in near real-time. Here, "small" may apply to both spatial extent as well as spectral change magnitude. However, dense time series of high-resolution optical data have a number of characteristics that pose a challenge to change detection applications. In addition to noise effects remaining after atmospheric correction and uncertainties in the geometric registration, these challenging characteristics include:

- Seasonality: Recurring seasonal patterns can be attributed to plant phenology and/or varying illumination conditions due to topography and solar angle. Considering that annual variations of temperature and rainfall also affect phenology, the seasonal patterns can vary between years.
- Irregular sampling interval: Satellites with a regular nadir acquisition scheme usually have a constant revisit cycle, for example, 5 days for the Sentinel-2A and B constellation. If images from overlapping orbits are integrated, though, the sampling interval becomes irregular. Data gaps due to masked clouds, cloud shadows, and snow also add to this irregularity.
- Presence of signal artefacts: Despite using state-of-the-art screening algorithms like Fmask [1], un-masked clouds, cloud shadows, and snow-covered areas remain in the pre-processed imagery. Corresponding observations have to be treated as invalid since the measured reflectance values neither represent the undisturbed land cover state nor a persistent change of it.

Algorithms can be divided based on how they deal with the time series characteristics described above. The different approaches used to handle these characteristics strongly affect the algorithms' suitability for monitoring changes in near real-time. Approaches that do not account for seasonality form a first group of algorithms. A review of change detection studies using Landsat time series concludes that many older studies focused on mapping changes only at annual or biannual time scales based on series of cloud-free composite images, which always represent the same season [2]. In this context, both the Vegetation Change Tracker (VCT) [3] as well as the LandTrendr approach [4] represent widely used algorithms, but they are not designed for near real-time mapping.

A second group of algorithms explicitly accounts for seasonality by using regression models based on trigonometric functions to capture the intra-annual variations (variations within one year, i.e., seasonality) of the spectral signatures independently for each pixel. With this approach, also frequently referred to as harmonic regression, periods of stable land cover are modelled as a deterministic, continuous function of time. Irregular sampling intervals and data gaps are therefore not a problematic issue, but the deterministic nature of the model does not allow inter-annual variations of the seasonal pattern (variations between different years, e.g., shifts in seasonality). The seasonal model represents an average of different conditions occurring within a stable period, e.g., dry and wet years, late and early leave outbreak. A widely used algorithm belonging to this group is the Breaks for Additive Season and Trend (BFAST) algorithm [5] and its evolution BFAST Monitor [6]. While the latter is tailored to near real-time mapping of new changes, the original version is intended for the analysis of historic time series. Both versions have been used in a variety of studies and can be applied to detect both abrupt and gradual changes. Concerning the robustness of BFAST to invalid observations, it has been stated that occasional signal artefacts are well handled, but temporally aggregated occurrences such as several consecutively un-masked clouds can be a source of error [7]. Also, additional pre-processing to eliminate artefacts was applied [8–10]. The second widely used implementation of the harmonic regression approach is the Continuous Change Detection and Classification algorithm (CCDC), where the original concept [11] is extended to include more types of land cover besides forest, as well as a classification framework [12]. From the beginning, CCDC was designed to work with dense Landsat time series and can handle seasonality, irregularly spaced observations, and signal artefacts to some extent. Both abrupt and gradual changes can be detected. Some further updates to the algorithm include (i) a mechanism to automatically adjust the complexity of the time series model based on the number of available clear observations, as well as (ii) a different method to estimate the model parameters which reduces overfitting [13]. A third algorithm employing harmonic regression utilizes residuals from the regression together with statistical quality control charts [14]. This approach comprises signal artefact detection with Shewhart X-bar charts. After the elimination of artefacts, both abrupt and gradual changes are indicated by exponentially weighted moving average (EWMA) charts in a near real-time manner. All of the described algorithms of the second group share certain basic concepts, but the individual implementations vary. They are designed to process large amounts of data in a highly automated way and therefore rely on data-driven statistical boundaries for detecting change, although the distinct nature and computation of these boundaries is quite different.

The third group of algorithms stands out by also taking inter-annual variations of the seasonal pattern into account. Structural time series models are set up in terms of components, such as trends and cycles, which have a direct interpretation. Their statistical treatment is based on the state space form and the Kalman filter and first described for time series analysis in econometrics [15]. Compared to the harmonic regression approach, the model is no longer deterministic. Kalman filtering denotes a versatile parameter estimation technique which yields optimal estimates in a statistical sense [16]. It is well established in many application fields and has been applied to numerous signal tracking problems [17]. The combination of structural time series models with the Kalman filter and the concept's suitability for remote sensing purposes has been investigated in a proof of concept study, where Landsat time series are used to detect storm damages in a small forest test site in Germany [18]. The Kalman filter has also been applied to time series of MODIS 8-day composites in order to detect insect-induced defoliation in near real-time at a forest test area in northern Sweden [19]. This study makes use of the CUSUM control chart [20] to indicate changes, but it does not combine structural time series models with the Kalman filter. Instead, the filter is used to derive a smoothed time series of the Normalized Difference Vegetation Index (NDVI) based on a global model trajectory.

With the advantages and limitations of existing algorithms in mind, this work combines the pixel-by-pixel modelling typical for existing harmonic regression algorithms with the Kalman filter's capability to dynamically adjust the model based on new observations. The main aim of this study is to present an innovative change detection algorithm for optical EO data which is based on a structural time series model and the Kalman filter. It is largely data-driven and designed especially for near real-time mapping in web- or cloud-based monitoring services. The algorithm presented in this paper accounts for seasonality and also allows inter-annual variations of the seasonal pattern, e.g., vegetation phenology. Furthermore, strategies for handling irregular sampling intervals and signal artefacts are presented. The algorithm is tested in a forest change detection application using time series of Sentinel-2 data (S-2). Forest disturbances are detected at two complex forest test sites in Austria and Malawi. The first test site is an alpine area in Austria characterized by frequent cloud cover, snow cover, strong topographic effects, and pronounced forest seasonality and phenology. The second test site is located in the dry tropical forests of Malawi, where forests show a strong and varying seasonality between dry and rainy seasons. In the Malawi test site, we also analyze and compare accuracy results for two different data scenarios: first, using S-2 images from only one orbit, and second, using all available S-2 imagery from two orbits. The aim of this analysis is to investigate if different viewing angle and inconsistent geo-location of the pixels resulting from the combination of two orbits decrease the overall change detection accuracies despite the boosted time series density.

## 2. Materials and Methods

### 2.1. Test Sites, Data, and Pre-Processing

For the forest change detection demonstration, we selected two test sites, one in Austria and one in Malawi. The location of the two sites is shown in Figure 1. The Austrian test site is located in the south-eastern part of the country. The test site is characterized by strong topography in the northern part (Alpine area), where coniferous forests dominate. The southern part of the test site is located in the foothills of the Alps with moderate topography and the forests are predominantly mixed forests composed of coniferous and deciduous trees. The annual temperature amplitude in the Alpine area is stronger than in the foothills; however, the Alpine coniferous forests have less pronounced phenological dynamics in comparison to the mixed and deciduous forests that dominate in the foothills. The eastern half of the study area is covered by two orbits, and the western half only by one orbit (see Figure 1). This is a typical data scenario encountered in practical applications.

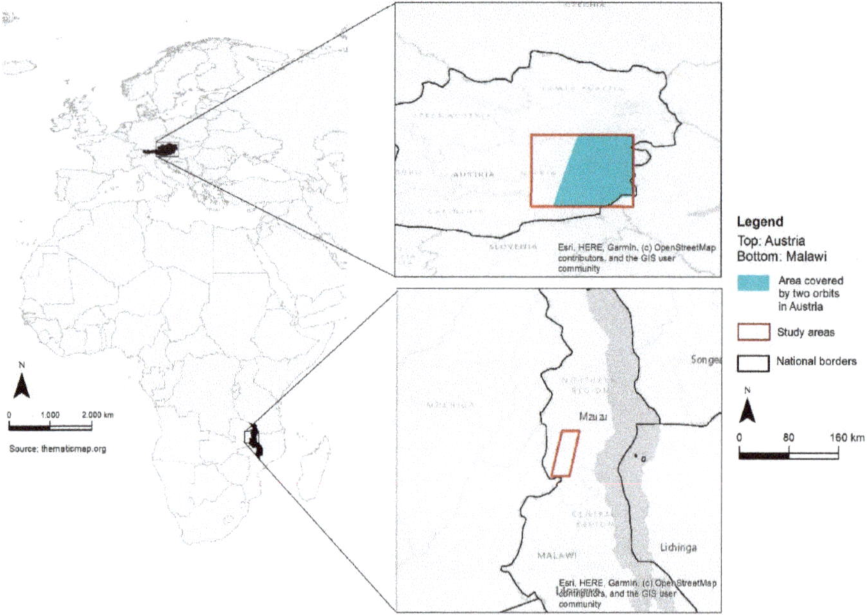


**Figure 1.** Location of the test sites.

The Malawi test site is characterized by flat to slightly hilly topography. The land cover is very heterogeneous and is subject to a precipitation gradient from east to west. As a result, the area is characterized by a vegetation phenology gradient from east to west. The forested areas differ in tree-cover density and tree-type composition and therefore, show very different spectral behavior. Dry tropical forests and the surrounding land use classes are more difficult to classify and monitor than humid evergreen forests, as they show a typical phenological development from highly vital in the rainy season to dry and leafless in the dry season [21]. Understory and grassland fires beneath the forest canopy can further complicate forest classification. Two S-2 orbits (relative orbits 92 and 135) cover the Malawi study area. The size of the test site was clipped to the overlap area of two orbits to investigate the effect of separately processing data from one and two orbits.

Both test sites are located in areas that show distinct seasonal patterns due to phenology. Austria has the typical European summer growing season with leaf-off time in winter for deciduous species due to low temperatures. The forests of Malawi also show strong phenological variation as water scarcity during the dry season (typically May–October) causes leaf-fall. The typical temporal NDVI signatures of different forest types are shown in Figure 2. Each time series represents two years of NDVI observations for a single pixel corresponding to a specific forest type and test site. In all cases, data from two orbits is used and cloud/snow masking has been applied as described below. Aside from the different seasonal patterns, several other characteristic properties of the data can be observed:

- The time series density achieved with the Sentinel-2 constellation and overlapping orbits has a high potential for near real-time monitoring applications.
- Larger data gaps occur during winter (snow cover in Austria) or the rainy season (frequent cloud cover in Malawi).
- The illustration also gives an impression of the smoothness of the time series. Significant short-term variance of consecutive observations caused by limited multi-temporal geometric registration accuracy [22], different viewing angles, and remaining atmospheric effects has to be expected.

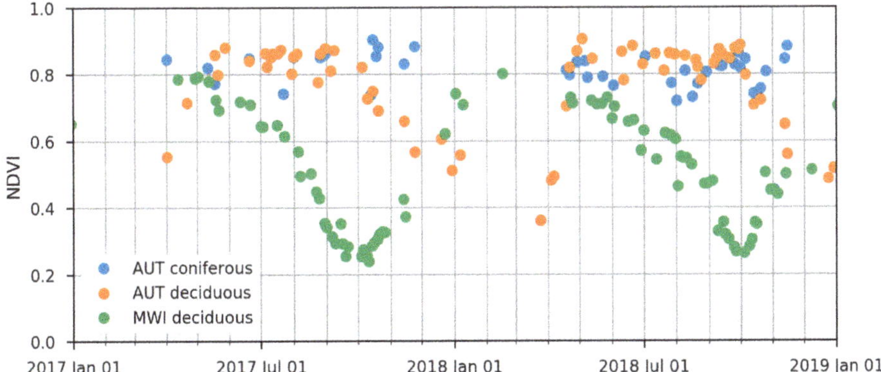

**Figure 2.** Typical temporal NDVI (Normalized Difference Vegetation Index) signatures over two years for single pixels corresponding to specific forest types and test sites. Larger data gaps occur in the winter months due to snow cover in Austria (AUT) or the rainy season in Malawi (MWI). Outside of these periods, the observation density is high enough to capture the seasonal patterns.

The current data quality report issued by the Sentinel-2 Mission Performance Centre (S2 MPC) gives statistics for the multi-temporal geometric registration performance [22]. For about half of the images, the co-registration error is larger than 0.5 pixels at 10 m resolution. Within homogenous land-cover areas, this can be treated as an additional noise component. The high geometric uncertainty becomes a larger problem at the border regions between land-cover classes, especially if a given pixel jumps between forested and non-forested states.

All available Sentinel-2 scenes with a nominal cloud cover below 90% were downloaded for the test sites (Table 1) at Level-1C and atmospherically corrected to surface reflectance (SR) values using the Sen2Cor processor version 2.5.5 [23]. We then resample the 20 m bands to 10 m spatial resolution and stack all bands to a 10-band output image. We calculate a combined cloud, cloud shadow, and snow mask with the FMask algorithm [24,25]. This mask is slightly altered by morphological operations (erode, expand) to fill cloud holes and the masked pixels are then removed from the pre-processed S-2 imagery by assigning no-data values to them. We also perform a topographic correction based on a modified Minnaert correction [26] using the Shuttle Radar Topography Mission (SRTM) model at 30 m spatial resolution as digital elevation model (DEM).

**Table 1.** Earth ObservationData Information.

| Test Site | Tile ID | No. of Images Used | Time Windows | |
| --- | --- | --- | --- | --- |
| | | | Initialization | Change Detection |
| Malawi—one orbit (R135) | 36LWM | 86 | 2016–2018 | 2019 |
| Malawi—two orbits (R135 and R092) | 36LWM | 134 | 2016–2018 | 2019 |
| Austria—two orbits (R122 and R079) | 33TWN | 160 | 2016–2018 | 2019 |

*2.2. Change Detection Method Using a Structural Time Series Model and the Kalman Filter*

The underlying assumption of the monitoring approach presented in this section is that the normal temporal trajectory of a given spectral band over the course of the year can be captured by a univariate structural time series model. Within the structural model, we can further distinguish between an observation sub-model and a dynamic sub-model. The observation sub-model on the one hand defines the relationship of the measurements to a set of state variables which cannot be observed directly. In a structural time series model, the state variables usually represent the series' additive decomposition into trend, seasonal, and long-term cyclical components. The dynamic sub-model on the other hand describes the expected temporal evolution of the state variables. By formulating the dynamic sub-model in continuous time, the problem of irregular sampling intervals is addressed. The Kalman filter is used to fit the model to the data and operates recursively from one point in time to the next. Each recursion may be divided into two steps. In the time-update step, the states' temporal evolution is predicted based on the dynamic sub-model. It is followed by the measurement update step, where the predicted state estimate is enhanced by incorporating newly available observations. Abrupt changes of the spectral signature, possibly linked to a forest disturbance, are indicated by statistically significant deviations between new observations and the Kalman filter predictions. The recursive operation of the filter further implies that prior knowledge about the initial state is required. Because the ability to distinguish anomalies from normal seasonal changes depends on the quality of the state estimates, a proper initialization is of high importance. Therefore, a robust least squares method is used to estimate the initial state from a historic time series covering at least one full year prior to the beginning of the monitoring period (see initialization time window in Table 1). The following sub-sections describe the implementation in more detail.

### 2.2.1. Time Series Model

Structural time series models are mathematically formulated using the discrete-time state space representation [15]. This concept assumes that a linear, time-variant system can be described by a set of state variables. Because these variables can usually not be observed directly, an observation sub-model linking the system state to a set of measurements is required. In case of univariate time series, the measurement equation is

$$z_k = \mathbf{h}_k \mathbf{x}_k + r_k, \tag{1}$$

where $\mathbf{h}_k$ is a row vector, $\mathbf{x}_k$ denotes the state vector, and $z_k$ is a scalar observation made at time $t_k$. In addition, $r_k$ represents serially uncorrelated observation noise with mean zero and variance $R_k$. We will refer to discrete points in time as epochs and define an epoch index $k \in \mathbb{N}$ to indicate a possible time-dependency of variables. The temporal evolution of the state vector is described by a dynamic sub-model using the transition equation:

$$\mathbf{x}_k = \mathbf{\Phi}_k \mathbf{x}_{k-1} + \mathbf{q}_k, \tag{2}$$

where $\mathbf{\Phi}_k$ denotes the transition matrix and $\mathbf{q}_k$ is a vector of serially uncorrelated process noise with mean zero and covariance matrix $\mathbf{Q}_k$. Note that no assumptions regarding the distributions of the observation and process noises are made at this point, but they are supposed to be uncorrelated with each other in all epochs. It is further assumed that the initial state $\mathbf{x}_0$ is known with a level of uncertainty characterized by the state error covariance matrix $\mathbf{P}_0$.

The structural time series model implemented in this study enables a decomposition of the time series into additive trend, seasonal, and irregular components. We assume a constant trend and directly introduce the state variable $\mu_k$, which represents the trend level at time $t_k$. The seasonal component on the other hand is constructed by a sum of $P$-periodic cosine waves, with $P$ denoting the fundamental duration of the seasonal cycle. Considering the nature of the time series investigated here, it is appropriate to measure time in days and thus set $P$ to 365.25. For each cosine wave added to the seasonal component, two more variables are appended to the state vector. Taking a wave with frequency $\omega_i$, the state variable $\gamma_{i,k}$ representing the cosine waves' level at time $t_k$ is added along with the variable $\gamma_{i,k}^*$, whose interpretation is not particularly important. The structural model for the Malawi test site features a seasonal component with frequencies corresponding to one and two periods per year. In case of the Austrian test site, we use just the fundamental frequency because of the less complex seasonal pattern. The measurement Equation (1) of the Malawi model becomes

$$z_k = \begin{bmatrix} 1 & 1 & 0 & 1 & 0 \end{bmatrix} \begin{bmatrix} \mu \\ \gamma_1 \\ \gamma_1^* \\ \gamma_2 \\ \gamma_2^* \end{bmatrix}_k + r_k \tag{3}$$

Note that $z_k$ represents the observed reflectance value at time $t_k$ and $\mathbf{h}$ is time-invariant. Using $\Delta t_k = t_k - t_{k-1}$ measured in days and $\omega_i = 2\pi i/365.25$, the transition Equation (2) of the Malawi model becomes

$$\begin{bmatrix} \mu \\ \gamma_1 \\ \gamma_1^* \\ \gamma_2 \\ \gamma_2^* \end{bmatrix}_k = \begin{bmatrix} 1 & 0 & 0 & 0 & 0 \\ 0 & \cos(\omega_1 \Delta t_k) & \sin(\omega_1 \Delta t_k) & 0 & 0 \\ 0 & -\sin(\omega_1 \Delta t_k) & \cos(\omega_1 \Delta t_k) & 0 & 0 \\ 0 & 0 & 0 & \cos(\omega_2 \Delta t_k) & \sin(\omega_2 \Delta t_k) \\ 0 & 0 & 0 & -\sin(\omega_2 \Delta t_k) & \cos(\omega_2 \Delta t_k) \end{bmatrix} \begin{bmatrix} \mu \\ \gamma_1 \\ \gamma_1^* \\ \gamma_2 \\ \gamma_2^* \end{bmatrix}_{k-1} + \begin{bmatrix} q_\mu \\ q_{\gamma_1} \\ q_{\gamma_1^*} \\ q_{\gamma_2} \\ q_{\gamma_2^*} \end{bmatrix}_k \tag{4}$$

In order to fully specify the state space model, two more quantities need to be defined: $R_k$ and $Q_k$. The first one will be addressed in the next sub-section. The covariance matrix $\mathbf{Q}_k$ quantifies the uncertainties within the dynamic sub-model and is defined depending on the nature of the process noise. Here, we assume that the state transition is affected by multivariate, continuous time white noise with constant variances $Q_c$ independently specified for the trend and seasonal component, that is:

$$\mathbf{Q}_k = \Delta t_k \begin{bmatrix} Q_c^{\text{trend}} & 0 & 0 & 0 & 0 \\ 0 & Q_c^{\text{seas}} & 0 & 0 & 0 \\ 0 & 0 & Q_c^{\text{seas}} & 0 & 0 \\ 0 & 0 & 0 & Q_c^{\text{seas}} & 0 \\ 0 & 0 & 0 & 0 & Q_c^{\text{seas}} \end{bmatrix} \tag{5}$$

Note that all off-diagonal elements of $\mathbf{Q}_k$ equal zero and we therefore assume that the different process noise components are uncorrelated. Implementing the dynamic model in continuous time means that the time series model is specified for arbitrary values of $\Delta t_k$ and irregular sampling intervals do not pose a problem. Finding appropriate values for $Q_c$ is one of the difficulties. Our experiments showed that setting individual values for each pixel and band proportional to the respective observation variance $R$ works quite well. Since $R$ is estimated from the data (see the next sub-section), the user has to specify two proportionality factors, one for the trend component and one for the seasonal component. We recommend to let $Q_c^{\text{seas}}$ be larger than $Q_c^{\text{trend}}$ in order to make the time series model more responsive to phenological variations.

### 2.2.2. Initial State Estimation

In order to obtain estimates of the initial state vector $\mathbf{x}_0$, its associated covariance matrix $\mathbf{P}_0$, and the observation variance $R$, the state space model outlined above is transformed to a linear regression model with the measurement equation:

$$
\begin{bmatrix} z_1 \\ z_2 \\ \vdots \\ z_m \end{bmatrix} = \begin{bmatrix} \mathbf{h}\Phi(t_1, t_0) \\ \mathbf{h}\Phi(t_2, t_0) \\ \vdots \\ \mathbf{h}\Phi(t_m, t_0) \end{bmatrix} \mathbf{x}_0 + \begin{bmatrix} r_1 \\ r_2 \\ \vdots \\ r_m \end{bmatrix} \tag{6}
$$

Equation (6) also illustrates the key difference between a structural time series model and a regression model. The latter does not distinguish between observation sub-model and dynamic sub-model and has no notion of process noise. Standard least squares methods may be used to obtain estimates of $\mathbf{x}_0$ and $\mathbf{P}_0$ from a batch of $m$ historic observations acquired before the monitoring period. We applied a robust parameter estimation approach known as iteratively reweighted least squares (IRLS). The technique belongs to the class of $M$-estimators [27] and the implementation follows [28]. We further use the mean squared error of the weighted least squares fit as an estimate for the observation variance $R$. If the number of valid historic observations is low, for example in areas with extremely high cloud probability, the observation variance can be severely underestimated. Existing studies applying harmonic regression also report this issue and we take up the suggestion to define a certain minimum value for $R$ that is used if the estimate is lower [12,29].

### 2.2.3. Kalman Filter

Once the time series model and the initial values $\mathbf{x}_0$ and $\mathbf{P}_0$ are defined, the discrete-time Kalman filter recursion can be used to obtain estimates for the state and its error covariance matrix in subsequent epochs. The time update step yields the predicted (a-priori) estimates $\tilde{\mathbf{x}}_k$ and $\tilde{\mathbf{P}}_k$ based on the dynamic model and the previous estimates at time $t_{k-1}$:

$$
\tilde{\mathbf{x}}_k = \Phi_k \hat{\mathbf{x}}_{k-1} \tag{7}
$$

$$
\tilde{\mathbf{P}}_k = \Phi_k \hat{\mathbf{P}}_{k-1} \Phi_k^T + \mathbf{Q}_k \tag{8}
$$

Then, the a-priori measurement residual $y_k$ and its variance $C_k$ is computed according to Equations (9) and (10). The residual represents the difference of the prediction to the actual measurement and is referred to as innovation, since it contains new information currently not present in the predicted state [17].

$$
y_k = z_k - \mathbf{h}\tilde{\mathbf{x}}_k \tag{9}
$$

$$
C_k = \mathbf{h}\tilde{\mathbf{P}}_k \mathbf{h}^T + R \tag{10}
$$

In the measurement update step of each recursion, new information is merged with the predictions to obtain improved (a-posteriori) estimates $\hat{\mathbf{x}}_k$ and $\hat{\mathbf{P}}_k$. The Kalman gain $\mathbf{k}_k$ (11) determines how much the newly acquired measurement will influence the a-posteriori estimates of the state and its error covariance and appears in both update Equations (12) and (13), where $\mathbf{I}$ represents an identity matrix.

$$
\mathbf{k}_k = \tilde{\mathbf{P}}_k \mathbf{h}^T C_k^{-1} \tag{11}
$$

$$
\hat{\mathbf{x}}_k = \tilde{\mathbf{x}}_k + \mathbf{k}_k y_k \tag{12}
$$

$$
\hat{\mathbf{P}}_k = (\mathbf{I} - \mathbf{k}_k \mathbf{h}_k) \tilde{\mathbf{P}}_k \tag{13}
$$

### 2.2.4. Signal Artefact Handling and Change Detection

When the sequence of measurements processed by the filter may contain artefacts, an additional anomaly detection step should be included before the measurement update. The properties of the innovations can be exploited to detect anomalous measurements by means of a statistical test. Provided that the underlying model assumptions are valid, and the observation noise is Gaussian, the innovations should be normally distributed with mean zero and variance $C_k$. The test statistic $\hat{T}_k$ given in (14) follows the $\chi^2$-distribution with a single degree of freedom. The hypotheses to be tested on a significance level $\alpha$ are stated in (15) and (16), respectively:

$$\hat{T}_k = \frac{y_k^2}{C_k}, \quad \hat{T}_k \sim \chi^2 \tag{14}$$

$$H_0 : \ y_k = 0 \text{ if } \hat{T}_k \leq \chi^2_{1,\,1-\alpha} \tag{15}$$

$$H_1 : \ y_k \neq 0 \text{ otherwise} \tag{16}$$

Considering that anomalous observations will cause large innovations, the null hypothesis will be rejected. In that case, the measurement update is not carried out so that the state estimate remains unbiased.

The change detection part of the algorithm is based on the cumulative sum control chart (or CUSUM) [20]. The key quantity here is again the sequence of filter innovations, which should have zero mean if the Kalman filter model assumptions reflect the truth closely enough. We use the CUSUM control chart to monitor if the innovations deviate from mean zero. Of course, the presence of artefacts like un-masked clouds presents a problem in this regard, because a single innovation corresponding to an artefact may shift the mean significantly and hence trigger a (inaccurate) change signal. The previously described anomaly test cannot distinguish between an artefact and an abrupt land cover change. To work around this limitation, a new quantity we call edited innovation $\check{y}_k$ is introduced. It represents the original innovation divided by its standard deviation, but also limited in magnitude based on the significance level $\alpha$ specified for the anomaly test, that is:

$$\check{y}_k = \begin{cases} \min\left(\frac{y_k}{\sqrt{C_k}}, \ \sqrt{\chi^2_{1,1-\alpha}}\right) \text{ if } y_k \geq 0 \\ \max\left(\frac{y_k}{\sqrt{C_k}}, \ -\sqrt{\chi^2_{1,1-\alpha}}\right) \text{ if } y_k < 0 \end{cases}. \tag{17}$$

Dividing by the standard deviation means that sequences of edited innovations should be close to having unit variance across different bands and pixels. The limit operation on the other hand ensures that the CUSUM control chart with $\check{y}_k$ as input is less sensitive to single statistical outliers. A temporal aggregation of edited innovations with the same sign is required to shift the mean of the sequence significantly. The CUSUM test statistic for a positive mean shift with respect to spectral band $i$ is implemented as:

$$\begin{aligned} S_i^+(0) &= 0 \\ S_i^+(k) &= \max(0, S_i^+(k-1) + \check{y}_i(k) - d) \end{aligned} \tag{18}$$

where $d$ is a drift parameter which generally compensates small deviations and also ensures that effects of occasional signal artefacts on the test statistic fade away over time. A mean shift is signaled if the test statistic crosses a predefined threshold. Instead of evaluating all processed bands separately, we decided to aggregate the test statistics of several bands by simple summation and then use a global threshold. We look for anomalous reflectance increases in the red, red edge, and short wave infrared

(SWIR) bands, as these have proven to be sensitive to vegetation changes [30]. The implemented criterion for detecting a forest disturbance at time $t_k$ is given in Equation (19) using S-2 band numbers.

$$\sum_i S_i^+(k) > change\ threshold \text{ where } i \in \{B04,\ B05,\ B11,\ B12\} \tag{19}$$

Both the change threshold in (19) and the drift parameter in (18) are user-defined tuning parameters. Appropriate values need to be determined empirically. Because of the limit operation described by (17), the maximum of a single edited innovation is a known constant and amounts to ~2.57 if $\alpha = 1\%$. This knowledge provides a helpful yardstick for setting both threshold and drift parameters. The statistical normalization applied to the edited innovations means that the same values can be used globally for all pixels. The flowchart depicted in Figure 3 illustrates how the methods discussed in the preceding sections are joined together in order to create a data-driven algorithm capable of detecting abrupt changes on the pixel level. A summary and some additional explanatory comments are given below.

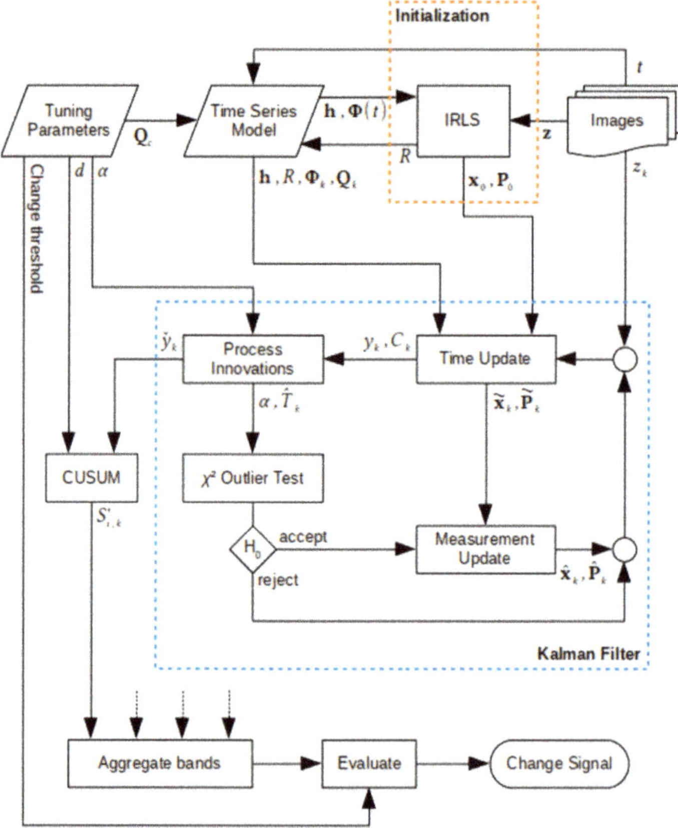


**Figure 3.** Flow chart of the Kalman filter approach.

1.  The initial state as well as the observation noise are estimated on the pixel level using the iteratively reweighted least squares (IRLS) method. Therefore, the user has to supply a stack of historic images as a training dataset. At least one year of observations is required, but, especially for models using two seasonal frequencies, we recommend two or three years to ensure a stable initial model.
2.  New images acquired in the monitoring period are processed one-at-a-time in a Kalman filter loop. A hypothesis test is used to identify anomalous observations showing significant deviations to the prediction.
3.  Whenever an observation is marked as anomalous, the measurement update step is bypassed in order to avoid adverse influence of signal artefacts on the state estimates.
4.  Spectral bands are processed in parallel and the aggregated CUSUM test statistic used for change detection is evaluated after processing each new image.

Figure 4 visualizes the method for a forest change pixel (10 by 10 m) in a deciduous forest in the Austrian test site. The first four sub-plots show the surface reflectance of Sentinel-2 bands 4 (red), 5 (red edge), 11 (SWIR1), and 12 (SWIR2) over time. The sub-plot at the bottom shows the value of the CUSUM test statistic defined in Section 2.2.4 over time. A phenological cycle that is typical for deciduous tree species can be observed. Higher reflectance values correspond to the leaf-off season during winter. The red line represents the Kalman fit and the grey area corresponds to the 90% confidence interval of the model forecast. While blue observations are considered in the model and measurement update step (compare Figure 3), the orange observations are flagged as anomalous and therefore ignored during the measurement update step. Please note that the level of significance for the anomaly test is $\alpha = 1\%$, hence blue observations may also appear outside of the plotted confidence interval. Occasional positive anomalies will cause a short-lived increase of the related CUSUM, but in most cases, it should not exceed the threshold (e.g., middle of 2016 or end of 2017 in Figure 4). Such anomalies occur if for example small cloud artefacts remain in the pre-processed data. A prolonged increase of the CUSUM is triggered by a persistent signal shift. At some point, the threshold is exceeded, and the pixel is flagged as changed, in this case, at the end of 2018. The vertical dashed line in Figure 4 marks the date on which the change is visible for the first time in the S-2 imagery. Note that anomalies with significant magnitude are first detected only in the SWIR bands, while the signal shift in the red and red edge band initially occurs within the 90% confidence interval; however, the aggregated CUSUM test statistic allows a timely detection of the change. After a threshold crossing, the test statistic is reset to zero. Repeated change alerts, as shown in this example, may thus occur. This could be used to increase the confidence about a detected change at the cost of delaying the detection; however, this aspect has not been investigated in detail in this work. For the resulting forest change maps at the two test sites, we first include all Sentinel-2 single pixel changes (10 by 10 m) that were detected by the described approach within the change detection time window. For the final forest change maps, a minimum mapping unit of 0.1 ha is applied to the forest change stratum, which relates to change areas represented by at least ten connected 10 m pixels. Detected forest change areas smaller than 10 pixels are removed from the final forest change maps.

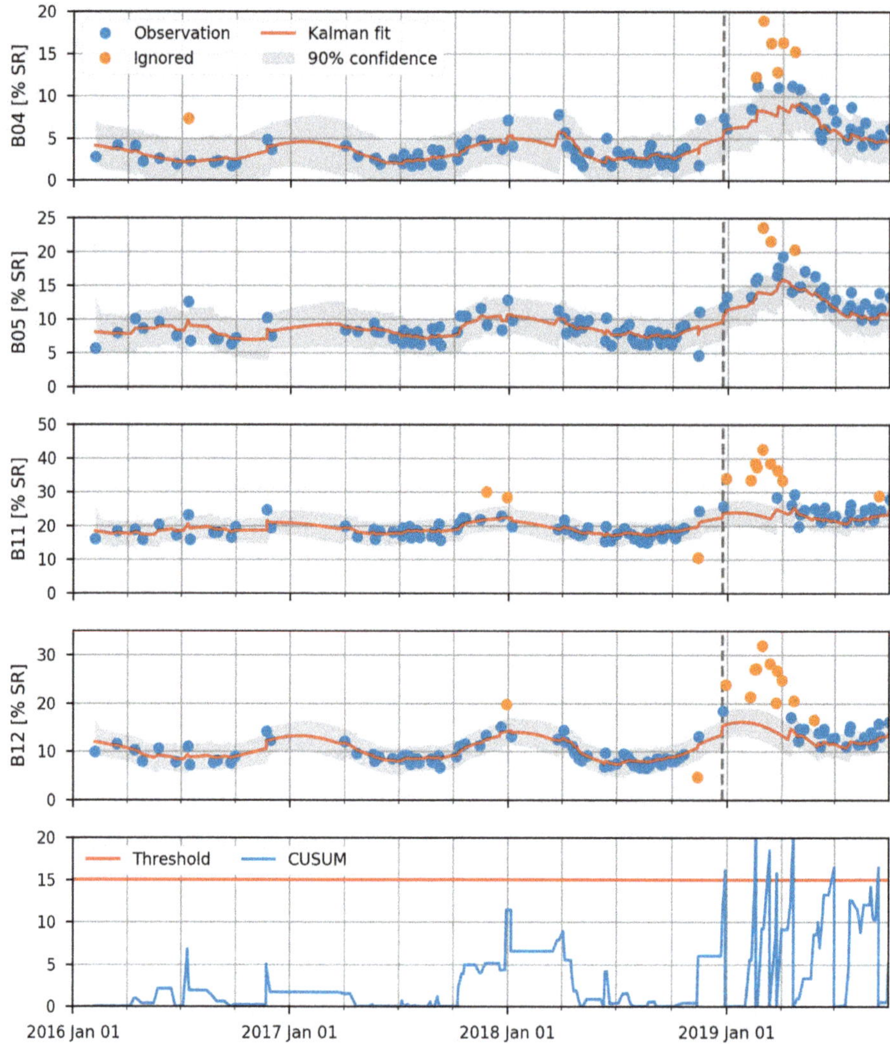



**Figure 4.** Example of the time series model for a deciduous forest pixel at the Austrian test site, where a forest change occurred at the end of 2018. The plots show the surface reflectance of Sentinel-2 bands 4, 5, 11, and 12. The vertical dashed line marks the date on which the change is visible for the first time in the S-2 image time series. The plot at the bottom shows the cumulative sum control chart (CUSUM) used for detecting the change event (coordinates: X 559045 m, Y 5235095 m, in EPSG 32633).

## 2.3. Evaluation Method

In order to test the operational application of the presented forest change detection method, a full area-based validation of the resulting change maps is performed in both test sites based on stratified random sampling points that are located within forest areas outlined by benchmark forest masks (status 2016 for Malawi and summer 2018 for Austria). The overall number of samples and samples per stratum are based on recommendations for land cover accuracy estimation [31,32]. Points that were found to be already deforested before the beginning of the change detection time window were excluded from the analysis since these cases correspond to errors of the initial forest masks.

For Malawi, 849 reference sample points are used for statistical analysis, with 735 sample points belonging to stratum forest and 114 sample points belonging to stratum change. During blind interpretation, we flagged all sample points that were located at or very close (~10 m) to the border of a forest change patch. This allows us to treat these points as correctly classified in a subsequent plausibility analysis. This approach is usually termed "plausibility analysis", because it is deemed plausible that the border point can also belong to the other stratum. This is especially true for the small-scale 10 by 10 m change assessment we use in this study considering that some of the spectral bands of Sentinel-2 used to derive the changes only have a nominal spatial resolution of 20 m. For Malawi, assessments are based on the following validation approaches and input data options:

- Blind versus plausibility validation approach.
- Input data from only one orbit versus combined input data from two orbits.

The plausibility analysis is only performed for the two-orbit input data. Thus, the comparison of the two input data options in Malawi is based only on the blind assessment approach.

For Austria, we interpreted a total of 1585 reference points, of which 21 points were removed as they were found to be non-forest already before the beginning of the change detection window. From the remaining 1564 points, 1212 belong to stratum forest and 352 to stratum change. The plausibility approach was carried out in the same manner as in Malawi. Please note that the Austrian test site is only partly covered by two orbits, which is a typical data scenario when working with Sentinel-2 data. Combined input data from two orbits is used where possible. For all assessments, we provide unbiased estimates of the mapped area proportions as well as the products', users', and producers' accuracies by applying Equations (1) and (6)–(8) from Reference [31].

## 3. Results

Table 2 gives a summary of key validation results for both test sites, different assessment approaches, and input data scenarios. The detailed results of the forest disturbance detection are shown in Tables 3–7, where the upper part presents the sample counts and the lower part presents the unbiased estimates of area proportion and accuracy measures. Overall accuracies are very high (96.4–99.3%). This is not surprising since the unchanged forest class accounts for 98.7% of the validation area at the Malawi test site and for 98.6% at the Austrian test site. It is also evidence for a low rate of false-positive change detections. For better comparison, Table 2 also lists the users' accuracies and producers' accuracies of the different validation approaches for the change class. The plausibility analysis increases the users' accuracy of the change class by 17.5% in Malawi and by 4.1% in Austria. Producers' accuracies show a strong increase of 31.1% in Malawi and 29.1% in Austria. Overall accuracies after plausibility analysis reach 99.3% at both test sites. Results show that combining data from two orbits leads to an increase in users' accuracy of 7.6% (Malawi—blind validation approach) compared to using data from only one orbit, while producers' accuracies remain the same. The detailed accuracy metrics for Malawi are shown in Tables 3–5, and for Austria, they are listed in Tables 6 and 7. For comparison between one and two orbits, please compare Tables 3 and 4. For comparison of blind and plausibility results, please compare Tables 4 and 5 for Malawi and Tables 6 and 7 for Austria.

**Table 2.** Summary of the accuracy measures for the forest change maps at both test sites for different assessment approaches (blind interpretation and plausibility analysis).

| Country | Validation Approach | Change Class | | Overall |
| --- | --- | --- | --- | --- |
| | | Users | Producers | |
| Malawi | Blind—1 orbit | 63.5% | 37.2% | 96.4% |
| Malawi | Blind—2 orbits | 71.1% | 37.1% | 98.0% |
| Malawi | Plausibility—2 orbits | 88.6% | 68.2% | 99.3% |
| Austria | Blind—2 orbits | 72.7% | 51.5% | 98.6% |
| Austria | Plausibility—2 orbits | 76.8% | 80.4% | 99.3% |

**Table 3.** Error matrix for Malawi, blind approach—1 orbit.

| | | Reference | | | $A_i$ (ha) | Users | Producers | Overall |
| --- | --- | --- | --- | --- | --- | --- | --- | --- |
| | | Forest | Change | Total | | | | |
| Map | Forest | 714 | 20 | 734 | 55,258 | | | |
| | Change | 42 | 73 | 115 | 1407 | | | |
| | Total | 756 | 93 | 849 | 56,665 | | | |
| | Forest | 0.949 | 0.027 | 0.975 | 54,266 | 97.3 ± 1.2% | 99.1 ± 0.2% | |
| | Change | 0.009 | 0.016 | 0.025 | 2399 | 63.5 ± 8.8% | 37.2 ± 10.7% | |
| | Total | 0.958 | 0.043 | 1.000 | 56,665 | | | 96.4 ± 1.2% |

Confidence interval of accuracy measures: 95%.

**Table 4.** Error matrix for Malawi, blind approach—2 orbits.

| | | Reference | | | $A_i$ (ha) | Users | Producers | Overall |
| --- | --- | --- | --- | --- | --- | --- | --- | --- |
| | | Forest | Change | Total | | | | |
| Map | Forest | 723 | 12 | 735 | 55,909 | | | |
| | Change | 33 | 81 | 114 | 756 | | | |
| | Total | 756 | 93 | 849 | 56,665 | | | |
| | Forest | 0.971 | 0.016 | 0.987 | 55,215 | 98.4 ± 0.9% | 99.6 ± 0.1% | |
| | Change | 0.004 | 0.009 | 0.013 | 1450 | 71.1 ± 8.4% | 37.1 ± 13.4% | |
| | Total | 0.974 | 0.026 | 1.000 | 56,665 | | | 98.0 ± 0.9% |

Confidence interval of accuracy measures: 95%.

**Table 5.** Error matrix for Malawi, plausibility approach—2 orbits.

| | | Reference | | | $A_i$ (ha) | Users | Producers | Overall |
| --- | --- | --- | --- | --- | --- | --- | --- | --- |
| | | Forest | Change | Total | | | | |
| Map | Forest | 731 | 4 | 734 | 55,909 | | | |
| | Change | 13 | 101 | 115 | 756 | | | |
| | Total | 744 | 105 | 849 | 56,665 | | | |
| | Forest | 0.981 | 0.005 | 0.987 | 55,691 | 99.5 ± 0.5% | 99.9 ± 0.1% | |
| | Change | 0.002 | 0.012 | 0.013 | 974 | 88.6 ± 5.9% | 68.2 ± 21.0% | |
| | Total | 0.983 | 0.017 | 1.000 | 56,665 | | | 99.3 ± 0.5% |

Confidence interval of accuracy measures: 95%.

**Table 6.** Error matrix for Austria, blind approach.

| | | Reference | | | $A_i$ (ha) | Users | Producers | Overall |
|---|---|---|---|---|---|---|---|---|
| | | Forest | Change | Total | | | | |
| Map | Forest | 1084 | 11 | 1095 | 426,192 | | | |
| | Change | 128 | 341 | 469 | 6253 | | | |
| | Total | 1212 | 352 | 1564 | 432,445 | | | |
| | Forest | 0.976 | 0.010 | 0.986 | 423,617 | $99.0 \pm 0.6\%$ | $99.6 \pm 0.1\%$ | |
| | Change | 0.004 | 0.010 | 0.014 | 8828 | $72.7 \pm 4.0\%$ | $51.5 \pm 14.8\%$ | |
| | Total | 0.980 | 0.020 | 1.000 | 432,445 | | | $98.6 \pm 0.6\%$ |

Confidence interval of accuracy measures: 95%.

**Table 7.** Error matrix for Austria, plausibility approach.

| | | Reference | | | $A_i$ (ha) | Users | Producers | Overall |
|---|---|---|---|---|---|---|---|---|
| | | Forest | Change | Total | | | | |
| Map | Forest | 1092 | 3 | 1095 | 426,192 | | | |
| | Change | 109 | 360 | 469 | 6253 | | | |
| | Total | 1201 | 363 | 1564 | 432,445 | | | |
| | Forest | 0.983 | 0.003 | 0.986 | 426,478 | $99.7 \pm 0.3\%$ | $99.6 \pm 0.1\%$ | |
| | Change | 0.003 | 0.011 | 0.014 | 5967 | $76.8 \pm 3.8\%$ | $80.4 \pm 17.8\%$ | |
| | Total | 0.986 | 0.014 | 1.000 | 432,445 | | | $99.3 \pm 0.3\%$ |

Confidence interval of accuracy measures: 95%.

Figure 5 shows some mapping examples of windthrow detection for the Austrian test site. The series of S-2 images illustrates the development of forest disturbances occurring at an alpine subset of the test site that was affected by windthrow late in the year 2018 (storm Vaia on 29/30 October 2018). Please note that the depicted sequence does not represent all available imagery, but a selection of cloud-free images of the area of interest. The first image (top left) shows the undisturbed state two weeks before the storm event. A large windthrow area can be identified in the second image (top middle; surrounding the red circle). Forest change detection is complicated by the fact that the timing of the storm exactly coincides with leaf discoloring and leaf-fall for the broadleaf trees at the site and by subsequent bad weather conditions also leading to snow cover. Remaining snow (blueish colored areas) can still be seen in early April 2019 (top right image). Harvesting of damaged forest areas leads to a continuous increase in deforested areas in the subsequent images. Typically, also areas adjacent to completely thrown forest areas are affected to some degree by wind damage, e.g., single tree throws or broken stems. In Austria, all storm damage affected areas are usually harvested directly after the storm event to prevent bark beetles from spreading. The last image in the series shows which pixels are flagged as changed by the algorithm and in which month the changes were detected. Due to frequent cloud and snow cover at the site from mid-November 2018 to April 2019, only few pixels were detected as changed shortly after the storm as only two usable post-storm observations were available until the end of 2018. Many detections are thus delayed until early spring 2019, when the time series of snow-free Sentinel-2 imagery continues and when the damaged forest areas are being cleared (removal of damaged and broken trees). This example shows both the capabilities, but also the limitations of the method when used in near real-time forest change mapping scenarios for Central European/Alpine forests.

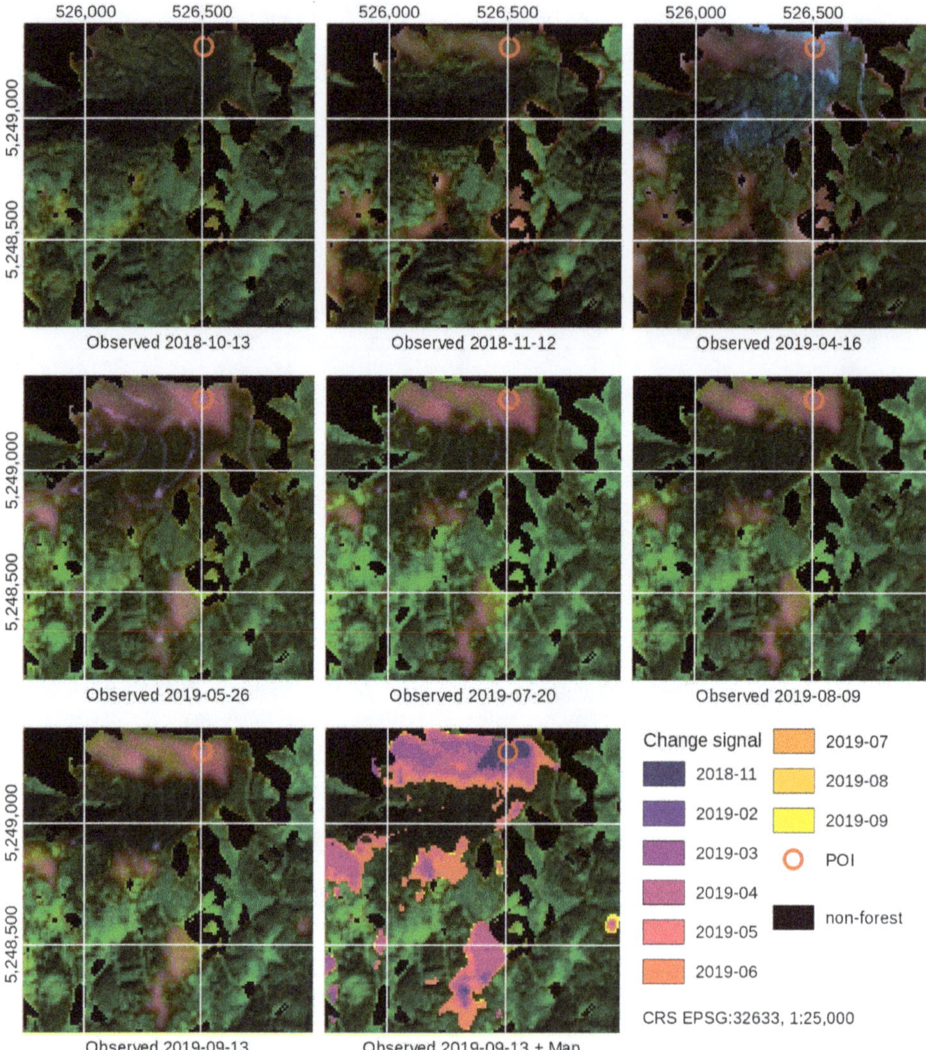



**Figure 5.** Examples of near real-time forest change detection in Austria. Storm "Vaia" windthrows on 29/30 October 2018 at an alpine forest site in Styria (~1400 m above sea level).

The software needed to create the presented mapping example and time series plots has been implemented in Python. This Python implementation is capable of processing smaller test sites (up to 10 MP) for development and demonstration purposes in reasonable time. For large-area processing, a performance-optimized C/C++ implementation is also available. However, both versions rely on certain in-house modules and libraries which are subject to licensing restrictions.

## 4. Discussion

When comparing our results to those of other studies on near real-time forest change detection, similarly high overall accuracy values can be observed. This is to be expected for a land cover class and test sites that are characterized by a large proportion of unchanged areas. In this case, users' and producers' accuracies of the change class are much better measures to determine the suitability and practical implementation potential of a monitoring approach. Many recent near real-time forest change monitoring approaches detect and evaluate changes at a minimum mapping unit (MMU) of at least one Landsat 8 pixel (~0.1 ha) as most related studies are still primarily based on Landsat 8 imagery with a lower spatial resolution than Sentinel-2 [29,33–35]. For better comparability, we have also applied a 0.1 ha MMU for the change class, but our change polygons can be of any shape representing at least ten connected Sentinel-2 change pixels.

There are a number of near real-time forest change studies and algorithms that provide accuracy statistics similar to those presented here. The near real-time humid tropical forest monitoring approach of Global Forest Watch was evaluated at national scale in Peru with a reported producer's accuracy of 75.4% and a user's accuracy of 92.2% [33]. However, a direct comparison of results is difficult, since 63% of the detected forest changes in this study were at least one hectare in size, while only 4% were the size of one single Landsat pixel (~0.1 ha). In our test sites, the vast majority of change polygons is smaller than 0.5 ha. Another change detection approach has recently been developed and tested for seasonal tropical forests in Myanmar. It uses a harmonic regression model to account for seasonality and a set of time series disturbance probabilities to detect forest changes [35]. The authors report overall disturbance detection accuracies of 78.3% for Landsat 8 data and 83.6% for Landsat 8 data combined with Sentinel-1 data. The reported users' accuracy for the disturbance class at the single Landsat 8 pixel level (~0.1 ha) is 84.1% and producers' accuracy is 78.6%, but disturbance detections are significantly delayed by 65 days on average. Very high users' and producers' accuracies of 88% and 89% were also reported for a forest change detection approach that combines Landsat 8, Sentinel-1, and ALOS-2 PALSAR-2 data at a dry tropical forest site in Bolivia characterized by distinct dry and wet seasons [34]. Instead of adapting the time series model in near real-time to the observed phenology, the authors apply a-priori spatial normalization to reduce the dry forest seasonality in the time series and then apply the change detection analysis on the normalized data [10]. The method is reported to perform well for extreme events, but we would expect that such a combined normalization and change detection approach could fail in years that behave significantly different from the average, e.g., years with extreme dryness (deviation in magnitude) or a very late start (deviation in time) of the wet season.

Our approach to use the Kalman filter is quite different in that it continuously accounts for inter-annual phenology variations and updates the time series model. The method proved to work properly at both test sites even for cases where the phenology curve significantly differs between years. Figure 6 shows three false-color Sentinel-2 images (band combination B11, B04, B03) of an unchanged forest area at the Malawi test site from three consecutive years, acquired almost on the same day of each year (mid-August). Significant differences in the phenological state can be observed both in the images as well as in the corresponding time series plot below (Figure 6). The time series plot shows the Sentinel-2 red band surface reflectance for a pixel at the center of the red circle in the images. Each year shows time periods with strong deviations from the average IRLS fit over all years. Year 2016 shows an early and stronger than average dry season, while the observations of year 2018 indicate a pronounced prolongation of the spring rainy season and thus late start of the dry season. The observed reflectances for August 2016 and August 2018 deviate by more than 7% of total surface reflectance. A widely used deterministic modelling approach based on harmonic regression (see IRLS fit represented by the green line in Figure 6) results in large and prolonged deviations between observations and the model curve which is disadvantageous for change detection. The Kalman filter is able to track the differences in plant vitality much better. In the given example, only one single observation in December 2018 was flagged as anomalous.

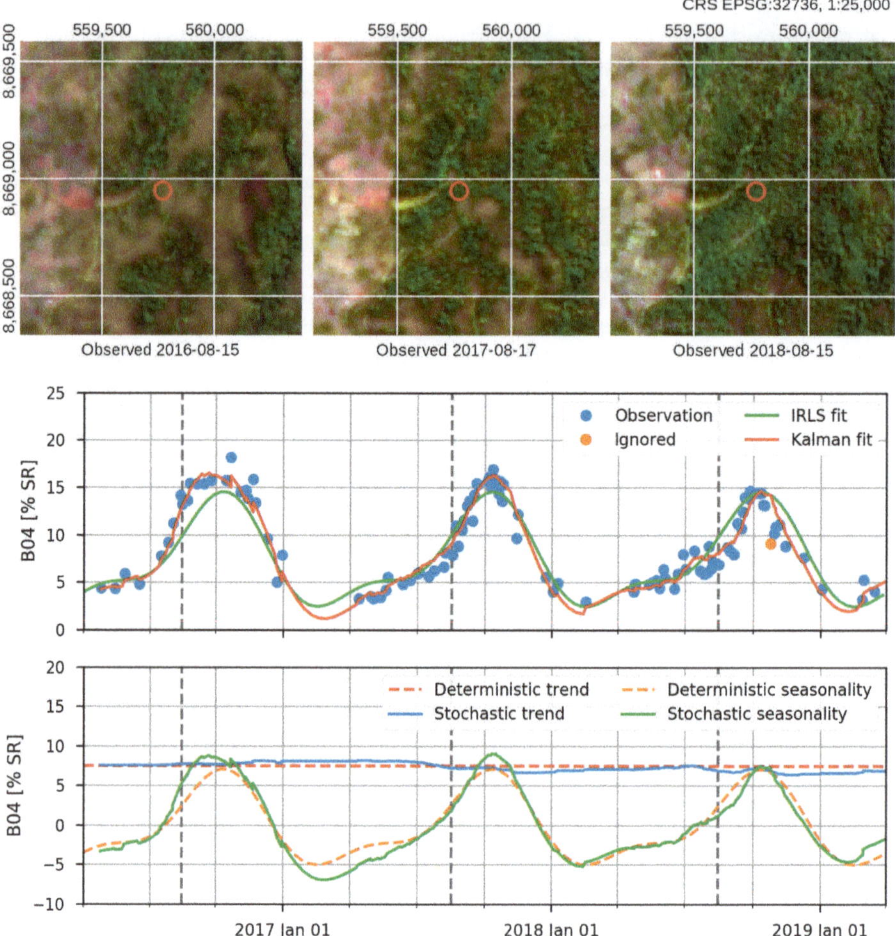

**Figure 6.** The illustration shows 3 false-color images (B11, B04, B03) acquired in three consecutive years in Malawi, almost on the same day (gray dashed lines). The upper time series plot shows the red band surface reflectance for a single pixel located in the center of the red circle together with a harmonic regression fit (IRLS—iteratively reweighted least squares) and the Kalman filtered trajectory. The lower plot shows the trend and seasonal components separately.

This seasonal phenological effect is further illustrated in Figure 7, which compares the residuals of the IRLS fit shown in Figure 6 with the corresponding Kalman filter innovations. The IRLS residuals show a much larger degree of unwanted systematic patterns left in the sequence, which is apparent when comparing the moving mean of the last 15 observations (dashed orange line in Figure 7). These systematic patterns represent the difference in phenology (both in time and magnitude) from the "average" fit. In combination with the CUSUM test, these systematic deviations of the IRLS fit would result in a larger amount of erroneous forest change detections. In case of the Kalman filter innovations, remaining non-random patterns appear when the model receives adjustments to the current phenology (especially the second half of 2018 in Figure 7). However, the drift parameter in Equation (18) avoids that small and short-lived deviations from the zero mean assumption accumulate to a change signal.

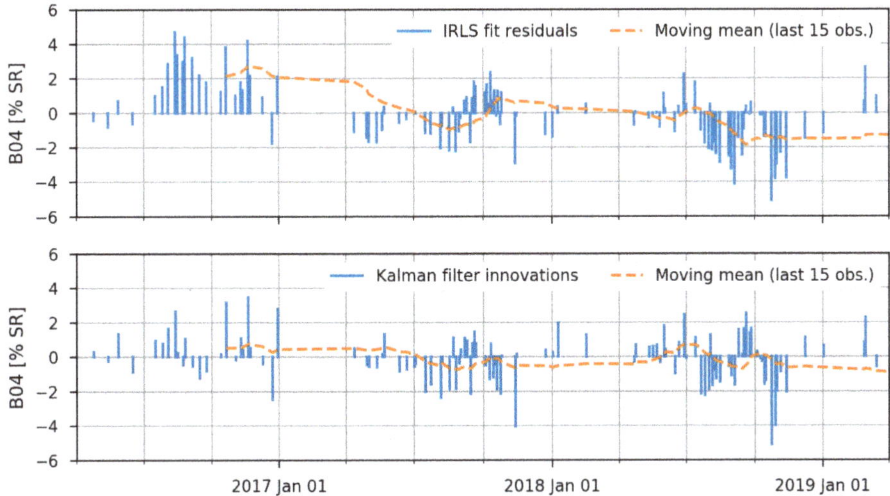

**Figure 7.** The residuals of the regression fit (IRLS) and the corresponding Kalman filter innovations for the same plot as in Figure 6. The orange dashed line represents the moving mean based on the last 15 observations.

We could show that the combination of Kalman filter innovations and the CUSUM test is suitable for rapid near real-time detection of abrupt forest changes as shown in Figure 4. With the CUSUM test, changes can be detected with a smaller time lag than with traditional change detection methods based on multiple confirmations [21,35 37]. Many existing algorithms require a fixed number of observed anomalies to signal a change, and some enforce the condition that anomalies also have to be detected consecutively. Certain statistical boundaries or change probabilities specific to the method have to be exceeded multiple times for a change to be recorded. When the spectral footprint of a change event is at the border of detectability, these restrictions likely lead to omission errors or a large time lag between change event and detection. Because of its cumulative nature, the proposed CUSUM test statistic is not restricted in the same away. Any post-change observation can add information to the test statistic, regardless of being flagged by the anomaly test or not. Depending on the spectral footprint of the change event, it may take two, three, or more post-change observations to detect the change. At this point, we cannot give a quantitative evaluation of the typical time lag because appropriate reference data were not available.

The reduction of the time lag between change event and change detection is of major importance to near real-time forest monitoring systems. For this reason, several recent studies have combined optical data with Synthetic Aperture Radar (SAR) data in order to reduce this time lag [34,35]. Regarding optical data, persistent cloud cover during rainy seasons represents a major limitation as the time gap between consecutive valid observations may become very large. Change detection methods that require multiple confirmed change detections will therefore show large time lags between the change event and its detection. Two studies carried out in tropical regions reported mean time lags of 63 days and 70 days for Myanmar and Bolivia respectively, if only optical Landsat 8 data is used [34,35].

At the Malawi site, we also tested if exploiting overlaps of two S-2 orbits delivers higher forest change detection accuracies. Using two orbits has the advantage of providing a much denser time series, but a potential disadvantage stems from the geometric shifts related to the different orbit viewing angles. Especially, the forest border is often slightly misplaced in data from two orbits as the DEM that is used by the European Space Agency (ESA) to orthorectify the imagery does not accurately account for tree height. At the 10 m spatial resolution of Sentinel-2, the effect of orbit-related geometric shifts on the spectral reflectance of single forest border pixels is much stronger than that, for example, in 30 m

Landsat 8 pixels. Elevation errors in the DEM may also lead to strong dislocations of up to 20 m at mountain ridges as observed in the alpine test areas in Austria. In Malawi, our test site is characterized by flat to slightly hilly terrain, so errors related to topography are negligible. The results show that for Malawi, the users' accuracy increases from 63.5% to 71.1% when using two orbits instead of only using one orbit. Especially, the commission error is much higher with only one orbit (see mapped areas of 1407 ha versus 756 ha in Tables 3 and 4, respectively). The main reason for this strong improvement may be found by looking at the inter-annual temporal distribution of valid observations. The time series density at the Malawi test site is quite inhomogeneous (see Figure 6). While many clear observations are available during the dry season, the time series is sparse in the rainy season from November to April due to frequent cloud cover. Using two orbits greatly increases the chance of including a few clear observations during the rainy season, which is very important for a proper initialization of the time series model. To our knowledge, such a comparative analysis on using Sentinel-2 data from one or two orbits as input to a change detection approach has not been performed before. Thus, we cannot assess whether our findings are in line with other research results. While our findings suggest that denser time series resulting from orbit overlaps lead to higher forest change mapping accuracies, it is not yet clear if these findings for the dry forests in Malawi are also valid for other forest types and for other regions. Further studies with different data input scenarios and at various forest test sites are still needed to fully answer this question.

The presented approach for detecting forest changes in near real-time using Sentinel-2 imagery has a high potential for operationalization of forest monitoring services, such as improved and automated REDD+ (Reducing Emissions from Deforestation and Forest Degradation) services in the tropics, and windthrow damage assessment or bark beetle monitoring in Central Europe. Future studies to improve the presented forest change detection method and similar near real-time approaches should focus on an integrated and automated separation of different biotic and abiotic forest disturbance agents. Additional developments should include a multivariate analysis of relevant reflectance bands and a spatiotemporal analysis of changes. The near real-time capability should be tested in operational scenarios and the time lag of change detection needs to be assessed with field data or EO data of very high geometric and temporal resolution (e.g., Planet data). For tropical forest monitoring and REDD+ activities, the method should be tested in combination with recent Sentinel-1 SAR forest change detection approaches, such as SAR shadow detection [38], backscatter composite differencing [39], or SAR backscatter thresholding based on the coefficient of variation [40]. With minor adaptations, the presented time series analysis approach might also be directly applicable to SAR data. The approach could also be used for other EO-based applications, such as phenology monitoring in agriculture or grassland monitoring (detection of mowing events).

## 5. Conclusions

In this paper, we presented a new algorithm designed for vegetation monitoring and change detection using optical EO data. The approach is largely data-driven and designed especially for near real-time mapping in web- or cloud-based monitoring services. Compared to existing algorithms employing harmonic regression, this study explored methodological improvements, mainly in two aspects:

- Seasonal patterns related to plant phenology can vary strongly between years because of different climate conditions, such as temperature and rainfall. With widely used harmonic regression models, it can be difficult to separate these normal variations from true disturbances due to the deterministic modelling approach. Our algorithm uses structural time series models in state space form which take into account that the trend and seasonal components in a time series can evolve over time. This class of stochastic models is typically used in conjunction with the Kalman filter, which also enables elegant handling of irregular sampling intervals and signal artefacts like un-masked clouds and cloud shadows. We showed that it is possible to track the phenology-related vegetation dynamics more closely without losing the ability to detect disturbances.

- Many existing algorithms require that specific change probabilities or statistical boundaries have to be exceeded multiple times for a change to be confirmed. In our algorithm, the sequence of Kalman filter innovations (i.e., the differences between the one-step-ahead model forecasts and corresponding observations) opens up alternative possibilities to characterize change. If the time series model assumptions are correct and no changes occur, the innovation sequence should be mostly free of temporal autocorrelation and it should have zero mean. We used the CUSUM control chart to monitor these properties and assumed that significant deviations indicate change. Additionally, a separate anomaly test intended to suppress commission errors due to signal artefacts was implemented. Compared to the multiple confirmation approach, our results indicate that the CUSUM approach has the potential to reduce the time lag between change event and detection and a better chance of detecting subtle and gradual changes because of its cumulative nature. However, a better understanding of different change types and especially reference data for validation are required to support this claim.

Two challenging test sites located in Austria and Malawi were selected to test the algorithm in a forest change detection scenario based on Sentinel-2 data. Both sites show pronounced and dynamic seasonal patterns in the Sentinel-2 time series due to plant phenology. The dominant change types in Malawi are deforestation and forest degradation. In Austria, we were mainly interested in changes caused by storm damages or bark beetle infestations, but of course, deforestation also occurs. The validation of the results was performed based on visually interpreted points derived by a stratified random sampling approach. For the forest change class, we reported users' accuracies of 76.8% (Austria) to 88.6% (Malawi), and producers' accuracies of 68.2% (Malawi) to 80.4% (Austria). Due to the low rate of commission errors and large proportion of stable forest, overall accuracies reached over 99%.

In the Malawi site, we further showed that a denser time series with data from two different orbits results in better change detection results compared to using data from only one orbit. The larger number of input images seems to outweigh the possible negative effects of spectral and geometric differences related to the different viewing angles, which especially occur at forest edges. The observed increase in users' accuracy when using two orbits amounted to 7.6%. However, further studies with different data inputs and at various forest test sites are required to confirm these results.

In summary, it can be stated that the combination of structural time series models and Kalman filtering represents an appropriate method for a variety of automated forest monitoring applications. Beside the possibility of detecting abrupt changes, for example caused by storm damages or deforestation, the method also shows a high potential to detect more subtle and continuous changes, such as bark beetle infestations, forest degradation, or drought stress. Regarding the detection of insect infestations, the interesting research question to be dealt with in future is whether an early detection of bark beetle green attack is feasible.

**Author Contributions:** Conceptualization, M.P., J.D., and M.H.; methodology, M.P.; software, A.W.; formal analysis, M.P., J.D., A.W., and U.S.; investigation, M.P. and A.W.; data curation, M.P. and A.W.; writing—original draft preparation, M.P., J.D., and M.H.; writing—review and editing, M.S.; visualization, M.P. and M.H.; funding acquisition, M.H., J.D., and M.S. All authors have read and agreed to the published version of the manuscript.

**Funding:** This research was funded by the European Union's Horizon 2020 research and innovation programme, grant numbers 633464 (project DIABOLO) and 685761 (project EOMonDis), and the Austrian Research Promotion Agency (FFG), grant numbers 859764 (project AlpMon) and 878891 (project BEAT IT!).

**Acknowledgments:** We greatly appreciate the free, full, and open data policy adopted by ESA for the COPERNICUS programme. We would also like to thank the reviewers for providing valuable feedback which helped to improve the manuscript.

**Conflicts of Interest:** The authors declare no conflict of interest. The funders had no role in the design of the study; in the collection, analyses, or interpretation of data; in the writing of the manuscript, or in the decision to publish the results.

## References


1. Zhu, Z.; Woodcock, C.E. Object-based cloud and cloud shadow detection in Landsat imagery. *Remote Sens. Environ.* **2012**, *118*, 83–94. [CrossRef]
2. Zhu, Z. Change detection using Landsat time series: A review of frequencies, preprocessing, algorithms, and applications. *ISPRS J. Photogramm. Remote Sens.* **2017**, *130*, 370–384. [CrossRef]
3. Huang, C.; Goward, S.N.; Masek, J.G.; Thomas, N.; Zhu, Z.; Vogelmann, J.E. An automated approach for reconstructing recent forest disturbance history using dense Landsat time series stacks. *Remote Sens. Environ.* **2010**, *114*, 183–198. [CrossRef]
4. Kennedy, R.E.; Yang, Z.; Cohen, W.B. Detecting trends in forest disturbance and recovery using yearly Landsat time series: 1. LandTrendr—Temporal segmentation algorithms. *Remote Sens. Environ.* **2010**, *114*, 2897–2910. [CrossRef]
5. Verbesselt, J.; Hyndman, R.; Zeileis, A.; Culvenor, D. Phenological change detection while accounting for abrupt and gradual trends in satellite image time series. *Remote Sens. Environ.* **2010**, *114*, 2970–2980. [CrossRef]
6. Verbesselt, J.; Herold, M.; Zeileis, A. Near real-time disturbance detection using satellite image time series. *Remote Sens. Environ.* **2012**, *123*, 98–108. [CrossRef]
7. DeVries, B.; Verbesselt, J.; Kooistra, L.; Herold, M. Robust monitoring of small-scale forest disturbances in a tropical montane forest using Landsat time series. *Remote Sens. Environ.* **2015**, *161*, 107–121. [CrossRef]
8. DeVries, B.; Pratihast, A.K.; Verbesselt, J.; Kooistra, L.; Herold, M. Characterizing forest change using community-based monitoring data and Landsat time series. *PLoS ONE* **2016**, *11*, e0147121. [CrossRef]
9. DeVries, B.; Decuyper, M.; Verbesselt, J.; Zeileis, A.; Herold, M.; Joseph, S. Tracking disturbance-regrowth dynamics in tropical forests using structural change detection and Landsat time series. *Remote Sens. Environ.* **2015**, *169*, 320–334. [CrossRef]
10. Hamunyela, E.; Verbesselt, J.; Herold, M. Using spatial context to improve early detection of deforestation from Landsat time series. *Remote Sens. Environ.* **2016**, *172*, 126–138. [CrossRef]
11. Zhu, Z.; Woodcock, C.E.; Olofsson, P. Continuous monitoring of forest disturbance using all available Landsat imagery. *Remote Sens. Environ.* **2012**, *122*, 75–91. [CrossRef]
12. Zhu, Z.; Woodcock, C.E. Continuous change detection and classification of land cover using all available Landsat data. *Remote Sens. Environ.* **2014**, *144*, 152–171. [CrossRef]
13. Zhu, Z.; Woodcock, C.E.; Holden, C.; Yang, Z. Generating synthetic Landsat images based on all available Landsat data: Predicting Landsat surface reflectance at any given time. *Remote Sens. Environ.* **2015**, *162*, 67–83. [CrossRef]
14. Brooks, E.B.; Wynne, R.H.; Thomas, V.A.; Blinn, C.E.; Coulston, J.W. On-the-fly massively multitemporal change detection using statistical quality control charts and Landsat data. *IEEE Trans. Geosci. Remote Sens.* **2014**, *52*, 3316–3332. [CrossRef]
15. Harvey, A.C. *Forecasting, Structural Time Series Models and the Kalman Filter*; Cambridge University Press: Cambridge, UK, 1989; ISBN 0-521-40573-4.
16. Kalman, R.E. A New Approach to Linear Filtering and Prediction Problems. *J. Basic Eng.* **1960**, *82*, 35–45. [CrossRef]
17. Gibbs, B.P. *Advanced Kalman Filtering, Least-Squares and Modeling*; John Wiley & Sons Inc.: Hoboken, NJ, USA, 2011; ISBN 9780470890042.
18. Puhm, M. Analysis of Landsat time series using state space models and Kalman filtering. Master's Thesis, Graz University of Technology, Graz, Austria, 2018. Available online: https://permalink.obvsg.at/tug/AC15229314 (accessed on 14 July 2020).
19. Olsson, P.O.; Lindström, J.; Eklundh, L. Near real-time monitoring of insect induced defoliation in subalpine birch forests with MODIS derived NDVI. *Remote Sens. Environ.* **2016**, *181*, 42–53. [CrossRef]
20. Page, E.S. Continuous Inspection Schemes. *Biometrika* **1954**, *41*, 100. [CrossRef]
21. Hirschmugl, M.; Deutscher, J.; Sobe, C.; Bouvet, A.; Mermoz, S.; Schardt, M. Use of SAR and optical time series for tropical forest disturbance mapping. *Remote Sens.* **2020**, *12*, 727. [CrossRef]
22. S2 MPC Team. L1C Data Quality Report. 2020. Available online: https://sentinel.esa.int/documents/247904/685211/Sentinel-2_L1C_Data_Quality_Report (accessed on 18 September 2020).

23. Louis, J.; Debaecker, V.; Pflug, B.; Main-Knorn, M.; Bieniarz, J.; Mueller-Wilm, U.; Cadau, E.; Gascon, F. Sentinel-2 Sen2Cor: L2A processor for users. In Proceedings of the Living Planet Symposium, Prague, Czech Republic, 9–13 May 2016; pp. 1–8.

24. Qiu, S.; He, B.; Zhu, Z.; Liao, Z.; Quan, X. Improving Fmask cloud and cloud shadow detection in mountainous area for Landsats 4–8 images. *Remote Sens. Environ.* **2017**, *199*, 107–119. [CrossRef]

25. Qiu, S.; Zhu, Z.; He, B. Fmask 4.0: Improved cloud and cloud shadow detection in Landsats 4–8 and Sentinel-2 imagery. *Remote Sens. Environ.* **2019**, *231*, 111205. [CrossRef]

26. Gallaun, H.; Schardt, M.; Linser, S. Remote sensing based forest map of Austria and derived environmental indicators. In Proceedings of the International Conference on Spatial Application Tools in Forestry (ForestSAT 2007), Montpellier, France, 5–7 November 2007.

27. Huber, P.J. Robust estimation of a location parameter. *Ann. Math. Stat.* **1964**, *35*, 73–101. [CrossRef]

28. Heiberger, R.M.; Becker, R.A. Design of an S Function for Robust Regression Using Iteratively Reweighted Least Squares. *J. Comput. Graph. Stat.* **1992**, *1*, 181–196. [CrossRef]

29. Zhu, Z.; Zhang, J.; Yang, Z.; Aljaddani, A.H.; Cohen, W.B.; Qiu, S.; Zhou, C. Continuous monitoring of land disturbance based on Landsat time series. *Remote Sens. Environ.* **2020**, *238*, 111116. [CrossRef]

30. Xue, J.; Su, B. Significant Remote Sensing Vegetation Indices: A Review of Developments and Applications. *J. Sensors* **2017**, *2017*, 1–17. [CrossRef]

31. Olofsson, P.; Foody, G.M.; Stehman, S.V.; Woodcock, C.E. Making better use of accuracy data in land change studies: Estimating accuracy and area and quantifying uncertainty using stratified estimation. *Remote Sens. Environ.* **2013**, *129*, 122–131. [CrossRef]

32. Olofsson, P.; Foody, G.M.; Herold, M.; Stehman, S.V.; Woodcock, C.E.; Wulder, M.A. Good practices for estimating area and assessing accuracy of land change. *Remote Sens. Environ.* **2014**, *148*, 42–57. [CrossRef]

33. Potapov, P.V.; Dempewolf, J.; Talero, Y.; Hansen, M.C.; Stehman, S.V.; Vargas, C.; Rojas, E.J.; Castillo, D.; Mendoza, E.; Calderón, A.; et al. National satellite-based humid tropical forest change assessment in Peru in support of REDD + implementation. *Environ. Res. Lett.* **2014**, *9*. [CrossRef]

34. Reiche, J.; Hamunyela, E.; Verbesselt, J.; Hoekman, D.; Herold, M. Improving near-real time deforestation monitoring in tropical dry forests by combining dense Sentinel-1 time series with Landsat and ALOS-2 PALSAR-2. *Remote Sens. Environ.* **2018**, *204*, 147–161. [CrossRef]

35. Shimizu, K.; Ota, T.; Mizoue, N. Detecting forest changes using dense Landsat 8 and Sentinel-1 time series data in tropical seasonal forests. *Remote Sens.* **2019**, *11*, 1899. [CrossRef]

36. Hansen, M.C.; Krylov, A.; Tyukavina, A.; Potapov, P.V.; Turubanova, S.; Zutta, B.; Ifo, S.; Margono, B.; Stolle, F.; Moore, R. Humid tropical forest disturbance alerts using Landsat data. *Environ. Res. Lett.* **2016**, *11*, 034008. [CrossRef]

37. Hirschmugl, M.; Sobe, C.; Deutscher, J.; Schardt, M. Combined Use of Optical and Synthetic Aperture Radar Data for REDD + Applications in Malawi. *Land* **2018**, *7*, 116. [CrossRef]

38. Bouvet, A.; Mermoz, S.; Ballère, M.; Koleck, T.; Le Toan, T. Use of the SAR shadowing effect for deforestation detection with Sentinel-1 time series. *Remote Sens.* **2018**, *10*, 1250. [CrossRef]

39. Rüetschi, M.; Small, D.; Waser, L.T. Rapid detection of windthrows using Sentinel-1 C-band SAR data. *Remote Sens.* **2019**, *11*, 115. [CrossRef]

40. Deutscher, J.; Gutjahr, K.; Perko, R.; Raggam, H.; Hirschmugl, M.; Schardt, M. Humid tropical forest monitoring with multi-temporal L-C and X-band SAR data. In Proceedings of the 9th International Workshop on the Analysis of Multitemporal Remote Sensing Images (MultiTemp), Brugge, Belgium, 27–29 June 2017; pp. 1–4.

*remote sensing*

Article

# Characterizing the Error and Bias of Remotely Sensed LAI Products: An Example for Tropical and Subtropical Evergreen Forests in South China

Yuan Zhao [1,2], Xiaoqiu Chen [1,*], Thomas Luke Smallman [2,3], Sophie Flack-Prain [2,3],
David T. Milodowski [2,3] and Mathew Williams [2,3]

[1] Laboratory for Earth Surface Processes of the Ministry of Education, College of Urban and
    Environmental Sciences, Peking University, Beijing 100871, China; zy581@pku.edu.cn
[2] School of GeoSciences, University of Edinburgh, Edinburgh EH9 3FF, UK; t.l.smallman@ed.ac.uk (T.L.S.);
    s.flack-prain@ed.ac.uk (S.F.-P.); d.t.milodowski@ed.ac.uk (D.T.M.); Mat.Williams@ed.ac.uk (M.W.)
[3] National Centre for Earth Observation, University of Edinburgh, Edinburgh EH9 3FF, UK
*   Correspondence: cxq@pku.edu.cn; Tel.: +86-010-62753976

Received: 31 August 2020; Accepted: 21 September 2020; Published: 23 September 2020

**Abstract:** Leaf area is a key parameter underpinning ecosystem carbon, water and energy exchanges via photosynthesis, transpiration and absorption of radiation, from local to global scales. Satellite-based Earth Observation (EO) can provide estimates of leaf area index (LAI) with global coverage and high temporal frequency. However, the error and bias contained within these EO products and their variation in time and across spatial resolutions remain poorly understood. Here, we used nearly 8000 in situ measurements of LAI from six forest environments in southern China to evaluate the magnitude, uncertainty, and dynamics of three widely used EO LAI products. The finer spatial resolution GEOV3 PROBA-V 300 m LAI product best estimates the observed LAI from a multi-site dataset ($R^2$ = 0.45, bias = $-0.54$ m$^2$ m$^{-2}$, RMSE = 1.21 m$^2$ m$^{-2}$) and importantly captures canopy dynamics well, including the amplitude and phase. The GEOV2 PROBA-V 1 km LAI product performed the next best ($R^2$ = 0.36, bias = $-2.04$ m$^2$ m$^{-2}$, RMSE = 2.32 m$^2$ m$^{-2}$) followed by MODIS 500 m LAI ($R^2$ = 0.20, bias = $-1.47$ m$^2$ m$^{-2}$, RMSE = 2.29 m$^2$ m$^{-2}$). The MODIS 500 m product did not capture the temporal dynamics observed in situ across southern China. The uncertainties estimated by each of the EO products are substantially smaller (3–5 times) than the observed bias for EO products against in situ measurements. Thus, reported product uncertainties are substantially underestimated and do not fully account for their total uncertainty. Overall, our analysis indicates that both the retrieval algorithm and spatial resolution play an important role in accurately estimating LAI for the dense canopy forests in Southern China. When constraining models of the carbon cycle and other ecosystem processes are run, studies should assume that current EO product LAI uncertainty estimates underestimate their true uncertainty value.

**Keywords:** remotely sensed LAI; field measured LAI; validation; magnitude; uncertainty; temporal dynamics

---

## 1. Introduction

The Leaf Area Index (LAI), the total leaf area per unit ground area, is a key biophysical variable playing an important role in global carbon, water, and energy cycles [1,2]. As such, it acts as an important parameter for several applications, such as land surface models [3], ecological models [4], and yield prediction models [5]. The amount of leaf area has a first-order control on photosynthesis, transpiration, and absorption of radiation, varying in both space and time. LAI seasonal dynamics provide information about phenological processes of canopy development, senescence, and plant traits [6]. Therefore, retrieving LAI over large areas and having a good knowledge of their yearly variations, errors, and bias is extremely important. Such information is central to accurately estimating primary productivity, understanding land surface-atmosphere exchanges, and detecting the response of terrestrial vegetation to climate change [7]. It is also beneficial for a large remote sensing community because it provides insights for the interpretation and correct usage of the LAI maps. Satellite-based Earth Observation (EO) offers the opportunity to retrieve information on LAI which is global in coverage and at increasing temporal and spatial resolution [8–12]. Robust estimates of uncertainties associated wth EO LAI estimates are needed to ensure their appropriate integration within further analyses such as data assimilation [6,13]. However, such robust evaluations are rarely possible due to the scarcity of in situ data at appropriate temporal and spatial resolutions [14,15].

LAI product uncertainty information represents the performance of the products' algorithm and reflects the uncertainties in the input data, model imperfections, and the inversion process [16–18], which could be called the theoretical uncertainties [19]. Evaluation studies typically assess the theoretical uncertainty of LAI based on the standard from Global Climate Observing System (GCOS) [20], which sets a target for absolute LAI uncertainty range to be within 0.5 $m^2$ $m^{-2}$ and a relative uncertainty of less than 20%. However, few studies validate the uncertainty estimates of EO LAI using in situ data [14] at appropriate scales. A key challenge for the validation process is the mismatch in the scale of satellite products (typically > 300 m) compared to that of the in situ data (<10 m). Heterogeneity of LAI at scales finer than the satellite resolution [21] means that simple comparison between measurements at different scales can be problematic [22].

The validation of LAI at temporal scales includes a more important component, the LAI temporal dynamics, besides the traditional absolute value checking [23]. The temporal dynamics of LAI time series contain useful ecological information (e.g., phenological infomation), which help study the relationship between plant phenology and climate [24] or assist in land cover classification [25]. However, the temporal dimension has been neglected for most of the LAI validation work [23]. To validate the seasonal variations of LAI products, higher field sampling rates are need, especially for the Evergreen Broadleaf Forests (EBF) across several regions (Africa, Eurasia, South America, Australia and Asia) [14].

Despite the important role played by tropical forests in regulating the global carbon [26,27] and energy balance [28], tropical phenology is highly uncertain. Large-scale monitoring of tropical forest LAI with satellites is hindered by several challenges, including signal saturation [23], poor observation conditions (cloud-aerosol contamination, etc.) [29], coarse spatial and temporal resolutions [30], and scarcity of both ground data and detailed validation tests [31–33]. Uncertainties are therefore particularly high in these regions [34]. Some progress has been made in the Amazon [35–39] and African forests [40–42]; however, the Southern China region has typically been neglected in remote sensing phenology studies [43–46]. Consequently, the phenological character of forests in this region remains uncertain, with additional complexity driven by fragmentation [47], high species diversity, and complex topography [48]. The lack of detailed validation studies in Southern China means that there is little information regarding the suitability of global EO LAI products for the forests in this region, which play a major role in the carbon sink across China, and span the climate gradient from the subtropics to tropics [49,50].

In this study, we take advantage of a network of 6 sites with ~8000 ground-based measurements of LAI for forests located across the tropical and subtropical Southern China region to fully evaluate three satellite-based remote sensed LAI products with global coverage (Moderate Resolution Imaging Spectroradiometer (MODIS) 500 m and PROBA-V GEOV2 1 km and V3 300 m). Specifically, we address the following questions:

(i)     How well do EO-based estimates of LAI capture temporal dynamics observed in ground measurements?

(ii)    How robust are EO LAI error estimates?

To answer these questions, we evaluate errors across multiple LAI retrieval algorithms and spatial resolutions and their associated temporal dynamics. We use additional fine resolution satellite data to evaluate the heterogeneity of canopy cover at the scales intermediate between in situ data and the satellite products at each location. Finally, we discuss the importance of robust error characterization for LAI products in the context of constraining models of the terrestrial carbon cycle and other ecosystem processes. We aim to characterize and understand errors and bias in the EO LAI products but do not make retrievals over the entire region. Whilst our scientific questions are focused on the Southern China region, the approach used can be applied elsewhere if in situ data is available. Assessing the uncertainties in LAI products through comparison with in situ measurements, i.e., direct validation is critical for their proper use in land surface models [51,52]. A better understanding of the uncertainties embedded in current LAI products will improve the assimilation of the LAI into land surface modeling studies [53,54] by providing a more robust error weighting [55].

## 2. Materials and Methods

### 2.1. Study Area

Site-level LAI observational data are collected from CERN (Chinese Ecosystem Research Network) [56]. The six sites included are Ban Na Forest (BNF), Ai Lao Forest (ALF), Gong Ga Forest (GGF), Ding Hu Forest (DHF), He Shan Forest (HSF), and Shen Nong Forest (SNF). These sites are spread across southern China between 21.9° and 31.3°N, covering major plant functional types from northern subtropical zones to the northern tropical zones in this region (Figure 1). The regional climate is characterized by a wet, warm summer and a dry, mild winter [57]. The site-specific Koeppen climate classifications [58] are Cwa (Temperate zone warm summer; dry winter) for ALF, BNF, HSF, and SNF. Cfa (Temperate zone hot summer; no dry season) for DHF and Dwb (Cold zone dry winter; cold summer) for GGF. The mean annual temperature ranges between 4.2 and 21.8 °C. Annual precipitation ranges between 1506 mm and 2175 mm. The sites vary in elevation (above sea level) between 70 and 3160 m (Table 1). Forest types include the tropical seasonal rainforests, mixed coniferous forest, evergreen and deciduous mixed broad-leaved forest. The sites used support publically forests which are natural, except for forests at HSF and DHF, which support planted forests over 40 years old. All forests have at least three vertical layers: woody, shrub, and herbs. The tree density (trees per hectare) varies widely across sites. The highest tree density reaches over 7000 and the lowest is around 600 (Table S1). The range of the mean diameter breath heights is 5–20 cm (Table S1). The dominant species composition differs between sites but mainly consists of the evergreen species (Table S1). An image of forests measured at each site is included in Figure S1.

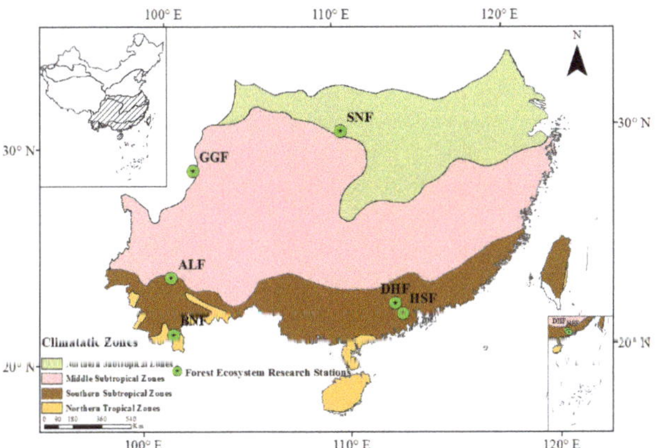


**Figure 1.** Map of the study sites in Southern China.

**Table 1.** Background information for each site. MAT: mean annual temperature; AP: annual precipitation.

| Sites | Latitude (N) | Longitude (W) | Altitude (m) | MAT (°C) | AP (mm) | Gradient | Forest Types |
|---|---|---|---|---|---|---|---|
| BNF | 21.91 | 101.2 | 730 | 21.8 | 1506 | 18–25° | Tropical seasonal rainforest |
| HSF | 22.67 | 112.89 | 70 | 21.7 | 1761 | 18–23° | Mixed coniferous forest |
| DHF | 23.16 | 112.53 | 300 | 21 | 1996 | 25–35° | Mixed coniferous forest |
| ALF | 24.54 | 101.01 | 2488 | 11 | 1931 | 5–25° | Natural wet evergreen broad-leaved forest |
| GGF | 29.57 | 101.98 | 3160 | 4.2 | 2175 | 30–35° | Subalpine Emei fir forest |
| SNF | 31.3 | 110.47 | 1650 | 10.6 | 1722 | 10–70° | Evergreen and deciduous mixed broad-leaved forest |

*2.2. Field LAI Data Measurements*

In each of the six sites, there are one to four subplots at which LAI is estimated, and each subplot area ranged from 400–10,000 m² (Figure S1). Field measurements of LAI were made each month using an LAI-2000 [59]. Field LAI measurement processing is consistent for each site and is based on the unified data collection and quality control protocols specified for CERN [60]. The sampling positions are distributed evenly and diagonally within the plot. The horizontal distance between each sampling points or plot boundary are to exceed 15 m and 10 m separately to avoid overlap sampling and reduce marginal effects. LAI-2000 were used to scan the canopy twice a day (8:00 am and 16:00 pm) to get the three-layer LAI (woody, shrub, and herbs) at each sampling point at the measurement day. The position for each sampling measurements and scan height is fixed. The median number of monthly measurements made at each site ranged from 6 to 80 (Table S2). BNF is the most intensively surveyed site, with between 203 and 360 measurements per year; therefore, BNF is particularly suited to evaluating the temporal dynamics of EO LAI products. For the other sites, only a subset of years and growth season months are measured (Table S2). A total of 795, 9 ground measurements from 2005 to 2017 were acquired across all the sites (Table S2).

*2.3. Earth Observation LAI Estimates*

In this study we evaluate three global EO LAI products. MODIS 500m product (MOD15A2H) provides estimates of LAI at 500 m spatial resolution with an 8-day interval from 2000-present [10]. GEOV2 1 km [11] and GEOV3 300 m LAI [12] were derived from the SPOT/VEGETATION sensor data at a 10-day interval, at spatial resolutions of 1/112° and 1/336°, respectively (approximately 1-km and 300 m at the equator), in the Plate Carrée projection. For the MODIS 500 m LAI product, we consider only the products derived from the main algorithm, which is based on the use of Look Up Tables (LUTs) built for six different plant functional types, with simulations from a three-dimensional radiative transfer model [61]. For the GEOV2 1 km and GEOV3 300 m LAI products, LAI is estimated using Neural Networks(NNTs) applied on Top-of-Canopy (TOC) input reflectance in the red, near-infrared, and shortwave infrared bands, at 1km resolution and 300m, respectively [11,12]. The period of MODIS 500 m LAI products and GEOV2 1 km LAI products is from 2005 to 2017 and 2014 to 2017 for the GEOV3 300 m LAI product (released from 2014).

The definition of uncertainty information varies between LAI products. For MODIS LAI, the standard deviation is used to measure the uncertainty of pixel LAI values, which is calculated over all acceptable solutions of a look-up table (LUT) retrieval method [17]. For GEOV2 1 km and GEOV3 300 m LAI products, the uncertainties are computed as the RMSE between the final decadal value and the daily NNTs estimates in the compositing period [12,62]. In this study, they are both treated as the LAI products' uncertainty. The quality assessment information is used to clean the LAI and inform uncertainty information at the pixel level for all of these LAI products. For MODIS 500 m products, data effected by cloud shadow, internal cloud masks, or aerosol are removed and only main retrievals were used in the analysis. For Copernicus LAI data (GEOV2 1 km and GEOV3 300 m), pixels that were filled or interpolated, or out of LAI range, were removed.

*2.4. Mapping Heterogeneity of Canopy Properties*

To evaluate the upscaling of field LAI measurements to satellite products, Landsat Level-2 Surface Reflectance products including Enhanced Vegetation Index (EVI) and Normalized Difference Vegetation Index (NDVI) derived from Landsat 5 (TM), Landsat 7 (ETM+) [63] and Landsat 8 (OLI) [64] scenes are also analyzed. EVI is calculated from red (R), near-infrared (NIR) and blue band (B) based on Equation (1) and incorporates parameters to adjust for canopy background, atmospheric resistance.

$$EVI \ = \ 2.5 * ((NIR - R)/(NIR + 6 * R - 7 * B + 1)) \tag{1}$$

NDVI is calculated as a ratio between the R and NIR values in a traditional fashion (Equation (2)).

$$NDVI \ = \ (NIR - R)/(NIR + R) \tag{2}$$

Landsat 5 (TM), Landsat 7 (ETM+), and Landsat 8 (OLI) Level-2 surface reflectance products were generated, using the Land Surface Reflectance Code (LaSRC) and the Landsat Ecosystem Disturbance Adaptive Processing System (LEDAPS) algorithms [65,66]. The LEDAPS and LaSRC surface reflectance algorithms correct for the temporally, spatially, and spectrally varying scattering and absorbing effects of atmospheric gases and aerosols, which is essential to derive the Earth's reflectance surface values. Landsat Spectral Indices are generated at 30 m spatial resolution on a Universal Transverse Mercator (UTM) mapping grid. The Spectral Indices could be obtained from the USGS Earth Resources Observation and Science (EROS) on-demand processing system.

The GTOPO 30 1 km elevation map [67] in the China region with Landsat scene footprints at each site is presented in Figure S2. In each of the six sites, there are one to four sampling subplots. The location of each site with the sampling subplots is shown on the Landsat scenes with path and row information (Figure S3) and on the 90 m SARTM elevation map [68], respectively (Figure S4). There are in total 778 Landsat 5–7 and Landsat 8 scenes with good quality (collection category is Tier 1) and cloud cover smaller than 20% of the total scene area from the year of 2005 to 2017 for all of these forests. Because there is no quality assessment information for the Landsat spectral indices, we use the pixel quality assurance band (pixel_qa) instead to obtain quality information. Pixels with cloud shadow, aerosols, etc., are filtered out, and we keep pixels with the best quality.

Landsat EVI and NDVI were extracted at the field sampling plots spatial level (20~100 m). Based on the Landsat scene quality information, we cleaned and kept just good quality Landsat pixel data. Then, daily EVI and NDVI were retrieved from the area-weighted averaged Landsat pixels inside the field sampling plot. We averaged the daily EVI and NDVI values by month. Finally, the monthly EVI and NDVI at the field sampling plots level were regressed with corresponding monthly in situ LAI values to calculate the statistics ($R^2$, etc.).

We investigated whether forest cover fraction in different LAI products' spatial resolution levels influenced product bias and error. We use the 30 m resolution GlobeLand30 land cover product for 2010 [69] to extract the forest cover fraction inside the relevant pixels for each of the LAI products at each site. We classify the land cover for each 30 m pixel on different spatial scales (LAI products' scale: 300 m, 500 m, and 1 km) at each site and plot. We only kept the 30 m pixels located entirely inside the LAI pixels. The total number of 30 m pixels and pixels classified as forests were both counted and summed to determine the total area and the forest areas at different spatial scales. Then we can obtain the forests cover proportion at different spatial scales for each plot and site. Finally, the forests' cover proportion was compared with the LAI products' bias at each site.

### 2.5. Statistical Analysis

Daily satellite LAI and uncertainty information were retrieved from the three LAI products at the location of each subplot of the six sites (Figure S1). If the field subplots overlapped with multiple pixels for a given LAI product, then the satellite-based value was estimated using the weighted sum of the contributing pixels, weighted by the area of overlap (Equation (3)). Data were cleaned based on the products' quality assessment information, before aggregation into monthly mean values, and comparison with the mean monthly ground-based LAI measurements at each site and plot. We calculate $R^2$, bias, and RMSE between the field and satellite LAI time series to evaluate the accuracy and error of the three LAI products.

$$LAI_R = \sum_{i=1}^{n} \frac{Ai}{A} * Pi, \tag{3}$$

where $LAI_R$ is the extracted remote sensing LAI value; $n$ represents the total number of the contributing pixels which overlapped with the field plots; $A$ is the total area for the plots, $Ai$ is the overlapped area between pixel and plots; $Pi$ is the remote sensing pixel value.

LAI products' uncertainty information was evaluated using the ratio of LAI products' uncertainty ($LAI_U$) with the bias against the field measurements, $r$, as follows (Equation (4)),

$$r = \frac{LAI_U}{|Bias|}, \tag{4}$$

where the bias (Equation (5)) is defined as:

$$Bias = LAI_{product} - LAI_{field}, \tag{5}$$

The ratio, $r$, can be used to evaluate whether or not the product uncertainty is robust. When $r \geq 1$, the bias in the LAI product is within the reported uncertainty range, indicating that the uncertainty estimate for LAI is robust; conversely, if $r < 1$, the reported uncertainty fails to capture the observed bias and therefore the uncertainty estimate is not robust. We calculate this ratio $r$ for each month at different plots and sites.

### 2.6. Calculation of Amplitude, Phases and Periods for LAI Time Series

We use metaCycle [70] to calculate the periodic characteristics (periods, phases, and amplitude) for the monthly field, MODIS, GEOV2 and V3 LAI time series. This method can calculate the periodic characteristics for time series with missing values [70], which frequently occurred for the satellite LAI times series. MetaCycle incorporates three methods: ARSER (Autoregressive spectral analysis) [71], JTK_CYCLE(Jonckheere-Terpstra-Kendall algorithm) [72], and Lomb-Scargle [73], for periodic signal detection, and it could output integrated analysis. The integrated period from MetaCycle is an arithmetic mean value of multiple periods, while phase integration based on the mean of circular quantities calculated from the above three mentioned methods.

MetaCycle recalculates the amplitude with the following model (Equation (6)):

$$Yi = B + TRE * \left( ti - \frac{\sum_{i=1}^{n} ti}{n} \right) + A * \cos\left( 2 * \pi * \frac{ti - PHA}{PER} \right), \tag{6}$$

where $Yi$ is the observed value at time $t_i$; $B$ represents the baseline value (mean value) of the LAI time series; $TRE$ is the trend level of the time-series profile; $A$ is the amplitude of the waveform. $PER$ and $PHA$ are the integrated period and phase, respectively. $PER$ represented the periods of the LAI time series; here, it is 12 months. $PHA$ represented the phases of the LAI time series at the period of 12 months. "$i$" is the time points of the time series. "$n$" represents the total number of the time points of the time series. In this model, only $B$, $TRE$, and $A$ are unknown parameters and can be calculated using ordinary least squares. Fisher's method is implemented in MetaCycle for integrating multiple $p$-values.

Relative amplitude (rAMP), which is the *A/B*, can be used to compare the amplitudes between time series with different baselines levels.

In this study, each LAI time series is decomposed into the baseline value (mean value), the trend, and the wavelength with specific amplitude, period, and phase. The integrated phases can be used to compare whether the LAI time series is lead, lag, or synchronous. The amplitude can be used to quantify the magnitude of seasonal fluctuations for LAI time series and the relative amplitude can be used to compare the LAI seasonal fluctuations from different sites and data source. We apply this method to the monthly averaged field LAI time series at BNF, and the monthly averaged retrieved satellite LAI at each site because only BNF has frequent enough field measurements over multiple months from the year 2005 to 2017.

### 2.7. Software

Remote-sensing data processing was carried out by Rv3.4.0 [74] using the package "raster" [75]. Amplitude, phases, and periods of LAI time series were calculated using the functions "meta2d" in the package 'MetaCycle' [70] in R. Figures and maps were produced in R and ArcGIS 10.4.

## 3. Results

### 3.1. Accuracy of the LAI Products Magnitude

Satellite derived mean LAI estimates were typically lower than the field measurements, which varied between 3.2 and 5.6 m$^2$ m$^{-2}$ (Table 2). GEOV3 300 m estimates were 10% lower than field measurements (mean = 4.50 m$^2$ m$^{-2}$), MODIS 500m were 30% lower (mean = 3.6 m$^2$ m$^{-2}$ and GEOV2 1 km were 40% lower (mean = 3.07 m$^2$ m$^{-2}$) (Figure 2; Table 2). LAI mean and relative uncertainty increased with product resolution: GEOV3 300m (0.22, 7%) < MODIS 500 m (0.37, 10%) < GEOV2 1 km (0.56, 19%)

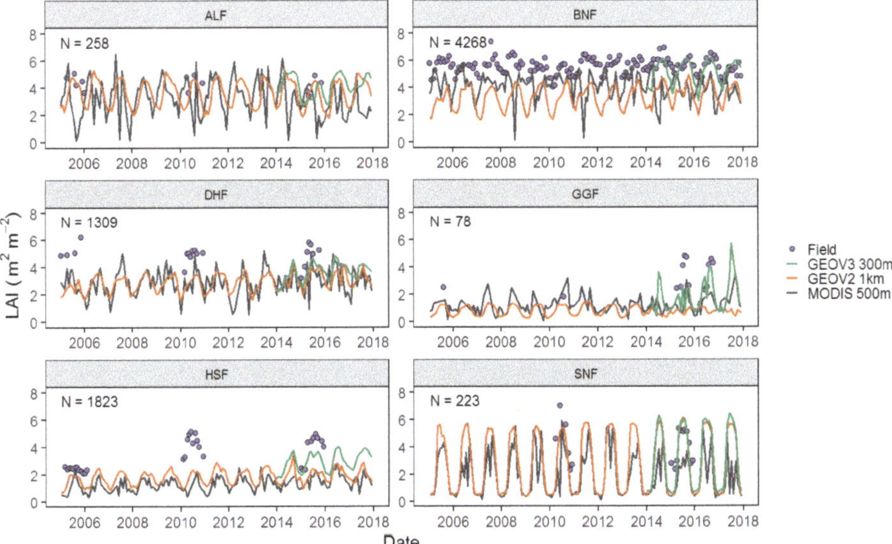


**Figure 2.** Time series of the LAI for the field, GEOV3 300 m, GEOV2 1 km, and the MODIS 500 m for different sites. The timespan of the LAI time series is from 2005 to 2017. Each field data point represented the mean value of the measurements from the sample subplots in each month. "N" denoted the number of total sample measurements for each site, including the repeat measurements in each month at each subplot.

**Table 2.** Mean site LAI values ($m^2\ m^{-2}$) and their statistical metrics calculated using months which have both field and EO LAI values from 2005 to 2017. "$r$" is median of the ratio between product uncertainty and bias for all months and subplots. $R^2$ was calculated from the linear regression between satellite products and field monthly measurements. "N" denotes the number of monthly values for all subplots at each site. All the statistics are calculated on a monthly scale.

| LAI Source | Metric Statics | ALF | BNF | DHF | GGF | HSF | SNF | Overall |
|---|---|---|---|---|---|---|---|---|
| Field (20~100 m) | Mean | 4.18 | 5.55 | 4.90 | 3.18 | 3.44 | 4.11 | 5.1 |
| | Measurements error (std) | 0.46 | 0.58 | 0.89 | 0.5 | 0.51 | 0.57 | 0.59 |
| | Relative error (std/mean) | 0.11 | 0.11 | 0.18 | 0.16 | 0.15 | 0.14 | 0.12 |
| | N | 38 | 571 | 57 | 14 | 96 | 17 | 793 |
| GEOV3 300 m | Mean | 3.28 | 5.15 | 3.82 | 2.38 | 2.94 | 3.34 | 4.50 |
| | Product uncertainty (RMSE) | 0.13 | 0.20 | 0.12 | 0.04 | 0.55 | 0.30 | 0.22 |
| | Relative uncertainty (RMSE/Mean) | 0.04 | 0.04 | 0.03 | 0.02 | 0.19 | 0.09 | 0.07 |
| | Bias (mean) | −0.58 | −0.40 | −0.89 | −1.18 | −0.94 | −0.66 | −0.54 |
| | $r$ (median) | 0.17 | 0.24 | 0.09 | 0.65 | 0.45 | 0.36 | 0.26 |
| | $R^2$ | 0.12 | 0.18 | 0.53 | 0.05 | 0.21 | 0.93 | 0.45 |
| | N | 20 | 191 | 23 | 11 | 30 | 11 | 286 |
| GEOV2 1 km | Mean | 3.3 | 3.35 | 2.92 | 0.82 | 1.72 | 2.96 | 3.07 |
| | Product uncertainty (RMSE) | 0.54 | 0.60 | 0.60 | 0.27 | 0.33 | 0.77 | 0.56 |
| | Relative uncertainty (RMSE/Mean) | 0.16 | 0.18 | 0.21 | 0.33 | 0.19 | 0.26 | 0.19 |
| | Bias(mean) | −0.88 | −2.20 | −1.97 | −2.36 | −1.72 | −1.16 | −2.04 |
| | $r$ (median) | 0.56 | 0.28 | 0.37 | 0.16 | 0.22 | 0.58 | 0.28 |
| | $R^2$ | 0.18 | 0.22 | 0.15 | 0.38 | 0.30 | 0.74 | 0.36 |
| | N | 38 | 571 | 57 | 14 | 96 | 17 | 793 |
| MODIS 500 m | Mean | 3.1 | 4.1 | 3.3 | 1.45 | 1.29 | 1.72 | 3.60 |
| | Product uncertainty (std) | 0.45 | 0.38 | 0.48 | 0.66 | 0.16 | 0.40 | 0.37 |
| | Relative uncertainty (std/mean) | 0.15 | 0.09 | 0.14 | 0.44 | 0.12 | 0.23 | 0.10 |
| | Bias (mean) | −0.99 | −1.35 | −1.54 | −1.94 | −2.09 | −2.38 | −1.47 |
| | $r$ (median) | 0.37 | 0.24 | 0.39 | 0.26 | 0.07 | 0.13 | 0.21 |
| | $R^2$ | 0.03 | 0.12 | 0.10 | 0.47 | 0.16 | 0.25 | 0.20 |
| | N | 33 | 506 | 52 | 7 | 85 | 17 | 700 |

Field mean LAI was best captured by GEOV3 300 m ($R^2$ = 0.45; Figure 3), followed by GEOV2 1 km ($R^2$ = 0.36), and MODIS 500 m ($R^2$ = 0.20) for all of these six forests in this region. This performance varied between forests and all of these products showed the highest $R^2$ (Table 2) for the forests at SNF, which showed more LAI seasonal variation (Figure 2). GEOV3 300 m LAI also had the lowest RMSE and bias (RMSE = 1.21, bias = −0.41) compared with GEOV2 1 km LAI (RMSE = 2.32, bias = −2.04) and MODIS 500m LAI (RMSE = 2.29, bias = −1.47) (Figure 3, Table 2).

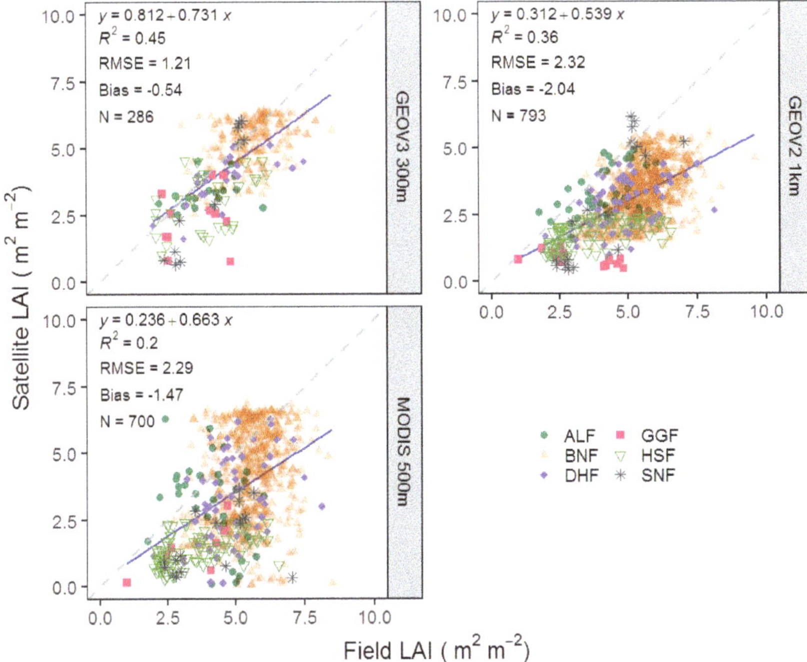

**Figure 3.** Regression between the field LAI and the satellite retrieved LAI. Each point represents the monthly LAI field measurement against the satellite LAI retrieval. The blue solid lines represent the regression between the field LAI against the satellite LAI for all the sites combined on a monthly scale. Point colors and shapes represent different sites.

## 3.2. Robustness of the LAI Products Uncertainty

The magnitude of the reported EO LAI uncertainty is typically smaller than the bias to in situ estimates (Figure 4), here shown as the ratio of uncertainty and reported bias ($r$), and this result is consistent across forests (Figure S5, Table 2). GEOV3 300 m has the smallest proportion of EO uncertainty estimates below the observed bias (86%) while MODIS 500m LAI and GEOV2 1 km have 87% and 91% of uncertainty estimates below the reported bias. The median ratio indicated similar average skill between both GEOV3 300 m (0.26) and GEOV2 1km LAI products (0.28) and MODIS 500 m LAI products (0.21) (Figure 4, Table 2).

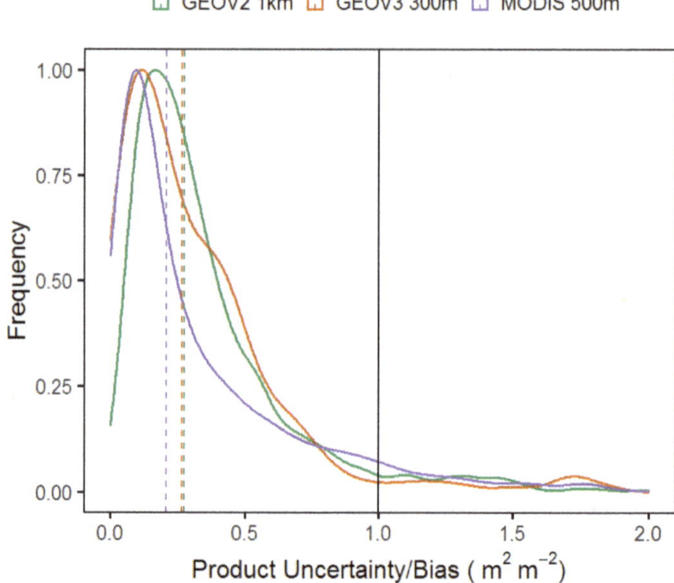



**Figure 4.** Distribution of the ratio of EO uncertainty estimates to bias, based on a comparison against in situ LAI measurements. Dashed lines represent the median values and the color represents different EO products. The vertical black line is the reference line where the ratio is equal to 1. A ratio larger than 1 indicates the uncertainty estimate is larger than the bias and is therefore considered robust.

### 3.3. Validation of the LAI Products Temporal Dynamics

The metaCycle analysis showed that a statistically significant period of 12 months ($p < 0.05$) is present in the monthly field LAI time series at BNF and in all monthly satellite LAI time series at all sites, except for MODIS 500 m LAI, which showed no significant 12-month periodic characteristics at BNF. Overall, LAI retrieved from GEOV3 300 m showed very close values of the base, phase, amplitude and relative amplitude with the field LAI at BNF (Figure 5). LAI retrieved from GEOV2 1km showed a lower base and amplitude, similar phase, but larger relative amplitude compared with GEOV3 300 m LAI for most sites and the field at BNF (Figure 5). LAI retrieved from MODIS 500 m product showed the lower base values, but had a high variation of phases with one underestimated phase for the forest at ALF compared with the other two LAI products (Figure 5b). The relative amplitude of LAI retrieved from MODIS 500 m is lower for the forest at GGF and SNF, but higher at HSF compared with the other two LAI products (Figure 5c).

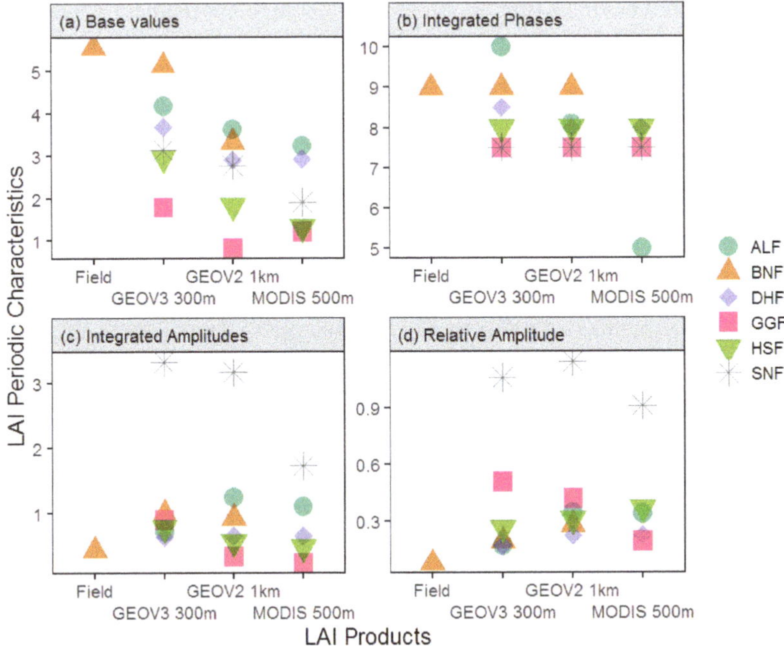



**Figure 5.** Periodic characteristics of the LAI time series from 2005 to 2017 for different sites. (**a**): Baseline values of the LAI time series; (**b**): Phase of the LAI time series; (**c**): Amplitude of the LAI time series; (**d**): Relative amplitude of the LAI time series. Point colors and shapes represent different sites. A detailed description of the calculation can be found in the Methods section of this paper.

### 3.4. Evaluation of Landcover Heterogeneity

The Landsat data provide a more spatially consistent comparison of surface reflectance against field data of LAI. The results (Figure S6) show that there are significant correlations between NDVI/EVI and LAI across sites, but the effect sizes are small ($R^2 = 0.1$). Thus, the relationships between remotely-sensed NDVI/EVI at the fine resolution and field-measured LAI are less significant than those between LAI products at the coarse resolution and field-measured LAI. The forest cover proportion at the pixel level varied between 38%–100% (Table S3). The landcover data show that LAI retrievals were degraded by the heterogeneity of landcover within product pixels. There was a clear pattern of LAI bias increasing with reductions in the proportions of forest cover at the pixel scale (Figure S7).

## 4. Discussion

The performance of three EO LAI products was evaluated at six forests across southern China. GEOV3 300 m LAI showed the best fit ($R^2 = 0.45$) and the smallest bias (bias = −0.54) (Figure 3) and the most similar LAI time series seasonal dynamic variation (phase and amplitude) compared with in situ LAI measurements (Figure 5). For the GEOV2 1 km LAI product, its performance had larger bias (Figure 3) and seasonality but similar phases of the LAI time series with the in situ observations compared with GEOV3 300 m LAI products (Figures 3 and 5)

The different performance of the two Copernicus products is likely due to the mismatch of the scaling between the EO pixel size and the field site. LAI is strongly non-linearly related to reflectance, making its estimation from remote sensing observations scale-dependent [76,77]. In contrast, the core operational algorithm (neural network techniques), data filtering and smoothing processes are similar for these two products. There are differences in the method used for temporal compositing, where temporal smoothing and gap filling using a climatology are used for GEOV2 1 km and interpolation applied in GEOV3 300 m) [78]. Differences in the applied gap-filling approach between these two products do not impact our conclusion that resolution is the primary driver of performance improvement at 300 m relative to 1 km, as all gap-filled and interpolated retrievals were removed in our study. The in situ measurements were usually conducted in protected mature forest areas with high plant densities, complex canopy stratums, and rich species diversities (Figure S1, Table S1). However, at coarser spatial resolutions, pixels integrate heterogeneity over a greater diversity of landscapes, habitats, and species, distributed across a variety of stages of growth and succession [79]. Thus, fine resolution GEOV3 300 m LAI product showed lower bias and similar seasonal dynamic variation to the in-situ measurements compared to the higher bias for the GEOV2 1 km LAI product.

For Collection 6 500 m MODIS LAI, our results indicate that this product does not capture the dynamics observed in situ, irrespective of the accuracy of estimated in situ LAI (Figure 3). Furthermore, it does not adequately capture the LAI seasonal dynamics (Figure 5). This result is consistent with two validation studies which both showed MODIS LAI had the poorest performance for the evergreen forests in the south of China [80,81]. The MODIS LAI values for tropical evergreen forests are severely impacted by atmospheric conditions, especially clouds during the growing season (around 42% data are influenced by the cloud in this study, Figure S8), which lead to strong noise in the input reflectance data and affect the retrieval [78]. Additionally, the reflectance saturation usually happened in dense canopies and the main algorithm is sensitive to uncertainties in atmospheric correction, particularly when red and NIR BRFs are saturated [82,83]. This means the reflectance does not contain sufficient information to estimate the LAI value [84] and leads to the instability of the LAI retrievals [14]. All of these may result in the poor performance of representation of evergreen forests in southern China for the MODIS 500 m LAI products. Furthermore, the product does not adequately capture LAI periodic characteristics in comparison to the field data and showed temporal inconsistency with GEOV2 and V3 products (Figure 5). Jiang (2017) also found the large temporal inconsistency between existing global LAI products at a longer time scale [85]. Cammalleri (2019) found GEOV2 fAPAR showed a systematic overestimation of the fAPAR anomalies compared with the MODIS fAPAR and proposed a two-step harmonization procedure to remove this discrepancy [86]. However, the homogenization may alter the magnitude of the original fAPAR time series in an undesirable way. These results highlight the need to validate the temporal consistency between different satellite products and explore more solutions to deal with such inconsistencies. In addition, geolocation uncertainty due to the spatial mismatch could also have influenced validation reliability, As our study sites are not homogeneous and sampling area is consistently smaller than the pixel of remote sensing data, mean or median remotely-sensed LAI values of surrounding $3 \times 3$ array of pixels [15,87,88] cannot be used for validation. To solve this problem, future field measurements should be conducted at a larger spatial scale and across more homogenous habitats.

Overall, the absolute and relative uncertainty of LAI products tends to be smaller for the fine resolution LAI products (Table 2). Collection 300 m GEOV3 has the smallest absolute uncertainty at around 0.22. Collection1 km GEOV2 has the largest absolute uncertainty at around 0.56. Uncertainty magnitude for Collection 6 500 m MODIS LAI is 0.37. This performance is consistent with the LAI relative uncertainty (Table 2). If compared with the absolute uncertainty requirements (±0.5) and relative uncertainty requirements (20%) set by the GCOS [12], all three products (Collection 300 m GEOV3, 0.22, 5%; Collection 6 500 m MODIS LAI, 0.37, 10%; Collection 1 km GEOV2 0.56, 18%) appear satisfactory. This is consistent with two global studies comparing different EO LAI products (MODIS, CYCLOPES, and GLOBCARBON) at two coarse spatial resolutions (5 km and 1 km). More pixels can meet the absolute uncertainty and relative uncertainty requirements at the spatial resolution of 1 km compared with the 5 km [19,34]. All of these results indicated that the LAI product uncertainty is scale-dependent. The algorithms will produce smaller uncertainty estimated for the LAI values at finer spatial scales. This could make the uncertainty estimates of the fine-resolution products become very conservative.

Changes in LAI error with spatial resolution could be due to vegetation heterogeneity, especially for the forests and scale-dependent reflectance values. However, it is still unclear how product uncertainty changes over different spatial scales. For MODIS, the product uncertainty is the standard deviation over all acceptable solutions of a look-up table (LUT) retrieval method [17]; for Collection 300 m GEOV3 and 1 km GEOV2 LAI products, the quantitative uncertainties (LAI_ERR) are computed as the RMSE between the final dekadal value and the daily NNTs estimates in the compositing period [12,62]. Standardizing different LAI products to the same spatial scale and averaging the LAI and uncertainty for different pixels during the upscaling and resampling process [19,89] could all generate errors [90]. When dealing with different spatial scales, the land surface properties (e.g., land cover proportion, soil properties) could change significantly. Thus, LAI retrieval algorithms are based on inherent empirical assumptions on the distribution of their parameters, which can depart significantly from the actual canopy and soil characteristics [14].

The ratio ($r$, in Equation (4)) of LAI product uncertainty to the bias of in situ measurements is dramatically lower than the reference value for all of these LAI products (Figure 4, Table 2). This result indicated that none of the three LAI uncertainty products tested here were able to provide a robust estimate for the in-situ measurements. The most promising product is the Collection 300 m GEOV3 LAI; however, while its accuracy is the best, its uncertainty is still too conservative to capture the offset from co-located field LAI measurements. This weakness reflects the complexity of this biome type and highlights that the algorithm used to generate the uncertainty information for these LAI products needs to be improved. For areas of high uncertainties, data producers may need to refine the algorithms and verify the information using additional data sources, particularly in situ data [34]. However, with respect to field data collection, more accurate measurements are also needed to test the results, such as destructive methods or using the litterfall traps. Direct measurement methods obtaining 'true' LAI values could also act as a reference to correct the bias and errors in indirect measurement techniques [31].

The accuracy of LAI data retrieved from remote sensing products is very important for a range of earth system models, for which LAI is a key internal variable. LAI uncertainty is particularly critical for model-data fusion in terrestrial carbon cycle studies [91]. The uncertainty produced for the existing remote sensing LAI products in the analysis would further lead to biases in vegetation productivity estimates [92–94]. For instance, the saturation of satellite-derived VIs generally results in an underestimation of GPP or NPP in areas with dense vegetation. Using finer resolution remote sensing data in carbon modeling studies could be an effective way to reduce the uncertainty in the estimates of C fluxes and/or stocks, but more robust errors are critical [95].

Uncertainty estimates of the observations are as important as the observations themselves because together they co-determine the outcome of the assimilation systems [96]. Bayesian approaches are commonly used data assimilation systems and the influence of observations are weighted by observational uncertainties. Uncertainties in data are critical as they often determine the outcome of analyses and forecasts [97]. The assimilation process requires clear error quantification for LAI to resolve model results and limit biases [98]. Besides, the presence of bias in a data stream will limit the utility of using multiple observation types in an assimilation framework. Therefore, it is imperative to characterize the error in the observations and understand better the error associated with the direct measurements of LAI to landscape pixel [99]. Our results indicated that none of these LAI uncertainty products are robust and therefore users should consider inflating the existing LAI product uncertainty when using these data sources in data assimilation frameworks [98,100,101] or in other studies for which the uncertainty of the product plays an important role.

One way of solving this spatial mismatch problem is to upscale the field measurements based on its relationship with the plant index retrieved from high-resolution images (e.g., 10–30 m resolution) [102], or using airborne Light Detection and Ranging (LiDAR) [103,104]. However, we failed to find a good relationship between the Landsat based EVI or NDVI and the field measurements (Figure S5). LAI-VI relationships are limited by a number of factors, including vegetation type, sun-surface-sensor geometry, leaf chlorophyll content, background reflectance, and atmospheric quality [31]. Geolocation uncertainties due to spatial mismatch and insufficient overlap data between in situ measurements and good quality Landsat scenes partly lead to this result. Saturation effects occur for both field measurement and spectral plant indices for very dense canopies, which also increases the challenge of upscaling field LAI to generate high-resolution LAI reference maps. In addition, the temporal mismatch between field and Landsat data also contributes to the poor relationship between field and Landsat data. This result may be also associated with the complex forest environments (such as terrain, tree densities, canopy structures, and species diversities) in southern China (Figure S4, Table S1). Thus, their LAI cannot be accurately modeled based on simple spectral indices, even with relatively high resolution (30 m) satellite data. More complex upscaling approaches or those utilizing very high-resolution sensors (e.g., LIDAR) to bridge the scaling gap may yield better results [102,105,106]. In particular, there is an exciting potential for unmanned aerial vehicles to provide critical upscaling information from field to satellite [21].

When using EO LAI products in the absence of reliable validation data, we recommend that the provided uncertainties are treated cautiously. Remote-sensing LAI products require improved algorithms with particular attention given to generating robust uncertainty estimates. New cloud and aerosol detection techniques based on time series and spatial analysis may help to improve the uncertainty products and LAI estimated for evergreen forests [107]. In addition, the new remote-sensing systems such as Global Ecosystem Dynamics Investigation (GEDI) [108], National Aeronautics and Space Administration's (NASA's) Ice, Cloud, and land Elevation Satellite-2 (ICESat-2) and Synthetic aperture radar (SAR) provide promising ways to solve cloud coverage issues [109]. Given the importance of data uncertainties (e.g., integration within terrestrial carbon modeling frameworks [92–94]), there is an urgent need for robust uncertainty characterization, based on links to in situ observations. The findings in this study are constrained to southern China. Future studies could explore the LAI and error products for other regions, or focus on the seasons where products show the largest uncertainties [19].

## 5. Conclusions

In this study, we validated three global LAI products and their uncertainty products using ~8000 in situ measurements of LAI from subtropical and tropical forests in the southern China region. The finest resolution product, Collection 300 m GEOV3, performed best, with the lowest RMSE and the lowest bias. It also best captured the temporal dynamics observed in the in-situ dataset, including the magnitude, amplitude, and phases of the LAI time series. Collection 6 500 m MODIS LAI showed the poorest performance and importantly it did not capture the temporal dynamics observed in situ for the forests in this specific region. Critically, for all three LAI products, the accompanying uncertainties were all far smaller than the bias compared to the in situ measurements, indicating that each product uncertainty estimate is not robust. Given the importance of LAI uncertainty in the context of constraining models of carbon cycling and other ecosystem processes, users should use the product's uncertainty with caution and consider inflating the existing LAI product uncertainty when using it within data assimilation frameworks or other studies in which the uncertainty plays an important role.

**Supplementary Materials:** The following are available online at http://www.mdpi.com/2072-4292/12/19/3122/s1, Table S1: Detailed forest community information in the southern China region, Table S2: The detailed information for field LAI measurements at each site, Table S3: Distribution of the ratio of EO uncertainty estimates to bias, based on a comparison against in situ LAI measurements, Figure S1: Forest appearance for each forest in this study in the southern China region, Figure S2: TOPO30 DEM (1 km) in the China region. Figure S3: Location of the sampling subplots and the GEOV2 1km LAI pixels showed on the Landsat EVI scenes for different sites, Figure S4: Location of the sampling subplots and the GEOV2 1 km LAI pixels showed on the SARTM 90 m DEM images for different sites. Figure S5: Frequency distribution of the ratio of LAI products uncertainty with LAI products' bias against the in situ measurements for different sites. Figure S6: Regression analyses between the Landsat EVI/NDVI values and the field observed LAI values at sampling subplots, Figure S7: Regression analyses between the proportion of forests cover (LAI pixel level) and the remotely sensed LAI bias against the field measured LAI for different LAI products, Figure S8: Summary of the dates for the MODIS LAI pixels which were influenced by the cloud in this study.

**Author Contributions:** Conceptualization, Y.Z. and M.W.; methodology, Y.Z., M.W., T.L.S., and D.T.M.; formal analysis, T.L.S. and S.F.-P. and D.T.M.; writing—original draft preparation, Y.Z. and M.W.; writing—review and editing, M.W., X.C., T.L.S., D.T.M., S.F.-P., and Y.Z. All authors have read and agreed to the published version of the manuscript.

**Funding:** This research was funded by the National Natural Science Foundation of China under grant number 41771049 and NCEO, CSSP Newton Fund, and China Scholarship Council.

**Acknowledgments:** The authors thank all the workers in the Chinese Ecosystem Research Network for field data collection and database establishment.

**Conflicts of Interest:** The authors declare no conflict of interest.

## References


1. Anav, A.; Murray-Tortarolo, G.; Friedlingstein, P.; Sitch, S.; Piao, S.; Zhu, Z. Evaluation of land surface models in reproducing satellite Derived leaf area index over the high-latitude northern hemisphere. Part II: Earth system models. *Remote Sens.* **2013**, *5*, 3637–3661. [CrossRef]

2. Mahowald, N.; Lo, F.; Zheng, Y.; Harrison, L.; Funk, C.; Lombardozzi, D.; Goodale, C. Projections of leaf area index in earth system models. *Earth Syst. Dynam.* **2016**, *7*, 211–229. [CrossRef]

3. Williams, M.; Richardson, A.D.; Reichstein, M.; Stoy, P.C.; Peylin, P.; Verbeeck, H.; Carvalhais, N.; Jung, M.; Hollinger, D.Y.; Kattge, J. Improving land surface models with FLUXNET data. *Biogeosciences* **2009**, *6*, 2785. [CrossRef]

4. Asner, G.P.; Scurlock, J.M.; Hicke, A.J. Global synthesis of leaf area index observations: Implications for ecological and remote sensing studies. *Glob. Ecol. Biogeogr.* **2003**, *12*, 191–205. [CrossRef]

5. Fang, H.; Liang, S.; Hoogenboom, G. Integration of MODIS LAI and vegetation index products with the CSM–CERES–Maize model for corn yield estimation. *Int. J. Remote Sens.* **2011**, *32*, 1039–1065. [CrossRef]

6. Richardson, A.D.; Dail, D.B.; Hollinger, D. Leaf area index uncertainty estimates for model–data fusion applications. *Agric. For. Meteorol.* **2011**, *151*, 1287–1292. [CrossRef]

7. Asaadi, A.; Arora, V.K.; Melton, J.R.; Bartlett, P. An improved parameterization of leaf area index (LAI) seasonality in the Canadian Land Surface Scheme (CLASS) and Canadian Terrestrial Ecosystem Model (CTEM) modelling framework. *Biogeosciences* **2018**, *15*, 6885–6907. [CrossRef]

8. Tum, M.; Günther, K.P.; Böttcher, M.; Baret, F.; Bittner, M.; Brockmann, C.; Weiss, M. Global gap-free MERIS LAI time series (2002–2012). *Remote Sens.* **2016**, *8*, 69. [CrossRef]

9. Gonsamo, A.; Chen, J.M. Improved LAI Algorithm Implementation to MODIS Data by Incorporating Background, Topography, and Foliage Clumping Information. *IEEE Trans. Geosci. Remote Sens.* **2014**, *52*, 1076–1088. [CrossRef]

10. Myneni, R.; Knyazikhin, Y.; Park, T. *MOD15A2H MODIS/Terra Leaf Area Index/FPAR 8-Day L4 Global 500 m SIN Grid V006* [Data set]. NASA EOSDIS Land Processes DAAC. 2015. Available online: https://doi.org/10.5067/MODIS/MCD15A2H.006 (accessed on 15 September 2020).

11. Verger, A.; Baret, F.; Weiss, M. Near Real-Time Vegetation Monitoring at Global Scale. *IEEE J. Sel. Top. Appl. Earth Obs. Remote Sens.* **2014**, *7*, 3473–3481. [CrossRef]

12. Baret, F.; Weiss, M.; Verger, A.; Smets, B. *Atbd for Lai, Fapar and Fcover From Proba-V Products at 300m Resolution (Geov3)*; INRA: Paris, France, 2016.

13. Williams, M.; Rastetter, E.B.; Shaver, G.R.; Hobbie, J.E.; Carpino, E.; Kwiatkowski, B.L. Primary production of an arctic watershed: An uncertainty analysis. *Ecol. Appl.* **2001**, *11*, 1800–1816. [CrossRef]

14. Garrigues, S.; Lacaze, R.; Baret, F.; Morisette, J.; Weiss, M.; Nickeson, J.; Fernandes, R.; Plummer, S.; Shabanov, N.; Myneni, R. Validation and intercomparison of global Leaf Area Index products derived from remote sensing data. *J. Geophys. Res. Biogeosci.* **2008**, *113*. [CrossRef]

15. Weiss, M.; Baret, F.; Garrigues, S.; Lacaze, R. LAI and fAPAR CYCLOPES global products derived from VEGETATION. Part 2: Validation and comparison with MODIS collection 4 products. *Remote Sens. Environ.* **2007**, *110*, 317–331. [CrossRef]

16. Baret, F.; Hagolle, O.; Geiger, B.; Bicheron, P.; Miras, B.; Huc, M.; Berthelot, B.; Niño, F.; Weiss, M.; Samain, O.; et al. LAI, fAPAR and fCover CYCLOPES global products derived from VEGETATION: Part 1: Principles of the algorithm. *Remote Sens. Environ.* **2007**, *110*, 275–286. [CrossRef]

17. Knyazikhin, Y. MODIS Leaf Area Index (LAI) and Fraction of Photosynthetically Active Radiation Absorbed by Vegetation (FPAR) Product (MOD 15) Algorithm Theoretical Basis Document. 1999. Available online: https://modis.gsfc.nasa.gov/data/atbd/atbd_mod15.pdf (accessed on 15 September 2020).

18. Pinty, B.; Andredakis, I.; Clerici, M.; Kaminski, T.; Taberner, M.; Verstraete, M.; Gobron, N.; Plummer, S.; Widlowski, J.L. Exploiting the MODIS albedos with the Two-stream Inversion Package (JRC-TIP): 1. Effective leaf area index, vegetation, and soil properties. *J. Geophys. Res. Atmos.* **2011**, *116*. [CrossRef]

19. Fang, H.; Jiang, C.; Li, W.; Wei, S.; Baret, F.; Chen, J.M.; Garcia-Haro, J.; Liang, S.; Liu, R.; Myneni, R.B.; et al. Characterization and intercomparison of global moderate resolution leaf area index (LAI) products: Analysis of climatologies and theoretical uncertainties. *J. Geophys. Res. Biogeosci.* **2013**, *118*, 529–548. [CrossRef]

20. GCOS. *Systematic Observation Requirements for Satellite-Based Products for Climate. 2011 Update Supplemetnatl Details to the Satellite 39 Based Component og the Implementation Plan for the Global Observing System for Climate in Support of the UNFCCC (2010 Update);* Technical Report; World Meteorological Organisation (WMO) 7 bis: Geneva, Switzerland, 2011.

21. Revill, A.; Florence, A.; MacArthur, A.; Hoad, S.; Rees, R.; Williams, M. Quantifying Uncertainty and Bridging the Scaling Gap in the Retrieval of Leaf Area Index by Coupling Sentinel-2 and UAV Observations. *Remote Sens.* **2020**, *12*, 1843. [CrossRef]

22. Williams, M.; Bell, R.; Spadavecchia, L.; Street, L.E.; Van Wijk, M.T. Upscaling leaf area index in an Arctic landscape through multiscale observations. *Glob. Chang. Biol.* **2008**, *14*, 1517–1530. [CrossRef]

23. Wang, Q.; Tenhunen, J.; Dinh, N.Q.; Reichstein, M.; Otieno, D.; Granier, A.; Pilegarrd, K. Evaluation of seasonal variation of MODIS derived leaf area index at two European deciduous broadleaf forest sites. *Remote Sens. Environ.* **2005**, *96*, 475–484. [CrossRef]

24. Wang, J.; Wu, C.; Wang, X.; Zhang, X. A new algorithm for the estimation of leaf unfolding date using MODIS data over China's terrestrial ecosystems. *ISPRS J. Photogramm. Remote Sens.* **2019**, *149*, 77–90. [CrossRef]

25. Kou, W.; Liang, C.; Wei, L.; Hernandez, A.J.; Yang, X. Phenology-based method for mapping tropical evergreen forests by integrating of MODIS and landsat imagery. *Forests* **2017**, *8*, 34.

26. Clark, D.A. Sources or sinks? The responses of tropical forests to current and future climate and atmospheric composition. *Philos. Trans. R. Soc. Lond. B Biol. Sci.* **2004**, *359*, 477–491. [PubMed]

27. Miller, S.D.; Goulden, M.L.; Hutyra, L.R.; Keller, M.; Saleska, S.R.; Wofsy, S.C.; Figueira, A.M.S.; da Rocha, H.R.; de Camargo, P.B. Reduced impact logging minimally alters tropical rainforest carbon and energy exchange. *Proc. Natl. Acad. Sci. USA* **2011**, *108*, 19431–19435. [CrossRef] [PubMed]

28. Tang, A.C.I.; Stoy, P.C.; Hirata, R.; Musin, K.K.; Aeries, E.B.; Wenceslaus, J.; Shimizu, M.; Melling, L. The exchange of water and energy between a tropical peat forest and the atmosphere: Seasonal trends and comparison against other tropical rainforests. *Sci. Total Environ.* **2019**, *683*, 166–175. [CrossRef]

29. Heiskanen, J.; Korhonen, L.; Hietanen, J.; Pellikka, P.K.E. Use of airborne lidar for estimating canopy gap fraction and leaf area index of tropical montane forests. *Int. J. Remote Sens.* **2015**, *36*, 2569–2583. [CrossRef]

30. Piao, S.; Liu, Q.; Chen, A.; Janssens, I.A.; Fu, Y.; Dai, J.; Liu, L.; Lian, X.; Shen, M.; Zhu, X. Plant phenology and global climate change: Current progresses and challenges. *Glob. Chang. Biol.* **2019**, *25*, 1922–1940.

31. Fang, H.; Baret, F.; Plummer, S.; Schaepman-Strub, G. An overview of global leaf area index (LAI): Methods, products, validation, and applications. *Rev. Geophys.* **2019**, *57*, 739–799.

32. Chhabra, A.; Panigrahy, S. Analysis of spatio-temporal patterns of leaf area index in different forest types of India using high temporal remote sensing data. *Int. Arch. Photogramm. Remote Sens. Spat. Inf. Sci.* **2011**, *38*, W20.

33. Olivas, P.C.; Oberbauer, S.F.; Clark, D.B.; Clark, D.A.; Ryan, M.G.; O'Brien, J.J.; Ordonez, H. Comparison of direct and indirect methods for assessing leaf area index across a tropical rain forest landscape. *Agric. For. Meteorol.* **2013**, *177*, 110–116.

34. Fang, H.; Wei, S.; Jiang, C.; Scipal, K. Theoretical uncertainty analysis of global MODIS, CYCLOPES, and GLOBCARBON LAI products using a triple collocation method. *Remote Sens. Environ.* **2012**, *124*, 610–621.

35. Huete, A.R.; Didan, K.; Shimabukuro, Y.E.; Ratana, P.; Saleska, S.R.; Hutyra, L.R.; Yang, W.; Nemani, R.R.; Myneni, R. Amazon rainforests green-up with sunlight in dry season. *Geophys. Res. Lett.* **2006**, *33*. [CrossRef]

36. Wagner, F.H.; Hérault, B.; Rossi, V.; Hilker, T.; Maeda, E.E.; Sanchez, A.; Lyapustin, A.I.; Galvão, L.S.; Wang, Y.; Aragao, L.E. Climate drivers of the Amazon forest greening. *PLoS ONE* **2017**, *12*, e0180932. [CrossRef] [PubMed]

37. Wu, J.; Kobayashi, H.; Stark, S.C.; Meng, R.; Guan, K.; Tran, N.N.; Gao, S.; Yang, W.; Restrepo-Coupe, N.; Miura, T. Biological processes dominate seasonality of remotely sensed canopy greenness in an Amazon evergreen forest. *New Phytol.* **2018**, *217*, 1507–1520. [CrossRef] [PubMed]

38. Wu, J.; Albert, L.P.; Lopes, A.P.; Restrepo-Coupe, N.; Hayek, M.; Wiedemann, K.T.; Guan, K.; Stark, S.C.; Christoffersen, B.; Prohaska, N. Leaf development and demography explain photosynthetic seasonality in Amazon evergreen forests. *Science* **2016**, *351*, 972–976. [CrossRef] [PubMed]

39. Tang, H.; Dubayah, R. Light-driven growth in Amazon evergreen forests explained by seasonal variations of vertical canopy structure. *Proc. Natl. Acad. Sci. USA* **2017**, *114*, 2640–2644. [CrossRef]

40. Yan, D.; Zhang, X.; Yu, Y.; Guo, W. A comparison of tropical rainforest phenology retrieved from geostationary (seviri) and polar-orbiting (modis) sensors across the congo basin. *IEEE Trans. Geosci. Remote Sens.* **2016**, *54*, 4867–4881. [CrossRef]

41. Adole, T.; Dash, J.; Atkinson, P.M. A systematic review of vegetation phenology in Africa. *Ecol. Inform.* **2016**, *34*, 117–128. [CrossRef]

42. Ryan, C.M.; Williams, M.; Grace, J.; Woollen, E.; Lehmann, C.E. Pre-rain green-up is ubiquitous across southern tropical Africa: Implications for temporal niche separation and model representation. *New Phytol.* **2017**, *213*, 625–633. [CrossRef]

43. Piao, S.; Fang, J.; Zhou, L.; Ciais, P.; Zhu, B. Variations in satellite-derived phenology in China's temperate vegetation. *Glob. Chang. Biol.* **2006**, *12*, 672–685. [CrossRef]

44. Wu, C.; Hou, X.; Peng, D.; Gonsamo, A.; Xu, S. Land surface phenology of China's temperate ecosystems over 1999–2013: Spatial–temporal patterns, interaction effects, covariation with climate and implications for productivity. *Agric. For. Meteorol.* **2016**, *216*, 177–187. [CrossRef]

45. Liu, Q.; Fu, Y.H.; Zeng, Z.; Huang, M.; Li, X.; Piao, S. Temperature, precipitation, and insolation effects on autumn vegetation phenology in temperate China. *Glob. Chang. Biol.* **2016**, *22*, 644–655. [CrossRef] [PubMed]

46. Ge, Q.; Dai, J.; Cui, H.; Wang, H. Spatiotemporal Variability in Start and End of Growing Season in China Related to Climate Variability. *Remote Sens.* **2016**, *8*, 433. [CrossRef]

47. Zhu, H. The Tropical Forests of Southern China and Conservation of Biodiversity. *Bot. Rev.* **2017**, *83*, 87–105. [CrossRef]

48. Wu, J.; Lin, W.; Peng, X.; Liu, W. A review of forest resources and forest biodiversity evaluation system in China. *Int. J. Res.* **2013**, *2013*. [CrossRef]

49. Piao, S.; Fang, J.; Ciais, P.; Peylin, P.; Huang, Y.; Sitch, S.; Wang, T. The carbon balance of terrestrial ecosystems in China. *Nature* **2009**, *458*, 1009–1013. [CrossRef]

50. Tang, X.; Zhao, X.; Bai, Y.; Tang, Z.; Wang, W.; Zhao, Y.; Wan, H.; Xie, Z.; Shi, X.; Wu, B. Carbon pools in China's terrestrial ecosystems: New estimates based on an intensive field survey. *Proc. Natl. Acad. Sci. USA* **2018**, *115*, 4021–4026. [CrossRef] [PubMed]

51. Justice, C.; Belward, A.; Morisette, J.; Lewis, P.; Privette, J.; Baret, F. 2000: Developments in the 'validation'of satellite sensor products for the study of the land surface. *Int. J. Remote Sens.* **2000**, *21*, 3383–3390.

52. Morisette, J.T.; Baret, F.; Privette, J.L.; Myneni, R.B.; Nickeson, J.E.; Garrigues, S.; Shabanov, N.V.; Weiss, M.; Fernandes, R.A.; Leblanc, S.G. Validation of global moderate-resolution LAI products: A framework proposed within the CEOS land product validation subgroup. *IEEE Trans. Geosci. Remote Sens.* **2006**, *44*, 1804–1817. [CrossRef]

53. Post, H.; Hendricks Franssen, H.J.; Han, X.; Baatz, R.; Montzka, C.; Schmidt, M.; Vereecken, H. Evaluation and uncertainty analysis of regional-scale CLM4.5 net carbon flux estimates. *Biogeosciences* **2018**, *15*, 187–208. [CrossRef]

54. Rüdiger, C.; Albergel, C.; Mahfouf, J.F.; Calvet, J.C.; Walker, J.P. Evaluation of the observation operator Jacobian for leaf area index data assimilation with an extended Kalman filter. *J. Geophys. Res. Atmos.* **2010**, *115*. [CrossRef]

55. Viskari, T.; Hardiman, B.; Desai, A.R.; Dietze, M.C. Model-data assimilation of multiple phenological observations to constrain and predict leaf area index. *Ecol. Appl.* **2015**, *25*, 546–558. [CrossRef]

56. Fu, B.; Li, S.; Yu, X.; Yang, P.; Yu, G.; Feng, R.; Zhuang, X. Chinese ecosystem research network: Progress and perspectives. *Ecol. Complex.* **2010**, *7*, 225–233. [CrossRef]

57. Yu, G.; Chen, Z.; Piao, S.; Peng, C.; Ciais, P.; Wang, Q.; Li, X.; Zhu, X. High carbon dioxide uptake by subtropical forest ecosystems in the East Asian monsoon region. *Proc. Natl. Acad. Sci. USA* **2014**, *111*, 4910–4915. [CrossRef] [PubMed]

58. Peel, M.C.; Finlayson, B.L.; McMahon, T.A. Updated world map of the Köppen-Geiger climate classification. *Eur. Geosci. Union* **2007**, *4*, 439–473.

59. Li-Cor, I. *LAI-2000 Plant Canopy Analyzer Instruction Manual*; LI-COR Inc.: Lincoln, NE, USA, 1992.

60. Wu, D.; Wei, W.; Zhang, S. *Protocols for Standard Biological Observation and Measurement in Terrestrial Ecosystems*; China Environmental Science Pres: Beijing, China, 2007.

61. Knyazikhin, Y.; Martonchik, J.; Myneni, R.B.; Diner, D.; Running, S.W. Synergistic algorithm for estimating vegetation canopy leaf area index and fraction of absorbed photosynthetically active radiation from MODIS and MISR data. *J. Geophys. Res. Atmos.* **1998**, *103*, 32257–32275. [CrossRef]

62. Verger, A.; Baret, F.; Weiss, M. Atbd for Lai, Fapar and Fcover from Proba-V Products Collection 1km Version 2; 2019, Issue I1.41. Available online: https://land.copernicus.eu/global/sites/cgls.vito.be/files/products/CGLOPS1_ATBD_LAI1km-V2_I1.41.pdf (accessed on 15 April 2020).

63. Masek, J.G.; Vermote, E.F.; Saleous, N.E.; Wolfe, R.; Hall, F.G.; Huemmrich, K.F.; Gao, F.; Kutler, J.; Lim, T.-K. A Landsat surface reflectance dataset for North America, 1990–2000. *IEEE Geosci. Remote Sens. Lett.* **2006**, *3*, 68–72. [CrossRef]

64. Vermote, E.; Justice, C.; Claverie, M.; Franch, B. Preliminary analysis of the performance of the Landsat 8/OLI land surface reflectance product. *Remote Sens. Environ.* **2016**, *185*, 46–56. [CrossRef]

65. Zanter, K.; Department of the Interior, U.S. Geological Survey. Landsat 4-7 Surface Reflectance (LEDAPS) Product Guide. Version 2.0. 2019, EROS, Sioux Falls, South Dakota. Available online: https://www.usgs.gov/media/files/landsat-4-7-surface-reflectance-code-ledaps-product-guide (accessed on 15 September 2020).

66. Zanter, K.; Department of the Interior, U.S. Geological Survey. Landsat 8 Surface Reflectance Code (LASRC) Product Guide. Version 2.0. 2019, EROS, Sioux Falls, South Dakota. Available online: https://www.usgs.gov/media/files/land-surface-reflectance-code-lasrc-product-guide (accessed on 15 September 2020).

67. USGS. GTOPO30: Global 30 Arc-Seconds Digital Elevation Model [Data Set]. Available online: https://www.usgs.gov/centers/eros/science/usgs-eros-archive-digital-elevation-global-30-arc-second-elevation-gtopo30?qt-science_center_objects=0#qt-science_center_objects (accessed on 15 September 2020).

68. Reuter, H.I.; Nelson, A.; Jarvis, A. An evaluation of void-filling interpolation methods for SRTM data. *Int. J. Geogr. Inf. Sci.* **2007**, *21*, 983–1008. [CrossRef]

69. Jun, C.; Ban, Y.; Li, S. Open access to Earth land-cover map. *Nature* **2014**, *514*, 434. [CrossRef]

70. Wu, G.; Anafi, R.C.; Hughes, M.E.; Kornacker, K.; Hogenesch, J.B. MetaCycle: An integrated R package to evaluate periodicity in large scale data. *Bioinformatics* **2016**, *32*, 3351–3353. [CrossRef]

71. Yang, R.; Su, Z. Analyzing circadian expression data by harmonic regression based on autoregressive spectral estimation. *Bioinformatics* **2010**, *26*, i168–i174. [CrossRef] [PubMed]

72. Hughes, M.E.; Hogenesch, J.B.; Kornacker, K. JTK_CYCLE: An efficient nonparametric algorithm for detecting rhythmic components in genome-scale data sets. *J. Biol. Rhythms* **2010**, *25*, 372–380. [CrossRef] [PubMed]

73. Glynn, E.F.; Chen, J.; Mushegian, A.R. Detecting periodic patterns in unevenly spaced gene expression time series using Lomb–Scargle periodograms. *Bioinformatics* **2006**, *22*, 310–316. [CrossRef] [PubMed]

74. R Core Team. *R: A Language and Environment for Statistical Computing*; R Foundation for Statistical Computing: Vienna, Austria, 2017; Available online: https://www.r-project.org/ (accessed on 3 December 2018).

75. Hijmans, R.J.; Van Etten, J.; Cheng, J.; Mattiuzzi, M.; Sumner, M.; Greenberg, J.A.; Lamigueiro, O.P.; Bevan, A.; Racine, E.B.; Shortridge, A. Package 'raster'. R package version 2.5-8 (2015).

76. Weiss, M.; Baret, F.; Smith, G.; Jonckheere, I.; Coppin, P. Review of methods for in situ leaf area index (LAI) determination: Part II. Estimation of LAI, errors and sampling. *Agric. For. Meteorol.* **2004**, *121*, 37–53. [CrossRef]

77. Garrigues, S.; Allard, D.; Baret, F.; Weiss, M. Influence of landscape spatial heterogeneity on the non-linear estimation of leaf area index from moderate spatial resolution remote sensing data. *Remote Sens. Environ.* **2006**, *105*, 286–298. [CrossRef]

78. Fuster, B.; Sánchez-Zapero, J.; Camacho, F.; García-Santos, V.; Verger, A.; Lacaze, R.; Weiss, M.; Baret, F.; Smets, B. Quality Assessment of PROBA-V LAI, fAPAR and fCOVER Collection 300 m Products of Copernicus Global Land Service. *Remote Sens.* **2020**, *12*, 1017. [CrossRef]

79. Nagendra, H. Using remote sensing to assess biodiversity. *Int. J. Remote Sens.* **2001**, *22*, 2377–2400. [CrossRef]

80. Liu, Z.; Shao, Q.; Liu, J. The Performances of MODIS-GPP and -ET Products in China and Their Sensitivity to Input Data (FPAR/LAI). *Remote Sens.* **2015**, *7*, 135–152. [CrossRef]

81. Li, X.; Lu, H.; Yu, L.; Yang, K. Comparison of the spatial characteristics of four remotely sensed leaf area index products over China: Direct validation and relative uncertainties. *Remote Sens.* **2018**, *10*, 148. [CrossRef]

82. Shabanov, N.V.; Huang, D.; Yang, W.; Tan, B.; Knyazikhin, Y.; Myneni, R.B.; Ahl, D.E.; Gower, S.T.; Huete, A.R.; Aragão, L.E.O. Analysis and optimization of the MODIS leaf area index algorithm retrievals over broadleaf forests. *IEEE Trans. Geosci. Remote Sens.* **2005**, *43*, 1855–1865. [CrossRef]

83. Yang, W.; Shabanov, N.; Huang, D.; Wang, W.; Dickinson, R.; Nemani, R.; Knyazikhin, Y.; Myneni, R. Analysis of leaf area index products from combination of MODIS Terra and Aqua data. *Remote Sens. Environ.* **2006**, *104*, 297–312. [CrossRef]

84. Yan, K.; Park, T.; Yan, G.; Chen, C.; Yang, B.; Liu, Z.; Nemani, R.R.; Knyazikhin, Y.; Myneni, R.B. Evaluation of MODIS LAI/FPAR Product Collection 6. Part 1: Consistency and Improvements. *Remote Sens.* **2016**, *8*, 359. [CrossRef]

85. Jiang, C.; Ryu, Y.; Fang, H.; Myneni, R.; Claverie, M.; Zhu, Z. Inconsistencies of interannual variability and trends in long-term satellite leaf area index products. *Glob. Chang. Biol.* **2017**, *23*, 4133–4146. [CrossRef] [PubMed]

86. Cammalleri, C.; Verger, A.; Lacaze, R.; Vogt, J. Harmonization of GEOV2 fAPAR time series through MODIS data for global drought monitoring. *Int. J. Appl. Earth Obs. Geoinf.* **2019**, *80*, 1–12. [CrossRef] [PubMed]

87. Pisek, J.; Chen, J.M.; Alikas, K.; Deng, F. Impacts of including forest understory brightness and foliage clumping information from multiangular measurements on leaf area index mapping over North America. *J. Geophys. Res. Biogeosci.* **2010**, *115*. [CrossRef]

88. Verger, A.; Baret, F.; Weiss, M. Performances of neural networks for deriving LAI estimates from existing CYCLOPES and MODIS products. *Remote Sens. Environ.* **2008**, *112*, 2789–2803. [CrossRef]

89. Claverie, M.; Vermote, E.F.; Weiss, M.; Baret, F.; Hagolle, O.; Demarez, V. Validation of coarse spatial resolution LAI and FAPAR time series over cropland in southwest France. *Remote Sens. Environ.* **2013**, *139*, 216–230. [CrossRef]

90. Tian, Y.; Wang, Y.; Zhang, Y.; Knyazikhin, Y.; Bogaert, J.; Myneni, R.B. Radiative transfer based scaling of LAI retrievals from reflectance data of different resolutions. *Remote Sens. Environ.* **2003**, *84*, 143–159. [CrossRef]

91. Bloom, A.A.; Exbrayat, J.-F.; Van Der Velde, I.R.; Feng, L.; Williams, M. The decadal state of the terrestrial carbon cycle: Global retrievals of terrestrial carbon allocation, pools, and residence times. *Proc. Natl. Acad. Sci. USA* **2016**, *113*, 1285–1290. [CrossRef]

92. Raupach, M.R.; Rayner, P.J.; Barrett, D.J.; DeFries, R.S.; Heimann, M.; Ojima, D.S.; Quegan, S.; Schmullius, C.C. Model–data synthesis in terrestrial carbon observation: Methods, data requirements and data uncertainty specifications. *Glob. Chang. Biol.* **2005**, *11*, 378–397. [CrossRef]

93. Xie, X.; Li, A.; Jin, H.; Tan, J.; Wang, C.; Lei, G.; Zhang, Z.; Bian, J.; Nan, X. Assessment of five satellite-derived LAI datasets for GPP estimations through ecosystem models. *Sci. Total Environ.* **2019**, *690*, 1120–1130. [CrossRef] [PubMed]

94. Liu, Y.; Xiao, J.; Ju, W.; Zhu, G.; Wu, X.; Fan, W.; Li, D.; Zhou, Y. Satellite-derived LAI products exhibit large discrepancies and can lead to substantial uncertainty in simulated carbon and water fluxes. *Remote Sens. Environ.* **2018**, *206*, 174–188. [CrossRef]

95. Xiao, J.; Chevallier, F.; Gomez, C.; Guanter, L.; Hicke, J.A.; Huete, A.R.; Ichii, K.; Ni, W.; Pang, Y.; Rahman, A.F. Remote sensing of the terrestrial carbon cycle: A review of advances over 50 years. *Remote Sens. Environ.* **2019**, *233*, 111383. [CrossRef]

96. Scholze, M.; Buchwitz, M.; Dorigo, W.; Guanter, L.; Quegan, S. Reviews and syntheses: Systematic Earth observations for use in terrestrial carbon cycle data assimilation systems. *Biogeosciences* **2017**, *14*, 3401–3429. [CrossRef]

97. Dietze, M.C. *Ecological Forecasting*; Princeton University Press: Princeton, NJ, USA, 2017.

98. López-Blanco, E.; Exbrayat, J.-F.; Lund, M.; Christensen, T.R.; Tamstorf, M.P.; Slevin, D.; Hugelius, G.; Bloom, A.A.; Williams, M. Evaluation of terrestrial pan-Arctic carbon cycling using a data-assimilation system. *Earth Syst. Dyn.* **2019**, *10*, 233–255. [CrossRef]

99. MacBean, N.; Peylin, P.; Chevallier, F.; Scholze, M.; Schuermann, G. Consistent assimilation of multiple data streams in a carbon cycle data assimilation system. *Geosci. Model. Dev.* **2016**, *9*, 3569–3588. [CrossRef]

100. De Kauwe, M.G.; Disney, M.; Quaife, T.; Lewis, P.; Williams, M. An assessment of the MODIS collection 5 leaf area index product for a region of mixed coniferous forest. *Remote Sens. Environ.* **2011**, *115*, 767–780. [CrossRef]
101. Chevallier, F. Impact of correlated observation errors on inverted CO2 surface fluxes from OCO measurements. *Geophys. Res. Lett.* **2007**, *34*. [CrossRef]
102. Xu, B.; Li, J.; Park, T.; Liu, Q.; Zeng, Y.; Yin, G.; Zhao, J.; Fan, W.; Yang, L.; Knyazikhin, Y. An integrated method for validating long-term leaf area index products using global networks of site-based measurements. *Remote Sens. Environ.* **2018**, *209*, 134–151. [CrossRef]
103. Stark, S.C.; Leitold, V.; Wu, J.L.; Hunter, M.O.; de Castilho, C.V.; Costa, F.R.; McMahon, S.M.; Parker, G.G.; Shimabukuro, M.T.; Lefsky, M.A. Amazon forest carbon dynamics predicted by profiles of canopy leaf area and light environment. *Ecol. Lett.* **2012**, *15*, 1406–1414. [CrossRef]
104. Jucker, T.; Hardwick, S.R.; Both, S.; Elias, D.M.; Ewers, R.M.; Milodowski, D.T.; Swinfield, T.; Coomes, D.A. Canopy structure and topography jointly constrain the microclimate of human-modified tropical landscapes. *Glob. Chang. Biol.* **2018**, *24*, 5243–5258. [CrossRef] [PubMed]
105. Li, X.; Liu, S.; Li, H.; Ma, Y.; Wang, J.; Zhang, Y.; Xu, Z.; Xu, T.; Song, L.; Yang, X. Intercomparison of six upscaling evapotranspiration methods: From site to the satellite pixel. *J. Geophys. Res. Atmos.* **2018**, *123*, 6777–6803. [CrossRef]
106. Shi, Y.; Wang, J.; Qin, J.; Qu, Y. An upscaling algorithm to obtain the representative ground truth of LAI time series in heterogeneous land surface. *Remote Sens.* **2015**, *7*, 12887–12908. [CrossRef]
107. Hilker, T.; Lyapustin, A.I.; Tucker, C.J.; Sellers, P.J.; Hall, F.G.; Wang, Y. Remote sensing of tropical ecosystems: Atmospheric correction and cloud masking matter. *Remote Sens. Environ.* **2012**, *127*, 370–384. [CrossRef]
108. Dubayah, R.; Blair, J.B.; Goetz, S.; Fatoyinbo, L.; Hansen, M.; Healey, S.; Hofton, M.; Hurtt, G.; Kellner, J.; Luthcke, S. The Global Ecosystem Dynamics Investigation: High-resolution laser ranging of the Earth's forests and topography. *Sci. Remote Sens.* **2020**, *1*, 100002. [CrossRef]
109. Narine, L.L.; Popescu, S.C.; Malambo, L. Using ICESat-2 to Estimate and Map Forest Aboveground Biomass: A First Example. *Remote Sens.* **2020**, *12*, 1824. [CrossRef]

*Article*

# Analysis of the Spatial Differences in Canopy Height Models from UAV LiDAR and Photogrammetry


Qingwang Liu [1,2,†], Liyong Fu [1,3,†], Qiao Chen [1,†], Guangxing Wang [4,*], Peng Luo [1], Ram P. Sharma [5], Peng He [6], Mei Li [1], Mengxi Wang [1] and Guangshuang Duan [1,7]

[1] Research Institute of Forest Resource Information Techniques, Chinese Academy of Forestry, Beijing 100091, China; liuqw@ifrit.ac.cn (Q.L.); fuly@ifrit.ac.cn (L.F.); chenq@ifrit.ac.cn (Q.C.); luopeng@ifrit.ac.cn (P.L.); limei@ifrit.ac.cn (M.L.); wangmx@ifrit.ac.cn (M.W.); duangs@ifrit.ac.cn (G.D.)
[2] Key Laboratory of Forestry Remote Sensing and Information System, National Forestry and Grassland Administration, Beijing 100091, China
[3] Key Laboratory of Forest Management and Growth Modeling, National Forestry and Grassland Administration, Beijing 100091, China
[4] School of Earth Systems and Sustainability, Southern Illinois University at Carbondale, Carbondale, IL 62901, USA
[5] Institute of Forestry, Tribhuwan Univeristy, Kritipur, Kathmandu 44600, Nepal; sharmar@fld.czu.cz
[6] Central South Inventory and Planning Institute, National Forestry and Grassland Administration, Changsha 410014, China; hepeng19880407@163.com
[7] College of Mathematics and Statistics, Xinyang Normal University, Xinyang 464000, China
[*] Correspondence: gxwang@siu.edu; Tel.: +1-618-453-6017
[†] These authors contributed equally to this work.


Received: 19 July 2020; Accepted: 1 September 2020; Published: 6 September 2020


**Abstract:** Forest canopy height is one of the most important spatial characteristics for forest resource inventories and forest ecosystem modeling. Light detection and ranging (LiDAR) can be used to accurately detect canopy surface and terrain information from the backscattering signals of laser pulses, while photogrammetry tends to accurately depict the canopy surface envelope. The spatial differences between the canopy surfaces estimated by LiDAR and photogrammetry have not been investigated in depth. Thus, this study aims to assess LiDAR and photogrammetry point clouds and analyze the spatial differences in canopy heights. The study site is located in the Jigongshan National Nature Reserve of Henan Province, Central China. Six data sets, including one LiDAR data set and five photogrammetry data sets captured from an unmanned aerial vehicle (UAV), were used to estimate the forest canopy heights. Three spatial distribution descriptors, namely, the effective cell ratio (ECR), point cloud homogeneity (PCH) and point cloud redundancy (PCR), were developed to assess the LiDAR and photogrammetry point clouds in the grid. The ordinary neighbor (ON) and constrained neighbor (CN) interpolation algorithms were used to fill void cells in digital surface models (DSMs) and canopy height models (CHMs). The CN algorithm could be used to distinguish small and large holes in the CHMs. The optimal spatial resolution was analyzed according to the ECR changes of DSMs or CHMs resulting from the CN algorithms. Large negative and positive variations were observed between the LiDAR and photogrammetry canopy heights. The stratified mean difference in canopy heights increased gradually from negative to positive when the canopy heights were greater than 3 m, which means that photogrammetry tends to overestimate low canopy heights and underestimate high canopy heights. The CN interpolation algorithm achieved smaller relative root mean square errors than the ON interpolation algorithm. This article provides an operational method for the spatial assessment of point clouds and suggests that the variations between LiDAR and photogrammetry CHMs should be considered when modeling forest parameters.

**Keywords:** digital surface model; digital terrain model; canopy height model; constrained neighbor interpolation; ordinary neighbor interpolation; point cloud density; stereo imagery

---

## 1. Introduction

Forest structure information is a prerequisite for forest resource inventories and forest ecosystem modeling [1–3]. Various techniques and methods can be used to obtain such information. Light detection and ranging (LiDAR) and photogrammetry have the ability to depict the three-dimensional canopy structure of forests and therefore can be used to monitor structural changes over time [4–8]. Unmanned aerial vehicle (UAV) LiDAR, hereafter called LiDAR, can be used to accurately measure the spatial distributions of forest canopies with a high density of point clouds [9–13]. Laser pulses can penetrate the gaps between branches and leaves of tree crowns and detect middle parts or areas under forest [14–17]. LiDAR point clouds are usually classified as vegetation, ground and other objects, which are used to generate digital surface models (DSMs) and digital terrain models (DTMs) using interpolation algorithms [10,18,19]. Generally, a DSM is created using the maximum algorithm in a regular grid at the given spatial resolution, which is determined according to the point cloud density [20–22]. A canopy height model (CHM) depicts the variations in the forest canopy height above the terrain, and these height variations can be determined by subtracting the DTM from a DSM [3,10,14,23–27]. Many algorithms that estimate the parameters at the individual tree or sample plot level are developed based on a CHM [14–16,23,26–32].

UAV photogrammetry, hereafter called photogrammetry, usually acquires images with high areas of overlap (usually 60–90% along-track and 30–60% across-track), which are processed to build the spatial structure of forest canopies using stereo imagery algorithms, such as structure from motion (SfM) [25,32–37] or semi-global matching [24,38–41]. Dense point clouds reconstructed from image pairs are used to create a DSM with a similar algorithm used for LiDAR point clouds [23,42]. The distribution of dense point clouds is substantially affected by the image along- and across-track overlap, flying height, terrain, and other factors [43–46]. It is quite difficult to reconstruct the terrain under dense forest due to mutual obscuration among tree crowns. A photogrammetry CHM (P-CHM) can be generated as the difference between a photogrammetry DSM (P-DSM) and LiDAR DTM (L-DTM) [23,33,35,47]. Forest attributes can be precisely estimated using the CHMs of either LiDAR or photogrammetry [23,25,48–50].

The observation geometry of LiDAR is obviously different from that of photogrammetry [23,46]. LiDAR can directly measure the spatial positions of branches and leaves at different heights, while photogrammetry obtains the surface envelope of the forest canopy using a stereo imagery algorithm. Thus, canopy heights estimated by LiDAR and photogrammetry show inherent spatial differences, and these differences have not been discussed in depth in the literature. The spatial resolutions of LiDAR and photogrammetry CHMs vary from centimetres to metres based on the different point cloud densities [9,14,15,23,25–27,37]. Detailed canopy structures tend to be suppressed with a coarse spatial resolution. The densities of LiDAR and photogrammetry point clouds are unevenly distributed due to various factors, such as flying height and crown shadows, and these variations produce varied points within different grid cells at a specified spatial resolution. A cell with one or more points is referred to as an effective cell (valid cell), while a cell without any points is referred to as a void cell (no data, missing or blank cell). The void cells can form variable holes in the grid. These holes are different from canopy gaps (canopy openings) created by the snapping and falling of trees, the impacts of insects or pathogens, or the mortality of single trees or small groups of trees; moreover, these gaps will eventually be closed in the ecological processes of forest ecosystems [14,23].

The size of the holes in a grid can be expressed as the number of continuous void cells, which varies at differing spatial resolutions. The same-sized hole will have more void cells at a fine resolution than at a coarse resolution. Some void cells in a hole will have effective neighboring cells, which are referred to as outer void cells, while the other void cells within a hole are referred to as inner void cells. The outer and inner void cells can be estimated or interpolated from the effective cells by using the ordinary neighbor (ON) interpolation algorithm. The interpolated inner void cells are subject to more uncertainties than the interpolated outer void cells considering the spatial relevance [9,14,51]. The number of effective neighboring cells can also affect the uncertainties of interpolated cells. The interpolated cells with many effective neighboring cells will have higher confidence than those with few effective neighboring cells. The ON algorithm should be constrained to obtain more reliable values with more effective neighboring cells.

The distribution of effective or void cells in a grid can reflect the homogeneity of the point cloud to some degree. Certain questions remain regarding the distribution of point clouds and estimated canopy heights, such as (1) how can the distribution of point clouds for a given spatial resolution be quantitatively described, (2) what spatial resolution is optimal for the specified point cloud, and (3) are spatial differences of the canopy heights depicted by LiDAR and photogrammetry affected by different spatial resolutions?

The purpose of this study was to assess the spatial distribution of point clouds and compare the differences between CHMs derived by LiDAR and photogrammetry. Three descriptors of the spatial distribution were introduced, namely, the effective cell ratio (ECR), point cloud homogeneity (PCH) and point cloud redundancy (PCR), to characterize the point clouds and the derived CHMs in the grid. The ON and constrained neighbor (CN) interpolation algorithms were used to fill the void cells of CHMs, and the CN algorithm was used to distinguish small and large holes. LiDAR CHMs were used as references to assess the photogrammetry CHMs that were created at different flying heights and with different image overlaps. This article will provide an operational method for point cloud assessment and spatial distribution analysis.

## 2. Materials and Methods

### 2.1. Study Site

The study site (114°05′ E, 31°52′ N) is located in the Jigongshan National Nature Reserve of Henan Province, Central China (Figure 1). This area belongs to a transition zone from a subtropical to temperate climate zone. The area covered is approximately 340 by 360 m, and the terrain elevation ranges from approximately 115 to 198 m. The forest is dominated by deciduous broad-leaved trees, including sawtooth oak (*Quercus acutissima* Carruth.), Chinese cork oak (*Quercus variabilis* Blume), Chinese sweet gum or Formosan gum (*Liquidambar formosana* Hance), and bald cypress (*Taxodium distichum* (L.) Rich.). Abundant shrubs are observed in the understory layer. This site has been used for studies on atmospheric nitrogen deposition in forest ecosystems [52,53].

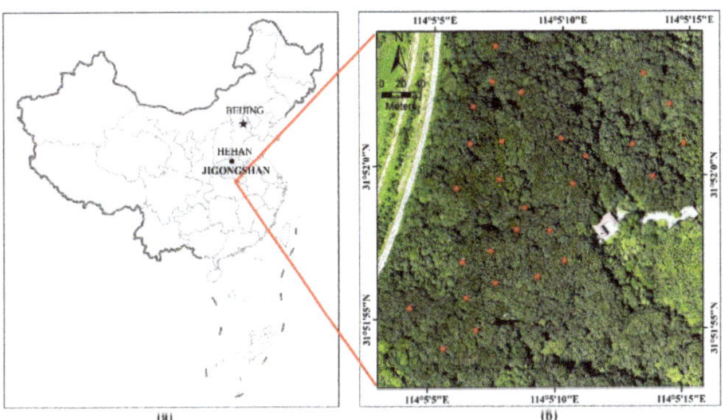


**Figure 1.** (**a**) Location of the study site in the Jigongshan Natural Reserve, Xinyang, Henan Province, China; and (**b**) true color map of the study site (red dots indicate the centres of ground plots).

The ground plots used were square with a size of 25 × 25 m (625 m$^2$), which were set up on the basis of the existing studies [52,53], and there were a total of 28 ground plots. The parameters of individual trees with a diameter at breast height (DBH) ≥5 cm, including the DBH (cm), tree height (m), crown radius in four directions (m) and species, were determined in August 2017 (Table A1). The number of stems within the ground plots varied from 30 to 96. Lorey's heights of the ground plots ranged from 15.7 to 31.2 m.

*2.2. UAV Data Sets*

The LiDAR data and photogrammetry images were acquired in August 2017. The LiDAR system used in this study is a Velodyne laser scanning system (VLP-16) with a high-precision global navigation satellite system and inertial measurement unit (GNSS and IMU) mounted on an eight-rotor aircraft [54,55]. The laser sensor has 16 channels with maximum scan angles of 30° and 360° in the along- and across-track directions, respectively. More detailed specifications of the LiDAR and photogrammetry system are shown in Table 1. The flying heights of the aircraft varied from 50 m to 55 m above the terrain or take-off point of the UAV at a speed of approximately 4.8 m s$^{-1}$.

**Table 1.** Light detection and ranging (LiDAR) and photogrammetry system specifications.

| LiDAR | | | |
|---|---|---|---|
| UAV model | GV1300 | Rotor | 8 |
| LiDAR model | VLP-16 | PRF | 300 kHz |
| Laser wavelength | 905 nm | Laser divergence | 3 mrad |
| Scan pattern | Rotate Mirror | Scan FOV | 30° × 360° |
| Echoes | 2 | Max Scan frequency | 20 Hz |
| Range | 2 m–120 m | Vertical Accuracy | <5 cm |
| **Photogrammetry** | | | |
| UAV model | GV1500 | Rotor | 6 |
| Camera model | EOS 5DS | Pixels | 50,320,896 |
| CMOS size | 36.0 × 24.0 mm | Image size | 8688 × 5792 pixels |
| FOV | 54.4° × 37.8° | Focal length | 35 mm |
| Pixel unit | 4.1 × 4.1 μm | Bands | R/G/B |

The LiDAR data acquired on 10 August 2017 did not cover the entire study area because the travel distances of some laser pulses exceeded the maximum range (120 m) of the LiDAR system. The areas with elevations below 130 m had almost no data, and they were mainly located in the west and southwest parts of the study site. The second LiDAR flight was designed to obtain supplementary point cloud data for the areas without data on 11 August 2017. The point clouds of the two flights were matched and combined as a single LiDAR data set (denoted as L-D55). Detailed information on the LiDAR data set is shown in Table A2. The point cloud density of LiDAR data varied from 0 to 1757 points m$^{-2}$, with a mean of 168 points m$^{-2}$ in the grid at a 1.0 m resolution (Figure A1).

The images were captured using a Canon EOS 5DS camera with a high-precision GNSS mounted on a six-rotor aircraft (see Table 1 for detailed specifications) [56,57], and the lens had a focal length of 35 mm. The size of the image consisted of 8688 × 5792 pixels, which corresponds to the size of the sensor of 36.0 × 24.0 mm. The photogrammetry system was operated at different heights varying from 80 to 300 m (the ground sample distance (GSD) varied from 1 cm to 4 cm) with image overlaps varying from 64% to 84% to obtain five data sets (denoted as P1-D300, P2-D150, P3-D150, P4-D150 and P5-D80) covering the same area as shown in Table A2. P1-D300 was acquired at a flying height of 300 m with a corresponding GSD of 4 cm. P2-D150, P3-D150 and P4-D150 were obtained at 150 m with a GSD of 2 cm using different across-track overlaps and flying speeds. P5-D80 was obtained at a flying height of 80 m with a corresponding GSD of 1 cm. The images were captured with an exposure time of 1/800 s or 1/1000 s on cloudy or sunny days using an automatic aperture and an ISO of 160. The images were processed to build a dense point cloud of the forest structure using the SfM method [58]. Several steps were involved to generate high-precision dense point clouds. The images were aligned with high accuracy using the camera positions obtained by static differential processing of high-precision GNSS data. The camera and lens parameters, including the focal length, principal point coordinates, affinity and skew (non-orthogonality) transformation coefficients, radial distortion coefficients, and tangential distortion coefficients, were optimized to minimize the camera position errors using the optimize camera alignment tool [58]. The dense points were built with high quality and mild depth filtering to maintain detailed characteristics of the forest canopies. The point cloud density of the photogrammetry data sets varied from 1 to 12,761 points m$^{-2}$, with means ranging from 257 to 2562 points m$^{-2}$ in the grid at a 1.0 m resolution (Figure A1).

### 2.3. Data Processing

The LiDAR point cloud was classified as ground, vegetation, building, and noise using TerraSolid [59]. The noise points were each carefully identified by visual evaluation. The ground algorithm was used to separate ground points from other objects with a maximum building size of 20 m, a terrain angle of 88°, an iteration angle of 6° to the plane and an iteration distance of 1.4 m to the plane. The ground points were modified by visual evaluation to eliminate false ground points. The points of artificial objects, such as water towers and buildings, were manually recognized. The remaining points were classified as vegetation after other objects were excluded. The L-DTM was generated from the ground points by a triangulated irregular network interpolation algorithm with spatial resolutions of 0.1, 0.2, 0.5, and 1.0 m [9,23,35,60,61]. The original LiDAR DSM (L-DSM) was created from the vegetation and ground points by using maximum algorithms with spatial resolutions of 0.1, 0.2, 0.5, and 1.0 m [9,14,28,61]. The original LiDAR CHM (L-CHM) was produced based on the difference between the original L-DSM and L-DTM. The void cells of the original L-DSM were partially interpolated by the CN interpolation algorithm to generate the interpolated L-DSM. The interpolated L-CHM was produced based on the difference between the interpolated L-DSM and L-DTM.

The displacements between the LiDAR and photogrammetry point clouds were adjusted using the reference points extracted from the buildings and open ground area. The horizontal displacement was calculated based on roof ridges and building edges. The vertical displacement was computed according to the ground control points after horizontal displacement was applied. The aligned data sets were generated after the vertical displacement was applied. The photogrammetry dense point cloud was used to generate the original P-DSM by using the maximum algorithm. The interpolated P-DSM was created by the CN interpolation algorithm. The original P-CHM and the interpolated P-CHM were created from the corresponding P-DSMs normalized by the L-DTM [23–25,32,33,37,48]. The DSMs and CHMs from LiDAR and photogrammetry were masked using the boundaries of the buildings and water towers to maintain consistency. The boundaries of the buildings were digitized according to the building points of the point cloud. Seventeen water towers with heights of approximately 38 m were distributed across the study site. The centres of the water towers were determined and used to create circular boundaries by buffering with a radius of 1.2 m.

### 2.4. Descriptors of Spatial Distribution

This study used three descriptors of the spatial distribution, namely, the ECR, PCH and PCR, to quantitatively depict the characteristics of the point cloud in the grid at the given spatial resolution. The original DSMs or CHMs obtained from LiDAR and photogrammetry may have void cells due to complex conditions, such as the laser scanning patterns, flight attitudes, and sunshine conditions [14,31,51]. The ECR is defined as the proportion of effective cells to total cells (Equation (1)) and used to depict the distribution patterns of effective and void cells. The ECR reflects the uneven characteristics of point clouds to some degree. A small ECR means that many points are clustered in a small area.

$$ECR = \frac{N_E}{N} \tag{1}$$

where $ECR$ is the effective cell ratio, $N_E$ is the number of effective cells, and $N$ is the number of total cells, including effective and void cells. If the $ECR$ equals 1, then the point cloud is evenly distributed over each cell of the grid.

Different densities of points may occur in grid cells for different data sets with the same ECR. To explain the homogeneity of the point cloud distributed in the grid, the PCH is defined in Equation (2) by introducing the number of points per cell, which can be calculated as the product of the point cloud density per square metre and the cell area.

$$PCH = \begin{cases} \left(\frac{ECR}{D_C}\right)^{1/D_C} & , \ if \ D_C < 1 \\ ECR^{D_C} & , \ if \ D_C \geq 1 \end{cases} \tag{2}$$

where $PCH$ is the point cloud homogeneity, $D_C$ is the mean number of points per cell by $D_C = n/N$, $n$ is the number of total points, and $N$ is the number of total cells. If $D_C$ equals 1, then one point is located in each cell and the $PCH$ is only determined by effective cells. If the $ECR$ equals 1, then the $PCH$ will be 1, regardless of how many points are located in each cell.

The PCR depicts the richness of points in each cell as defined in Equation (3). A lower PCH corresponds to a higher PCR, which indicates that some cells hold more unnecessary points. An evenly distributed data set should have a higher PCH and lower PCR.

$$PCR = 1 - \frac{ECR}{D_C} \tag{3}$$

where $PCR$ is the point cloud redundancy, $ECR$ is the effective cell ratio, and $D_C$ is the mean number of points per cell. If the $ECR$ and $D_C$ both equal 1, then the points are ideally distributed evenly in each cell and the $PCR$ equals 0. If the $D_C$ is much greater than the $ECR$, then the $PCR$ approximately approaches 1.

The ECR, PCH, and PCR of the point cloud could also be calculated based on the interpolated DSM or CHM. The interpolation introduced some degree of variation in these descriptors of the spatial distribution.

### 2.5. Constrained Neighbor Interpolation

The original DSM or CHM had void cells in the grid at a specified spatial resolution due to the uneven distribution of point clouds. The void cells can be filled using the effective neighboring cells. The number of effective neighboring cells (i.e., not including void neighboring cells) for one void cell varies from 1 to 8, which affects the spatial variation in the filled cells [9,14]. The CN interpolation algorithm is used to calculate the interpolated values for the void cells (Equation (4)).

$$C_i = \frac{\sum_{j=1}^{k} NC_j}{k}, \ k \leq 8, \ if \ k \geq Q \tag{4}$$

where $C_i$ (m) is the interpolated value of the $i$th void cell, $NC_j$ (m) is the value of the $j$th effective neighboring cell of the $i$th void cell, k is the number of effective neighboring cells of the $i$th void cell, and Q is the threshold of effective neighboring cells of the void cell. The $i$th void cell will be interpolated if the number of its effective neighboring cells is not less than the threshold ($Q$).

The value of the threshold $Q$ varies from 1 to 8. The void cell will be interpolated if there is at least one effective neighboring cells; that is, $Q = 1$. The ON interpolation algorithm will similarly interpolate the void cells with any effective neighboring cells, which can be regarded as the special case ($Q = 1$) of the CN interpolation algorithm. The CN and ON interpolation algorithms use an iterative process to interpolate the void cells. Some void cells with effective neighboring cells can be interpolated in the first loop, while other void cells may be interpolated in the second or more loops.

The void cells of different holes have a varied number of effective neighboring cells (Figure 2). For example, the maximum number of effective neighboring cells of the void cells in Figure 2a is 4, which will not be interpolated if the threshold $Q$ equals 5. Eight void cells will be interpolated in the first loop and 4 void cells interpolated in the second loop if the threshold $Q$ equals 4. In practice, any holes could be interpolated by iterative loops if the threshold $Q$ is less than 5. The runs of iterative loops will vary from several times to dozens of times depending on the number of void cells. The CN algorithm will continue to loop until every hole is either eliminated or reduced to a specific size as depicted in Figure 2. The double width cross pattern of 12 cells would be the minimum hole if $Q$ equals 5 under normal circumstances in Figure 2a. Another rarely occurring case would be that the interleaved chess pattern in Figure 2e could not be interpolated. The small holes in Figure 2b–d will not be interpolated if the threshold $Q$ equals 6, 7, or 8, while those holes can be filled if the threshold $Q$ equals 5. Therefore, the threshold of $Q = 5$ is chosen for filling void cells in this study.

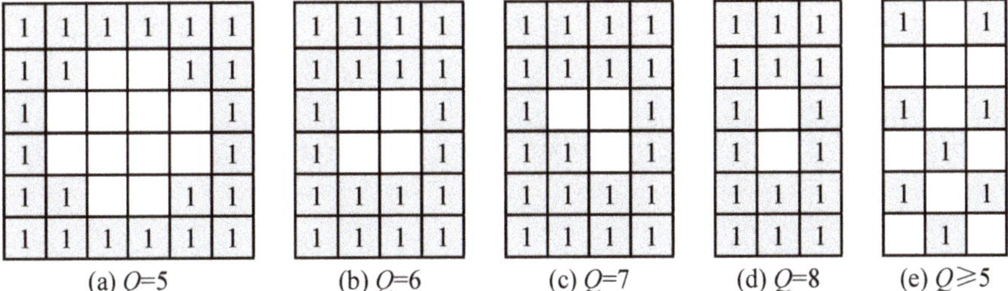

**Figure 2.** Minimum holes left by the constrained neighbor (CN) interpolation algorithm with different threshold values ($Q$) (void cells are white, and effective cells are grey): (**a**) $Q = 5$; (**b**) $Q = 6$; (**c**) $Q = 7$; (**d**) $Q = 8$; and (**e**) $Q \geq 5$.

The hole in Figure 2a is larger than the other holes in Figure 2, and it is referred to as the diagnostic hole, which is used as the criteria to distinguish small and large holes. Small holes have one or more void cells, and the numbers of columns and rows of void cells were not less than 4. Large holes have twelve or more void cells, and both the numbers of columns and rows were equal to or greater than 4. The actual area of the diagnostic hole was determined based on the spatial resolution, for example, the area was 0.12 or 0.48 m$^2$ with a spatial resolution of 0.1 or 0.2 m, respectively.

## 2.6. Analysis of Spatial Resolution

The spatial resolution will affect the number of points in each cell of the original DSM or CHM. A fine spatial resolution tends to result in many void cells in the grid, while a coarse spatial resolution will discard many points in the effective cells of the grid for a given point cloud, which represents a compromise for determining the spatial resolution [9,14].

The spatial distribution characteristics of point clouds depend on the spatial resolution. The ECR will increase as the number of void cells decreases at a coarse spatial resolution. The interpolated void cells by the CN algorithm will also, to some degree, cause the ECR to increase. The variation in the ECR can be used to analyze the optimal spatial resolution for the DSM or CHM, in which as much information about the point cloud is retained as possible so that the void cells can be confidently interpolated in the grid. Series of ECR pairs of original DSMs and interpolated DSMs by the CN algorithm can be calculated at different spatial resolutions. The difference between the ECR of the interpolated DSM and that of the original DSM is referred to as the ECR pair difference, which is used as an indicator to select the spatial resolution. The ECR pair differences might decrease from fine to coarse spatial resolution when the void cells of the grid are interpolated by the CN algorithm. The optimal spatial resolution could be determined as the corresponding spatial resolution if the ECR pair differences do not become greater than the specified threshold (for example, 0.10). The series of spatial resolutions used as candidates is 0.1 m, 0.2 m, 0.5 m, and 1.0 m in this study. The optimal spatial resolution is considered relative to the other spatial resolution candidates.

### 2.7. Data Set Assessment

The estimated canopy heights obtained from the LiDAR data set were used as references to compare the estimated canopy heights obtained from five different photogrammetry data sets with spatial resolutions of 0.1, 0.2, 0.5, and 1.0 m. The overall mean difference ($\bar{d}$) (Equation (5)) and mean absolute difference ($\overline{|d|}$) (Equation (6)) were calculated for this purpose [14,23]. The canopy heights were divided into bins at 1.0 m intervals and used as a stratified variable to calculate a series of mean differences for analyzing the potential trends of canopy heights.

$$\bar{d} = \frac{\sum d_i}{N} = \frac{\sum LH_i - PH_i}{N}, \tag{5}$$

$$\overline{|d|} = \frac{\sum \left| d_i - \bar{d} \right|}{N}, \tag{6}$$

where $\bar{d}$ (m) is the overall mean difference, $d_i$ (m) is the canopy height difference between LiDAR and photogrammetry for the $i^{th}$ effective cell, $LH_i$ (m) and $PH_i$ (m) are the canopy heights of LiDAR and photogrammetry for the $i^{th}$ effective cell (i = 1,...,N), N is the total number of effective cells, and $\overline{|d|}$ (m) is the mean absolute difference.

The original and interpolated CHMs obtained from the LiDAR data set (response) were linearly regressed and analyzed with the corresponding CHMs of the five photogrammetry data sets (predictor). Three widely used statistical criteria, namely, the coefficient of determination ($R^2$) (7), root mean square error (*RMSE*) (8), and relative *RMSE* (*rRMSE*) (9), were used to assess the model accuracy [41]. The *RMSE* combines the mean error and error variance to provide a robust measure of overall accuracy [62].

$$R^2 = 1 - \frac{\sum \left( LH_i - L\hat{H}_i \right)^2}{\sum \left( LH_i - \overline{LH} \right)^2}, \tag{7}$$

$$RMSE = \sqrt{\bar{e}^2 + \sigma_e^2}, \tag{8}$$

$$rRMSE = \frac{RMSE}{\overline{LH}}, \tag{9}$$

where $LH_i$ and $L\hat{H}_i$ are the reference and estimated forest canopy heights (m) for the $i$th effective cell (i = 1,...,N), N is the total number of the reference canopy heights, $\overline{LH}$ is the mean of the reference canopy heights, $\bar{e}$ is the mean error calculated by $\bar{e} = \sum e_i / N = \sum \left( LH_i - L\hat{H}_i \right)/N$, $e_i$ is the $i$th error and $\sigma_e^2$ is the error variance calculated by $\sigma_e^2 = \sum (e_i - \bar{e})^2 / (N - 1)$.

## 3. Results

### 3.1. CHMs of LiDAR and Photogrammetry

The original L-DSM of the L-D55 data set with a spatial resolution of 0.1 m had many holes within and between crowns. The interpolation of the L-DSM by the CN algorithm filled all the small holes with areas less than 1.2 m². The hole-free L-DSM was generated by the ON algorithm. The original and interpolated L-CHMs were calculated based on the corresponding L-DSMs and L-DTM. The small holes of the original L-CHM with a spatial resolution of 0.1 m (Figure 3a) were interpolated by the CN algorithm (Figure 3b). All of the holes of the original L-CHM were filled by the ON algorithm except for the holes representing the water tower and building areas (Figure 3c).

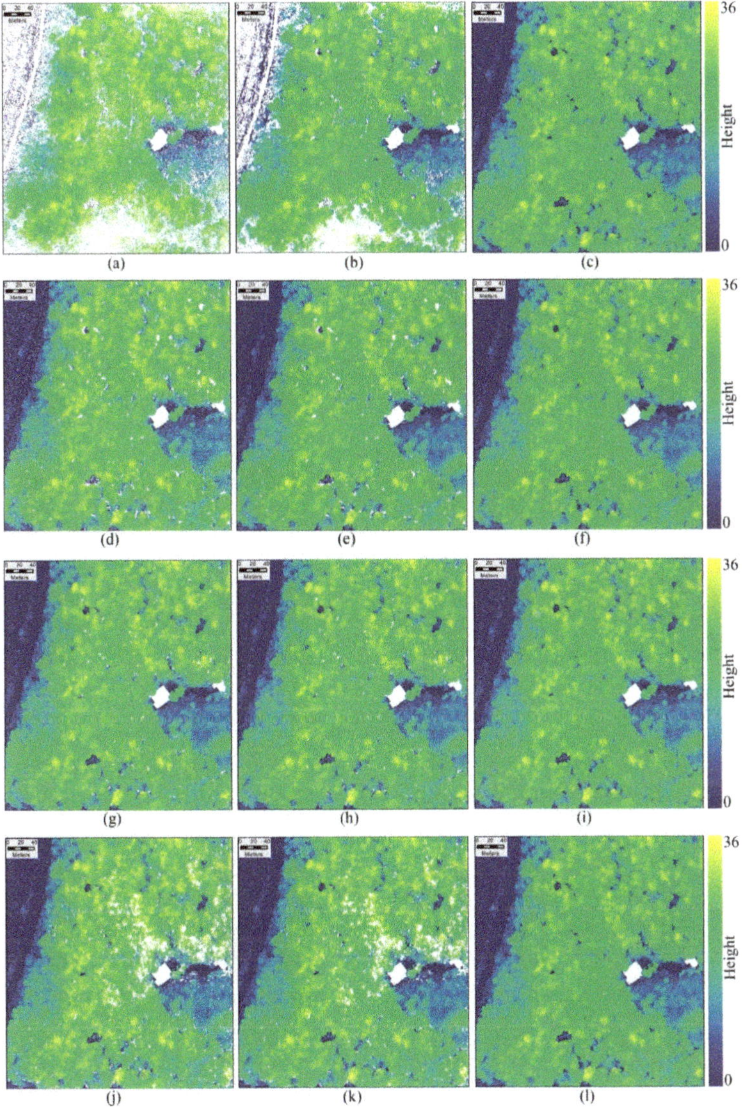

**Figure 3.** Original canopy height models (CHMs) and interpolated CHMs by the constrained neighbor (CN) and ordinary neighbor (ON) algorithms from the LiDAR and photogrammetry data sets with a spatial resolution of 0.1 m (no data areas are white): (**a**) original CHM of L-D55; (**b**) CHM of L-D55 interpolated by the CN algorithm; (**c**) CHM of L-D55 interpolated by the ON algorithm; (**d**) original CHM of P1-D300; (**e**) CHM of P1-D300 interpolated by the CN algorithm; (**f**) CHM of P1-D300 interpolated by the ON algorithm; (**g**) original CHM of P3-D150; (**h**) CHM of P3-D150 interpolated by the CN algorithm; (**i**) CHM of P3-D150 interpolated by the ON algorithm; (**j**) original CHM of P1-D80; (**k**) CHM of P1-D80 interpolated by the CN algorithm; and (**l**) CHM of P1-D80 interpolated by the ON algorithm. The color describes the lowest to highest height from blue to green. The CHMs of P2-D150 and P4-D150 are similar to that of P3-D150 at the same flying height of 150 m and thus are not shown here.

The P-DSMs of the five photogrammetry data sets were processed similarly to those of the LiDAR data set to generate original P-DSMs and those interpolated by the CN and ON algorithms, and the corresponding P-CHMs were generated by subtracting the L-DTM (Figure 3d–l). The original P-CHM of the P1-D300 data set also had some small holes within crowns and large holes between the crowns, and the distributions of the holes were similar to the distribution of the holes in the L-CHM. The original P-CHMs of the P2-D150, P3-D150 and P4-D150 data sets had few small holes within crowns but large holes existed between the crowns. The original P-CHM of the P5-D80 data set had some large holes within the crowns and much larger holes between the crowns than any other data set. The large holes were mainly distributed in the shadow area of tree crowns.

*3.2. Spatial Distribution of Point Clouds*

The spatial distribution of the point clouds was described using the ECR, PCH and PCR. The calculated ECRs, PCHs and PCRs of the original and interpolated L-CHMs with spatial resolutions of 0.1, 0.2, 0.5, and 1.0 m were shown in Figure 4. The ECRs of the original and interpolated L-CHMs increased as the spatial resolution changed from 0.1 to 1.0 m. The ECRs of the original L-CHM with a spatial resolution of 1.0 m were still less than 1.0, which indicated that some areas had sparsely distributed points due to weak ability of backscattering. The CHMs interpolated by the ON algorithm achieved the ideal ECR of 1.0 for any spatial resolution, although the method introduced great spatial variations and decreased the accuracy of the estimated canopy heights.

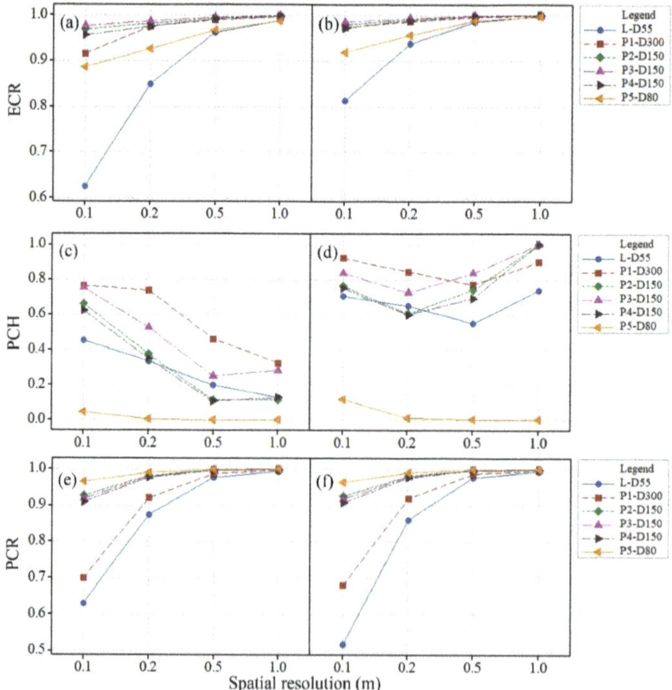

**Figure 4.** Effective cell ratios (ECRs), point cloud homogeneities (PCHs), and point cloud redundancies (PCRs) of the original CHMs and interpolated CHMs by the constrained neighbor (CN) algorithm from LiDAR and photogrammetry data sets: (**a**) ECRs of the original CHMs; (**b**) ECRs of the CHMs interpolated by the CN algorithm; (**c**) PCHs of the original CHMs; (**d**) PCHs of the CHMs interpolated by the CN algorithm; (**e**) PCRs of the original CHMs; and (**f**) PCRs of the CHMs interpolated by the CN algorithm.

The ECRs, PCHs and PCRs of P-CHMs are shown in Figure 4. All the ECRs of the original P-CHMs were greater than those of the L-CHM at spatial resolutions less than 1.0 m, which indicated that the photogrammetry data sets had fewer holes than the LiDAR data set. The ECRs of the original and interpolated P-CHMs also increased as the spatial resolution decreased.

The PCH of the original L-CHM was less than that of the original P-CHMs of most photogrammetry data sets except for P-D80 with a spatial resolution of 0.1 m, which showed that the dense point cloud of photogrammetry was more evenly distributed than the LiDAR point cloud. The low PCH of the original P-CHM of the P-D80 data set was caused by many large holes due to the failed reconstruction of the dense point cloud. The PCHs of the original L- and P-CHMs decreased continuously as the spatial resolution decreased. The PCRs of the original P-CHMs were all higher than those of the L-CHM, indicating that there were more redundant points in the photogrammetry data sets.

The difference between the ECRs of the original L-CHM and interpolated L-CHM by CN at the spatial resolution of 0.1 m was the largest (0.19) compared with those at other spatial resolutions. The ECR pair differences decreased as the spatial resolution decreased. The optimal spatial resolution of L-CHM was 0.2 m if the threshold of the ECR pair difference was set as 0.10. The optimal spatial resolution of P-CHMs was 0.1 m when the same threshold of the ECR pair difference as that of L-CHM was used.

### 3.3. Differences between L-CHMs and P-CHMs

The differences between the original and interpolated L- and P-CHMs were calculated on a cell-by-cell basis. The void cells of the holes might not have corresponding cells among different CHMs. These cells were ignored if there was a void cell in either of the CHM pairs used to calculate the differences. The mean differences between the original L-CHM and original P-CHMs with a spatial resolution of 0.1 m varied from −0.1 to −0.5 m, while the corresponding mean absolute differences varied from 0.9 to 1.1 m (Table A3). This result indicated that the original L-CHM was lower overall than the original P-CHMs and large positive and negative differences occurred within the original CHMs of different data sets. The mean differences between the original L-CHM and P-CHMs increased from negative to positive as the spatial resolution became coarser because low values within cells were ignored when CHMs with a coarse spatial resolution were generated using the maximum algorithm. The mean absolute differences between the original L-CHM and original P-CHMs decreased as the spatial resolution increased, which indicated that the spatial variation in canopy heights was suppressed at a coarse spatial resolution.

The mean differences and mean absolute differences between the LiDAR and photogrammetry data sets slightly increased when the CHMs were interpolated by the CN algorithm compared with the original CHMs. The mean differences and mean absolute differences further increased using the CHMs that were interpolated by the ON algorithm compared with the CHMs interpolated by the CN algorithm.

The stratified mean differences between the original L-CHM and P-CHMs within the 1.0 m bin of the canopy heights were calculated to analyze the spatial variations among canopy heights (Figure 5). Three mark heights of mean differences were observed along the canopy heights, including the minimum mark height, negative mark height and positive mark height. The minimum mark height of the mean difference between CHMs at a spatial resolution of 0.1 m was the 3.0 m bin of the canopy height corresponding to a minimum mean difference of 3.3 m. All the mean differences between CHMs with a spatial resolution of 0.1 m were negative when the canopy height was less than the 21.0 m bin and were positive when the canopy height was greater than the 26.0 m bin, which were referred to as negative and positive mark heights, respectively. The mean differences increased from negative to positive as the canopy height increased above the 3.0 m bin. This result indicated that photogrammetry tends to overestimate the low canopy heights and underestimate the high canopy heights compared to those obtained from LiDAR data.

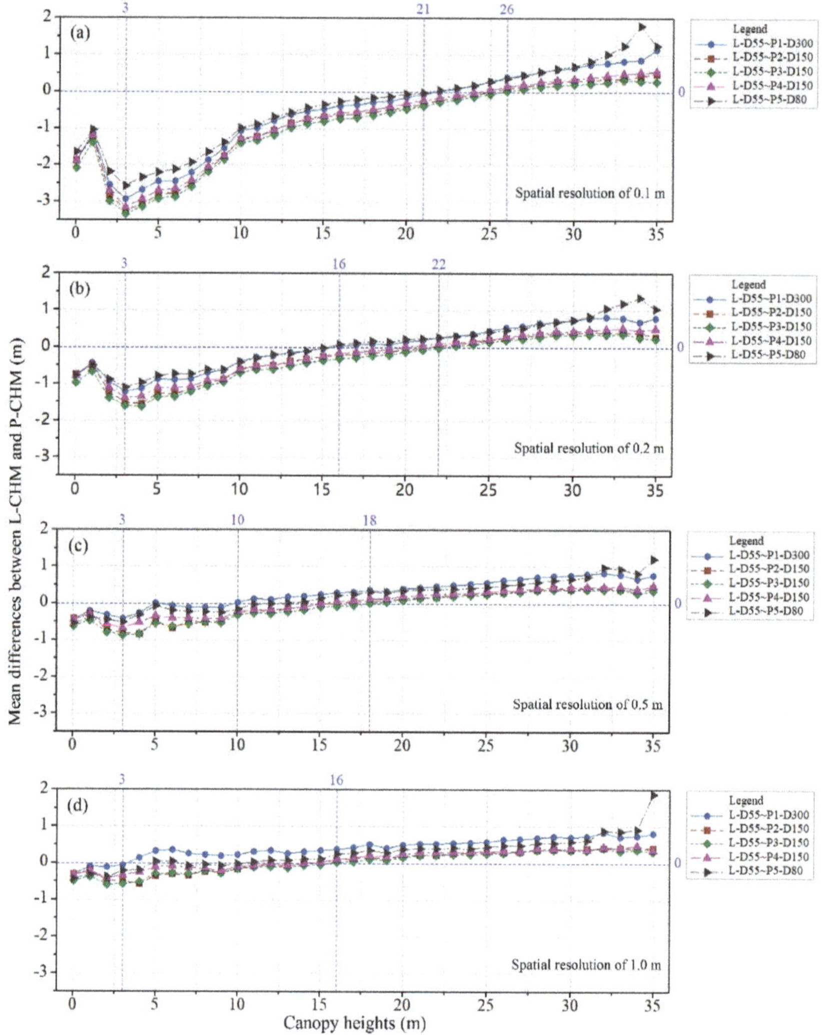

**Figure 5.** Stratified mean differences between the original LiDAR CHM and photogrammetry CHMs within 1.0 m bins of the canopy heights: (**a**) mean differences between CHMs at a spatial resolution of 0.1 m; (**b**) mean differences between CHMs at a spatial resolution of 0.2 m; (**c**) mean differences between CHMs at a spatial resolution of 0.5 m; and (**d**) mean differences between CHMs at a spatial resolution of 1.0 m.

The minimum mean differences increased from −3.3 to −0.6 m as the spatial resolution changed from 0.1 to 1.0 m. The negative and positive mark heights both shifted from high bins to low bins as the spatial resolution became coarser.

### 3.4. Correlation between L-CHMs and P-CHMs

The original and interpolated L-CHMs (responses) were regressed with the corresponding original and interpolated P-CHMs (predictors) (see Figure 6). High correlations were observed between the LiDAR and photogrammetry data sets at different spatial resolutions as shown by the $R^2$, which varied from 0.87 to 0.98. The original L-CHMs were better correlated with P-CHMs at a coarse spatial resolution than a fine spatial resolution. The *RMSE* between the original L-CHMs and P-CHMs decreased as the spatial resolution changed from 0.1 to 1.0 m.

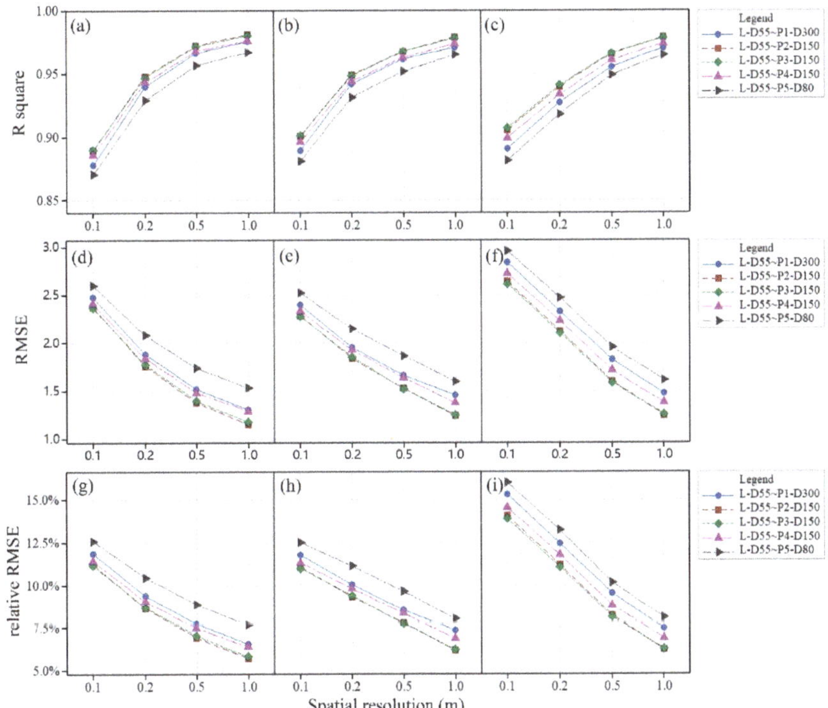

**Figure 6.** Coefficient of determination ($R^2$), *RMSE*, and relative *RMSE* between the CHMs from LiDAR and photogrammetry data sets: (**a**) $R^2$ between the original CHMs; (**b**) $R^2$ between the CHMs interpolated by the CN algorithm; (**c**) $R^2$ between the CHMs interpolated by the ON algorithm; (**d**) *RMSE* between the original CHMs; (**e**) *RMSE* between the CHMs interpolated by the CN algorithm; (**f**) *RMSE* between the CHMs interpolated by the ON algorithm; (**g**) relative *RMSE* between the original CHMs; (**h**) relative *RMSE* between the CHMs interpolated by the CN algorithm; and (**i**) relative *RMSE* between the CHMs interpolated by the ON algorithm.

The L-CHMs and P-CHMs interpolated by the CN algorithm had slight effects on the $R^2$ and root mean square errors (RMSEs) compared with those between the original L-CHM and P-CHMs. The RMSEs between the interpolated L-CHM and P-CHMs by the ON algorithm were higher than those between the interpolated CHMs by the CN algorithm, which was expected. The relative RMSEs between the interpolated L-CHM and P-CHMs by the ON algorithm increased by approximately 3.1% compared with those between the original L-CHM and P-CHMs with a spatial resolution of 0.1 m. This result indicated that the ON interpolation algorithm would introduce obvious spatial variation into the CHM at fine spatial resolution. The relative RMSE decreased as the spatial resolution became coarser.

## 4. Discussion

### 4.1. Spatial Distribution of Point Clouds

The spatial distribution of point clouds is usually uneven across a survey area due to various conditions, such as the LiDAR or photogrammetry system configuration, flying height, crown shadows, or terrain conditions [43–45,63]. At different locations, some points are closely clustered while other points are sparsely distributed. The mean point density can describe only the overall characteristics of point clouds and does not reflect the features of uneven distributions. The ECR, PCH and PCR are used to describe the spatial distribution of point clouds based on regular grids.

The spatial resolution determines the number of points located within each cell of the grid and affects the values of these spatial descriptors. The coverage, homogeneity and redundancy of point clouds are dependent on the spatial resolution of the grid. The coverage of the point clouds will increase at a coarse spatial resolution if the ECR is less than 1.0, the homogeneity of the point clouds will decrease at a coarse spatial resolution, while the redundancy of the point clouds will increase at a coarse spatial resolution.

The relative correlation of the spatial distribution of different data sets might remain constant. For example, P3-D150 had the best coverage, L-D55 had the worst coverage, P1-D300 was the most evenly distributed data set, P5-D80 had the worst homogeneity, L-D55 had the lowest redundancy, and P5-D80 had the most redundant points, and these observations were all based on CHMs with the same spatial resolution (Figure 4). The ideal data set should have the maximum coverage, best homogeneity and lowest redundancy. No data set fulfilled these criteria in this study case.

### 4.2. Effects of the CN and ON Algorithms on the CHM

The void cells of the CHM could be filled by a neighbor interpolation algorithm. The ON algorithm continuously interpolates void cells from the outside cells to inside cells. On the other hand, the CN algorithm interpolates only void cells that meet the criteria of neighboring cells. For example, if the threshold of effective neighboring cells ($Q$) is 5, the void cells along the edges of a large hole will not be filled while the void cells at the corners will be conditionally filled by the CN algorithm.

The interpolated void cells of the CHM by the CN algorithm varied as the spatial resolution changed and produced different ECR values (Figure 4). The ON algorithm interpolated all void cells and obtained the same ECR at all spatial resolutions. If the dimension of the spatial resolution was much smaller than the distance of the point cloud, then the number of the effective neighboring cells of void cells was smaller than the threshold of effective neighboring cells ($Q$) and no void cells were interpolated by the CN algorithm. If the dimension of the spatial resolution was larger than the size of the maximum hole, then the ECR was equal to 1.0. The CN algorithm had no effect on the ECR of the CHM according to these two mentioned cases.

The CN algorithm had a weak effect on the mean absolute differences between the L-CHM and P-CHMs at the spatial resolution of 0.1 m. The mean absolute differences between the original L-CHM and original P-CHMs at the spatial resolution of 0.1 increased by 0.2 m compared with that between the L-CHM and P-CHMs interpolated by the ON algorithm (Table A3). The effects of the CN and ON algorithms on the mean absolute differences both became weaker at a coarser spatial resolution.

The relative RMSEs between the L-CHMs and P-CHMs interpolated by the ON algorithm were larger than those between the CHMs interpolated by the CN algorithm at all spatial resolutions. The CN algorithm had the ability to control the spatial variation introduced by the ON algorithm. Thus, the CN algorithm is recommended for interpolating the void cells of a CHM rather than the ON algorithm.

### 4.3. Distinguishing Small and Large Holes within a CHM

The holes within a CHM with a specified spatial resolution consist of one or several continuous void cells. Small holes in an L-CHM are usually caused by the fluctuation of the UAV platform, mutual obscuration of crowns, gaps within and between crowns, lost echoes of laser pulses, etc. Large holes in an L-CHM are mainly affected by the large incident angles on crowns and the loss of many backscattering signals. The influencing factors of small and large holes of the P-CHM include the reconstruction algorithm of dense point clouds, the quality of images, illumination conditions, texture of crowns, pits of crowns, gap between crowns, and swinging of crowns due to wind. Large holes in the P-CHM are mostly due to weak texture and poor lighting (extremely bright or dim). The light conditions of different photogrammetry data sets are very different and will cause different shadows to affect the spatial distribution of the reconstructed point cloud.

The coverage of holes is the ratio of void cells to all cells expressed as 1-ECR. The original CHM with a fine spatial resolution might include small and large holes, while the CHM interpolated by the CN algorithm would include only large holes. For example, the area covered by holes of the original L-CHM with a spatial resolution of 0.1 m was 38% of the whole area. The large holes ($\geq$0.12 m$^2$) of the interpolated L-CHM occupied 19% of the whole area. This result indicated that 50% of the holes were large holes. The percent of large holes ($\geq$0.48 m$^2$) decreased to 40% of the holes when the spatial resolution of the original L-CHM was 0.2 m. The minimum size of large holes further increased, and the percent of large holes further decreased when the spatial resolution of the L-CHM became coarser.

### 4.4. Optimal Spatial Resolution of the CHM

The spatial resolution of a CHM is usually determined by considering the mean point cloud density. Many void cells will exist within the CHM at this spatial resolution if the point cloud is unevenly distributed across the survey area. Small holes of the CHM can be interpolated by the CN algorithm. The interpolated cells reflect the potential space that would become effective cells at a coarse spatial resolution. The optimal spatial resolution may be determined by the potential space, which is referred to as the potential space criteria. The potential space is calculated as the difference between the ECRs of the original CHM and the CHM interpolated by the CN algorithm.

The optimal spatial resolution can also be determined by the ECR of the CHM interpolated by the CN algorithm, which is referred to as the ECR criteria. If the ECR approaches 1 and is greater than the specified threshold (for example, 0.90), then the corresponding spatial resolution will be regarded as optimal. The ECR criteria are easily affected by very large holes, whose sizes are far greater than the mean space of the point cloud. Additional criteria can be used to select the optimal spatial resolution. We recommend potential space criteria that will not be affected by large holes. Moreover, the optimal spatial resolution can be simply determined by the ECR pair differences among predefined spatial resolutions, which should be calculated by automatic methods in future studies.

### 4.5. Effects of Flying Configuration on the CHM

Five photogrammetry data sets at three different flying heights were used in this study. P1-D300 had the highest flying height and a highly efficient capability of data acquisition when compared to the other data sets collected at lower flying heights. The optimal spatial resolution of P1-D300 was 0.1 m if the threshold of the difference between the ECRs of the original CHM and the CHM interpolated by the CN algorithm was 0.1. Therefore, an optimal flying height of approximately 300 m was recommended to generate a CHM at a spatial resolution of 0.1 m using the similar photogrammetry system in this study.

The P5-D80 data set had the lowest flying height among the photogrammetry data sets, although the ECRs of the original P-CHM of P5-D80 were smaller than those of the other photogrammetry data sets. Very large holes exist within the bright crown area in the original P-CHM of the P5-D80 data set. This finding indicated that some tree crowns failed to be reconstructed using images with very high spatial resolution.

Different shadows occurred in the photogrammetry data sets with different light conditions, and these differences would affect the distribution patterns of void cells in the P-CHMs. The overall effects of light conditions on the reconstructed point cloud and the P-CHMs were weaker than the effects of the flying height in this study. The ECRs of the original P-CHMs with a spatial resolution of 0.1 m had slight differences under different light conditions at the same flying height of 150 m. The differences among the ECRs of the P-CHMs with a spatial resolution of 0.1 m at flying heights of 80, 150 and 300 m were greater than those among the ECRs of the P-CHMs at the same flying height of 150 m.

The mean differences between the combined pairs of photogrammetry data sets were smaller and fluctuated from −0.5 m to 0.5 m when the canopy heights were less than 30 m. Obviously large variations occurred when the canopy heights were greater than 30 m. This variation might be due to the swinging of leaves and twigs of tall trees affected by wind. The mean differences between the combined pairs of the P2-D150, P3-D150 and P4-D150 data sets at the same flying heights were smaller than those of other pairs at different flying heights for all canopy height bins.

The mean differences between the P2-D150 and P4-D150 data sets with different image overlaps varied from 0 m to 0.1 m, which meant that image overlap (>68%) had a minor effect on the estimated canopy height. However, the point density of the photogrammetry data set was substantially affected by image overlap (Figure A1). The mean differences between the P2-D150 and P3-D150 data sets with different flying speeds varied from −0.2 to 0 m, indicating that canopy height was only slightly affected by flying speed (<8 m s$^{-1}$).

### 4.6. Effects of Gaps within and between Crowns on the CHM

The CHMs obtained from LiDAR and photogrammetry data sets for the given spatial resolution were affected by the gaps within and between crowns [14,23,27]. Small gaps within crowns could be depicted in the L-CHM with fine spatial resolution, while the P-CHM tended to ignore such small gaps [23]. A number of deep gaps were observed between the tall tree crowns, which were much darker than the surrounding crowns under the variable illumination conditions. It was difficult to reconstruct the spatial structure in these deep gaps [14,38].

The spatial resolution will affect the height values of gaps within and between crowns. The cells of a CHM are often calculated as the highest value if there are numerous points within the corresponding cell. This condition will suppress the low values of gaps and promote high values of crowns, thus reducing the spatial variation in the canopy heights. LiDAR can provide more detailed structural information than photogrammetry on a fine scale and more details will be lost on a coarse scale. The CHMs between LiDAR and photogrammetry tend to have more consistency as the spatial resolution decreases.

## 5. Conclusions

This study compared the canopy heights obtained from UAV LiDAR and UAV photogrammetry and interpolated by two spatial interpolation algorithms CN and ON. The comparison was conducted based on three proposed spatial distribution descriptors of point clouds: the ECR, PCH and PCR quantifying the unevenness, homogeneity and redundancy characteristics, respectively, of point clouds in the grid at a given spatial resolution. The stratified mean differences revealed that there existed an inherent trend between the estimated canopy heights from LiDAR and photogrammetry, which changed from negative to positive as the canopy heights increased. The LiDAR CHM strongly correlated with the photogrammetry CHM. More importantly, the CN algorithm had the ability to distinguish small and large holes and determine the optimal spatial resolution according to the ECR pair differences, while the ON algorithm did not have this ability. Compared with the ON algorithm, the CN algorithm apparently reduced the spatial variation in the CHM, led to smaller RMSE values, and could be recommended to obtain reliable CHM values.

Some large holes in the CHMs interpolated by the CN interpolation algorithm still occurred at very high point densities, which were typically distributed around the deep gaps between tall tree crowns. Precisely measuring such deep gaps is quite challenging. Photogrammetry tends to ignore gaps within the crowns, overestimate low canopy heights and underestimate high canopy heights; these issues need to be considered while canopy cover, canopy closure, and forest stand height are modelled. Overall, this article provides an operational method for the spatial assessment of point clouds and suggests that the differences between LiDAR and photogrammetry derived CHMs should be considered when forest parameters are estimated. In particular, further study is necessary to enhance understanding the quality measures of the point cloud spatial distribution, optimal spatial

*Remote Sens.* **2020**, *12*, 2884

resolution, small and large holes within CHMs, gaps within and between crowns, and their effects on estimation of forest parameters using additional data sets.

**Author Contributions:** Conceptualization, Q.L. and L.F.; data curation, Q.C., P.L., P.H., M.L. and M.W.; formal analysis, M.L., M.W. and G.D.; funding acquisition, Q.L. and L.F.; methodology, Q.L. and L.F.; supervision, G.W.; visualization, Q.C., P.L. and P.H.; writing—original draft, Q.L.; writing—review and editing, G.W. and R.P.S. All authors contributed to interpreting results and the improvement of the article. All authors have read and agreed to the published version of the manuscript.

**Funding:** This research was partially funded by the National Key R&D Program of China under grant number 2017YFD0600904, the Central Public-interest Scientific Institution Basal Research Fund (Grant No. CAFYBB2019QD003 and CAFYBB2016SZ003).

**Acknowledgments:** We thank Xincheng Shi, Denglong Ha, Yixi Ju, Qiuyan Wang, et al. for assisting field work. We thank the support of the Jigongshan National Nature Reserve, Xinyang, Henan Province, China and the Xinyang Normal University, Xinyang, Henan Province, China for experiments. We thank the anonymous reviewers for their constructive comments and recommendations, which we used to significantly improve our article.

**Conflicts of Interest:** The authors declare no conflict of interest.

## Appendix A

The statistics of the plot parameters are shown in Table A1.

**Table A1.** The statistics of the plot parameters.

| Plot Parameters | Mean | Stand Deviation | Minimum | Maximum |
|---|---|---|---|---|
| Number of stems | 63 | 21 | 30 | 96 |
| Basal area ($m^2$) | 2.4 | 0.9 | 1.0 | 4.4 |
| Lorey's height (m) | 23.8 | 4.4 | 15.7 | 31.2 |
| Mean crown width (m) | 5.4 | 0.6 | 4.3 | 6.8 |

The acquisition information of the LiDAR and photogrammetry data sets is shown in Table A2.

**Table A2.** LiDAR and photogrammetry data sets.

| Data Set | Date(y/m/d) | Local Time (h:m) | Flying Height (m) [1] | Interval (m)/Overlap (%) | | Flying Speed (m s$^{-1}$) | Footprint Size/GSD (cm) |
|---|---|---|---|---|---|---|---|
| | | | | Along-Track | Across-Track | | |
| L-D55 | 2017/8/10 | 11:40–11:50 | 55 | / | 55 | 4.8 | 17 |
| | 2017/8/11 | 10:13–10:24 | 50 [2] | / | 55 | 4.8 | 15 |
| P1-D300 | 2017/8/9 | 13:38–13:56 | 300 | 40/80% | 50/84% | 4.8 | 4 |
| P2-D150 | 2017/8/14 | 15:14–15:32 | 150 | 20/80% | 30/80% | 4.8 | 2 |
| P3-D150 | 2017/8/14 | 16:39–16:51 | 150 | 20/80% | 30/80% | 8.0 | 2 |
| P4-D150 | 2017/8/15 | 11:08–11:20 | 150 | 20/80% | 50/68% | 4.8 | 2 |
| P5-D80 | 2017/8/15 | 15:33–15:51 | 80 | 10/81% | 30/64% | 4.8 | 1 |

[1] The flying height is the relative height above the take-off point (194 m above sea level) of the UAV, which is near the highest point of the study site (198 m above sea level); [2] LiDAR flight with a height of 50 m above the terrain other than the take-off point of the UAV.

The calculated mean differences and mean absolute differences between the L-CHMs and P-CHMs are listed in Table A3.

The point density distributions of the LiDAR and photogrammetry data sets are shown in Figure A1.

**Table A3.** The mean differences and mean absolute differences between the CHMs of LiDAR and photogrammetry data sets.

| Data Set | L-D55 | | | | | | Spatial Resolution (m) |
|---|---|---|---|---|---|---|---|
| | Original CHM | | Interpolated CHM by the CN Algorithm [1] | | Interpolated CHM by the ON Algorithm [2] | | |
| | $\bar{d}$ (m) | $\overline{|d|}$ (m) | $\bar{d}$ (m) | $\overline{|d|}$ (m) | $\bar{d}$ (m) | $\overline{|d|}$ (m) | |
| P1-D300 | −0.1 | 1.0 | −0.2 | 1.1 | −0.3 | 1.3 | |
| P2-D150 | −0.4 | 0.9 | −0.5 | 1.0 | −0.6 | 1.1 | |
| P3-D150 | −0.5 | 0.9 | −0.5 | 1.0 | −0.6 | 1.1 | 0.1 |
| P4-D150 | −0.3 | 1.0 | −0.4 | 1.0 | −0.5 | 1.2 | |
| P5-D80 | −0.1 | 1.1 | −0.1 | 1.2 | −0.3 | 1.4 | |
| P1-D300 | 0.2 | 0.8 | 0.1 | 0.9 | 0.0 | 1.0 | |
| P2-D150 | −0.1 | 0.7 | −0.1 | 0.7 | −0.2 | 0.8 | |
| P3-D150 | −0.1 | 0.7 | −0.2 | 0.8 | −0.3 | 0.9 | 0.2 |
| P4-D150 | 0.0 | 0.8 | −0.1 | 0.8 | −0.2 | 0.9 | |
| P5-D80 | 0.2 | 0.9 | 0.1 | 1.0 | 0.1 | 1.1 | |
| P1-D300 | 0.4 | 0.8 | 0.4 | 0.8 | 0.3 | 0.8 | |
| P2-D150 | 0.1 | 0.6 | 0.1 | 0.6 | 0.1 | 0.6 | |
| P3-D150 | 0.1 | 0.6 | 0.0 | 0.6 | 0.0 | 0.6 | 0.5 |
| P4-D150 | 0.1 | 0.7 | 0.1 | 0.7 | 0.1 | 0.7 | |
| P5-D80 | 0.3 | 0.8 | 0.3 | 0.9 | 0.3 | 0.9 | |
| P1-D300 | 0.5 | 0.7 | 0.5 | 0.7 | 0.5 | 0.8 | |
| P2-D150 | 0.2 | 0.5 | 0.1 | 0.5 | 0.1 | 0.5 | |
| P3-D150 | 0.1 | 0.5 | 0.1 | 0.5 | 0.1 | 0.5 | 1.0 |
| P4-D150 | 0.2 | 0.6 | 0.2 | 0.6 | 0.2 | 0.6 | |
| P5-D80 | 0.3 | 0.7 | 0.3 | 0.7 | 0.3 | 0.8 | |

[1] CN means the constrained neighbor interpolation algorithm; [2] ON means the ordinary neighbor interpolation algorithm.

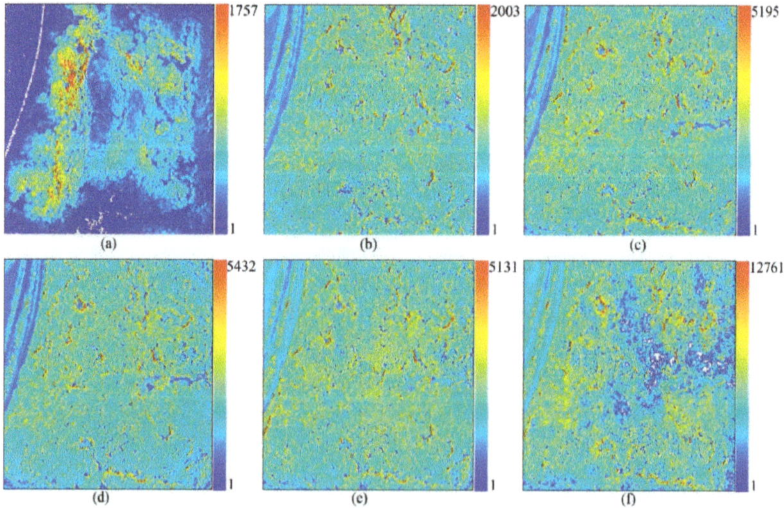

**Figure A1.** Point density of the 1 m resolution grid (no data areas are white; the grid is stretched with a percent clip of 0.5): (**a**) L-D55 point cloud with a mean of 168 points m$^{-2}$; (**b**) P1-D300 point cloud with a mean of 303 points m$^{-2}$; (**c**) P2-D150 point cloud with a mean of 1319 points m$^{-2}$; (**d**) P3-D150 point cloud with a mean of 1191 points m$^{-2}$; (**e**) P4-D150 point cloud with a mean of 1050 points m$^{-2}$; (**f**) P5-D80 point cloud with a mean of 2529 points m$^{-2}$.

## References


1. Lefsky, M.A.; Cohen, W.B.; Acker, S.A.; Parker, G.G.; Spies, T.A.; Harding, D. Lidar Remote Sensing of the Canopy Structure and Biophysical Properties of Douglas-Fir Western Hemlock Forests. *Remote Sens. Environ.* **1999**, *70*, 339–361. [CrossRef]
2. Junttila, V.; Kauranne, T.; Finley, A.O.; Bradford, J.B. Linear Models for Airborne-Laser-Scanning-Based Operational Forest Inventory With Small Field Sample Size and Highly Correlated LiDAR Data. *IEEE Trans. Geosci. Remote Sens.* **2015**, *53*, 5600–5612. [CrossRef]
3. White, J.C.; Coops, N.C.; Wulder, M.A.; Vastaranta, M.; Hilker, T.; Tompalski, P. Remote Sensing Technologies for Enhancing Forest Inventories: A Review. *Can. J. Remote Sens.* **2016**, *42*, 619–641. [CrossRef]
4. Suárez, J.C.; Ontiveros, C.; Smith, S.; Snape, S. Use of airborne LiDAR and aerial photography in the estimation of individual tree heights in forestry. *Comput. Geosci.* **2005**, *31*, 253–262. [CrossRef]
5. White, J.C.; Stepper, C.; Tompalski, P.; Coops, N.C.; Wulder, M.A. Comparing ALS and Image-Based Point Cloud Metrics and Modelled Forest Inventory Attributes in a Complex Coastal Forest Environment. *Forests* **2015**, *6*, 3704–3732. [CrossRef]
6. Rahlf, J.; Breidenbach, J.; Solberg, S.; Næsset, E.; Astrup, R. Digital aerial photogrammetry can efficiently support large-area forest inventories in Norway. *Forestry* **2017**, *90*, 710–718. [CrossRef]
7. Marinelli, D.; Paris, C.; Bruzzone, L. A Novel Approach to 3-D Change Detection in Multitemporal LiDAR Data Acquired in Forest Areas. *IEEE Trans. Geosci. Remote Sens.* **2018**, *56*, 3030–3046. [CrossRef]
8. Liu, Q.; Fu, L.; Wang, G.; Li, S.; Li, Z.; Chen, E.; Pang, Y.; Hu, K. Improving Estimation of Forest Canopy Cover by Introducing Loss Ratio of Laser Pulses Using Airborne LiDAR. *IEEE Trans. Geosci. Remote Sens.* **2020**, *58*, 567–585. [CrossRef]
9. Wallace, L.; Lucieer, A.; Watson, C.S. Evaluating Tree Detection and Segmentation Routines on Very High Resolution UAV LiDAR Data. *IEEE Trans. Geosci. Remote Sens.* **2014**, *52*, 7619–7628. [CrossRef]
10. Brede, B.; Lau, A.; Bartholomeus, H.M.; Kooistra, L. Comparing RIEGL RiCOPTER UAV LiDAR Derived Canopy Height and DBH with Terrestrial LiDAR. *Sensors* **2017**, *17*, 2371. [CrossRef]
11. Liu, K.; Shen, X.; Cao, L.; Wang, G.; Cao, F. Estimating forest structural attributes using UAV-LiDAR data in Ginkgo plantations. *ISPRS J. Photogramm. Remote Sens.* **2018**, *146*, 465–482. [CrossRef]
12. Wang, Y.; Lehtomäki, M.; Liang, X.; Pyörälä, J.; Kukko, A.; Jaakkola, A.; Liu, J.; Feng, Z.; Chen, R.; Hyyppä, J. Is field-measured tree height as reliable as believed—A comparison study of tree height estimates from field measurement, airborne laser scanning and terrestrial laser scanning in a boreal forest. *ISPRS J. Photogramm. Remote Sens.* **2019**, *147*, 132–145. [CrossRef]
13. Wang, Y.; Pyörälä, J.; Liang, X.; Lehtomäki, M.; Kukko, A.; Yu, X.; Kaartinen, H.; Hyyppä, J. In Situ biomass estimation at tree and plot levels: What did data record and what did algorithms derive from terrestrial and aerial point clouds in boreal forest. *Remote Sens. Environ.* **2019**, *232*, 111309. [CrossRef]
14. Vepakomma, U.; St-Onge, B.; Kneeshaw, D. Spatially explicit characterization of boreal forest gap dynamics using multi-temporal lidar data. *Remote Sens. Environ.* **2008**, *112*, 2326–2340. [CrossRef]
15. Kane, V.; Gersonde, R.; Lutz, J.; Mcgaughey, R.; Bakker, J.; Franklin, J. Patch dynamics and the development of structural and spatial heterogeneity in Pacific Northwest forests. *Can. J. For. Res.* **2011**, *41*, 2276–2281. [CrossRef]
16. Wang, Y.; Hyyppä, J.; Liang, X.; Kaartinen, H.; Yu, X.; Lindberg, E.; Holmgren, J.; Qin, Y.; Mallet, C.; Ferraz, A.; et al. International Benchmarking of the Individual Tree Detection Methods for Modeling 3-D Canopy Structure for Silviculture and Forest Ecology Using Airborne Laser Scanning. *IEEE Trans. Geosci. Remote Sens.* **2016**, *54*, 5011–5027. [CrossRef]
17. Wang, X.; Zheng, G.; Yun, Z.; Moskal, L.M. Characterizing Tree Spatial Distribution Patterns Using Discrete Aerial Lidar Data. *Remote Sens.* **2020**, *12*, 712. [CrossRef]
18. Næsset, E. Vertical Height Errors in Digital Terrain Models Derived from Airborne Laser Scanner Data in a Boreal-Alpine Ecotone in Norway. *Remote Sens.* **2015**, *7*, 4702–4725. [CrossRef]
19. Okojie, J.A.; Okojie, L.O.; Effiom, A.E.; Odia, B.E. Relative canopy height modelling precision from UAV and ALS datasets for forest tree height estimation. *Remote Sens. Appl. Soc. Environ.* **2020**, *17*, 100284. [CrossRef]
20. Järnstedt, J.; Pekkarinen, A.; Tuominen, S.; Ginzler, C.; Holopainen, M.; Viitala, R. Forest variable estimation using a high-resolution digital surface model. *ISPRS J. Photogramm. Remote Sens.* **2012**, *74*, 78–84. [CrossRef]

21. Sofonia, J.J.; Phinn, S.; Roelfsema, C.; Kendoul, F.; Rist, Y. Modelling the effects of fundamental UAV flight parameters on LiDAR point clouds to facilitate objectives-based planning. *ISPRS J. Photogramm. Remote Sens.* **2019**, *149*, 105–118. [CrossRef]

22. Wang, X.-H.; Zhang, Y.-Z.; Xu, M.-M. A Multi-Threshold Segmentation for Tree-Level Parameter Extraction in a Deciduous Forest Using Small-Footprint Airborne LiDAR Data. *Remote Sens.* **2019**, *11*, 2109. [CrossRef]

23. White, J.C.; Tompalski, P.; Coops, N.C.; Wulder, M.A. Comparison of airborne laser scanning and digital stereo imagery for characterizing forest canopy gaps in coastal temperate rainforests. *Remote Sens. Environ.* **2018**, *208*, 1–14. [CrossRef]

24. Granholm, A.-H.; Lindgren, N.; Olofsson, K.; Nyström, M.; Allard, A.; Olsson, H. Estimating vertical canopy cover using dense image-based point cloud data in four vegetation types in southern Sweden. *Int. J. Remote Sens.* **2017**, *38*, 1820–1838. [CrossRef]

25. Hawryło, P.; Tompalski, P.; Wężyk, P. Area-based estimation of growing stock volume in Scots pine stands using ALS and airborne image-based point clouds. *Forestry* **2017**, *90*, 686–696. [CrossRef]

26. Sullivan, F.B.; Ducey, M.J.; Orwig, D.A.; Cook, B.; Palace, M.W. Comparison of lidar- and allometry-derived canopy height models in an eastern deciduous forest. *Forest Ecol. Manage.* **2017**, *406*, 83–94. [CrossRef]

27. Senécal, J.-F.; Doyon, F.; Messier, C. Tree Death Not Resulting in Gap Creation: An Investigation of Canopy Dynamics of Northern Temperate Deciduous Forests. *Remote Sens.* **2018**, *10*, 121. [CrossRef]

28. Jakubowski, M.K.; Li, W.; Guo, Q.; Kelly, M. Delineating Individual Trees from Lidar Data: A Comparison of Vector- and Raster-based Segmentation Approaches. *Remote Sens.* **2013**, *5*, 4163–4186. [CrossRef]

29. Duan, Z.; Zhao, D.; Zeng, Y.; Zhao, Y.; Wu, B.; Zhu, J. Assessing and Correcting Topographic Effects on Forest Canopy Height Retrieval Using Airborne LiDAR Data. *Sensors* **2015**, *15*, 12133–12155. [CrossRef]

30. Chen, C.; Wang, Y.; Li, Y.; Yue, T.; Wang, X. Robust and Parameter-Free Algorithm for Constructing Pit-Free Canopy Height Models. *ISPRS Int. J. Geo-Inf.* **2017**, *6*, 219. [CrossRef]

31. Alexander, C.; Korstjens, A.H.; Hill, R.A. Influence of micro-topography and crown characteristics on tree height estimations in tropical forests based on LiDAR canopy height models. *Int. J. Appl. Earth. Obs. Geoinf.* **2018**, *65*, 105–113. [CrossRef]

32. Dietmaier, A.; McDermid, G.J.; Rahman, M.M.; Linke, J.; Ludwig, R. Comparison of LiDAR and Digital Aerial Photogrammetry for Characterizing Canopy Openings in the Boreal Forest of Northern Alberta. *Remote Sens.* **2019**, *11*, 1919. [CrossRef]

33. Dandois, J.P.; Ellis, E.C. High spatial resolution three-dimensional mapping of vegetation spectral dynamics using computer vision. *Remote Sens. Environ.* **2013**, *136*, 259–276. [CrossRef]

34. Jayathunga, S.; Owari, T.; Tsuyuki, S. The use of fixed-wing UAV photogrammetry with LiDAR DTM to estimate merchantable volume and carbon stock in living biomass over a mixed conifer-broadleaf forest. *Int. J. Appl. Earth. Obs. Geoinf.* **2018**, *73*, 767–777. [CrossRef]

35. Ni, W.; Sun, G.; Pang, Y.; Zhang, Z.; Liu, J.; Yang, A.; Wang, Y.; Zhang, D. Mapping Three-Dimensional Structures of Forest Canopy Using UAV Stereo Imagery: Evaluating Impacts of Forward Overlaps and Image Resolutions With LiDAR Data as Reference. *IEEE J. Sel. Top. Appl. Earth Obs. Remote Sens.* **2018**, *11*, 3578–3589. [CrossRef]

36. Yurtseven, H.; Akgul, M.; Coban, S.; Gulci, S. Determination and accuracy analysis of individual tree crown parameters using UAV based imagery and OBIA techniques. *Measurement* **2019**, *145*, 651–664. [CrossRef]

37. Moe, K.T.; Owari, T.; Furuya, N.; Hiroshima, T. Comparing Individual Tree Height Information Derived from Field Surveys, LiDAR and UAV-DAP for High-Value Timber Species in Northern Japan. *Forests* **2020**, *11*, 223. [CrossRef]

38. Baltsavias, E.; Gruen, A.; Eisenbeiss, H.; Zhang, L.; Waser, L.T. High-quality image matching and automated generation of 3D tree models. *Int. J. Remote Sens.* **2008**, *29*, 1243–1259. [CrossRef]

39. Nurminen, K.; Karjalainen, M.; Yu, X.; Hyyppä, J.; Honkavaara, E. Performance of dense digital surface models based on image matching in the estimation of plot-level forest variables. *ISPRS J. Photogramm. Remote Sens.* **2013**, *83*, 104–115. [CrossRef]

40. Ullah, S.; Dees, M.; Datta, P.; Adler, P.; Koch, B. Comparing Airborne Laser Scanning, and Image-Based Point Clouds by Semi-Global Matching and Enhanced Automatic Terrain Extraction to Estimate Forest Timber Volume. *Forests* **2017**, *8*, 215. [CrossRef]

41. Tompalski, P.; Rakofsky, J.; Coops, N.C.; White, J.C.; Graham, A.N.V.; Rosychuk, K. Challenges of Multi-Temporal and Multi-Sensor Forest Growth Analyses in a Highly Disturbed Boreal Mixedwood Forests. *Remote Sens.* **2019**, *11*, 2102. [CrossRef]

42. Leberl, F.; Irschara, A.; Pock, T.; Meixner, P.; Gruber, M.; Scholz, S.; Wiechert, A. Point Clouds: Lidar versus 3D Vision. *Photogramm. Eng. Remote Sens.* **2010**, *76*, 1123–1134. [CrossRef]

43. Chasmer, L.; Hopkinson, C.; Smith, B.; Treitz, P. Examining the Influence of Changing Laser Pulse Repetition Frequencies on Conifer Forest Canopy Returns. *Photogramm. Eng. Remote Sens.* **2006**, *72*, 1359–1367. [CrossRef]

44. Goodwin, N.R.; Coops, N.C.; Culvenor, D.S. Assessment of forest structure with airborne LiDAR and the effects of platform altitude. *Remote Sens. Environ.* **2006**, *103*, 140–152. [CrossRef]

45. Hopkinson, C. The influence of flying altitude, beam divergence, and pulse repetition frequency on laser pulse return intensity and canopy frequency distribution. *Can. J. Remote Sens.* **2007**, *33*, 312–324. [CrossRef]

46. White, J.C.; Wulder, M.A.; Vastaranta, M.; Coops, N.C.; Pitt, D.; Woods, M. The Utility of Image-Based Point Clouds for Forest Inventory: A Comparison with Airborne Laser Scanning. *Forests* **2013**, *4*, 518–536. [CrossRef]

47. St-Onge, B.; Vega, C.; Fournier, R.A.; Hu, Y. Mapping canopy height using a combination of digital stereo-photogrammetry and lidar. *Int. J. Remote Sens.* **2008**, *29*, 3343–3364. [CrossRef]

48. Vastaranta, M.; Wulder, M.; White, J.; Pekkarinen, A.; Tuominen, S.; Ginzler, C.; Kankare, V.; Holopainen, M.; Hyyppä, J.; Hyyppä, H. Airborne laser scanning and digital stereo imagery measures of forest structure: Comparative results and implications to forest mapping and inventory update. *Can. J. Remote Sens.* **2013**, *39*, 382–395. [CrossRef]

49. Pitt, D.G.; Woods, M.; Penner, M. A Comparison of Point Clouds Derived from Stereo Imagery and Airborne Laser Scanning for the Area-Based Estimation of Forest Inventory Attributes in Boreal Ontario. *Can. J. Remote Sens.* **2014**, *40*, 214–232. [CrossRef]

50. Salach, A.; Bakuła, K.; Pilarska, M.; Ostrowski, W.; Górski, K.; Kurczyński, Z. Accuracy Assessment of Point Clouds from LiDAR and Dense Image Matching Acquired Using the UAV Platform for DTM Creation. *ISPRS Int. J. Geo-Inf.* **2018**, *7*, 342. [CrossRef]

51. Ben-Arie, J.R.; Hay, G.J.; Powers, R.P.; Castilla, G.; St-Onge, B. Development of a pit filling algorithm for LiDAR canopy height models. *Comput. Geosci.* **2009**, *35*, 1940–1949. [CrossRef]

52. Yan, J.; Zhang, W.; Wang, K.; Qin, F.; Wang, W.; Dai, H.; Li, P. Responses of CO2, N2O and CH4 fluxes between atmosphere and forest soil to changes in multiple environmental conditions. *Glob. Chang. Biol.* **2014**, *20*, 300–312. [CrossRef] [PubMed]

53. Zhang, W.; Shen, W.; Zhu, S.; Wan, S.; Luo, Y.; Yan, J.; Wang, K.; Liu, L.; Dai, H.; Li, P.; et al. CAN Canopy Addition of Nitrogen Better Illustrate the Effect of Atmospheric Nitrogen Deposition on Forest Ecosystem? *Nature* **2015**, *5*, 11245. [CrossRef] [PubMed]

54. Beijing GreenValley Technology Co. Ltd. User Guide GV1300. 2017.

55. Velodyne LiDAR Inc. VLP-16 User Manual. Available online: https://velodynelidar.com/wp-content/uploads/2019/12/63-9243-Rev-E-VLP-16-User-Manual.pdf (accessed on 3 September 2020).

56. Beijing GreenValley Technology Co. Ltd. User Guide GV1500. 2017.

57. Canon, U.S.A. Inc. Canon EOS 5DS. Available online: https://www.usa.canon.com/internet/portal/us/home/products/details/cameras/eos-dslr-and-mirrorless-cameras/dslr/eos-5ds (accessed on 6 June 2020).

58. Agisoft LLC. Agisoft PhotoScan User Manual Professional Edition Version 1.2. Available online: https://www.agisoft.com/pdf/photoscan-pro_1_2_en.pdf (accessed on 3 September 2020).

59. Terrasolid Ltd. TerraScan User's Guide. Available online: http://terrasolid.com/download/tscan.pdf (accessed on 3 September 2020).

60. Bogawski, P.; Grewling, Ł.; Dziób, K.; Sobieraj, K.; Dalc, M.; Dylawerska, B.; Pupkowski, D.; Nalej, A.; Nowak, M.; Szymańska, A.; et al. Lidar Derived Tree Crown Parameters: Are They New Variables Explaining Local Birch (Betula sp.) Pollen Concentrations? *Forests* **2019**, *10*, 1154. [CrossRef]

61. Pyörälä, J.; Saarinen, N.; Kankare, V.; Coops, N.C.; Liang, X.; Wang, Y.; Holopainen, M.; Hyyppä, J.; Vastaranta, M. Variability of wood properties using airborne and terrestrial laser scanning. *Remote Sens. Environ.* **2019**, *235*, 111474. [CrossRef]

62. Hyndman, R.J.; Koehler, A.B. Another look at measures of forecast accuracy. *Int. J. Forecast.* **2006**, *22*, 679–688. [CrossRef]
63. Honkavaara, E.; Markelin, L.; Rosnell, T.; Nurminen, K. Influence of solar elevation in radiometric and geometric performance of multispectral photogrammetry. *ISPRS J. Photogramm. Remote Sens.* **2012**, *67*, 13–26. [CrossRef]

 *remote sensing*

*Article*

# Prediction of Individual Tree Diameter and Height to Crown Base Using Nonlinear Simultaneous Regression and Airborne LiDAR Data

Zhaohui Yang [1,2,†], Qingwang Liu [1,†], Peng Luo [1], Qiaolin Ye [3], Guangshuang Duan [1,4], Ram P. Sharma [5], Huiru Zhang [1,2], Guangxing Wang [6] and Liyong Fu [1,2,*]

[1]  Research Institute of Forest Resource Information Techniques, Chinese Academy of Forestry, Beijing 100091, China; yzh@ifrit.ac.cn (Z.Y.); liuqw@ifrit.ac.cn (Q.L.); luopeng@ifrit.ac.cn (P.L.); duangs@ifrit.ac.cn (G.D.); huiru@ifrit.ac.cn (H.Z.)
[2]  Key Laboratory of Forest Management and Growth Modeling, National Forestry and Grassland Administration, Beijing 100091, China
[3]  College of Information Science and Technology, Nanjing Forestry University, Nanjing 210037, China; yqlcom@njfu.edu.cn
[4]  College of Mathematics and Statistics, Xinyang Normal University, Xinyang 464000, China
[5]  Institute of Forestry, Tribhuwan Univeristy, Kritipur, Kathmandu-44600, Nepal; sharmar@fld.czu.cz
[6]  Department of Geography and Environmental Resources, Southern Illinois University at Carbondale, Carbondale, IL 62901, USA; gxwang@siu.edu
*  Correspondence: fuly@ifrit.ac.cn; Tel.: +86-10-62889179; Fax: +86-10-62888315
†  These authors contributed equally to this work.

Received: 18 June 2020; Accepted: 29 June 2020; Published: 13 July 2020

**Abstract:** The forest growth and yield models, whichor are used as important decision-support tools in forest management, are commonly based on the individual tree characteristics, such as diameter at breast height (DBH), crown ratio, and height to crown base (HCB). Taking direct measurements for DBH and HCB through the ground-based methods is cumbersome and costly. The indirect method of getting such information is possible from remote sensing databases, whichor can be used to build DBH and HCB prediction models. The DBH and HCB of the same trees are significantly correlated, and so their inherent correlations need to be appropriately accounted for in the DBH and HCB models. However, all the existing DBH and HCB models, including models based on light detection and ranging (LiDAR) have ignored such correlations and thus failed to account for the compatibility of DBH and HCB estimates, in addition to disregarding measurement errors. To address these problems, we developed a compatible simultaneous equation system of DBH and HCB error-in-variable (EIV) models using LiDAR-derived data and ground-measurements for 510 *Picea crassifolia* Kom trees in northwest China. Four versatile algorithms, such as nonlinear seemingly unrelated regression (NSUR), two-stage least square (2SLS) regression, three-stage least square (3SLS) regression, and full information maximum likelihood (FIML) were evaluated for their estimating efficiencies and precisions for a simultaneous equation system of DBH and HCB EIV models. In addition, two other model structures, namely, nonlinear least squares with HCB estimation not based on the DBH (NLS and NBD) and nonlinear least squares with HCB estimation based on the DBH (NLS and BD) were also developed, and their fitting precisions with a simultaneous equation system compared. The leave-one-out cross-validation method was applied to evaluate all estimating algorithms and their resulting models. We found that only the simultaneous equation system could illustrate the effect of errors associated with the regressors on the response variables (DBH and HCB) and guaranteed the compatibility between the DBH and HCB models at an individual level. In addition, such an established system also effectively accounted for the inherent correlations between DBH with HCB. However, both the NLS and BD model and the NLS and NBD model did not show these properties. The precision of a simultaneous equation system developed using NSUR appeared the best among all the evaluated algorithms. Our equation system does not require the stand-level information as input, but it does require the information of tree height, crown width, and crown projection area, all of whichor can be readily derived from LiDAR imagery using the delineation algorithms and ground-based DBH measurements. Our results indicate that NSUR is a more reliable and quicker algorithm for developing DBH and HCB models using large scale LiDAR-based datasets. The novelty of this study is that the compatibility problem of the DBH model and the HCB EIV model was properly addressed, and the potential algorithms were compared to choose the most suitable one (NSUR). The presented method and algorithm will be useful for establishing similar compatible equation systems of tree DBH and HCB EIV models for other tree species.

**Keywords:** *Picea crassifolia* Kom; compatible equation; nonlinear seemingly unrelated regression; error-in-variable modeling; leave-one-out cross-validation

---

## 1. Introduction

A tree crown is characterized by crown height, crown width, crown density, leaf area, and crown ratio, and their measurements are useful for forest management and research. The crown ratio is considered a reliable indicator of the vigor and potential growth of a tree [1–4]. Height to crown base (HCB) is an important tree measure to derive crown ratio and is also regarded as an indicator of log quality. HCB is usually understood as the vertical height from the ground to the bottom of live whorled branch on the bole of a tree [5]. The ground-based measurement of HCB is a time-consuming and labor-intensive process; thus, it is rarely done during field inventory [6,7]. Most researchers have obtained the HCB value by establishing linear or nonlinear HCB models with other variables as predictors, such as DBH, tree height, basal area, basal area larger than a target tree, the sum of basal area of all trees with diameter bigger than a target tree, crown competition factor, climate, and site index [8–12]. Tree diameter at breast height (DBH) is also an important tree attribute that is used as a main predictor in forest growth and yield, taper, and biomass models. In general, the measurement of DBH is very common in ground-based inventory; however, field-inventory data could have a low accuracy, and their measurement needs more time and cost, especially measurements required for extensive forest areas. Therefore, methods of HCB data collection have been transformed from the traditional forest field inventory to modeling and prediction based on remote sensing technology [13–16].

Light detection and ranging (LiDAR) can accurately determine the geographical position of surface objects by transmitting and receiving laser pulses. Laser pulses travel down the forest canopy, and detailed information on the three-dimensional structures of the forest canopy and understory topography can be obtained [17]. Many tree attributes, such as tree height and crown dimensions [18] can be obtained based on the LiDAR data. The study approaches based on HCB prediction may be divided into two categories: direct and indirect approaches. The direct approaches refer to those derived from HCB with various geometrical shapes of the crown [12,19–23] or predicting HCB according to descriptive statistics of the LiDAR-based data distribution [4,24]. Direct approaches do not require any ground-measured HCB data, whichor are costly and time-consuming, as they only require point-cloud data processing and analysis including tree detection and the determination of crown base positions. In addition, this approach could also cause considerable uncertainties in determining the base of the first normal green branch as a part of the crown. Therefore, its application is quite limited to estimating HCB. The indirect approach, on the other hand, refers to predicting HCB through the application of statistical modeling [22,25–28]. This approach requires field-measured HCB data to establish the models for the prediction of HCB. The models for the accurate prediction of individual tree HCB can be built using LiDAR-based information, and so this method has been frequently used in recent years [22,25–28].

The application of ordinary least square (OLS) regression to estimate the parameters of LiDAR-based DBH and HCB models is not generally preferred, but it is still used [16,29]. This estimation method usually assumes that (i) regressors are random variables with errors, (ii) regressors are fixed variables without errors, and (iii) the associated error is subject to normal distribution with zero mean and constant variance [30]. Any violation of the second assumption leads to the substantially biased estimation of the models [30], whichor eventually reduces the prediction accuracy.

The prediction accuracy of the developed HCB and DBH models uses the LiDAR-based tree height, crown width and crown area may not be always satisfactory for a couple reasons. Firstly, LiDAR-based tree height, crown width, and crown area have random or systematic errors caused by LiDAR system configuration and parameter estimation. Any error involved in the variables could increase the residual variance of the model and also lead to invalid statistical tests [31,32]. Secondly, the estimated DBH from a LiDAR-based DBH estimation model contains non-ignorable or inevitable errors [33]. If such erroneous DBH is used as a predictor in a LiDAR-based HCB model, substantial bias would occur due to error transfers [34]. In addition, estimating with a LiDAR-based DBH model and a LiDAR-based HCB model separately or independently using OLS disregards the inherent correlations of HCB with DBH and thus fails to account for the compatibility of the estimated HCB and DBH. Thus, estimating the parameters of both model types independently with OLS may create a remarkable problem, especially in the condition when errors are associated with both the regressors and response variables. An appropriate settlement of this problem is to apply error-in-variable (EIV) modeling, whichor takes the errors into consideration and can guarantee compatibility between HCB and DBH [29,35–37].

Fuller [35] first introduced the theory on the development and application of linear EIV models, and, later on, Carroll et al. [32] applied this concept on the nonlinear EIV modeling in detail. Kangas [31] investigated the effects of EIV on the parameters of the diameter growth model and applied the simulation extrapolation algorithm to adjust the errors in the estimated parameters. Lindely [38] proved that validation data from the same population as the fitting data resulted in predictions that were usually unbiased, even though the regressors were subject to error. Tang and Zhang [36] developed an EIV model to investigate the unbiased parameter estimates. Tang and Wang [39] proposed the two-stage EIV method to estimate the model parameters. In their study, the EIV concept was introduced into forest attribute modeling, whichor provides a theoretical basis for studying the influence of errors on stand growth and harvest models. Li and Tang [40] compared three methods, namely simulation extrapolation, regression calibration, and EIV to estimate the models and found a better performance with EIV with smaller variances compared to other two methods.

Few studies have been carried out with DBH EIV modeling using remote sensing data. For example, Fu et al. [33] developed an individual tree DBH and above-ground biomass (AGB) EIV model with LiDAR-based tree height and crown projection area as predictors with the application of the two-stage error-in-variable modeling (TSEM) and nonlinear seemingly-unrelated regression (NSUR) to estimate model parameters. Both TSEM and NSUR explain the correlations of DBH with AGB and also effectively explain the errors in DBH on the prediction of AGB. Zhang et al. [29] reported that the DBH EIV model developed with errors associated with both response and regressor variables through the application of the maximum likelihood method was most appropriate. To the authors' knowledge, no studies have been carried out on developing LiDAR-based HCB EIV models that were attributed to compatibility.

This study thus aimed (a) to develop a compatible simultaneous equation system of DBH and HCB EIV models based on the LiDAR data at the individual tree level for *Picea crassifolia* Kom forests in northwest China, (b) to evaluate the compatibility of two different nonlinear OLS-based DBH and HCB models with the leave-one-out cross validation method, and (c) to compare various unbiased fitting algorithms including NSUR. To simplify the proposed simultaneous equation system and to guarantee its application in the future, only response variables (HCB and DBH) were assumed as the error-in-variables [39], and predictor variables were regarded as error-out variables [33]. The presented compatible simultaneous equation system of DBH and HCB models will be applicable to other *Picea* species whose growth and stand conditions are very much similar to the basis of our studied species. This tree species is crucial to the economic and social development of the rural population, as well as regional carbon storage and cycling, and the maintenance of the structures and functions of the forest ecosystems in northwest China. This article is mainly concerned with the methodology employed in this study, whichor is clearly described in the Methods section; additionally, the major strengths and weaknesses of the methodologies, along with the main findings of the study, are thoroughly discussed while the potential contribution of the study is highlighted.

## 2. Methods

### 2.1. Data Collection

The study site is located at the Xishui forest farm of the Su'nan Yuguzu autonomous county, Gansu province (38°29′–38°35′N, 100°12′–100°20′E) (Figure 1a) with *Picea crassifolia* Kom as the dominate tree species. The climate in this field is a temperate semi-arid zone. It is covered by mountainous forests. Slopes with south-facing aspect are covered by grass, and the slopes with north-facing aspect are covered by natural secondary pure forests with one dominating tree species of *Picea crassifolia* Kom. The ground is covered by a moss floor, and the average elevation here is around 2993 m. The typical soil type is sandy loam. Along the hill, we established a permanent sample plot (PSP) with 100 m long and 100 m wide in 2008, and the PSP was divided into sixteen sub-plots that were 25 m long and 25 m wide. The PSP designed in this study was very representative of the entire forest of the study area and was mainly used for the carbon flux observation and dynamic monitoring of forest quality.

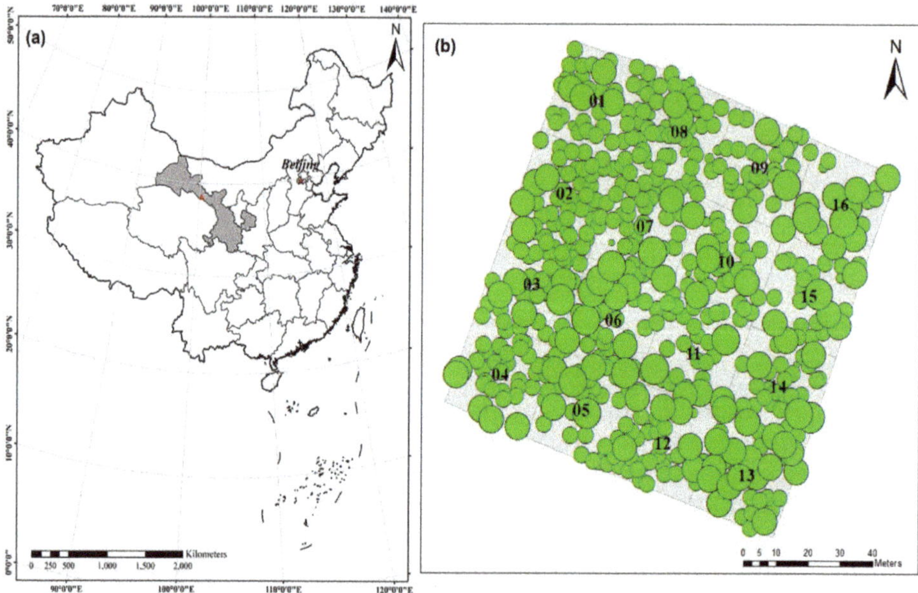

**Figure 1.** (a) Location of the study site: Xishui forest farm located in the Su'nan Yuguzu autonomous county of the Gansu Qilian Mountains National Nature Reserve, Western China and (b) tree positions within 16 sub-sample plots nested within a permanent sample plot of 100 × 100 m.

Airborne LiDAR data were acquired by the LiteMapper 5600 system with laser scanner—Riegl LMS-Q560 by a specification of 50 kH pulse repetition frequency, a 49 HZ scanning frequency, and a 30° maximum scanning angle [41]. The LiDAR data were collected on 23 June 2008, and field-measured data were collected on 1 June through 13 June 2008. The wavelength was 1550 nm, and the pulse length and laser beam divergence were 3.5 ns and 0.5 mrad, respectively. The average flight height was 3699 m, and the average flight speed was 230 km h$^{-1}$. The scanner's pulse repetition frequency, scanning frequency, and maximum scanning angle were 50 kHZ, 49 Hz, and 30°, respectively, and the mean density of point cloud was 4.34 m$^{-2}$. The spatial distribution of neighbor smoothed 510 *Picea crassifolia* Kom trees is shown in Figure 1b. Data summary is presented in Table 1, and the relationships of HCB with DBH, LiDAR-derived tree height (LH), LiDAR-derived crown width (LCW), and LiDAR-derived crown projection area (LCA) are shown in Figure 2.

**Table 1.** Descriptive statistics of tree measurements (SD, standard deviation).

| Variable | Min. | Max. | Mean | SD |
|---|---|---|---|---|
| LH (m) | 4.62 | 22.15 | 13.84 | 3.20 |
| HCB (m) | 0.90 | 10.20 | 4.80 | 3.52 |
| LCW (m) | 2.00 | 7.50 | 4.20 | 0.95 |
| DBH (cm) | 3.60 | 81.10 | 22.57 | 8.54 |
| LCA (m) | 3.19 | 38.63 | 12.21 | 5.47 |

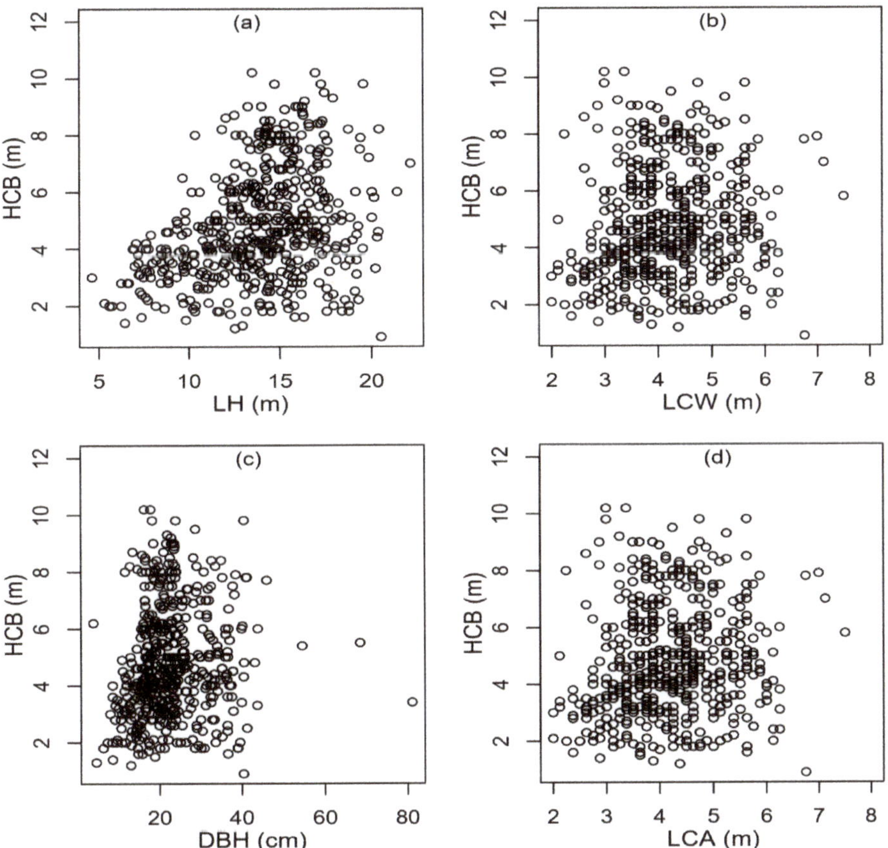

**Figure 2.** The relationships of height to crown base (HCB) with other tree variables: (**a**) light detection and ranging (LiDAR)-derived tree height (LH), (**b**) crown width (LCW), (**c**) diameter at breast height (DBH), and (**d**) crown projection area (LCA) for *Picea crassifolia* Kom.

The correlation analysis of LiDAR-derived tree attributes and ground-measured tree attributes are shown in Figure 3, whichor indicates that these LiDAR-derived tree attributes are highly correlated with ground-measure tree attributes. Thus, these LiDAR-derived tree attributes in the sample could be used for our modeling study. Figure 4 presents the Y coordinate value versus the predicted HCB value showing the vertical profile of the LiDAR product.

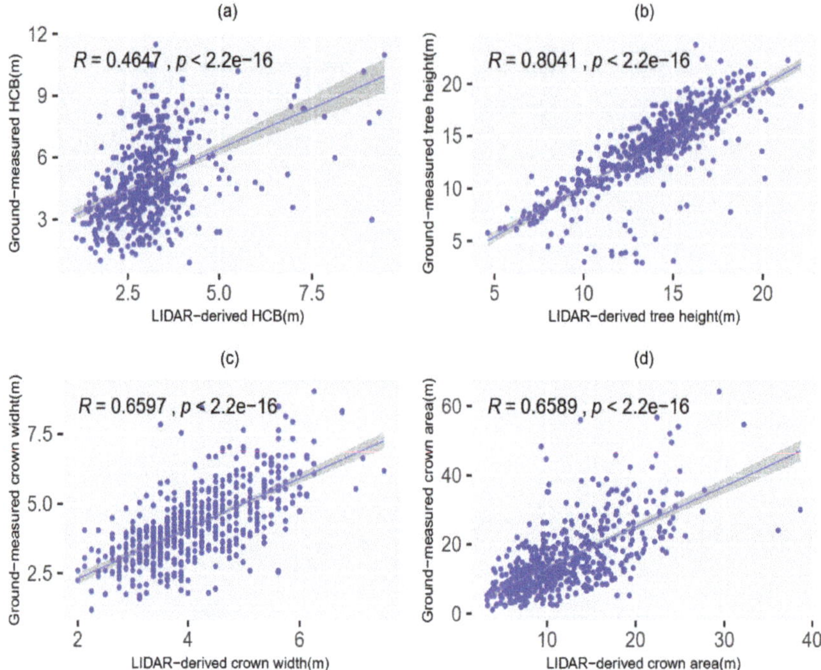


**Figure 3.** Correlations between LiDAR-derived HCB and ground-measured HCB (**a**), correlation between LiDAR-derived tree height and ground-measured tree height (**b**), correlation between LiDAR-derived crown with and ground-measured crown width (**c**), and correlation between LiDAR-derived crown area and ground measured crown area (**d**).

The point cloud was created from LiDAR waveforms by the data provider [41]. By applying the algorithm of TerraScan 005 (Terrasolid, Helsinki, Finland), the ground points were classified, and this was used to create the digital elevation model (DEM) (Figure 5a) with a 0.25 m resolution. With ground and vegetation points, the digital surface model (DSM) with a 0.25 m resolution was created using the Highest hit z algorithm of TerraScan 005. A canopy height model (CHM) (Figure 5b) with a resolution of 0.25 m and a window size of 3 × 3 m was obtained by subtracting the DSM and DEM [42–44]. The pits in the CHM were smoothed by neighbor smoothing algorithm [45]. Using the local maximum method with a window size of 2.0 m to detect the crown top from the CHM, the LH values were estimated as the values of detected crown tops with a prediction accuracy of 0.65. Using the region growing algorithm proposed by Liu et al. [46], the LCW of each tree was estimated to be the average of the horizontal ranges of the identified crown from west to east and north to south [45]. After determining canopy boundary, the LCA was obtained. Ground measurements were done for various tree attributes including individual tree DBH, HCB, crown width, and total tree height (H) of 16 sub-plots for a total of 510 *Picea crassifolia* Kom trees.

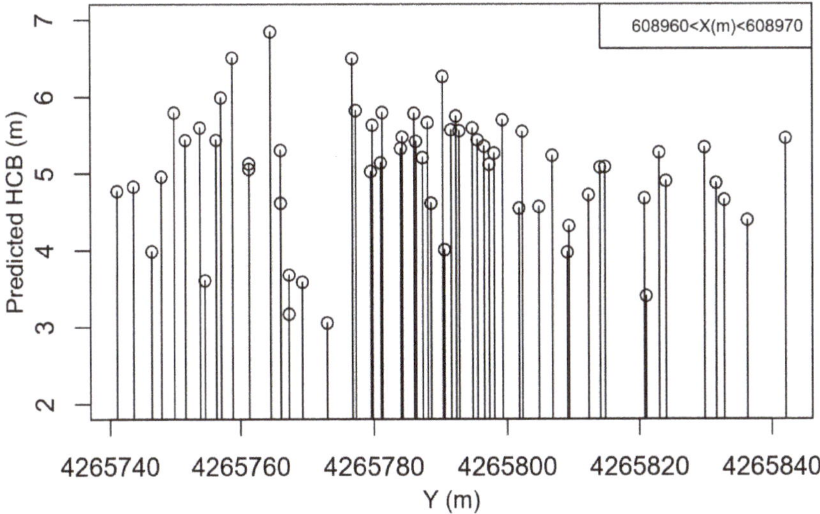



**Figure 4.** Predicted HCB (m) of trees against y-coordinate Y (m) located within coordinate X (608,960–608,970 m).

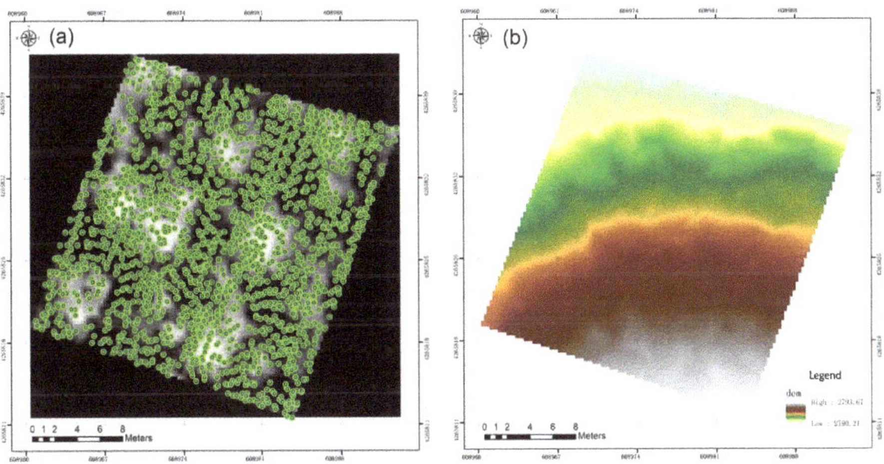



**Figure 5.** Digital elevation model (DEM) of sample plot (**a**) and footprint of laser pulse (**b**).

## 2.2. Base Model

### 2.2.1. LiDAR–DBH Base Model

Fu et al. [33] developed an exponential LiDAR-based DBH model using LH and LCA as predictors for *Picea crassifolia* Kom and found a significantly higher prediction accuracy than other three candidate LiDAR-based DBH model forms (linear, Richards, and logistic). Our preliminary analyses exhibited the biggest $R^2$ and the smallest root mean square error (RMSE) of the exponential LiDAR-based DBH model form, indicating its greatest suitability according to our data characteristics, and its prediction accuracy could be further improved by including LH and LCW as predictors (Equation (1)):

$$DBH = \beta_1 \exp(-\beta_2 LH - \beta_3 LCW) + \varepsilon_{DBH} \tag{1}$$

where $\beta_1, \beta_2, \beta_3$ are parameters to be estimated, and $\varepsilon_{DBH}$ is a residual error.

### 2.2.2. LiDAR–HCB Base Model

Similar to Walters and Hann [40], we used the logistic model as a LiDAR–HCB base model, whichor had DBH and LCA as predictors in this study.

$$HCB = \frac{LH}{1 + \exp(\gamma_1 DBH + \gamma_2 LCA)} + \varepsilon_{HCB} \tag{2}$$

where $\gamma_1$ and $\gamma_2$ are parameters to be estimated, and $\varepsilon_{HCB}$ is a residual error.

### 2.3. A Compatible Individual Tree DBH and HCB EIV Equation System

A compatible equation system consisting of tree-based DBH and HCB EIV models (Equation (3)) was built by integrating both the LiDAR–DBH base model (Equation (1)) and the LiDAR–HCB base model (Equation (2)) by following the methods suggested by existing modeling studies [36,37,47].

$$\begin{cases} dbh_i = \beta_1 \exp(-\beta_2 LH_i - \beta_3 LCW_i) \\ hcb_i = LH_i / (1 + \exp(\gamma_1 DBH_i + \gamma_2 LCA_i) \\ DBH_i = dbh_i + \varepsilon_{DBH_i} \\ HCB_i = hcb_i + \varepsilon_{DBH_i} \\ \varepsilon_i = \varepsilon_{DBH_i} + \varepsilon_{HCB_i} \end{cases} \tag{3}$$

where $DBH_i$ (cm) and $HCB_i$ (m) ($i = 1,2\ldots,N$) are the ground-measured diameter at breast height with errors and height to crown base with errors of the $i^{th}$ tree, respectively; $dbh_i$ and $hcb_i$ are true values (with the assumption of no errors) of $DBH_i$ and $HCB_i$, respectively; $\varepsilon_{DBH_i}$ and $\varepsilon_{HCB_i}$ represent the errors of $DBH_i$ and $HCB_i$, respectively. Error $\varepsilon_i$ is a two-dimensional vector that is assumed to be normally distributed with zero means and variance–covariance matrix $\Sigma$; $LH_i$, $LCW_i$, and $LCA_i$ are the LiDAR-derived tree height (m), crown width (m), and crown projection area (m$^2$) of the $i^{th}$ tree, respectively. In this simultaneous equation system (Equation (3)), both DBH and HCB are the EIV, while LH, LCW, and LCA are regarded as error-free variables. The other parameters and variables are the same as defined above. The elements in the variance–covariance matrix $\Sigma$ were applied to account for the inherent correlations of DBH with HCB.

It was assumed that the simultaneous equation system (Equation (3)) with an error term $\varepsilon_{ti}$ (t = DBH and HCB; and i = 1, … , N) that were not correlated among the observations but were contemporaneously correlated across the sub-models. For each observation, we assumed that:

$$\Sigma = \begin{pmatrix} \sigma_{DBH \times DBH} & \sigma_{DBH \times HCB} \\ \sigma_{HCB \times DBH} & \sigma_{HCB \times HCB} \end{pmatrix} \tag{4}$$

where $\sigma_{DBH \times DBH}$, $\sigma_{HCB \times HCB}$, $\sigma_{DBH \times HCB}$, and $\sigma_{HCB \times DBH}$ are the variance and covariance related elements for both DBH and HCB.

The covariance matrix of the stacked error terms ($\varepsilon = (\varepsilon_{DBH}^T, \varepsilon_{HCB}^T)^T (\varepsilon_t = (\varepsilon_{t1}, \ldots, \varepsilon_{tN})^T, t = DBH$ and HCB) would be $R = \Sigma \otimes I_N$.

### 2.4. Parameter Estimation

Four commonly used algorithms, such as NSUR, two-stage least square (2SLS), three-stage least square (3SLS), and full information maximum likelihood (FIML) were applied to estimate the parameters $\mathbf{B} = (\boldsymbol{\beta}^T_{DBH} = (\beta_1, \beta_2, \beta_3), \boldsymbol{\gamma}^T_{HCB} = (\gamma_1, \gamma_2))$ in our simultaneous equation system (Equation (3)). We briefly describe these algorithms in a methodological flow chart (Figure 6), and the details are given in the sub-sections below.

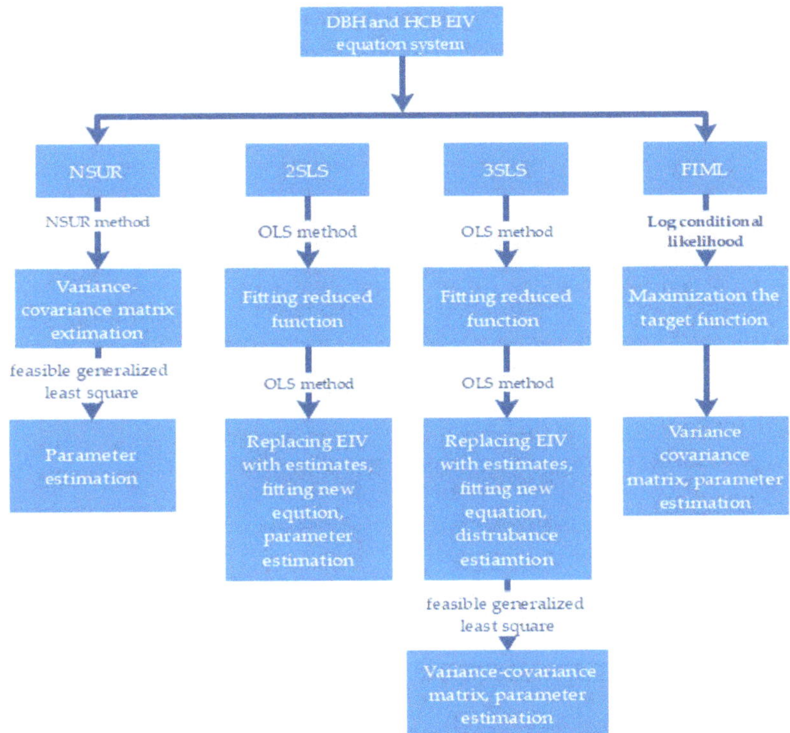


**Figure 6.** A flow chart depicting a brief description of four algorithms (NSUR, nonlinear seemingly unrelated regression; 2SLS, two-stage least square; 3SLS, three-stage least square; and FIML, full information maximum likelihood) used to estimate a DBH and HCB error-in-variable (EIV) equation system.

(1) NSUR algorithm

The NSUR algorithm considers the disturbance across the two equations as a linkage of the equation system but assumes that disturbances are uncorrelated across the observations; thus, this algorithm is known as a seemingly unrelated regression. The estimation of parameters in the simultaneous equation system (Equation (3)) was done using the NSUR algorithm [34,48–50] with the feasible generalized least square regression method described as follows:

Step 1: Two sub-models (Equations (1) and (2)) in the simultaneous equation system (Equation (3)) were fitted with the NSUR algorithm, and the resulting residuals $\hat{\varepsilon}_t$ ($t = DBH$ *and* $HCB$) were used for estimating the variance–covariance matrix, $\Sigma$. The residuals of each sub-model were estimated with OLS using following formula:

$$\hat{\sigma}_{ij} = \frac{1}{N}\sum_{i=1}^{N} \varepsilon_{it}\varepsilon_{jt} \tag{5}$$

The estimated variance–covariance matrix $\Sigma$ is given by:

$$\hat{\Sigma} = \begin{pmatrix} \hat{\sigma}_{11}\hat{\sigma}_{12}\dots\hat{\sigma}_{1n} \\ \hat{\sigma}_{21}\hat{\sigma}_{22}\dots\hat{\sigma}_{2n} \end{pmatrix} \tag{6}$$

Step 2: Based on the estimated $\hat{\Sigma}$, a covariance matrix R was defined as $\hat{R} = \hat{\Sigma} \otimes I_N$. The parameters in the simultaneous equation system (Equation (3)) were estimated using a feasible generalized least square method.

$$\hat{B} = \left(X^T(\hat{\Sigma}^{-1} \otimes I_N)X\right)^{-1}(\hat{\Sigma}^{-1} \otimes I_N)y \tag{7}$$

(2) 2SLS algorithm

The 2SLS algorithm was applied using the following steps [51]:

Step 1: By composing the reduced function (Equation (8)) for all the error-in variables on the right side of an equation, the estimated error-in variables were obtained using OLS:

$$Y_i = \Pi_i X_i + \varepsilon_i \tag{8}$$

where $X_i$ is the vector of the error-free variables and $\Pi_i$ is the $i^{\text{th}}$ parameter vector for X.

Step 2: The error-in variables on the right side were replaced with estimated error-in variables $\hat{Y}_i$, and the parameters were estimated using OLS.

$$\text{cov}(\hat{B}) = \left(X^T(diag(\Sigma^{-1}) \otimes I_N)X\right)^{-1} \tag{9}$$

(3) 3SLS algorithm

The 3SLS algorithm considers the correlation of disturbance terms among different equations. It was carried out with following steps [51]:

Step 1: The same as step 1 in 2SLS.

Step 2: The same as step 2 in 2SLS; in addition, the disturbance $\varepsilon_i$ was estimated.

Step 3: The covariance $\sigma_{ij}$ and $\hat{\varepsilon}_i$ was estimated in step 2, and then an estimated variance–covariance matrix, $\hat{\Sigma}$, was obtained. $\Phi = \Sigma \otimes I_N$ was defined, and parameters were estimated with a feasible generalized least squares regression.

$$\hat{B} = \left(X^T(\hat{\Sigma}^{-1} \otimes I_N)X\right)^{-1}(\hat{\Sigma}^{-1} \otimes I_N)y \tag{10}$$

$$\text{cov}(\hat{B}) = \left(X^T(diag(\hat{\Sigma}^{-1}) \otimes V_N)X\right)^{-1} \tag{11}$$

where **V** is a matrix of the instrumental variables and **I** is an identity matrix.

(4) FIML algorithm

Instead of only making use of the reduced function information, FIML makes full use of all the information by estimating all the parameters in the simultaneous equation system at the same time. There must be equal number of error-in-variables and sub-models in this equation system, whichor we used. Otherwise, if the number of endogenous variables is more than that of the sub-models, the limited information maximum likelihood method needs to be applied. The FIML maximizes the following conditional log-likelihood function [52]:

$$Q_n(B,\Sigma) = -\frac{M}{2}\ln(2\pi) - \frac{1}{2}\ln(|\Sigma|) - \frac{1}{2}\sum_{i=1}^{m}(Y_i - y_i)^T\Sigma^{-1}(Y_i - y_i) \tag{12}$$

where $M$ is the number of sub-models; the other parameters and variables are the same as defined above.

### 2.5. Other Model Structures for Comparison

#### 2.5.1. Nonlinear Least Squares with HCB Estimation not Based on DBH (NLS and NBD)

The DBH in the LiDAR–HCB base model (Equation (2)) was substituted by the LiDAR–DBH base model (Equation (1)). Therefore, in this case, the HCB estimation was independent of DBH. The HCB based on DBH model is given by:

$$
\begin{aligned}
HCB_i &= LH_i/(1 + \exp(\gamma_1 DBH_i + \gamma_2 LCA_i)) + \varepsilon_{HCB_i} \\
&= LH_i/(1 + \exp(\mu_1 \exp(-\mu_2 LH_i - \mu_3 LCW_i) + \mu_4 LCA_i) + \widetilde{\varepsilon}_{HCB_i}
\end{aligned}
\tag{13}
$$

where $\mu_1 = \beta_1\gamma_1$, $\mu_2 = \beta_2$, $\mu_3 = \beta_3$, and $\mu_4 = \gamma_2$ are parameters in the model, and the error of the $HCB_i$ was changed into:

$$
\widetilde{\varepsilon}_{HCB_i} = f(LH_i, LCW_i)\varepsilon_{DBH_i} + \varepsilon_{HCB_i}
\tag{14}
$$

$$
f(LH_i, LCW_i) = \beta_1\gamma_1 \exp(-\beta_2 LH_i - \beta_3 LCW_i)
\tag{15}
$$

With this method, $DBH_i$ was estimated by the LiDAR–DBH base model (Equation (1)), and $HCB_i$ was estimated by the NLS and NBD model (Equation (13)). It should be noted that the inherent correlations of HCB with DBH could not be addressed for this method. In addition, the compatibility between the estimated DBH and HCB could not be achieved.

#### 2.5.2. Nonlinear Least Squares with HCB Estimation Based on DBH (NLS and BD)

The LiDAR–DBH base model (Equation (1)) and the LiDAR–HCB base model (Equation (2)) were fitted separately based on the database by the NLS and BD. This method was applied to quantify the consequences in the HCB estimation by using predicted DBH to take place of an actual value while ignoring its error. The NLS and BD approach could explain the compatibility between DBH and HCB, but it failed to account for the effect of the errors in the estimated DBH on HCB estimation. The estimated values of $DBH_i$ and $HCB_i$ ($i = 1, 2 \ldots, N$) are, respectively, given by:

$$
D\hat{B}H_i = \hat{\beta}_1 \exp(-\hat{\beta}_2 LH_i - \hat{\beta}_3 LCW_i)
\tag{16}
$$

$$
H\hat{C}B_i = LH_i/(1 + \exp(\hat{\gamma}_1 DBH_i + \hat{\gamma}_2 LCA_i))
\tag{17}
$$

where $\beta_1, \beta_2, \beta_3, \gamma_1$, and $\gamma_2$ are the model parameters; the other parameters and variables are the same as defined above.

### 2.6. Comparison and Evaluation of Models

We only had 510 observations, whichor was not enough to divide a full data set into fitting and validation sets. As such, we applied the leave-one-out cross validation (LOOCV) method [53,54] for the validation of the models. Each time, one tree from the full dataset was deleted, and the fitting data set was formed by the remaining trees. A fitting data set was used to fit the DBH and HCB models and estimated their parameters. Using the estimated parameter values, the deleted tree's DBH and HCB were predicted, and commonly used prediction statistics, such as mean bias($\bar{e}$), variance of bias ($\sigma_{\bar{e}}^2$), RMSE, and mean absolute error (MAE) (Equations (18)–(21)) were computed with the difference obtained from the predicted and observed values. Then, we put the tree back in place, deleted another tree, and performed the same model-fitting and prediction processes. This procedure was performed on all the trees in the full data set. We present the LOOCV computational codes with NSUR algorithm as an example in Appendix A, and we used these codes to evaluate the equation system.

Finally, the prediction performance of the simultaneous equation system (Equation (3)) was estimated with each of the six different methods: NSUR [33,47,55], 2SLS, 3SLS, FIML, NLS and BD [33], and NLS and NBD [33] were evaluated by three statistics including mean bias, bias variance, and root mean square error that were calculated with Equations (18)–(21). The model with the smallest $\bar{e}$,

$\sigma_e^2$, RMSE, and MAE were defined as the final model to predict DBH and HCB. We performed all computations with R software version 3.4.4 [56].

$$\bar{e} = \sum_{i=1}^{N} e_i/N = \sum_{i=1}^{N} (y_i - \hat{y}_i)/N \tag{18}$$

$$\sigma_e^2 = \sum_{i=1}^{N} (e_i - \bar{e})^2/(N-1) \tag{19}$$

$$RMSE = \sqrt{\bar{e}^2 + \sigma_e^2} \tag{20}$$

$$\text{MAE} = |\sum_{i=1}^{N} e_i|/N = |\sum_{i=1}^{N} (y_i - \hat{y}_i)|/N \tag{21}$$

where $y_i$ and $\hat{y}_i$ are the measured and estimated height to crown base or DBH for the $i^{th}$ observation, N is the number of observations, $\bar{e}$ is the mean bias, $\sigma_e^2$ is the variance of bias, *RMSE* is the root mean square error, and MAE is the mean absolute error.

## 3. Results

For the DBH model, the RMSE of NSUR was identical to the NLS and BD model and smaller than that of 2SLS, 3SLS, and FIML. For the HCB model, the RMSE of NSUR was smaller than that of the NLS and BD model, 2SLS, 3SLS, and FIML. The MAE of NSUR for the HCB model was the smallest.

### 3.1. Parameters Estimation

All the parameters in the LiDAR–DBH base model (Equation (1)), the LiDAR–HCB base model (Equation (2)), and the simultaneous equation system (Equation (3)) were estimated with four different methods, namely NSUR, 2SLS, 3SLS, and FIML using all the data. Most of the parameter estimates were significantly different from zero, and their magnitudes and signs could meet biological logics, except for parameters $\mu_1, \mu_2$, and $\mu_3$ for NLS and NBD and $\gamma_1$ for both 2SLS and 3SLS, whichor were not significant ($p < 0.05$) (Table 2).

**Table 2.** Parameter estimates of the LiDAR–DBH base model (Equation (1)), the LiDAR–HCB base model (Equation (2)), and the NLS and BD model (Equation (13)), as well as the simultaneous equation system (Equation (3)). The first three models were estimated using ordinary least squares regression, and the last one was estimated using NSUR, 2SLS, 3SLS, and FIML.

| Model | Method | Parameters | Estimates | Standard Error | t-Value |
|---|---|---|---|---|---|
| LiDAR–DBH base model (Equation (1)) | NLS | $\beta_1$ | 5.7161 | 0.3599 | 15.882 |
| | | $\beta_2$ | −0.0567 | 0.0061 | −9.344 |
| | | $\beta_3$ | −0.1264 | 0.0178 | −7.108 |
| | | $\sigma^2$ | 37.22 | | |
| LiDAR–HCB base model (Equation (2)) | NLS | $\gamma_1$ | 0.0053 | 0.0031 | 1.68 |
| | | $\gamma_2$ | 0.0367 | 0.0057 | 6.4340 |
| | | $\sigma^2$ | 3.37 | | |
| HCB based on DBH model (Equation (13)) NLS and BD | NLS | $\mu_1$ | −0.0003 | 0.0547 | −0.00 |
| | | $\mu_2$ | 0.2472 | 0.1987 | −1.24 |
| | | $\mu_3$ | 0.0800 | 0.0520 | 1.54 |
| | | $\mu_4$ | 0.0283 | 0.0137 | 2.07 |
| | | $\sigma^2$ | 3.33 | | |

**Table 2.** *Cont.*

| Model | Method | Parameters | Estimates | Standard Error | t-Value |
|---|---|---|---|---|---|
| Simultaneous equation system (Equation (3)) | NSUR | $\beta_1$ | 5.6992 | 0.3615 | 15.77 |
| | | $\beta_2$ | −0.0563 | 0.0061 | −9.27 |
| | | $\beta_3$ | −0.1283 | 0.0180 | −7.12 |
| | | $\gamma_1$ | 0.0059 | 0.0032 | 1.87 |
| | | $\gamma_2$ | 0.0361 | 0.0058 | 6.26 |
| | | $\sigma^2_{DBH}$ | 37.22 | | |
| | | $\sigma^2_{HCB}$ | 3.32 | | |
| | | $\sigma_{DH}$ | 0.608 | | |
| Simultaneous equation system (Equation (3)) | 2SLS | $\beta_1$ | 5.2496 | 0.3586 | 14.64 |
| | | $\beta_2$ | −0.0567 | 0.0064 | −8.83 |
| | | $\beta_3$ | −0.1451 | 0.0186 | −7.80 |
| | | $\gamma_1$ | 0.0054 | 0.0051 | 1.04 |
| | | $\gamma_2$ | 0.0370 | 0.0091 | 4.08 |
| | | $\sigma^2_{DBH}$ | 37.93 | | |
| | | $\sigma^2_{HCB}$ | 3.32 | | |
| | | $\sigma_{DH}$ | 0.551 | | |
| Simultaneous equation system (Equation (3)) | 3SLS | $\beta_1$ | 5.2641 | 0.3590 | 14.66 |
| | | $\beta_2$ | −0.0564 | 0.0064 | −8.78 |
| | | $\beta_3$ | −0.1457 | 0.0186 | −7.83 |
| | | $\gamma_1$ | 0.0052 | 0.0051 | 1.01 |
| | | $\gamma_2$ | 0.0373 | 0.0091 | 4.12 |
| | | $\sigma^2_{DBH}$ | 37.88 | | |
| | | $\sigma^2_{HCB}$ | 3.32 | | |
| | | $\sigma_{DH}$ | 0.553 | | |
| Simultaneous equation system (Equation (3)) | FIML | $\beta_1$ | 5.7036 | 0.3469 | 16.44 |
| | | $\beta_2$ | −0.0561 | 0.0047 | −11.84 |
| | | $\beta_3$ | −0.1289 | 0.0149 | −8.66 |
| | | $\gamma_1$ | 0.0074 | 0.0042 | 1.76 |
| | | $\gamma_2$ | 0.0335 | 0.0076 | 4.39 |
| | | $\sigma^2_{DBH}$ | 37.72 | | |
| | | $\sigma^2_{HCB}$ | 3.31 | | |
| | | $\sigma_{DH}$ | 0.65 | | |

The elements of the variance–covariance matrix were significantly different from each other ($p < 0.05$) in the simultaneous equation system (Equation (3)), whichor was estimated using NSUR, 2SLS, 3SLS, and FIML, implying that correlation of DBH with HCB was highly significant.

### 3.2. Model Prediction

The LOOCV was carried out for the LiDAR–DBH base model (Equation (1)), the LiDAR–HCB base model (Equation (2)), the NLS and NBD model (Equation (13)), and the NLS and BD model (Equation (16)) estimated using the nonlinear OLS, as well as the simultaneous equation system (Equation (3)) estimated using NSUR, 2SLS, 3SLS, FIML, and TSEM. The evaluations and comparisons of all these models were carried out using $\bar{e}$, $\sigma^2_e$, and *RMSE* (Table 3).

A compatible DBH and HCB EIV equation system fitted with NSUR showed a better prediction ability than those fitted with other alternative methods (Table 3). For the DBH model, the $\sigma^2_e$ of NSUR was identical to that of NSL and BD, as well as 0.37%, 0.37%, and 0.18% smaller than that of 2SLS, 3SLS, and FIML, respectively. The RMSE of NSUR was identical to that of NLS and BD, and it was 0.02%, 0.01%, and 0.08% smaller than that of 2SLS, 3SLS, and FIML, respectively. For the HCB model, the $\sigma^2_e$ of NSUR was 0.35%, 0.021%, 0.006%, and 0.17% smaller than that of NLS and BD, 2SLS, 3SLS, and FIML, respectively. The RMSE of NSUR was 2.75%, 0.022%, 0.011%, and 0.082% smaller than that of NLS and BD, 2SLS, 3SLS, and FIML, respectively. The MAE of NSUR for HCB was the smallest.

The residuals of six different alternative models and equation systems were calculated based on a full dataset. This analysis indicated that the mean residuals of the NLS method for HCB were higher than other alternative methods, among whichor NSUR showed the smallest mean residual for HCB (Table 4).

**Table 3.** Prediction statistics of the models: the DBH-based model (Equation (1)), the HCB-based model (Equation (2)) fitted with NLS, the HCB based on DBH model (Equation (13)) fitted with NLS, and the simultaneous equation system (Equation (3)) fitted with the NSUR, 2SLS regression, 3SLS regression, and FIML algorithms.($\bar{e}$, mean bias; $\sigma_e^2$, bias variance; and *RMSE*, root mean square error. All other acronyms are the same as defined in Table 2).

| Fitting Method | Variables | $\bar{e}$ | $\sigma_e^2$ | RMSE | MAE |
|---|---|---|---|---|---|
| NLS and NBD | DBH | −0.0426 | 37.6572 | 6.1367 | 3.6998 |
|  | HCB | −0.0367 | 3.3387 | 1.8276 | 1.4619 |
| NLS and BD | HCB | −0.4186 | 3.3530 | 1.8784 | 1.4619 |
| NSUR | DBH | −0.0458 | 37.6577 | 6.1368 | 3.7008 |
|  | HCB | −0.0276 | 3.3414 | 1.8281 | 1.4606 |
| 2SLS | DBH | 0.0093 | 37.7985 | 6.1481 | 3.6984 |
|  | HCB | −0.0336 | 3.3421 | 1.8285 | 1.4612 |
| 3SLS | DBH | 0.0026 | 37.7977 | 6.1480 | 3.7004 |
|  | HCB | −0.0341 | 3.3416 | 1.8283 | 1.4613 |
| FIML | DBH | −0.0457 | 37.6598 | 6.1369 | 3.7012 |
|  | HCB | −0.0195 | 3.3470 | 1.8296 | 1.4613 |

**Table 4.** Descriptive statistics of residuals of the LiDAR– DBH base model (Equation (1)), LiDAR–HCB base model (Equation (2)), and model (Equation (13)), and simultaneous equation system (Equation (3)). The first three models were estimated using ordinary least squares regression and last one was estimated using NSUR, 2SLS, 3SLS, and FIML. (SD, standard deviation).

| Model | Method | Response Variable | Min. of Residuals | Max. of Residuals | Mean of Residuals | SD of Residuals |
|---|---|---|---|---|---|---|
| LiDAR–DBH base model (Equation (1)) | NLS | DBH | −60.0992 | 21.6027 | 0.0423 | 6.1365 |
| LiDAR–HCB base model (Equation (2)) | NLS | HCB | −6.1705 | 4.3487 | 0.0367 | 1.8272 |
| HCB based on DBH model, NLS and BD (Equation (13)) | NLS | HCB | −6.1708 | 4.3495 | 0.0342 | 1.8300 |
| Simultaneous equation system (Equation (3)) | NSUR | DBH | −60.0883 | 21.5831 | 0.0458 | 6.1366 |
|  |  | HCB | −6.2068 | 4.4099 | 0.0276 | 1.8279 |
| Simultaneous equation system (Equation (3)) | 2SLS | DBH | −60.1132 | 22.2219 | −0.0093 | 6.1481 |
|  |  | HCB | −6.2193 | 4.4061 | 0.0336 | 1.8282 |
| Simultaneous equation system (Equation (3)) | 3SLS | DBH | −60.0874 | 22.1638 | −0.0026 | 6.1480 |
|  |  | HCB | −6.2150 | 4.3999 | 0.0341 | 1.8280 |
| Simultaneous equation system (Equation (3)) | FIML | DBH | −60.0819 | 21.5710 | 0.0457 | 6.1368 |
|  |  | HCB | −6.2407 | 4.4668 | 0.0295 | 1.8295 |

The prediction accuracy of the simultaneous equation system (Equation (3)) fitted with all four fitting algorithms appeared almost identical (Figure 7), indicating that each of the fitting algorithms were able to produce almost equally unbiased estimations and prediction accuracies. The prediction accuracy of DBH seemed to be much higher than that of HCB.

The inherent correlations between the ground-measured DBH, model-estimated DBH, ground-measured HCB, and model estimated-HCB were all significantly high (Figure 8). The inherent correlation between DBH and HCB was substantially high. This figure suggested that all models and all fitting algorithms were appropriately suited to our data.

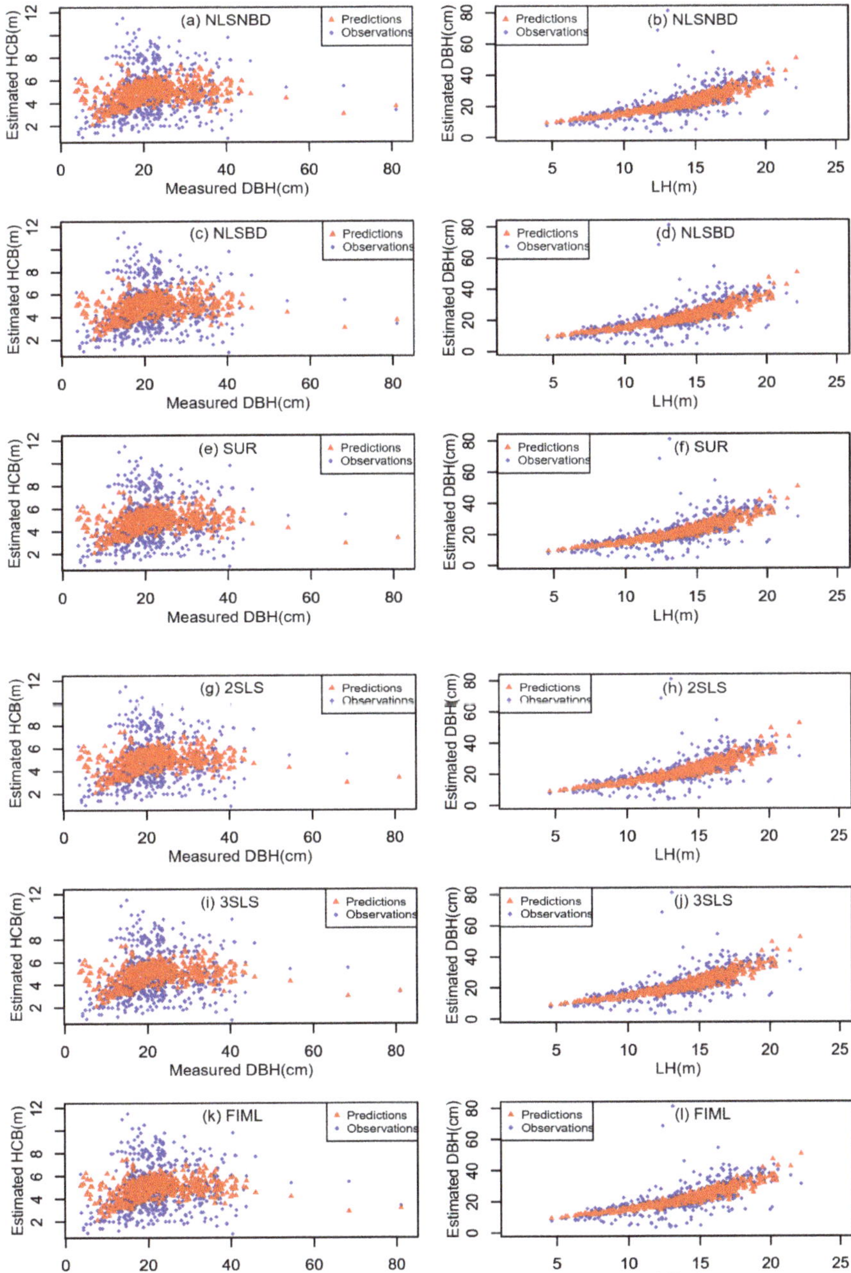

**Figure 7.** Scattered plots of estimated values of HCB versus ground-measured DBH for nonlinear least squares with HCB estimation not based on DBH (NLS and NBD) (**a**), nonlinear least squares with HCB estimation based on the DBH (NLS and BD) (**c**), NSUR (**e**), 2SLS regression (**g**), 3SLS regression (**i**), and FIML (**k**). Scattered plots of estimated DBH versus LH for NLS and NBD (**b**), NLS and BD (**d**), NSUR (**f**), 2SLS (**h**), 3SLS (**j**), and FIML (**l**).

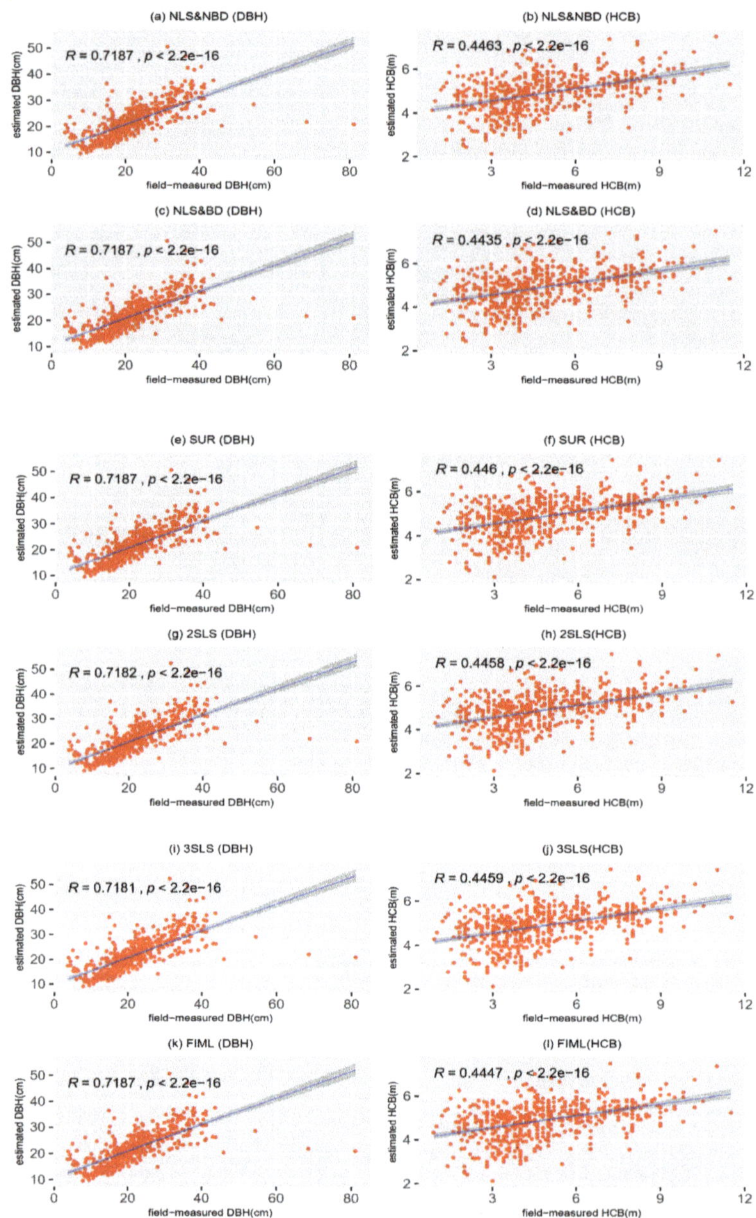

**Figure 8.** Correlations between the ground-measured tree diameter at breast height (DBH) and estimated DBH for nonlinear least squares with height to crown base (HCB) estimation not based on DBH (NLS and NBD) (**a**), nonlinear least squares with HCB estimation based on the DBH (NLS and BD) (**c**), NSUR (**e**), 2SLS regression (**g**), 3SLS regression (**i**), and FIML (**k**), as well as correlations between ground-measured tree HCB and estimated HCB from NLS and NBD (**b**), NLS and BD (**d**), NSUR (**f**), 2SLS (**h**), 3SLS (**j**), and FIML (**l**). R = Pearson's correlation coefficient.

## 4. Discussion

HCB is an important tree attribute to assess tree productivity and tree vigor. DBH is commonly used to predict HCB model, but DBH estimated with LiDAR-based attributes contains unignorable errors. In addition, the compatibility between DBH and HCB needs to be considered when estimating HCB. In this study, we investigated four algorithms to estimate DBH and HCB in an EIV equation system—NSUR, 2SLS, 3SLS, and FIML—that were compared with two model structures. The prediction accuracy of the four EIV equation system algorithms and two model structures were reflected by RMSE and MAE. The results showed that the impacts of measurement error of DBH on HCB and the compatibility between DBH and HCB were well accounted for by the NSUR algorithm.

HCB is an important indicator for tree vigor and tree stem form, as well as an indispensable measure for retrieving the crown ratio. However, measuring in-situ HCB is quite labor-intensive and costly, especially when conducted for large forest areas. In this situation, an efficient method of obtaining precise HCB is necessary, whichor can be possible with the HCB prediction model developed from the LiDAR-derived variables, such as tree height, crown projection area, crown width, and ground-measured DBH. The first three variables can be relatively more accurately and easily measured by applying the advanced remote sensing techniques. The HCB can be estimated from the established HCB model, whichor may also contain DBH as a predictor [11,28]. The DBH estimation model can also be developed using the LiDAR-derived information [33]. The estimation of HCB and DBH from their corresponding prediction models would be substantially biased if separately developed models were used, i.e., DBH model and HCB models developed independently from each other from the same tree data. In order to overcome such a bias, developing a compatible simultaneous equation system is the most appropriate solution. However, this equation system of DBH and HCB models is still unavailable in forest modeling literature. As mentioned in the introduction, other compatible simultaneous equation systems developed through the EIV modeling approach are available, e.g., a system of equations of DBH and individual tree above-ground biomass models [33]. Considering the knowledge gap, we developed the simultaneous equation system of DBH and HCB models using the tree-level predictors (LH, LCW, and LCA), the information of whichor was derived from the LiDAR imagery. Four different algorithms (NSUR, 2SLS, 3SLS, and FIML) were used to estimate this equation system.

The data used in our study originated from the *Picea crassifolia* Kom forest, whichor is crucial to the economic and social benefits to the rural population, as well as regional carbon storage, regional carbon cycling, and the maintenance of the balanced-functions of forest ecosystems in northwest China. Two different model structures (the NLS and NBD model and the NLS and BD model) built by assuming errors associated with all the regressors and response variables were found to be inappropriate because this approach did not account for the inherent correlations of DBH with HCB and all the estimated parameters and variances were biased.

Generally, the structural estimators or fitting algorithms (NSUR, 2SLS, 3SLS, and FIML) should always be preferred to the NLS, as each of them effectively accounted for the errors in variables in an appropriate way. However, surprisingly, we found that NLS could sometimes provide a closer estimation of the structural estimators applied in this study, and it was the same for NLS and NBD. The NLS and NBD model had a smaller bias variance, so it has possibility to produce a smaller RMSE. However, NLS standard errors are, in all the likelihoods, not useful for inference purposes [57]. The prediction accuracy of the NLS and BD model was the worst with the highest $\sigma_e^2$ and the biggest RMSE, thus, in this case, the EIV modeling approach clearly displayed the advantage over NLS. In general, individual tree DBH and HCB models based on the LiDAR data and field-measurements contain errors that exist in image capture, image processing, and the extraction of the information processes, and they are therefore very hard to completely avoid [29,33,58].

The NLS and NBD could neither address the compatibility problem of DBH and HCB nor account for their inherent correlations. However, a simultaneous equation system (Equation (3)) can effectively address these issues. Among the four algorithms used in fitting simultaneous equation system (Equation (3)), NSUR and 2SLS are classified into the limited information estimators, while 3SLS and FIML are the full information estimators. The former two estimators can make use of the reduced model information, while the latter two estimators can make use of full information from the model [33,34]. Based on the model validation results with LOOCV, the prediction accuracy of NSUR was slightly better than that of the other algorithms (2SLS, 3SLS, and FIML). This was probably because NSUR has a better ability to address the error transfers caused by DBH in the simultaneous equation system of the DBH and HCB models. Potentially because of this, Parresol [49] applied NSUR to develop the additive tree biomass models in a pioneer modeling study about a simultaneous equation system in forestry. The prediction accuracy of 3SLS was slightly better than 2SLS, confirming the findings of Tang et al. [34], who found that when errors in across equations were correlated, 3SLS outperformed 2SLS, and—when errors involved across equations were uncorrelated—2SLS outperformed 3SLS.

Our HCB equation system developed in this study was based on the most attractive fit statistics of the base model among the five frequently used HCB base candidate models [10,59,60]. The analysis of correlations between the regressors and HCB showed strong connections among LCA, LH, DBH, and HCB. In other words, these tree characteristics strongly influenced HCB variations. Our DBH base model, whichor replaced LCPA with LCW in the models of Fu et al. [33], showed a better fitting performance with a smaller RMSE. Both the HCB model applied with all the LiDAR-based data (except for DBH data, whichor were obtained from ground measurement) and the DBH model were developed by LiDAR data, and this enabled the DBH–HCB-compatible EIV models, suggesting the high possibility of the equation system's application to an extensive forest area. The validation results based on the LOOCV for NSUR, 2SLS, 3SLS, and FIML were almost identical, even though NSUR slightly outperformed others; however, the prediction difference was still insignificant (Table 3). In this study, we only considered DBH and HCB as error-in-variables; however, other regressors may contain various errors including measurement errors, tree crown delineation errors, and errors of parameter estimation. Ignoring all these errors can cause the complex uncertainties while developing models. Future researchers should focus on these issues. Therefore, readers need to be cautious when considering the conclusion of this study.

As mentioned in the introduction section, this study was based on a novel methodology, whichor resulted in a system of compatible simultaneous equations of DBH and HCB models in whichor various LiDAR-derived tree attributes were used. The measurement errors of both DBH and HCB were simultaneously taken into consideration to address the problem of compatibility between DBH and HCB models and to account for inherent correlations between these tree variables through a simultaneous modeling approach. The presented equation system of DBH and HCB models can fulfill the gaps of the unavailability of such an HCB EIV model system in forest modeling literature. A compatible simultaneous equation system of the DBH and HCB models developed using the information of the tree-level predictors (LH, LCW, and LCA) derived from LiDAR imagery and ground-based measurements confirmed the accurate prediction of HCB and DBH. Compared to any of the previously developed HCB models using only ground measurements [11] and those based on LiDAR-derived databases [22,25–28], the presented equation system in this article will be interesting and useful to both researchers and forest managers, as this system is able to accurately predict HCB. Furthermore, the presented modeling approach and algorithm in this article will be useful for establishing similar compatible equation systems of DBH and HCB EIV models for other tree species and other tree variables that have inherent correlations between themselves.

## 5. Conclusions

This study developed a compatible simultaneous equation system of DBH and HCB EIV models on the basis of LiDAR-derived and ground-measured data of *Picea crassifolia* Kom trees in northwest China. Four different algorithms—NSUR, 2SLS regression, 3SLS regression, and FIML—were used to estimate the parameters in an equation system. The NLS used for estimating both the LiDAR–DBH base model (Equation (1)) and the LiDAR–HCB base model (Equation (2)) produced biased results, while the other fitting algorithms used for estimating a simultaneous equation system (Equation (3)) produced unbiased results with similar SSE, MSE, and RMSE. Two additional model structures—nonlinear least squares with HCB estimation not based on DBH (NLS and NBD) and nonlinear least squares with HCB estimation based on the DBH (NLS and BD)—were also developed for comparison. All the fitting algorithms and their resulting models were assessed by a leave-one-out cross validation method. This study indicates that only EIV modeling method can effectively account for the effects of errors associated with the regressors on the response variables and can guarantee the compatibility between DBH model and HCB model at the individual tree level. However, neither the NLS and BD model nor the NLS and NBD model exhibited these advantages. Among the various evaluated algorithms and models, NSUR showed a slightly better performance than the others. The results showed that the methodology proposed in this article is a reliable and efficient, and it can estimate individual tree DBH and HCB from LiDAR-based data over the extensive forest area. In addition, the presented simultaneous equation system (Equation (3)) does not need measurement of any stand-level variable, whichor would require an additional cost. The presented modeling approach and algorithm will be useful for establishing similar compatible equation systems of DBH and HCB EIV models for other tree species and other tree variables that have inherent correlations between themselves.

**Author Contributions:** Conceptualization, Z.Y., L.F., Q.L., Q.Y., H.Z., and G.D.; Formal analysis and writing-original draft preparation Z.Y., L.F., G.D., R.P.S., P.L., and G.W. All authors contributed to interpreting results and the improvement of the paper. All authors have read and agreed to the published version of the manuscript.

**Funding:** This research was funded by the Thirteenth Five-year Plan Pioneering project of High Technology Plan of the National Department of Technology (No. 2017YFC0504101), the Central Public interest Scientific Institution Basal Research Fund under (Grant No. CAFYBB2019QD003) and the Chinese National Natural Science Foundations (Grant Nos. 31570627 and 31570628).

**Acknowledgments:** We thank the National Program on Key Basic Research Project (973 Program) (No. 2007CB714400) for data support. We also appreciate the valuable comments and constructive suggestions from two anonymous referees and the Associate Editor.



## Appendix A

An R program for leave-one-out cross validation (LOOCV) using the SUR fitting algorithm is illustrated on full data set.

```
library("openxlsx")
library(systemfit)
mydata<- read.xlsx("sample.xlsx")
LOOCV<-function(mydata) {
N<-nrow(mydata)
EstD<-array(dim=N)
EstHCB<-array(dim=N)
start.values<-c(a0=5,a1=-0.1,a2=-0.1,b0=0.2, b1=0.1)
eqD<-DBH~(a0*exp(-a1*LH-a2*LCW))
eqHCB<-HCB~LH/(1+exp(b0*DBH+b1*LCA))
model<-list(eqD,eqHCB)
for (i in 1: N) {
Temp1<-mydata[-i,]
Temp2<-mydata[i,]
try (sur<-nlsystemfit ("SUR", model, start values, data=mydata), TRUE)
if(class(sur)=="try-error")
{EstD[i]<-"NA"
EstHCB[i]<-"NA"}
else {
```

EstD[i]<-sur$b[1]*exp(-sur$b[2]*Temp2$LH-sur$b[3]*Temp2$LCW))
EstHCB[i]<-Temp2$LH/(1+exp(sur$b[4]*Temp2$DBH+sur$b[5]*Temp2$LCA))}
return (list (EstD,EstHCB))}

## References


1. Daniels, R.F.; Burkhart, H.E. *Simulation of Individual Tree Growth and Stand Development in Managed Loblolly Pine Plantations*; FWS-5-75; Division of Forestry and Wildlife Resources, Virginia Polytechnic and State University: Blacksburg, VA, USA, 1975; 69p.
2. Navratil, S. Wind damage in thinned stands. In Proceedings of the A Commercial Thinning Workshop, Whitecourt, AB, Canada, 17–18 October 1997; pp. 29–36.
3. Ancelin, P.; Courbaud, B.; Fourcaud, T. Development of an individual tree-based mechanical model to predict wind damage within forest stands. *For. Ecol. Manag.* **2004**, *203*, 101–121. [CrossRef]
4. Dean, T.J.; Cao, Q.V.; Roberts, S.D.; Evans, D.L. MeaNSURing heights to crown base and crown median with LiDAR in a mature, even-aged loblolly pine stand. *For. Ecol. Manag.* **2009**, *257*, 126–133. [CrossRef]
5. Scott, J.H.; Reinhardt, D. *Assessing Crown Fire Potential by Linking Models of Surface and Crown Fire Potential*; USDA Forest Service, Rocky Mountain Research: Washington, WA, USA, 2001.
6. Hynynen, J. Predicting tree crown ratio model for Austrian forests. *Can. J. For. Res.* **1995**, *25*, 57–62. [CrossRef]
7. Vauhkonen, J. Estimating crown base height for Scots pine by means of the 3D geometry of airborne laser scanning data. *Int. J. Remote Sens.* **2010**, *31*, 1213–1226. [CrossRef]
8. Michael, E.D.; Burkhart, H.E. Compatible crown ratio and crown height models. *Can. J. For. Res.* **1987**, *17*, 572–574.
9. Ritchie, M.W.; Hann, D.W. *Equations for Predicting Height to Crown Base for Fourteen Tree Species in Southwest Oregon*; Research Paper; Oregon State University, Forestry Research Laboratory: Corvallis, OR, USA, 1987.
10. Baburam, R.; Aaron, R.; Weiskittel, J.A.; Kershaw, J. Development of height to crown base models for thirteen tree species of the North American Acadian Region. *For. Chron.* **2012**, *88*, 60–73. [CrossRef]
11. Sharma, R.P.; Vacek, Z.; Vacek, S.; Podrázský, V.; Jansa, V. Modelling individual tree height to crown base of Norway spruce (*Picea abies* (L.) Karst.) and European beech (*Fagus sylvatica* L.). *PLoS ONE* **2017**, *12*, e0186394. [CrossRef] [PubMed]
12. Fu, L.Y.; Sharma, R.P.; Hao, K.J.; Tang, S.Z. A generalized interregional nonlinear mixed-effects crown width model for Prince Rupprecht larch in northern China. *For. Ecol. Manag.* **2017**, *389*, 364–373. [CrossRef]
13. Popescu, S.C. Estimating biomass of individual pine trees using airborne LiDAR. *Biomass Bioenergy* **2007**, *31*, 646–655. [CrossRef]
14. Broadbent, E.N.; Asner, G.P.; Peña-Claros, M.; Palace, M.; Soriano, M. Spatial partitioning of biomass and diversity in a lowland Bolivian forest: Linking field and remote sensing measurements. *For. Ecol. Manag.* **2008**, *255*, 2602–2616. [CrossRef]
15. Heurich, M. Automatic recognition and measurement of single trees based on data from airborne laser scanning over the richly structured natural forests of the Bavarian Forest National Park. *For. Ecol. Manag.* **2008**, *255*, 2416–2433. [CrossRef]
16. Bi, H.; Fox, J.C.; Li, Y.; Lei, Y.; Pang, Y. Evaluation of nonlinear equations for predicting diameter from tree height. *Can. J. Remote Sens.* **2012**, *42*, 789–806. [CrossRef]
17. Lefsky, M.A.; Cohen, W.B.; Acker, S.A.; Parker, G.G.; Spies, T.A.; Harding, D. LiDAR Remote Sensing of the Canopy Structure and Biophysical Properties of Douglas-Fir Western Hemlock Forests. *Remote Sens. Environ.* **1999**, *70*, 339–361. [CrossRef]
18. Næsset, E. Predicting forest stand characteristics with airborne scanning laser using a practical two-stage procedure and field data. *Remote Sens. Environ.* **2002**, *80*, 88–99. [CrossRef]
19. Pyysalo, U.; Hyyppä, H. Reconstructing tree crowns from laser scanner data for feature extraction. Int. Archives of Photogram. *Remote Sens. Spat. Inf. Sci. XXXIV Part 3B* **2002**, *34*, 218–221.
20. Holmgren, J.; Persson, Å.; Söderman, U. Species identification of individual trees by combining high resolution LIDAR data with multi-spectral images. *Int. J. Remote Sens.* **2008**, *29*, 1537–1552. [CrossRef]
21. Popescu, S.C.; Zhao, K. A voxel-based LiDAR method for estimating crown base height for deciduous and pine trees. *Remote Sens. Environ.* **2008**, *112*, 767–781. [CrossRef]

22. Maltamo, M.; Bollandsås, O.M.; Vauhkonen, J.; Breidenbach, J.; Gobakken, T.; Næsset, E. Comparing different methods for prediction of mean crown height in Norway spruce stands using airborne laser scanner data. *Forestry* **2010**, *83*, 257–268. [CrossRef]

23. Luo, L.; Zhai, Q.; Su, Y.; Ma, Q.; Kelly, M.; Guo, Q. Simple method for direct crown base height estimation of individual conifer trees using airborne LiDAR data. *Opt. Express.* **2018**, *26*, A562–A578. [CrossRef]

24. Solberg, S.; Næsset, E.; Bollandsås, O.M. Single tree segmentation using airborne laser scanning data in a structurally heterogeneous spruce forest. *Photogramm. Eng. Remote Sens.* **2006**, *72*, 1369–1378. [CrossRef]

25. Næsset, E.; Økland, T. Estimating tree height and tree crown properties using airborne scanning laser in a boreal nature reserve. *Remote Sens. Environ.* **2002**, *79*, 105–115. [CrossRef]

26. Andersen, H.E.; McGaughey, R.J.; Reutebuch, S.E. Estimating forest canopy fuel parameters using LiDAR data. *Remote Sens. Environ.* **2005**, *94*, 441–449. [CrossRef]

27. Maltamo, M.; Hyyppä, J.; Malinen, J. A comparative study of the use of laser scanner data and field measurements in the prediction of crown height in boreal forests. *Scand. J. Forest Res.* **2006**, *21*, 231–238. [CrossRef]

28. Maltamo, M.; Karjalainen, T.; Repola, J.; Vauhkonen, J. Incorporating tree- and stand-level information on crown base height into multivariate forest management inventories based on airborne laser scanning. *Silva ennica* **2018**, *52*, 1–18. [CrossRef]

29. Zhang, W.; Ke, Y.; Quackenbush, L.J.; Zhang, L. Using error-in-variable (EIV) regression to predict tree diameter and crown width from remotely sensed imagery. *Can. J. For. Res.* **2010**, *40*, 1095–1108. [CrossRef]

30. Rechenr, A.C.; Schaalje, G.B. *Linear Models in Statistics*, 2nd ed.; Woley: New York, NY, USA, 2008.

31. Kangas, A.S. Effect of errors-in-variables on coefficients of a growth model and on prediction of growth. *For. Ecol. Manag.* **1998**, *102*, 203–212. [CrossRef]

32. Carroll, R.J.; Ruppert, D.; Stefanski, L.A.; Crainiceanu, C.M. *Measurement Error in Nonlinear Models: A Modern Perspective*; Taylor & Francis Group LLC: New York, NY, USA, 2006; p. 438.

33. Fu, L.Y.; Liu, Q.W.; Sun, H.; Wang, Q.Y.; Li, Z.Y.; Chen, E.X.; Pang, Y.; Song, X.Y.; Wang, G.X. Development of a system of compatible individual tree diameter and aboveground biomass prediction models using error-in-variable regression and airborne LiDAR data. *Remote Sens.* **2018**, *10*, 325. [CrossRef]

34. Tang, S.; Li, Y.; Fu, L.Y. *Statistical Foundation for Biomathematical Models*, 2nd ed.; Higher Education Press: Beijing, China, 2015; p. 435. ISBN 978-7-04-042303-7.

35. Fuller, W.A. *Meansureement Error Models*; John Wiley and Sons: New York, NY, USA, 1987.

36. Tang, S.; Zhang, S. Measurement error models and their applications. *J. Biomath.* **1998**, *13*, 161–166.

37. Tang, S.; Li, Y.; Wang, Y. Simultaneous equations, errors-invariable models, and model integration in systems ecology. *Ecol. Model.* **2001**, *142*, 285–294. [CrossRef]

38. Lindley, D.V. Regression lines and the linear functional relationship. *J. R. Stat. Soc. B* **1947**, *9*, 218–244. [CrossRef]

39. Tang, S.; Wang, Y. A parameter estimation program for the errors-in-variable model. *Ecol. Model.* **2002**, *156*, 225–236. [CrossRef]

40. Walters, D.K.; Hann, D.W. *Taper Equations for Six Conifer Species in Southwest Oregon*; Forest Research Laboratory, Oregon State University: Corvallis, OR, USA, 1986; p. 41.

41. Pang, Y.; Chen, E.; Liu, Q.; Xiao, Q.; Zhong, K.; Li, X.; Ma, M. WATER: Dataset of airborne LiDAR mission at the super site in the Dayekou watershed flight zone on Jun. 23, 2008. In *Chinese Academy of Forestry; Institute of Remote Sensing Applications, Chinese Academy of Sciences*; Cold and Arid Regions Environmental and Engineering Research Institute, Chinese Academy of Sciences; Heihe Plan Science Data Center: Lanzhou, China, 2008.

42. Koch, B.; Heyder, U.; Weinacker, H. Detection of Individual Tree Crowns in Airborne Lidar Data. *Photogramm. Eng. Remote Sens.* **2006**, *72*, 357–363. [CrossRef]

43. Parent, J.R.; Volin, J.C. Assessing the potential for leaf–off LiDAR data to model canopy closure in temperate deciduous forests. *Photogramm. Eng. Remote Sens.* **2014**, *95*, 134–145. [CrossRef]

44. Liu, Q.; Fu, L.; Wang, G.; Li, S.; Li, Z.; Chen, E.; Pang, Y.; Hu, K. Improving estimation of forest canopy cover by introducing loss ratio of laser pulses using airborne LiDAR. *IEEE Trans. Geosci. Remote.* **2020**, *58*, 567–584. [CrossRef]

45. Liu, Q. Study on the Estimation Method of Forest Parameters Using Airborne Lidar. Ph.D. Thesis, Chinese Academy of Forestry, Beijing, China, 2009. (In Chinese).

46. Liu, Q.; Li, Z.; Chen, E.; Pang, Y.; Wu, H. Extracting individual tree heights and crowns using airborne LIDAR data. *J. Beijing For. Univ.* **2008**, *30*, 83–89.

47. Fu, L.; Lei, Y.; Wang, G.; Bi, H.; Tang, S.; Song, X. Comparison of seemingly unrelated regressions with multivariate errors-in-variables models for developing a system of nonlinear additive biomass equations. *Trees* **2016**, *30*, 839–857. [CrossRef]

48. Parresol, B.R. Assessing tree and stand biomass: A review with examples and critical comparisons. *For. Sci.* **1999**, *45*, 573–593.

49. Parresol, B.R. Additivity of nonlinear biomass equations. *Can. J. For. Res.* **2011**, *31*, 865–878. [CrossRef]

50. Judge, G.G.; Hill, R.C.; Griffifiths, W.E.; Lutkepohl, H.; Lee, T.C. *Introduction to the Theory and Ractice of Econometrics*, 2nd ed.; Wiley: New York, NY, USA, 1988.

51. Zellner, A.; Theil, H. Three Stage Least Squares: Simultaneous Estimation of Simultaneous Equations. *Econometrica* **1962**, *30*, 54–78. [CrossRef]

52. Fumio, H. *Econometrics*; Shanghai University of Finance and Economics Press: Shanghai, China, 2005; p. 536. ISBN 7-81098-499-3.

53. Nord-Larsen, T.; Meilby, H.; Skovsgaard, J.P. Site-specific height growth models for six common tree species in Denmark. *Scand. J. For. Res.* **2009**, *24*, 194–204. [CrossRef]

54. Timilsina, N.; Staudhammer, C.L. Individual tree-based diameter growth model of slash pine in Florida using nonlinear mixed modeling. *For. Sci.* **2013**, *59*, 27–31. [CrossRef]

55. Zhao, D.; Lynch, T.B.; Westfall, J.; Coulston, J.; Kane, M.; Adams, D.E. Compatibility, development and estimation of taper and volume equation systems. *For. Sci.* **2019**, *65*, 1–13. [CrossRef]

56. R Core Team. *R: A Language and Environment for Statistical Computing*; R Foundation for Statistical Computing: Vienna, Austria, 2018; Available online: https://www.R-project.org/ (accessed on 15 March 2020).

57. Greene, W.J. *Econometric Analysis*, 7th ed.; Pearson Education, Inc.: New York, NY, USA, 2001; p. 1188. ISBN 0-13-139538-6.

58. Curran, P.J.; Hay, A.M. The importance of mesauement error for certain procedures in remote sensing at optical wavelengths. *Photogramm. Eng. Remote Sens.* **1986**, *52*, 229–241.

59. Wykoff, W.R. A basal area increment model for individual conifers in the northern Rocky Mountains. *For. Sci.* **1990**, *136*, 1077–1104.

60. Van Deusen, P.C.; Biging, G.S. *STAG, A Stand Generator for Mixed Species Stands*; Research Note; Northern 55 California Forest Yield Cooperative, Department of Forestry and Resource Management, University of California: Berkeley, CA, USA, 1985; Volume 11, 25p.

Article

# Land Use/Land Cover Mapping Using Multitemporal Sentinel-2 Imagery and Four Classification Methods—A Case Study from Dak Nong, Vietnam

**Huong Thi Thanh Nguyen [1], Trung Minh Doan [1], Erkki Tomppo [2,3,* ] and Ronald E. McRoberts [4]**

[1]   Department of Forest resource & Environment management (Frem), Faculty of Agriculture and Forestry, Tay Nguyen University, Le Duan Str. 567, Buon Ma Thuot City 63000, Daklak Province, Vietnam; nguyenthithanhhuong@ttn.edu.vn (H.T.T.N.); doantrung@ttn.edu.vn (T.M.D.)

[2]   Department of Electronics and Nanoengineering, School of Electrical Engineering, Aalto University, P.O. Box 11000, 00076 Aalto, Finland

[3]   Department of Forest Sciences, University of Helsinki, Latokartanonkaari 7, P.O. Box 27 FI-00014 Helsinki, Finland

[4]   Raspberry Ridge Analytics, 15111 Elmcrest Avenue North, Hugo, MN 55038, USA; mcrob001@umn.edu

*   Correspondence: erkki.tomppo@aalto.fi

Received: 23 March 2020; Accepted: 24 April 2020; Published: 26 April 2020

**Abstract:** Information on land use and land cover (LULC) including forest cover is important for the development of strategies for land planning and management. Satellite remotely sensed data of varying resolutions have been an unmatched source of such information that can be used to produce estimates with a greater degree of confidence than traditional inventory estimates. However, use of these data has always been a challenge in tropical regions owing to the complexity of the biophysical environment, clouds, and haze, and atmospheric moisture content, all of which impede accurate LULC classification. We tested a parametric classifier (logistic regression) and three non-parametric machine learning classifiers (improved k-nearest neighbors, random forests, and support vector machine) for classification of multi-temporal Sentinel 2 satellite imagery into LULC categories in Dak Nong province, Vietnam. A total of 446 images, 235 from the year 2017 and 211 from the year 2018, were pre-processed to gain high quality images for mapping LULC in the 6516 km$^2$ study area. The Sentinel 2 images were tested and classified separately for four temporal periods: (i) dry season, (ii) rainy season, (iii) the entirety of the year 2017, and (iv) the combination of dry and rainy seasons. Eleven different LULC classes were discriminated of which five were forest classes. For each combination of temporal image set and classifier, a confusion matrix was constructed using independent reference data and pixel classifications, and the area on the ground of each class was estimated. For overall temporal periods and classifiers, overall accuracy ranged from 63.9% to 80.3%, and the Kappa coefficient ranged from 0.611 to 0.813. Area estimates for individual classes ranged from 70 km$^2$ (1% of the study area) to 2200 km$^2$ (34% of the study area) with greater uncertainties for smaller classes.

**Keywords:** classification; Sentinel 2; land use land cover; improved k-NN; logistic regression; random forest; support vector machine

---

## 1. Introduction

### 1.1. Motivation

Most Vietnamese forests are classified as tropical with natural forest accounting for more than 70% of the total forest area [1]. Dak Nong province has the most abundant natural forest resources in Vietnam. The great diversity of this resource is primarily owing to a wide variety of environmental and climatic factors, most of which are governed by latitude and topography [2]. However, Dak Nong's natural forests are being lost at an alarming rate owing to factors that include expanding agriculture, conversion to commercial and plantation forest types, and increasing human population. For many years, the Highland Plateau, which includes Dak Nong, has been a major "hot spot" for conversion of forest to agriculture in Vietnam. During the 1990s and early 2000s, forest was lost at an average annual rate of 15,000 ha per year [3], with forest cover declining from 75% in 1985 to 60% in 2009. During this time, the annual rate of deforestation in the Highland Plateau was the greatest of all regions, accounting for 46.3% of the entire national forest area lost.

The Highland Plateau is characterized by a complex topography with mountains, highlands, valleys, deltas, and diversified soil types. Approximately 1.3 million ha are fertile soils, rich in organic matter and nutrients, that facilitate development of high value industrial perennial crops such as coffee, rubber, pepper, cashew, and fruit trees. Additionally, the distinct rainy and dry seasons in the south of Vietnam cause differences in the rates of plant growth. Finally, climatic differences from north to south in Vietnam cause vegetation to vary in physiognomy and lead to morphological differences among land cover types, particularly between semi-evergreen and deciduous Dipterocarp forests.

Current, accurate, and detailed land cover information that reflects these unique topographic and climatic conditions, particularly for natural forest types, is crucial for land managers, decision makers, and policy makers tasked with developing forest management strategies and policies [4–6]. Forest resource decision-making is characterized by a large degree of uncertainty regarding the outcomes of alternative choices. The result is a wide variety of opinions regarding the different options that impedes agreement on a clear way forward. Although there is usually agreement on general objectives such as sustainable forest use, biodiversity conservation, and the alleviation of rural poverty, conflicts among stakeholders over the best course of action for achieving these objectives almost always arise. New issues or new actors may appear and influence discussions, external events may unexpectedly require the revision of agreed policy proposals, and deadlocks can exist for long periods, all continuing until pressing circumstances lead to settlements and decisions [7].

### 1.2. Remotely Sensed Data

Remote sensing offers a unique environmental capability for monitoring extensive geographical areas in a cost-efficient manner, while simultaneously producing information related to the Earth's land, atmosphere, and oceans [8]. Land cover mapping represents one of the most common uses of remotely sensed data [9–11], with satellite imagery serving as one of the most important data sources [11].

As previously, Dak Nong presents unique challenges for the construction of accurate remote sensing-based land use land cover (LULC) maps [12]. The variation in vegetation owing to the rainy/dry seasonal variation affects the spectral reflectance properties of vegetation. For example, deciduous dipterocarp forests have spectral properties in the dry season that are similar to those of other cover types such as industrial coffee and rubber crops, whereas the respective spectral properties are quite different in the rainy season. Only a few studies have accommodated this kind of seasonal variation when constructing satellite image-based land cover classifications. Sothe et al. (2017) [13] combined multi-spectral fall and spring season images when mapping land cover with Landsat-8 data and found that the inclusion of additional band data considerably improved classifications when compared with the use of fall spectral bands alone. For both classifiers used by Sothe et al. (2017) [13], there were meaningful increases in classification accuracy, by 4.8% and 2.9% for the random forests and support vector machine classifiers, respectively, when the "spring" spectral bands were added.

Dak Nong's seasonal growth variation, varying vegetation spectral signatures, and varying topography suggest that Sentinel-2 satellite spectral data, with its fine spatial resolution (10–60 m), fine temporal resolution (five days), and fine spectral resolution (13 spectral bands), may be particularly well-suited for land cover classification purposes in the province. Although data from the Sentinel-2 sensor have been investigated for a variety of vegetation monitoring [14,15], terrestrial monitoring [16], and forest classification [16] applications, only a few studies have used Sentinel-2 for land cover mapping [17–19]. Therefore, additional studies that evaluate the utility of this imagery for land cover classification for regions with extremely diverse conditions such as those in Dak Nong are well-justified [20].

*1.3. Classification Techniques*

Factors that affect classification accuracy include sensor type, sources of training and accuracy assessment data, the number of classes, and the classification method [21–23]. Of these factors, the selection of a suitable algorithm that achieves acceptable classification accuracy with minimal processing time can be crucial [24]. Many methods have been proposed for constructing satellite image-based land cover maps [25,26], including both unsupervised and supervised methods and both parametric and non-parametric methods. Although unsupervised algorithms such as IsoData and K-Means clustering have been widely used for many years, general purpose clustering algorithms are cumbersome and difficult to develop [27]. Parametric supervised algorithms such as linear discriminant analysis [28–31] and multinomial logistic regression (MLR) have also been broadly used [32,33] and are often considered standards for comparison purposes [29,30,34]. In the last decade, non-parametric methods including support vector machine (SVM) [35–37], k-nearest neighbors (k-NN) [38,39], and random forests (RF) [40–42] have gained attention for remote sensing-based land cover classification. However, both SVM and RF require the selection of values for multiple parameters that affect their efficacy, and both are computationally intensive [6,35]. For k-NN, Naidoo et al. (2012) [43] reported difficulty in selecting the optimal value of k and that the genetic algorithms recommended for optimization can be computationally intensive [44,45]. Finally, object-based classification has been shown to be an effective method for classifying fine resolution imagery [46,47]. Object-based methods have been used with both fuzzy sets [48,49] and neural networks [48,50] to map land cover using satellite imagery. Although object-based classification methods have been shown to increase accuracy for some land cover mapping applications, fine spatial resolution remotely sensed imagery remains the most frequently used data source for these applications [51].

Because of the unique features of each study and study area including definitions, sample sizes, and numbers and characteristics of the classes, comparisons of methods with respect to accuracy among studies is difficult. Even so, not much effort has been committed to comparing methods with respect to accuracy for diverse tropical forest regions such as Dak Nong. Meyfroidt et al., 2013 [52] used MLR with Landsat data to assess classes of forest change and reported land cover classification accuracies of 0.64 to 0.69. Use of RF for land cover classification has been reported for multiple studies in Vietnam. Bourgoin et al. (in press) [53] used RF with both Landsat and Sentinel-2 data for multiple land cover and land cover change classes and reported an overall accuracy of 0.81. Nguyen et al. (2018) [6] used RF and Landsat data for 10 classes including multiple forest classes in Vietnam. Overall accuracy was approximately 0.90. Ha et al. (2018) [54] used RF and Landsat data for seven land cover classes including forest land and reported overall accuracies greater than 0.90. Finally, Phan and Kappas (2018) [20] reported that SVM was more accurate than RF for classifying six types of land cover types including one forest class in the North of Vietnam (Red River Delta) using Sentinel-2 data. In summary, although only a few studies using only a few methods have been used for the classification of forest land in Vietnam, the reported accuracies are relatively large. Thus, there is merit in a more comprehensive evaluation of classification methods, particularly for diverse tropical regions such as in Dak Nong province, Vietnam, with their distinct seasonal effects.

*1.4. Objectives*

The overall objective was to evaluate the utility of multi-seasonal Sentinel-2 spectral data for land cover classification and mapping in Dak Nong province, Vietnam. A subordinate objective was to compare the parametric MLR and non-parametric ik-NN, SVM, and RF classification methods with respect to both overall and class-level accuracies and with respect to whether the methods exploited the beneficial effects of the multi-seasonal Sentinel-2 data. Google Earth Engine was used for collecting and pre-processing both training and accuracy assessment data. A second subordinate objective was rigorous statistical estimation of the ground area of each land cover class.

## 2. Materials and Methods

### 2.1. Overview

The structure of this section has multiple components. First, the Dak Nong study area is described in Section 2.2, the Sentinel-2 satellite imagery and its separation into temporal periods are described in Section 2.2, and the land cover data from multiple sources and their separation into training and validation subsets are described in Section 2.3. Next, brief descriptions of the four classifiers are provided in Section 2.4, including descriptions of their statistical properties, details on their required input parameters, and procedures for optimizing their performance. Finally, in Section 2.5, the two analytical components used to compare all combinations of the four temporal image periods and the four classifiers are described. The first component focuses on map accuracy assessment, while the second component focuses on estimating LULC class areas and their corresponding uncertainties. The overall research approach is summarized in Figure 1.

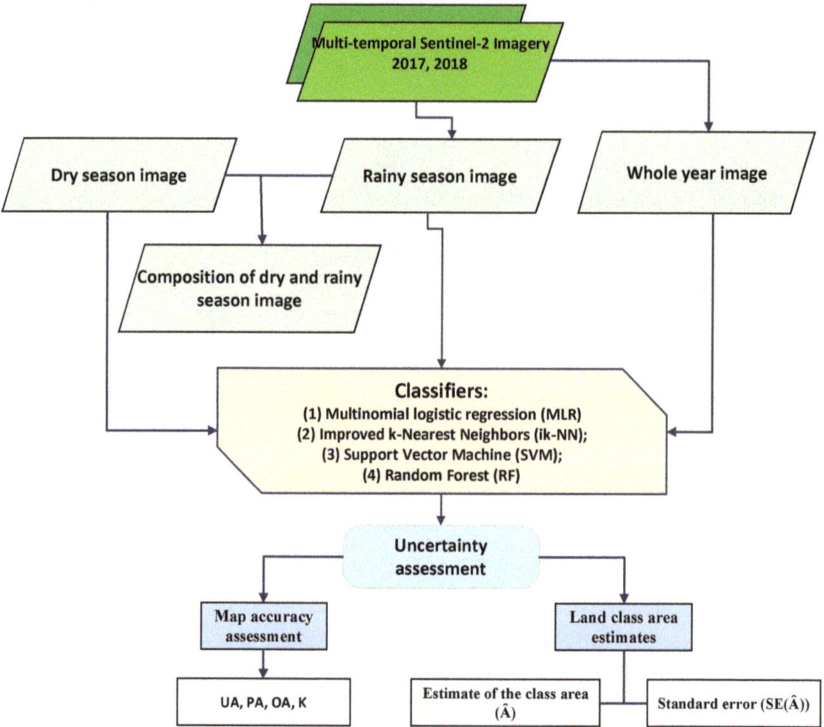


**Figure 1.** Research approach as a flowchart. The Sentinel-2 2017 and 2018 data were collected for different seasons: dry, rainy, whole year, and rainy and dry composited images. The MLR, ik-NN, SVM, and RF classifiers were tested with the resulting uncertainty assessments used as criteria for comparing combinations of seasonal datasets and classifiers. OA, overall accuracy; K, Kappa coefficient; PA, producer's accuracy; UA, user's accuracy.

## 2.2. Study Area

The study was conducted in Dak Nong Province in the Central Highlands of Vietnam (Figure 2). The average elevation is between 600 and 700 m above sea level. The mean temperature is 24 degrees Celsius. The climate conditions produce general characteristics of a subequatorial tropical monsoon climate. The province has characteristics of humid tropical highland climate and is affected by dry-hot southwest monsoons. There are two distinct annual seasons: the rainy season starts in April and ends in November, and the dry season starts in December and ends in March the following year. The average annual rainfall is 2500 mm, of which 90% occurs during the rainy season. The study area extends over 6516 km$^2$ and is characterized by substantial fragmentation, thereby making LULC classification particularly challenging. Natural forest consists of patches of natural evergreen broad-leaved, mixed bamboo, deciduous dipterocarp, and semi-deciduous forest with different levels of disturbance, mainly human in origin.

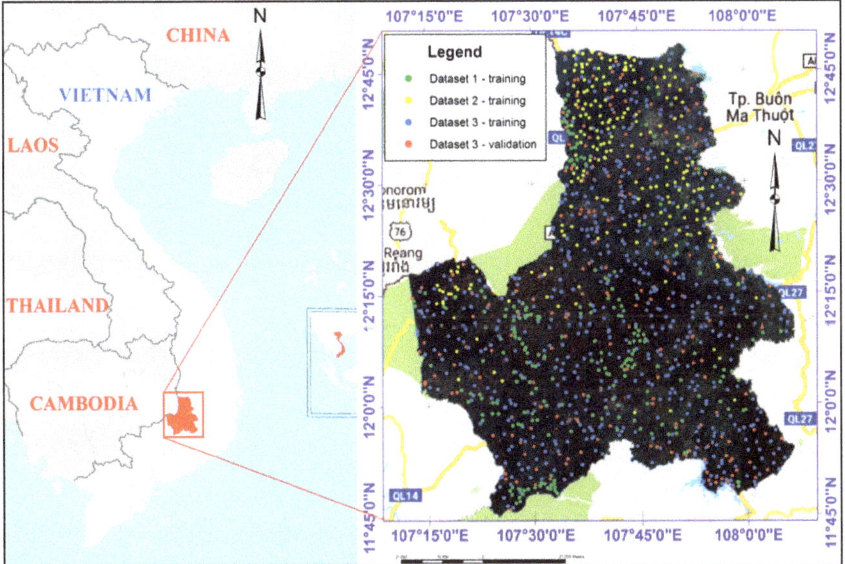


**Figure 2.** The study area in Dak Nong province, Vietnam, with sample unit locations.

## 2.3. Data

### 2.3.1. Sentinel-2 Imagery

Sentinel-2 MSI (multi-spectral instrument, Level-1C) remotely sensed data were used for LULC classification. The Sentinel-2 mission was developed by the European Space Agency (ESA) as a part of the Copernicus Programme [55]. The mission's wide swath, fine spatial resolution (10 m–60 m), multi-spectral features (13 spectral bands), and frequent revisit time (10 days at the equator with one satellite, and 5 days with two satellites) support monitoring vegetation changes within a growing season, forest monitoring, land cover change detection, and natural disaster management [56]. The spectrum characteristics of the Sentinel 2 images are described in Table 1.

**Table 1.** Basic characteristics of Sentinel 2 multi-spectral instrument (MSI).

| Name | Min | Max | Scale | Resolution | Wavelength | Description |
|------|-----|-----|-------|------------|------------|-------------|
| B1 | 0 | 10,000 | 0.0001 | 60 Meters | 443 nm | Aerosols |
| B2 | 0 | 10,000 | 0.0001 | 10 Meters | 490 nm | Blue |
| B3 | 0 | 10,000 | 0.0001 | 10 Meters | 560 nm | Green |
| B4 | 0 | 10,000 | 0.0001 | 10 Meters | 665 nm | Red |
| B5 | 0 | 10,000 | 0.0001 | 20 Meters | 705 nm | Red Edge 1 |
| B6 | 0 | 10,000 | 0.0001 | 20 Meters | 740 nm | Red Edge 2 |
| B7 | 0 | 10,000 | 0.0001 | 20 Meters | 783 nm | Red Edge 3 |
| B8 | 0 | 10,000 | 0.0001 | 10 Meters | 842 nm | Near infrared (NIR) |
| B8a | 0 | 10,000 | 0.0001 | 20 Meters | 865 nm | Red Edge 4 |
| B9 | 0 | 10,000 | 0.0001 | 60 Meters | 940 nm | Water vapor |
| B10 | 0 | 10,000 | 0.0001 | 60 Meters | 1375 nm | Cirrus |
| B11 | 0 | 10,000 | 0.0001 | 20 Meters | 1610 nm | Short-wave infrared (SWIR) 1 |
| B12 | 0 | 10,000 | 0.0001 | 20 Meters | 2190 nm | SWIR 2 |
| QA10 | | | | 10 Meters | | Always empty |
| QA20 | | | | 20 Meters | | Always empty |
| QA60 | | | | 60 Meters | | Cloud mask |

The difference in solar illumination geometry during image acquisition between the two seasons was considered in the present study. Although vegetation in the study area presents reduced climatic and phonological seasonality, the observed reflectance varies by season owing to changes in the solar illumination geometry caused by the Earth's translation movement [13]. Therefore, seasonal image datasets were separately classified to evaluate these influences. Accordingly, the scenes of interest included the following: (i) a collection of Sentinel-2 MSI scenes in the study area during the dry season of 2017 and 2018 (1 January 2017 to 31 March 2017, and 1 December 2017 to 31 March 2018), designated imagery 1 (IMG 1) (Table 2); (ii) a collection of Sentinel-2 MSI scenes during the rainy season of 2017 and 2018 (from 1 April 2017 to 30 November 2017 and 1 April 2018 to 30 June 2018), designated IMG 2; (iii) a collection of Sentinel 2 MSI scenes for all months of 2017 (from 1 January 2017 to 31 December 2017), designated IMG 3; and (iv) a combination of all bands for the dry and rainy seasons (combination of IMG 1 and IMG 2, designated IMG 4.

**Table 2.** Seasonal satellite images used in the classification.

| Image Name | Time | Acquisition Date | Number of Images Involved | Number of Bands |
|------------|------|------------------|---------------------------|-----------------|
| IMG 1 | Dry season, 2017–2018 | 01/01/2017–03/31/2017 and 12/01/2017–03/31/2018 | 169 | 10 |
| IMG 2 | Rainy season, 2017–2018 | 04/01/2017–11/30/2017 and 04/01/2018–06/30/2018 | 277 | 10 |
| IMG 3 | All for year 2017 | 01/01/2017–12/31/2017 | 265 | 10 |
| IMG 4 | Combination of all bands for both 2017 and 2018 (IMG 1 + IMG 2) | Dry season 2017–2018 + Rainy season 2017–2018 | 446 | 20 |

The different seasonal image datasets represented for each season were considered for the analyses based on the median value of the collection. The multi-spectral bands in the study included Blue (B2), Green (B3), Red (B4), Red Edge 1 (B5), Red Edge 2 (B6), Red Edge 3 (B7), near infrared (NIR) (B8), Red Edge 4 (B8A), short-wave infrared (SWIR) 1 (B11), and SWIR 2 (B12). In addition to these spectral bands, the normalized difference vegetation index (NDVI) and a digital elevation model (DEM) were added to the seasonal image data (IMG 1–4) with the objective of increasing classification accuracy, as reported from previous studies [22,57]. These bands, including NDVI [58,59] and DEM, were resampled at 10 m resolution. Image information is described in Table 2 below.

To conduct the analyses, the JavaScript API Code Editor in the Google Earth Engine (GEE) was used to collect data for a large number of images. GEE provides most freely available image data and an application programming interface (API) to analyze and visualize the data [60,61]. Surface reflectance (SR) images for 2017 were not available, and for 2018 images, approximately 50% of the study area was covered by clouds. Hence, a top of atmosphere (TOA) dataset acquired for 2017 and 2018 was used for the study. The set of collected images was pre-processed to reduce the effects of topography and bidirectional reflectance distribution function (BRDF). At the same time, cloud areas were masked out and shadows were removed during this process.

All images underwent pixel-wise cloud and cloud shadow masking using the Google cloudScore algorithm for cloud masking and temporal dark outlier mask (TDOM) for cloud shadows, both of which built on ideas from Landsat TDOM and cloudScore algorithm. The original concept was written by Carson Stam, adapted by Ian Housman, currently documented in [60], and described and evaluated in a forthcoming paper [62]. This study implemented the correction of reflectance spectral values by BRDF based on the method described by Roy et al. 2017a,b [63,64]. Topographic correction is the process to account for diffuse atmospheric irradiance caused by slope, aspect, and elevation effects. The sun-canopy-sensor + C (SCSc) correction method based on the C-correction, as described by Soenen et al. [65], was applied for topographic correction in this study. The median function was then applied to create an image object (single image) representing the median value of all images in the filtered collection [66,67]. The median lies closer to the majority of values and is insensitive to extreme values and has exactly half the values smaller and half the values greater than the median, as elaborately applied by [68]. The post-processed images were then resampled to a spatial resolution of 10 m using the nearest neighbor method [69], and subsetted to the study area. The entire pre-processing was implemented on GEE based on the script available on "Open Geo Blog - Tutorials, Code snippets and examples to handle spatial data" [61,70].

### 2.3.2. Training and Validation Data

Within the study area, 11 LULC classes were distinguished: (1) dense evergreen broadleaved forest (the forest has been slightly impacted); (2) open evergreen broadleaved forest (the forest has been moderately to heavily disturbed); (3) semi-evergreen forest (the forest that consists of a mixture of evergreen and deciduous dipterocarp tree species); (4) deciduous dipterocarp forest; (5) plantation forest; (6) mature rubber ($\geq$3 years old); (7) perennial industrial plants; (8) croplands (annual crop land); (9) residential area; (10) water surface; and (11) other lands including, but not limited to, other types of grassland, shrubs, bare land, and abandoned land.

Acquiring adequate training and validation data is often challenging in tropical regions. Sothe et al., 2017 [13] and Teluguntla et al., 2018 [71] both obtained good results using sample data from a combination of sources including field investigations, very fine spatial resolution Google Earth imagery, current Landsat and Sentinel imagery, and other sources such as maps. A similar approach was used for this study for which three sets of sample data were acquired in 2017 and 2018: (1) field observations for a purposive sample of size 232; (2) visual interpretations of fine and very fine resolution imagery from sources that included Google Earth for a purposive sample size of 214; and (3) visual interpretations of fine and very fine resolution imagery from sources that included Google Earth and Sentinel- 2A imagery for a simple random sample size of 800. For the latter sample dataset, field observations and data from the 2016 Dak Nong Forest Inventory Map were used to clarify and refine interpretations for the LULC classes such as semi-evergreen forest, plantation forest, and some perennial industrial crops that were difficult to distinguish in the imagery.

To obtain the probability sample necessary for validation, a systematic sample of the probability-based third dataset was selected. The plots in the third dataset were first sorted by their east and north coordinates, and then a systematic sample was selected from within each LULC class. For each class, the proportion selected was arbitrary, but was guided by the desire for a minimum sample size of 15, where possible, while yet retaining a sufficient sample size for training purposes.

For the eleven LULC classes, the proportions were, in order, as follows: 0.20, 0.20, 0.50, 0.67, 1.00, 0.50, 0.11, 0.50, 0.67, 0.50, 0.20. Because the third dataset was selected using a simple random sample, and it was systematically subsampled, each subsample can also be considered a simple random sample and, therefore, can be used for validation. The remaining portion of the third dataset was used for training purposes. The result was a sample size of 1036 for training and a sample size of 208 for validation. The number of validation plots by LULC category was considered sufficient and generally complied with the recommendation of Särndal et al. (1992) [72]. A summary of the training and validation datasets is shown in Table 3 with the spatial distribution of sample locations shown in Figure 2.

**Table 3.** Training and validation data.

| Dataset | Use | Land Cover Class | | | | | | | | | | | Total |
| | | 1 | 2 | 3 | 4 | 5 | 6 | 7 | 8 | 9 | 10 | 11 | |
|---|---|---|---|---|---|---|---|---|---|---|---|---|---|
| 1 | Training | 77 | 6 | 15 | 13 | 29 | 34 | 0 | 13 | 32 | 4 | 9 | 232 |
| 2 | Training | 6 | 8 | 52 | 33 | 11 | 14 | 19 | 21 | 20 | 4 | 25 | 213 |
| 3 | Training | 99 | 97 | 22 | 9 | 0 | 17 | 234 | 20 | 8 | 11 | 74 | 591 |
| Total | Training | 182 | 111 | 89 | 55 | 40 | 65 | 253 | 54 | 60 | 19 | 108 | 1036 |
| 3 | Validation | 25 | 25 | 22 | 17 | 7 | 17 | 28 | 20 | 16 | 12 | 19 | 208 |

*2.4. Classifiers*

The MLR, ik-NN, RF, and SVM supervised classification algorithms were used to classify the satellite image data as described above. The training areas for each LULC type were selected based on Google Earth, field data, and prior knowledge, as well as available data. The models were used as supervised classifiers to classify pixels based on their spectral signatures.

2.4.1. Multinomial Logistic Regression (MLR)

With MLR, the probability of class c for the $i^{th}$ plot, c=1,..., C, is estimated as follows:

$$p\left(y_i = c\middle|x_i\right) = \frac{\exp(\beta_c \cdot x_i)}{1 - \sum_{m=1}^{C-1} \exp(\beta_m \cdot x_i)} + \varepsilon_i, \text{ for } c = 1, \ldots, C-1 \tag{1}$$

and

$$p\left(y_i = C\middle|x_i\right) = \frac{1}{1 - \sum_{m=1}^{C-1} \exp(\beta_m \cdot x_i)} + \varepsilon_i, \tag{2}$$

where C is the number of the LULC classes, $x_i$ is the vector of predictor variable observations for the $i^{th}$ population unit, and $\beta_c$ is the vector of regression coefficients associated with LULC class c. The class with the greatest probability is selected as the prediction for the $i^{th}$ population unit. Optimal estimates for $\{\beta_c : c = 1, \ldots, C\}$ can be obtained using any of multiple statistical software packages.

2.4.2. Improved k-Nearest Neighbors (ik-NN)

In the terminology of nearest neighbors techniques, the auxiliary or predictor variables are designated feature variables and the space defined by the feature variables is designated the feature space; the set of sample units for which observations of both response and feature variables are available is designated the reference set; and the set of population units for which predictions of response variables are desired is designated the target set (Chirici et al., 2016) [73]. All population units for both the reference and target sets are assumed to have complete sets of observations for all feature variables.

For the $i^{th}$ target unit, all forms of nearest neighbors algorithms entail selecting the k-nearest or most similar neighbors, $\left\{y_j^i : j = 1, 2, \ldots, k\right\}$, from the reference set with respect to a distance metric, d, formulated as a function of the feature variables. For categorical response variables such as land cover classes, the prediction, $\hat{y}_i$, for the $i^{th}$ target unit is the most heavily weighted class among the k-nearest neighbors, a weighted median or mode in case of ordinal scale variables, or a mode in the case of nominal variables. Implementation of nearest neighbors techniques requires multiple selections: (i) the distance metric, d, to assess nearness or similarity in the feature space; (ii) a scheme for weighting the predictor variables in the distance metric; (iii) a scheme for weighting individual neighbors when calculating predictions; and (iv) a number, k, of nearest neighbors [73].

For this study, the distance metric was a simplified version of the metric proposed by Tomppo and Halme (2004) [44], as used in the operational Finnish multi-source national forest inventory (MS-NFI),

$$d_{ij} = \sqrt{\sum_{m=1}^{p} w_m^2 \cdot \left(x_{im} - x_{jm}\right)^2}, \tag{3}$$

where i denotes a target unit; j denotes a reference unit; $d_{ij}$ is the distance between the units i and j; m indexes the feature variables; $x_{im}$ and $x_{jm}$ are observations of the $m^{th}$ feature variable for the $i^{th}$ target unit and $j^{th}$ reference unit, respectively; and $w_m$ is a feature variable weight. Neighbor weights are typically formulated as powers, $t \in [0, 2]$, of distances between target and reference units. Often, the necessary selections to implement a nearest neighbor algorithm are made in an arbitrary method, whereas improved k-NN (ik-NN) entails optimized selection of the weights, $w_m$, using a technique such as genetic algorithms [44,45,74]

### 2.4.3. Support Vector Machine (SVM)

The principle behind the SVM classifier is a hyperplane that separates the data for different classes. The main focus is construction of the hyperplane by maximizing the distance from the hyperplane to the nearest data point of either class. These nearest data points are known as support vectors [75].

According to Huang et al. (2002) [35] (p. 734), by mapping the input data into a high-dimensional space, the kernel function converts non-linear boundaries in the original data space into linear boundaries in the high-dimensional space, which can then be located using an optimization algorithm. Therefore, selection of the kernel function and appropriate values for corresponding kernel parameters, referred to as the kernel configuration, can affect the performance of the SVM.

The radial basis function kernel (RBF kernel) is one of the most popular kernels used to implement the support vector machine algorithm and was used for this study. The squared Euclidean distance metric was used to construct completely non-linear hyperplanes. The RBF kernel of the SVM classifier is commonly used and has performed well. Polynomial kernels, especially high-order kernels, took far more time to train than RBF kernels [35].

Meyer et al. (2002) [76] stated that, for classification tasks, C-classification is most likely used with the RBF kernel because of its good general performance and the small number of parameters (only two, C and $\gamma$). Therefore, the two parameters that must be defined for this classification algorithm are the cost parameter (C) and the kernel width parameter ($\gamma$). According to Knorn et al. (2009) [77] (p.960), C is a regularization parameter that controls the trade-off between maximizing the margin and minimizing the training error. C is a preset penalty value for misclassification errors, while $\gamma$ describes the kernel width, which affects the smoothing of the shape of the class-dividing hyperplane.

The authors of LIBSVM suggest trying small and large values for C, such as 1 to 1000, then using cross-validation to decide which is optimal for the data, and finally trying several $\gamma$'s for the optimal C's. A small C-value tends to emphasize the margin while ignoring the outliers in the training data, while a large C-value may overfit the training data [77] (p.960). Optimal selection of C and $\gamma$ parameters was done by testing C parameters in the range $2^{-1}$ to $2^8$ and $\gamma$ parameters in the range 0.1 to 2.0.

### 2.4.4. Random Forests (RF)

The RF classifier developed by Breiman (2001) [78] requires selection of three parameters: ntree (number of trees to grow), mtry (the number of variables to split each node), and variable importance (the number of variables/bands influences model performance). Liaw & Wiener (2002) [79] recommend using the square root of the number of input variables as the default value for mtry. A large value for ntree produces a stable result for variable importance, which is estimated using two indicators: (i) mean decrease accuracy (MDA) and ii) mean decrease gini (MDG). MDA is the accuracy associated with each predictor variable based on the out-of-bag error rate (OOB). Gini impurity is a measure of how often a randomly chosen element from the set would be incorrectly labeled if it is randomly labeled according to the distribution of labels in the subset. For this study, MDA values were investigated to determine the importance of variables. Nguyen et al. (2018) [6] indicated that within the range $1 \leq$ ntree $\leq 500$, ntree = 300 produced the best fit. In addition, Breiman (2001) [78] stated that using more than the required number of trees may be unnecessary, albeit not harmful, because the relationship between accuracy and ntree is asymptotic. The 'RandomForest' package in the R environment developed by Liaw and Wiener (version 4.6–14 in 2018) was used in present study. Optimal values of mtry, ntree, and variable importance were selected based on the smallest OOB error. The optimal variable importance depended on the MDA value and accuracy of the model.

### 2.5. Analyses

#### 2.5.1. Accuracy Assessment

Accuracy assessment is an important step before accepting a classification result [21]. The classification accuracy of a map product is estimated by constructing a confusion matrix between reference and classified pixels. Classification accuracy was assessed using criteria such as overall accuracy (OA), Kappa coefficient (K), producer's accuracy (PA), and user's accuracy (UA). Congalton and Green (1999) [80] assert that analysis of the causes of differences in the confusion matrix can be one of the most important and interesting steps in the construction of a map from remotely sensed data.

The objectives of the study included comparing the performance of classifiers as well as assessing the effects of Sentinel-2 satellite images for different seasons, as described in Table 2. The number of seasonal bands used with the four classifiers is reported in Table 4.

**Table 4.** Classifiers and seasonal bands. Ik-NN, improved k-nearest neighbors; MLR, multinomial logistic regression; SVM, support vector machine; RF, random forests.

| Classification Algorithm | Image Set | Number of Bands |
| --- | --- | --- |
| ik-NN | IMG 1 | 10 |
|  | IMG 2 | 10 |
|  | IMG 3 | 10 |
|  | IMG 4 | 20 |
| MLR | IMG 1 | 10 |
|  | IMG 2 | 10 |
|  | IMG 3 | 10 |
|  | IMG 4 | 20 |
| SVM | IMG 1 | 10 |
|  | IMG 2 | 10 |
|  | IMG 3 | 10 |
|  | IMG 4 | 20 |
| RF | IMG 1 | 10 |
|  | IMG 2 | 10 |
|  | IMG 3 | 10 |
|  | IMG 4 | 20 |

#### 2.5.2. Land Cover Class Area Estimation

For each land cover class, an estimate of the class area and the corresponding standard error were calculated using a combination of confusion matrices and stratified estimators [81,82]. For each class,

C, a confusion matrix was constructed for two classes: (i) class C and (ii) the aggregation of all other classes into a single class designated ~C (Table 5). Using estimates of proportions and corresponding variances as indicated in Table 5, the stratified estimate of the area of class C was as follows:

$$\hat{A}_C = A_{tot} \cdot \left(wt_1 \cdot \hat{p}_1 + wt_2 \cdot \hat{p}_2\right), \tag{4}$$

with standard error,

$$SE\left(\hat{A}_C\right) = A_{tot} \cdot \left[wt_1^2 \cdot \hat{Var}\left(\hat{p}_1\right) + wt_2^2 \cdot \hat{Var}\left(\hat{p}_2\right)\right] \tag{5}$$

where $wt_1$ is the proportion of the total map area in class C, $wt_2 = 1 - wt_1$, and $A_{tot}$ is the total area of interest. Approximate 95% confidence intervals for the class areas can be estimated as follows:

$$\hat{A}_C \pm 2 \cdot SE\left(\hat{A}_C\right). \tag{6}$$

**Table 5.** Confusion matrix.

| Map Class | Reference Class | | Total | UA * | $\hat{p}_h$ | $\hat{Var}(\hat{p}_h)$ |
|---|---|---|---|---|---|---|
| | C | ~C | | | | |
| C | $n_{11}$ | $n_{12}$ | $n_{1.} = n_{11} + n_{12}$ | $ua_1 = \frac{n_{11}}{n_{1.}}$ | $\hat{p}_1 = \frac{n_{11}}{n_{1.}}$ | $\hat{Var}\left(\hat{p}_1\right) = \frac{\hat{p}_1 \cdot (1-\hat{p}_1)}{n_{1.}}$ |
| ~C | $n_{21}$ | $n_{22}$ | $n_{2.} = n_{21} + n_{22}$ | $ua_2 = \frac{n_{22}}{n_{2.}}$ | $\hat{p}_2 = \frac{n_{21}}{n_{2.}}$ | $\hat{Var}\left(\hat{p}_2\right) = \frac{\hat{p}_2 \cdot (1-\hat{p}_2)}{n_{2.}}$ |
| Total | $n_{.1} = n_{11} + n_{21}$ | $n_{.2} = n_{12} + n_{22}$ | | | | |
| PA * | $pa_1 = \frac{n_{11}}{n_{.1}}$ | $pa_2 = \frac{n_{22}}{n_{.2}}$ | | | | |

* UA = user's accuracy, * PA = producer's accuracy.

## 3. Results

### 3.1. Classifiers

#### 3.1.1. Multinomial Logistic Regression (MLR)

The parameters of the multinomial logistic regression model (Equations (1) and (2)) were estimated using the *multinom* function of the R packages [83]. All variables (spectral values of all bands of the image set in the analysis) were included in the model. The log-likelihood stabilized after 100 iterations. The importance of the variables was quite similar among different Sentinel-2 datasets. For the dry season image (IMG 1) and for the all-month image (IMG 3), the most important variables were Blue and SWIR 2, otherwise, the variable importance values were approximately the same. For the rainy season image (IMG 2), the results were also similar, although the Blue and SWIR 2 importance values were slightly less than for IMG 1 and IMG 3. Differences among importance values were small for the two-season image (IMG 4).

#### 3.1.2. Improved k-NN (ik-NN)

The improved k-NN (ik-NN) algorithm was applied as described in [45], except that only overall accuracy was used in the fitness function. The value of k = 10 was used. For the genetic algorithm, the number of the generations was 60, the population and medi-population sizes were 50, and the maximum number of the random iterations was 4000. Otherwise, the genetic algorithm parameters were as reported by Tomppo et al. (2009) [45]. Because genetic algorithms as a heuristic optimization method may select a local optimum, several trial runs were used to find a near optimal solution. Pixel-level estimates can be readily calculated with ik-NN when the weights of the variables have been optimized. The importance for the different variables was similar for ik-NN and MLR. However, in the case of predictor variables with large correlations, caution should be used when drawing conclusions with either method.

### 3.1.3. Support Vector Machine (SVM)

With the SVM algorithm using the RBF kernel, determination of the optimal cost (C) and Gamma parameters for the model is important. Following Qian et al. (2015) [84] and using our actual dataset, the R function 'tune()' was used to select these two SVM parameters. The optimal cost (C) value was determined from the values: $2^{-1}, 2^0, 2^1, 2^2, 2^3, 2^4, 2^5, 2^6, 2^7, 2^8$, and the Gamma ($\gamma$) value was a free parameter set from 0.1 to 2. The optimal parameters were determined based on classification error. Figure 3 describes the performance of the SVM model using the different cost and Gamma parameters. The darker the blue area, the better the performance of the model presented.

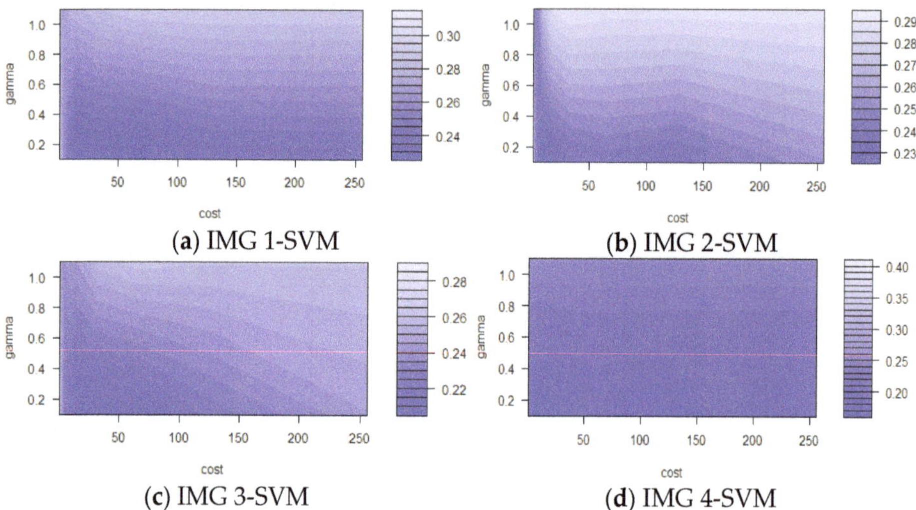

**Figure 3.** Parameter tuning of support vector machine: (**a**) IMG1-SVM; (**b**) IMG 2-SVM; (**c**) IMG 3-SVM; (**d**) IMG 4-SVM.

For the IMG 2-SVM combination and the IMG 4-SVM combination, the optimal C of $2^3$ and Gamma of 0.1 produced classification errors of 0.2283 and 0.1525, respectively, while for the IMG 1-SVM and IMG 3-SVM combinations, the optimal C of $2^5$ and $2^6$, both with Gamma of 0.1, produced classification errors of 0.2212 and 0.2145, respectively.

### 3.1.4. Random Forests (RF)

The three RF parameters, ntree, mtry, and variable importance, play important roles in classification. The algorithm assesses the importance of each variable in the classification process by means of a specific measure. 'Importance()' and 'varImplot()' functions were used to determine the MDA values and to select the potential variables that were actually needed for the optimal RF models. Figure 4 shows the variable importance ranked in the direction of decreasing MDA from right to left for the four seasonal images. The selection of variables was based on MDA using the backward selection method [85]. With this method, the algorithm starts with all predictor variables and then sequentially removes some variables until the greatest accuracy is achieved. Accordingly, the least MDAs were attributed to two bands of IMG 1, 2, and 3: specifically, B6 and B8 for IMG 1 and B6 and B7 for both IMG 2 and IMG 3. For IMG 4, the five bands were included: B5a, B2a, B2b, B7a, and B8b. In addition, the NIR band reduced the accuracy for all images.

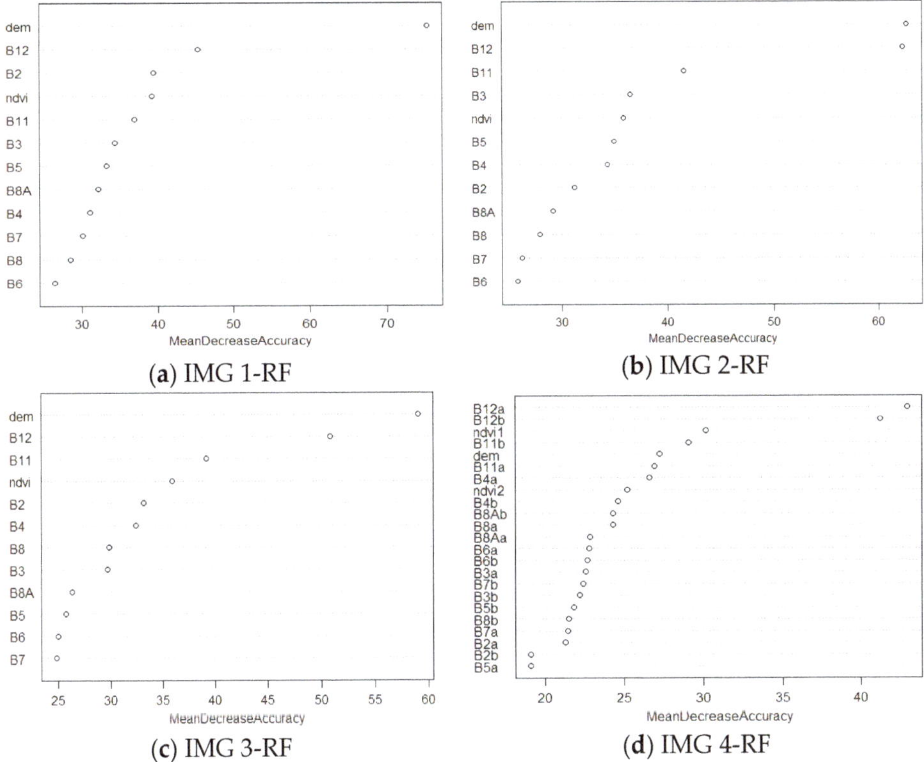

(**a**) IMG 1-RF

(**b**) IMG 2-RF

(**c**) IMG 3-RF

(**d**) IMG 4-RF

**Figure 4.** Ranking the variable importance measurement (bands): (**a**) IMG 1-RF; (**b**) IMG 2-RF; (**c**) IMG 3-RF; (**d**) IMG 4-RF.

The number of variables used for splitting at each node (mtry) was determined using the tuneRF function based on the variable importance and the number of trees (ntree). On the basis of the OOB error estimation, the optimum ntree and mtry parameters were chosen for the models.

Figure 5 shows the OOB errors when the model was run with ntree ranging from 1 to 500 trees. The smallest OOB errors were obtained for ntree = 500 trees for all seasonal image combinations.

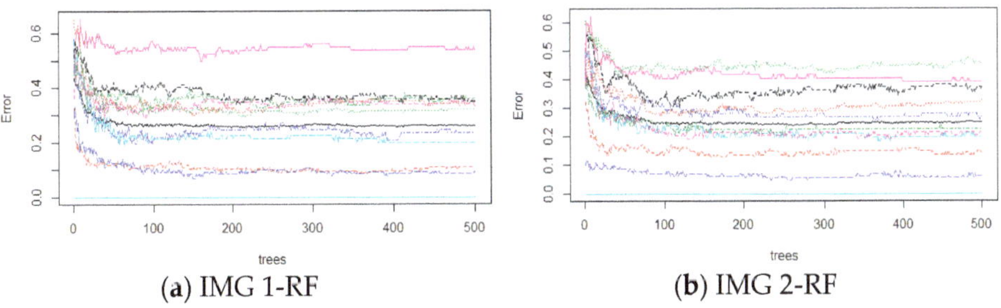

(**a**) IMG 1-RF

(**b**) IMG 2-RF

**Figure 5.** *Cont.*

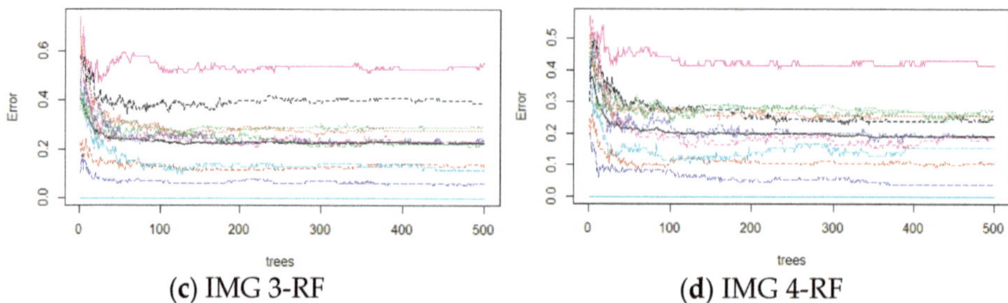

**(c)** IMG 3-RF    **(d)** IMG 4-RF

**Figure 5.** Out-of-bag error rate (OOB) errors versus ntree values: (**a**) IMG 1-RF; (**b**) IMG 2-RF; (**c**) IMG 3-RF; (**d**) IMG 4-RF.

The OOB errors associated with different mtry values are shown in Figure 6. The smallest OOB error was obtained with mtry = 3 for the IMG 2/RF and IMG 3/RF combinations, with mtry = 6 for the IMG 1/RF combination, and with mtry = 4 for the IMG4/RF combination.

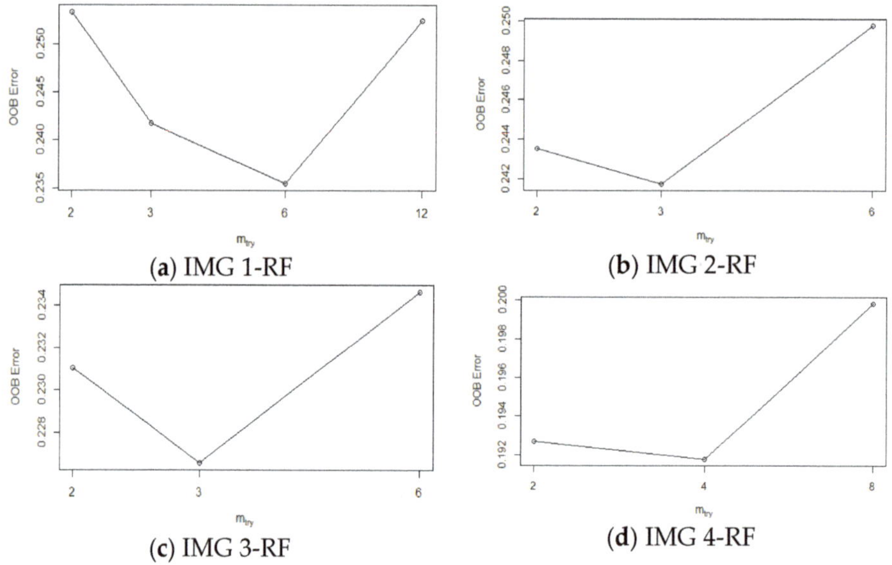

**(a)** IMG 1-RF    **(b)** IMG 2-RF

**(c)** IMG 3-RF    **(d)** IMG 4-RF

**Figure 6.** The OOB error of the model based on mtry parameter: IMG 1-RF; (**b**) IMG 2-RF; (**c**) IMG 3-RF; (**d**) IMG 4-RF.

*3.2. Analyses*

### 3.2.1. Accuracy of Classification Results

OA, K, PA, and UA for each class for the different combinations of image groups and classifiers are reported Table 6 and a comparison of the results is reported in Figure 7. In general, relatively large accuracies were found with OA >60% and K >0.6. Of interest, the IMG 4 composite of rainy and dry images produced the greatest accuracies for all classifiers. By contrast, the rainy season IMG 2 images produced the smallest accuracies. Classification accuracies for IMG 1 and IMG 3 were less than for IMG 4, but greater than for IMG 2.

**Table 6.** Accuracy estimates: OA = overall accuracy, K = Kappa, PA = producer's accuracy, UA = user's accuracy, $\hat{A}$ = class area estimate (km²), SE($\hat{A}$) = standard error of class area estimate.

| Image | Classifier | OA | Kappa | Accuracy | Land Class * | | | | | | | | | | |
|---|---|---|---|---|---|---|---|---|---|---|---|---|---|---|---|
| | | | | | 1 | 2 | 3 | 4 | 5 | 6 | 7 | 8 | 9 | 10 | 11 |
| IMG 1 | MLR | 68.3 | 0.657 | PA | 97.60 | 35.50 | 70.00 | 39.40 | 0.000 | 47.70 | 91.40 | 47.10 | 48.60 | 100.00 | 89.30 |
| | | | | UA | 58.50 | 66.70 | 66.70 | 80.00 | 0.000 | 68.80 | 75.90 | 70.00 | 100.00 | 85.70 | 61.50 |
| | | | | $\hat{A}$ | 821.72 | 932.59 | 502.16 | 378.25 | 241.3 | 456.51 | 1923.88 | 502.16 | 202.17 | 91.3 | 469.56 |
| | | | | SE($\hat{A}$) | 104.35 | 156.52 | 97.82 | 97.82 | 84.78 | 110.87 | 195.65 | 123.91 | 45.65 | 13.04 | 71.74 |
| | Ik-NN | 72.1 | 0.732 | PA | 98.20 | 55.40 | 81.90 | 45.10 | 32.50 | 38.30 | 93.60 | 28.20 | 31.60 | 100.00 | 87.70 |
| | | | | UA | 80.00 | 65.20 | 80.80 | 80.00 | 60.00 | 84.60 | 68.60 | 91.70 | 91.70 | 92.30 | 62.10 |
| | | | | $\hat{A}$ | 886.94 | 808.68 | 404.34 | 541.29 | 202.17 | 463.04 | 1878.23 | 456.51 | 189.13 | 97.82 | 593.47 |
| | | | | SE($\hat{A}$) | 78.26 | 143.48 | 78.26 | 123.91 | 71.74 | 130.43 | 208.69 | 123.91 | 52.17 | 6.52 | 104.35 |
| | RF | 67.7 | 0.67 | PA | 92.20 | 64.00 | 62.10 | 42.10 | 46.90 | 18.90 | 89.90 | 38.40 | 42.00 | 100.00 | 87.80 |
| | | | | UA | 77.40 | 63.00 | 85.70 | 81.80 | 66.70 | 100.00 | 59.50 | 85.70 | 91.70 | 92.30 | 58.10 |
| | | | | $\hat{A}$ | 886.94 | 763.03 | 502.16 | 515.21 | 182.61 | 697.81 | 1689.10 | 397.82 | 215.21 | 104.35 | 567.38 |
| | | | | SE($\hat{A}$) | 104.35 | 123.91 | 104.35 | 123.91 | 58.69 | 163.04 | 215.21 | 97.82 | 52.17 | 6.52 | 104.35 |
| | SVM | 73.2 | 0.748 | PA | 94.90 | 54.30 | 73.60 | 47.10 | 26.80 | 36.20 | 94.40 | 42.80 | 46.50 | 100.00 | 84.80 |
| | | | | UA | 76.70 | 64.00 | 87.00 | 100.00 | 60.00 | 78.60 | 70.60 | 92.30 | 92.90 | 92.30 | 63.00 |
| | | | | $\hat{A}$ | 880.42 | 886.94 | 404.34 | 456.51 | 189.13 | 397.82 | 1956.49 | 463.04 | 169.56 | 104.35 | 613.03 |
| | | | | SE($\hat{A}$) | 91.3 | 163.04 | 84.78 | 110.87 | 71.74 | 117.39 | 215.21 | 104.35 | 52.17 | 6.52 | 110.87 |
| IMG 2 | MLR | 63.9 | 0.611 | PA | 67.80 | 37.70 | 71.80 | 33.70 | 8.80 | 96.30 | 85.30 | 39.50 | 41.40 | 100.00 | 86.20 |
| | | | | UA | 54.50 | 58.80 | 66.70 | 53.30 | 10.00 | 84.20 | 67.90 | 70.00 | 90.90 | 92.30 | 64.30 |
| | | | | $\hat{A}$ | 854.33 | 978.24 | 469.56 | 371.73 | 189.13 | 228.26 | 1813.01 | 645.64 | 280.43 | 104.35 | 586.95 |
| | | | | SE($\hat{A}$) | 150.00 | 189.13 | 71.74 | 84.78 | 71.74 | 26.09 | 215.21 | 136.95 | 65.22 | 6.52 | 104.35 |
| | Ik-NN | 54.3 | 0.673 | PA | 90.90 | 36.50 | 61.20 | 44.90 | 56.30 | 42.40 | 84.50 | 38.80 | 38.10 | 86.00 | 82.00 |
| | | | | UA | 74.20 | 80.00 | 62.10 | 85.70 | 83.30 | 85.70 | 51.20 | 81.30 | 92.30 | 91.70 | 63.00 |
| | | | | $\hat{A}$ | 854.33 | 1317.37 | 417.38 | 404.34 | 104.35 | 365.21 | 1643.45 | 436.95 | 182.61 | 91.3 | 704.33 |
| | | | | SE($\hat{A}$) | 104.35 | 202.17 | 91.3 | 84.78 | 39.13 | 110.87 | 228.26 | 110.87 | 58.69 | 13.04 | 123.91 |
| | RF | 57.5 | 0.712 | PA | 86.30 | 39.40 | 65.90 | 66.30 | 53.10 | 58.60 | 85.00 | 42.00 | 28.00 | 100.00 | 80.200 |
| | | | | UA | 78.60 | 68.80 | 75.00 | 91.70 | 62.50 | 87.50 | 58.30 | 100.00 | 91.70 | 85.70 | 56.700 |
| | | | | $\hat{A}$ | 913.03 | 1180.41 | 404.34 | 319.56 | 104.35 | 293.47 | 1760.84 | 547.82 | 195.65 | 91.3 | 717.38 |
| | | | | SE($\hat{A}$) | 110.87 | 202.17 | 84.78 | 45.65 | 39.13 | 84.78 | 228.26 | 117.39 | 65.22 | 13.04 | 136.95 |

**Table 6.** *Cont.*

| Image | Classifier | OA | Kappa | Accuracy | Land Class * | | | | | | | | | | |
|---|---|---|---|---|---|---|---|---|---|---|---|---|---|---|---|
| | | | | | 1 | 2 | 3 | 4 | 5 | 6 | 7 | 8 | 9 | 10 | 11 |
| | SVM | 68.4 | 0.717 | PA | 83.10 | 38.60 | 66.00 | 52.10 | 41.80 | 66.00 | 85.90 | 43.80 | 42.10 | 100.00 | 89.40 |
| | | | | UA | 67.70 | 64.70 | 72.00 | 81.80 | 80.00 | 88.20 | 63.60 | 100.00 | 92.90 | 85.70 | 64.30 |
| | | | | Â | 815.2 | 1180.41 | 436.95 | 345.65 | 130.43 | 247.82 | 1871.7 | 534.77 | 136.95 | 91.3 | 723.9 |
| | | | | SE(Â) | 104.35 | 208.69 | 91.3 | 65.22 | 45.65 | 78.26 | 228.26 | 123.91 | 52.17 | 13.04 | 117.39 |
| IMG 3 | MLR | 64.2 | 0.611 | PA | 73.20 | 36.00 | 69.70 | 29.40 | 6.90 | 96.20 | 84.70 | 50.80 | 48.40 | 100.00 | 85.40 |
| | | | | UA | 54.50 | 58.80 | 66.70 | 53.30 | 10.00 | 84.20 | 67.90 | 70.00 | 90.90 | 92.30 | 64.30 |
| | | | | Â | 939.11 | 971.72 | 430.43 | 384.78 | 221.74 | 254.34 | 1663.01 | 717.38 | 319.56 | 117.39 | 502.16 |
| | | | | SE(Â) | 156.52 | 182.61 | 71.74 | 97.82 | 78.26 | 26.09 | 202.17 | 136.95 | 71.74 | 6.52 | 97.82 |
| | Ik-NN | 66.9 | 0.684 | PA | 88.80 | 40.30 | 87.70 | 67.50 | 38.90 | 18.40 | 90.10 | 38.50 | 37.10 | 86.80 | 87.80 |
| | | | | UA | 75.90 | 60.00 | 69.00 | 78.60 | 100.00 | 54.50 | 60.00 | 81.30 | 85.70 | 91.70 | 70.80 |
| | | | | Â | 919.55 | 965.2 | 319.56 | 280.43 | 169.56 | 723.9 | 1754.32 | 443.47 | 182.61 | 97.82 | 658.68 |
| | | | | SE(Â) | 117.39 | 182.61 | 45.65 | 45.65 | 58.69 | 182.61 | 228.26 | 117.39 | 58.69 | 13.04 | 104.35 |
| | RF | 69.5 | 0.721 | PA | 90.90 | 51.20 | 84.80 | 56.10 | 46.80 | 13.70 | 91.40 | 39.60 | 45.50 | 100.00 | 100.00 |
| | | | | UA | 82.10 | 61.50 | 74.10 | 90.90 | 100.00 | 80.00 | 60.50 | 85.70 | 92.90 | 92.30 | 67.90 |
| | | | | Â | 913.03 | 893.46 | 313.04 | 352.17 | 176.08 | 743.46 | 1754.32 | 502.16 | 150 | 104.35 | 613.03 |
| | | | | SE(Â) | 104.35 | 163.04 | 45.65 | 78.26 | 52.17 | 176.08 | 221.74 | 117.39 | 45.65 | 6.52 | 78.26 |
| | SVM | 71.2 | 0.743 | PA | 92.20 | 42.60 | 86.00 | 68.40 | 50.60 | 25.80 | 94.50 | 37.10 | 43.20 | 88.90 | 97.70 |
| | | | | UA | 77.40 | 59.10 | 83.30 | 92.30 | 100.00 | 85.70 | 65.80 | 81.30 | 100.00 | 91.70 | 66.70 |
| | | | | Â | 886.94 | 1043.46 | 358.69 | 280.43 | 136.95 | 658.68 | 1819.53 | 469.56 | 169.56 | 117.39 | 586.95 |
| | | | | SE(Â) | 104.35 | 189.13 | 45.65 | 39.13 | 45.65 | 150 | 208.69 | 117.39 | 52.17 | 13.04 | 78.26 |
| IMG 4 | MLR | 65.9 | 0.611 | PA | 69.30 | 39.40 | 78.70 | 26.60 | 1.20 | 98.70 | 85.50 | 31.30 | 58.80 | 100.00 | 83.30 |
| | | | | UA | 54.50 | 58.80 | 66.70 | 53.30 | 10.00 | 84.20 | 67.90 | 70.00 | 90.90 | 92.30 | 64.30 |
| | | | | Â | 854.33 | 1030.42 | 560.86 | 345.65 | 195.65 | 456.51 | 1799.97 | 449.99 | 273.91 | 71.74 | 489.12 |
| | | | | SE(Â) | 156.52 | 189.13 | 84.78 | 84.78 | 71.74 | 45.65 | 215.21 | 130.43 | 45.65 | 6.52 | 97.82 |
| | Ik-NN | 74.3 | 0.781 | PA | 91.10 | 52.60 | 69.50 | 44.90 | 39.30 | 55.00 | 94.70 | 35.20 | 47.00 | 100.00 | 87.20 |
| | | | | UA | 75.00 | 81.30 | 90.00 | 92.90 | 57.10 | 87.50 | 67.60 | 93.30 | 92.90 | 92.30 | 70.80 |
| | | | | Â | 867.38 | 1036.94 | 449.99 | 449.99 | 123.91 | 280.43 | 2034.75 | 384.78 | 143.48 | 97.82 | 639.12 |
| | | | | SE(Â) | 110.87 | 176.08 | 84.78 | 130.43 | 52.17 | 84.78 | 228.26 | 104.35 | 45.65 | 6.52 | 104.35 |

**Table 6.** *Cont.*

| Image | Classifier | OA | Kappa | Accuracy | Land Class * | | | | | | | | | | |
|---|---|---|---|---|---|---|---|---|---|---|---|---|---|---|---|
| | | | | | 1 | 2 | 3 | 4 | 5 | 6 | 7 | 8 | 9 | 10 | 11 |
| | RF | 80 | 0.802 | PA | 89.40 | 61.90 | 78.40 | 68.90 | 46.00 | 77.20 | 95.10 | 34.20 | 33.20 | 100.00 | 98.10 |
| | | | | UA | 85.20 | 69.20 | 83.30 | 92.30 | 62.50 | 100.00 | 82.10 | 83.30 | 100.00 | 92.30 | 69.20 |
| | | | | Â | 945.63 | 952.16 | 391.3 | 280.43 | 130.43 | 189.13 | 2204.31 | 482.6 | 176.08 | 91.3 | 678.25 |
| | | | | SE(Â) | 110.87 | 163.04 | 58.69 | 52.17 | 52.17 | 32.61 | 195.65 | 136.95 | 58.69 | 13.04 | 84.78 |
| | SVM | 80.3 | 0.813 | PA | 89.40 | 63.30 | 77.80 | 70.20 | 42.20 | 81.00 | 95.10 | 39.60 | 39.30 | 100.00 | 97.50 |
| | | | | UA | 82.10 | 73.90 | 90.50 | 93.30 | 71.40 | 100.00 | 80.60 | 86.70 | 92.30 | 85.70 | 69.20 |
| | | | | Â | 932.59 | 971.72 | 391.3 | 358.69 | 123.91 | 234.78 | 2223.87 | 710.86 | 378.25 | 97.82 | 639.12 |
| | | | | SE(Â) | 113.48 | 163.04 | 58.69 | 104.35 | 52.17 | 39.13 | 208.69 | 182.61 | 143.48 | 13.04 | 84.78 |

* Class 1: dense evergreen; 2: open evergreen; 3: semi-evergreen; 4: dipterocarp; 5: plantation; 6: rubber; 7: industrial plants; 8: crop land; 9: residential. 10: water surface; 11: other land.

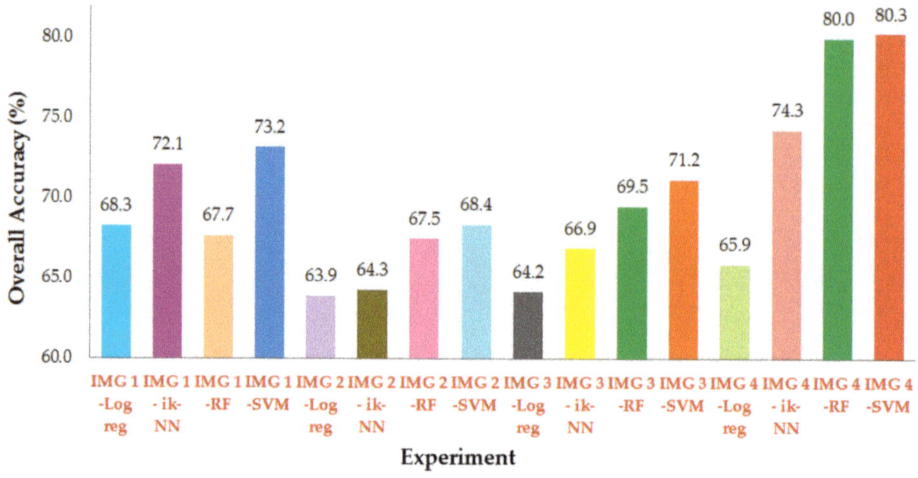


**Figure 7.** Overall accuracy for all combinations.

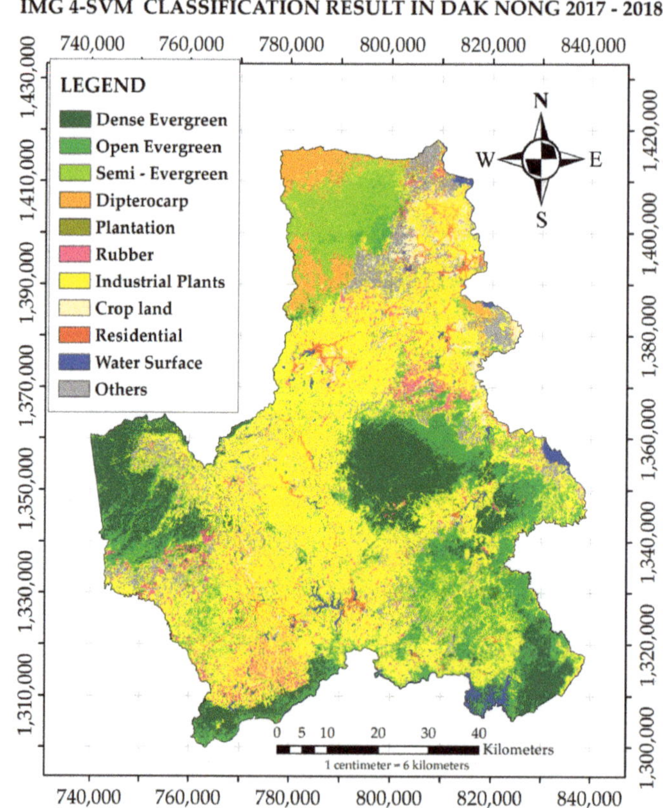


**Figure 8.** Land use and land cover (LULC) map produced for the most accurate classification combination.

The greatest accuracy was achieved for the composite two-season IMG 4 using the SVM classifier, specifically OA = 80.3% and Kappa = 0.813. The smallest accuracy was for IMG 2 with the MLR classifier. The differences between the greatest and smallest accuracies were 16.4% percentage points for OA and 0.202 for K. With respect to the classification algorithms, differences between the greatest and smallest accuracies for the four algorithms were 14.4% percentage points for OA and 0.202 for K. For SVM, the differences were 11.9% percentage points and 0.096 for K; for ik-NN, the differences were 10% percentage points for OA and 0.108 for K. The final LULC map was constructed using the classification for the most accurate IMG 4/SVM combination and is shown in Figure 8.

Average UA and PA estimates were greater than 60%, apart from a few cases, but differed by LULC class. For the forest cover classes, dense forest (1) had the greatest accuracy, while open forest (2) had the smallest accuracies for the four methods for most seasons (Figure 9).

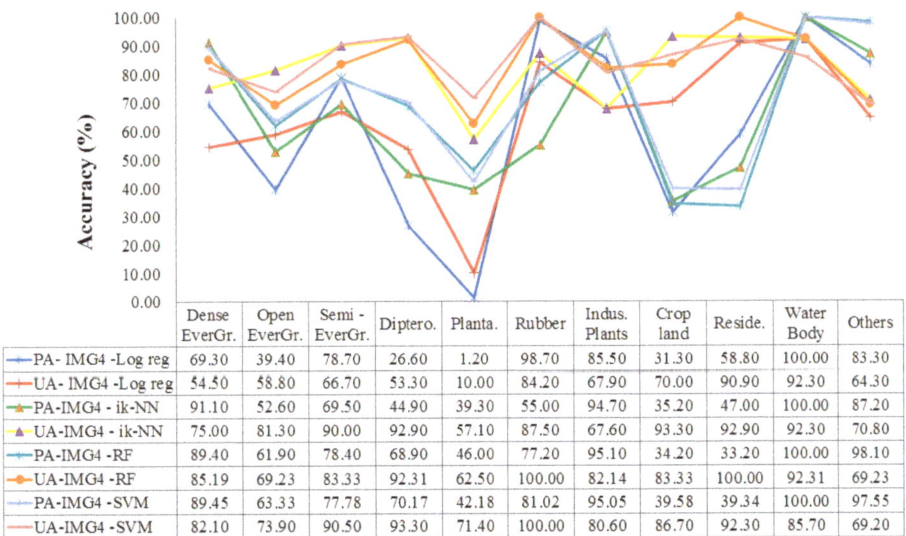

| | Dense EverGr. | Open EverGr. | Semi - EverGr. | Diptero. | Planta. | Rubber | Indus. Plants | Crop land | Reside. | Water Body | Others |
|---|---|---|---|---|---|---|---|---|---|---|---|
| PA- IMG4 -Log reg | 69.30 | 39.40 | 78.70 | 26.60 | 1.20 | 98.70 | 85.50 | 31.30 | 58.80 | 100.00 | 83.30 |
| UA- IMG4 -Log reg | 54.50 | 58.80 | 66.70 | 53.30 | 10.00 | 84.20 | 67.90 | 70.00 | 90.90 | 92.30 | 64.30 |
| PA-IMG4 - ik-NN | 91.10 | 52.60 | 69.50 | 44.90 | 39.30 | 55.00 | 94.70 | 35.20 | 47.00 | 100.00 | 87.20 |
| UA-IMG4 - ik-NN | 75.00 | 81.30 | 90.00 | 92.90 | 57.10 | 87.50 | 67.60 | 93.30 | 92.90 | 92.30 | 70.80 |
| PA-IMG4 -RF | 89.40 | 61.90 | 78.40 | 68.90 | 46.00 | 77.20 | 95.10 | 34.20 | 33.20 | 100.00 | 98.10 |
| UA-IMG4 -RF | 85.19 | 69.23 | 83.33 | 92.31 | 62.50 | 100.00 | 82.14 | 83.33 | 100.00 | 92.31 | 69.23 |
| PA-IMG4 -SVM | 89.45 | 63.33 | 77.78 | 70.17 | 42.18 | 81.02 | 95.05 | 39.58 | 39.34 | 100.00 | 97.55 |
| UA-IMG4 -SVM | 82.10 | 73.90 | 90.50 | 93.30 | 71.40 | 100.00 | 80.60 | 86.70 | 92.30 | 85.70 | 69.20 |

**Figure 9.** User's and producer's accuracies by class for the four classifiers using the IMG 4 combination of the rainy and dry Sentinel 2 satellite imagery.

### 3.2.2. Land Cover Class Area Estimates

Land cover class area estimates with corresponding standard errors are shown by class for the 16 seasonal image and prediction technique combinations in Table 6. Class area estimates ranged from slightly greater than 70 km$^2$ for class 10 for the IMG 4 and MLR combination to slightly greater than 2200 km$^2$ for class 7 for the IMG 4 and SVM combination. Standard errors ranged from approximately 6 km$^2$ to approximately 230 km$^2$, with larger standard errors associated with larger area estimates (Figure 10), although smaller ratios of standard errors to area estimates were associated with larger area estimates. Regardless of the seasonal image and prediction technique combination, the three classes with the greatest areas, in order, were class 7: perennial industrial plants, class 2: open evergreen broadleaved forest, and class 1: dense evergreen broadleaved forest. For IMG 1, IMG 2, and IMG 3, the sums of the estimated areas for these three classes as proportions of the total area ranged from slightly more than 0.50 to approximately 0.63, but with larger estimates for IMG 4.

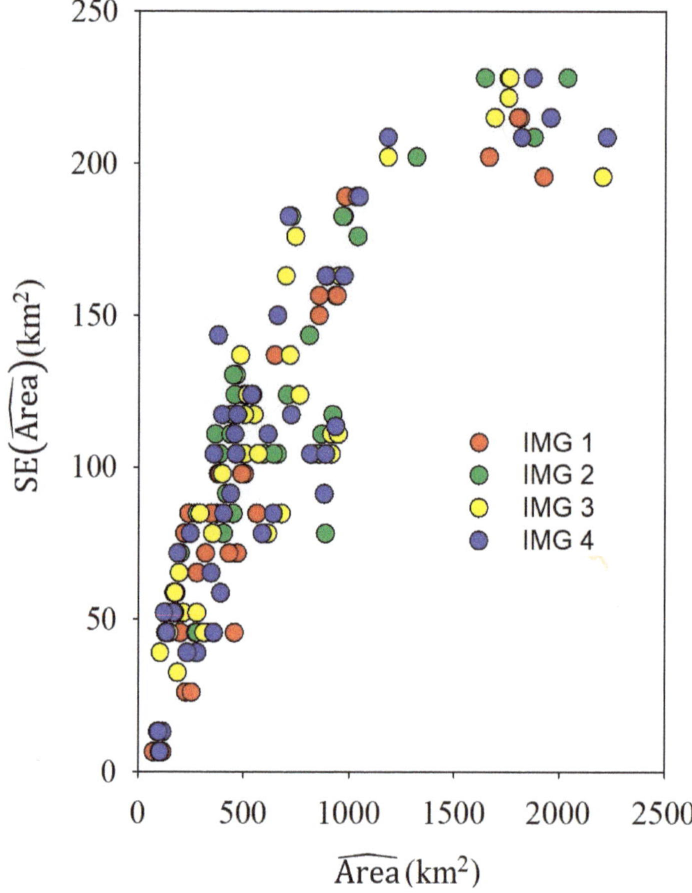



**Figure 10.** Standard errors (SE) (km²) of class area estimates versus area estimates (km²).

## 4. Discussion

Errors are present in any classification, estimation, or prediction [21,86–88]. Comparison of the results of this study and those of earlier studies is not straightforward because the numbers and definitions of the vegetation classes differ by study. Thus, optimality differs by study and user [21,86–88]. There are also no generally accepted limits on how accurate a classification should be to be characterized as reliable, because different users may have different concerns about accuracy. They may, for example, be interested in the accuracy for a specific class or in accuracy for areal estimates [89]. In addition, multiple factors influence classification accuracy: image quality, classifier, image composition, number and details of classes, and sample size.

Andersen et al. (1976) [90] recommended that accuracies of 85% for mapping land cover are acceptable. However, as Foody (2008) [91] noted, for many contemporary mapping applications, the challenge may be more difficult than assessed by Anderson et al. (1976) [90], particularly when attempting to distinguish among a large number of relatively detailed classes at a relatively local, large cartographic scale. Consequently, in such applications, the use of the 85% target suggested by Anderson et al. (1976) [90] may be inappropriate, as it may be unrealistically large.

Indeed, many studies have been conducted to select the most accurate classifier, either among those simultaneously evaluated or with classifiers evaluated in other studies. Such works reach no consensus, because the performance of a classifier always depends on the specific site characteristics, on the type and quality of the remotely sensed data, and also on the number and general aspects of the classes of interest [13]. Using the RF, SVM, maximum likelihood, and neural network classification algorithms to discriminate among four individual land cover classes based on two Landsat-8 OLI scenes, Lowe and Kulkarni (2015) [40] reported overall classification accuracies of 96.25%, 86.88%, 83.13%, and 76.87%, respectively. Kennedy et al. (2015) [41] used RF to classify Landsat time-series data from 1198 training patches for four classes (agriculture, forest, urbanization, and stream) and reported OA greater than 80%, but most successfully for the numerically well-represented forest management class. Meanwhile, Franco-Lopez (2001) [38] used k-NN to map 13 types of land cover using Landsat TM and achieved OA = 64%. Tomppo et al. (2008) [92] reported OA between 70% and 80% for classifying dominant tree species in one boreal forest test site in Finland when using two adjacent Landsat 7 ETM+ scenes and the ik-NN method. Pelletier et al. (2016) [18] used RF and SVM algorithms to classify SPOT-4 imagery and Landsat-8 HR-SITS images in southern France. The authors reported an OA of 83.3% for RF and 77.1% for SVM. Research by Phan and Kappas (2017) [20] showed different results among RF, SVM, and k-NN classifiers used to discriminate six types of LULC using Sentinel-2 image data in the Red River Delta of Vietnam. This research reported that SVM produced the greatest OA (95.29%) with the least sensitivity to the training sample sizes, followed consecutively by RF (94.44%) and k-NN (94.13%). These results indicate that no standard of accuracy is appropriate for all cases, because accuracy relevance depends on both the objective and the user.

Spatial information including remotely sensed data has been an excellent source of information for decision makers in forest management, albeit in conjunction with an understanding of classification uncertainties, whereby the probabilities of non-optimal and infeasible decisions are reduced. For this study, OA ranged from 63.9% to 80.3% (Figure 7) when using Sentinel 2 data to classify 11 LULC classes, with SVM producing the greatest accuracies. The difference between accuracies for the most accurate SVM classifier and the least accurate MLR classifier was approximately 14.4%. Although the results for SVM and RF were relatively similar, some authors recommend RF because training is less time-consuming and parameter selection is easier [18], a recommendation that was confirmed in our study.

Producer's and user's accuracies among the 11 LULC classes differed considerably (Figure 9). In general, the open evergreen forest classes were confused more than the other forest cover classes. This result is attributed to the heterogeneous conditions of natural tropical forests. In addition, forests in the study area have been disturbed to different degrees [21]. Among the forest classes, deciduous dipterocarp and semi-evergreen forest are considered the most challenging for remote sensing classification because of the seasonal deciduous characteristics of these forest types in the dry season [93]. However, this problem may be solved by using the combination of dry and rainy season images, as investigated in the present study.

The Sentinel-2 images acquired for different seasons (plant growth stages) produced different results. The greatest accuracies were for the composite rainy and dry season IMG 4; by contrast, the lowest accuracies were for the rainy season IMG 2. The observed reflectance varied by season owing to changes in the solar illumination geometry caused by the Earth's translation movement. In addition, the vegetation in the study area varies depending on the season, owing to the substantial rainfall differences for the two seasons. Sothe et al. (2017) [13] assert that differences in classification accuracies for the dry and rainy seasons can be attributed to the differences in solar illumination geometry between the two seasons. For images acquired in the dry season, the incident sun radiation arrives in a more perpendicular direction to the Earth's surface, thus reducing the shadow effect caused by topography and variations in the forest canopy height, and leading to greater pixel illumination. For the current study, there was a substantial increase in classification accuracies when using a composite of dry and rainy Sentinel 2 images (IMG 4). For the ik-NN, RF, and SVM classifiers, the greatest accuracies were obtained for the combined rainy and dry IMG 4 relative to the rainy or dry season alone (Table 6). The accuracy increase for the composite image may be explained by the fact that different seasons contain different information for the same kind of land cover (e.g., dipterocarp forest is deciduous in dry seasons and green in rainy seasons). Combining the two season's image bands captures additional information on land cover.

Among all combinations of images, classification algorithms, and land classes, the smallest SE for area estimates was for the water surface class owing to its stability, whereas the largest SE was for the industrial plant class. In fact, because cultivation characteristics of industrial plants in the study area are quite complex with a variety of species such as coffee, pepper, and cashew, all with uneven ages, large SEs are inevitable. This complexity also explains the large difference among area estimates for this class, ranging from 1643.45 km$^2$ to 2223.87 km$^2$, or from 25% to 34% of the total area (Figure 10).

Although classification accuracies for vegetation classes were not particularly large, the classifications are still useful for complex tropical rain forests that have been disturbed to different degrees such as in the Central Highlands of Vietnam. The area estimates and spatial distributions of the LULC classes produced from the current study will assist local authorities, managers, and other stakeholders in decision-making and planning regarding forest land cover and uses. The usual practice is for the Institution of Forest inventory and Planning (FIPI) to conduct a forest inventory and construct a forest map every five years. Local forest units such as Dak Nong receive the maps and update them manually. However, the accuracy of the map has usually not been announced, and inaccuracies and errors have been detected only by local forest staff when patrolling in the field. Moreover, LULC changes, particularly for industrial land, occur quickly and easily owing to factors such as unstable crop markets and increasing population resulting from migration. Thus, the results of this study will not only provide authorities with updated information on current conditions, but will also serve as a recommendation regarding methods for proactively updating LULC maps in a timely and costly manner. Specifically, timely and updated maps assist authorities by serving as a basis for formulating suitable solutions and policies for managing LULC including forest cover.

## 5. Conclusions

This research showed the utility of combining Sentinel-2, multi-spectral, and dry and rainy season band data when mapping LULCs in Dak Nong Province, Vietnam. The greatest accuracies were achieved for the composite IMG 4 obtained by combining dry and rainy season image sets using the SVM classifier.

Among the classifiers, SVM produced the greatest accuracies, although RF, which had similar accuracies, was simpler to train and apply, and was less computationally intensive. For IMG 4, the greatest accuracies with SVM were OA = 80.3% and Kappa index = 0.813; for RF, the greatest accuracies were OA = 80.0% and K = 0.802. Thus, the combination of dry and rainy season imagery used with the SVM or RF may contribute to potentially new ways for classifying the complex tropical forest of Vietnam and similar areas. The area estimates and spatial distributions of the LULC classes produced from the current study will assist local authorities, managers, and other stakeholders in decision-making and planning regarding forest land cover and uses.

In conclusion, the two-season, multi-spectral Sentinel-2 images provided useful data for classifying LULC classes in areas with substantial fragmentation, especially for natural forests that have been disturbed and degraded at different levels such as in Dak Nong, Vietnam. The SVM and RF machine learning algorithms were both accurate classifiers when used with the Sentinel 2 imagery. The methods developed for this study are applicable to boreal and temporal forests with different classes in addition to the tropical forests for the current study. However, the model parameters always need to be re-estimated for each application.

**Author Contributions:** Conceptualization, H.T.T.N., E.T., and R.E.M.; methodology, H.T.T.N., E.T., and R.E.M.; software, H.T.T.N., T.M.D., E.T., and R.E.M.; validation, H.T.T.N., E.T., R.E.M., and T.M.D.; formal analysis, H.T.T.N., R.E.M., E.T., and T.M.D.; investigation, T.M.D. and H.T.T.N.; resources, H.T.T.N.; data curation, H.T.T.N. and T.M.D.; writing—original draft preparation, H.T.T.N. and T.M.D.; writing—review and editing, R.E.M. and E.T.; visualization, T.M.D.; supervision, H.T.T.N.; project administration, H.T.T.N.; funding acquisition, H.T.T.N. All authors have read and agreed to the published version of the manuscript.

*Remote Sens.* **2020**, *12*, 1367

**Funding:** This research was funded by UNITED STATES AGENCY FOR INTERNATIONAL DEVELOPMENT, grant number AID-OAA-A-11-00012.

**Acknowledgments:** This work is part of the research project under the PEER program (Partnerships for Enhanced Engagement in Research), a U.S. government program to fund scientific research in developing countries. This is a program sponsored by USAID in partnership with several other U.S. Government agencies and administered by the U.S. National Academy of Sciences (NAS). The authors would like to thank all of the people involved in collecting field data for classification and validation. The authors thank also the editor and three anonymous reviewers for the constructive comments.

**Conflicts of Interest:** The authors declare no conflict of interest.

## References


1. Ministry of Agriculture and Rural Development. *Decision on the Declaration of Forest Status of the Country in 2018*; Decision No. 911/QD-BNN-LN; Ministry of Agriculture and Rural Development: Hanoi, Vietnam, 2019.
2. Thai, V.T. *Vietnamese Forest Vegetation*, 1st ed.; Science and Technique Publishing House: Hanoi, Vietnam, 1978.
3. Hoang, M.H.; Do, T.H.; van Noordwijk, M.; Pham, T.T.; Palm, M.; To, X.P.; Doan, D.; Nguyen, T.X.; Hoang, T.V.A. *An Assessment of Opportunities for Reducing Emissions from All Land Uses–Vietnam Preparing for REDD. Final National Report*; ASB Partnership for the Tropical Forest Margins: Nairobi, Kenya, 2010; p. 85.
4. Burkhard, B.; Kroll, F.; Nedkov, S.; Müller, F. Mapping ecosystem service supply, demand and budgets. *Ecol. Indic.* **2012**, *21*, 17–29. [CrossRef]
5. Gebhardt, S.; Wehrmann, T.; Ruiz, M.A.M.; Maeda, P.; Bishop, J.; Schramm, M.; Kopeinig, R.; Cartus, O.; Kellndorfer, J.; Ressl, R.; et al. MAD-MEX: Automatic wall-to-wall land cover monitoring for the Mexican REDD-MRV program using all Landsat data. *Remote Sens.* **2014**, *6*, 3923–3943. [CrossRef]
6. Nguyen, T.T.H.; Doan, M.T.; Volker, R. Applying random forest classification to map land use/land cover using landsat 8 OLI. *Int. Soc. Photogramm. Remote Sens.* **2018**, *XLII-3/W4*, 363–367. [CrossRef]
7. Arnold, F.E.; van der Werf, N.; Rametsteiner, E. *Strengthening Evidence-Based Forest Policy-Making: Linking Forest Monitoring With National Forest Programmes*; Forestry Policy and Institutions Working; FAO: Rome, Italy, 2014; p. 33.
8. Khatami, R.; Mountrakis, G.; Stehman, S.V. A meta-analysis of remote sensing research on supervised pixel-based land cover image classification processes: General guidelines for practitioners and future research. *Remote Sens. Environ.* **2016**, *177*, 89–100. [CrossRef]
9. Jensen, J.R.; Cowen, D.C. Remote sensing of urban/suburban infrastructure and socioeconomic attributes. *Photogramm. Eng. Remote Sens.* **1999**, *65*, 611–622.
10. Deka, J.; Tripathi, O.P.; Khan, M.L. Study on land use/land cover change dynamics through remote sensing and GIS–A case study of Kamrup District, North East India. *J. Remote Sens. GIS* **2014**, *5*, 55–62.
11. Topaloğlu, H.R.; Sertel, E.; Musaoğlu, N. Assessment of classification accuracies of SENTINEL-2 and LANDSAT-8 data for land cover/use mapping. *Int. Arch. Photogramm. Remote Sens. Spat. Inf. Sci.* **2016**, *XLI-B8*, 1055–1059.
12. Gomez, C.; White, J.C.; Wulder, M.A. Optical remotely sensed time series data for land cover classification: A review. *Int. Soc. Photogramm.* **2016**, *116*, 55–72. [CrossRef]
13. Sothe, C.; Almeida, C.M.; Liesenberg, V.; Schimalski, M.B. Evaluating sentinel-2 and landsat-8 data to map sucessional forest stages in a subtropical forest in Southern Brazil. *Remote Sens.* **2017**, *9*, 838. [CrossRef]
14. Addabbo, P.; Focareta, M.; Marcuccio, S.; Votto, C.; Ullo, S.L. Contribution of sentinel-2 data for applications in vegetation monitoring. *Acta IMEKO* **2016**, *5*, 44–54. [CrossRef]
15. Song, X.; Yang, C.; Wu, M.; Zhao, C.; Yang, G.; Hoffmann, W.C.; Huang, W. Evaluation of sentinel-2a satellite imagery for mapping cotton root rot. *Remote Sens.* **2017**, *9*, 906. [CrossRef]
16. Li, J.; Roy, D.P. A global analysis of sentinel-2a, sentinel-2b and landsat-8 data revisit intervals and implications for terrestrial monitoring. *Remote Sens.* **2017**, *9*, 902.
17. Pirotti, F.; Sunar, F.; Piragnolo, M. Benchmark of machine learning methods for classification of a sentinel-2 image. *Int. Arch. Photogramm. Remote Sens. Spat. Inf. Sci.* **2016**, *XLI-B7*, 335–340. [CrossRef]
18. Pelletier, C.; Valero, S.; Inglada, J.; Champion, N.; Dedieu, G. Assessing the robustness of Random Forests to map land cover with high resolution satellite image time series over large areas. *Remote Sens. Environ.* **2016**, *187*, 156–168. [CrossRef]

19. Sharma, R.C.; Hara, K.; Tateishi, R. High-resolution vegetation mapping in japan by combining sentinel-2 and landsat 8 based multi-temporal datasets through machine learning and cross-validation approach. *Land* **2017**, *6*, 50. [CrossRef]

20. Phan, T.N.; Kappas, M. Comparison of random forest, k-nearest neighbor, and support vector machine classifiers for land cover classification using sentinel-2 imagery. *Sensors* **2018**, *18*, 20.

21. Nguyen, T.T.H. *Forestry Remote Sensing: Multi-Source Data in Natural Evergreen Forest Inventory in the Central Highlands of Vietnam*, 1st ed.; Lambert Academic Publishing: Saarbruecken, Germany, 2011; p. 165.

22. Manandhar, R.; Odeh, I.O.A.; Ancev, T. Improving the accuracy of land use and land cover classification of landsat data using post-classification enhancement. *Remote Sens.* **2009**, *1*, 330–344. [CrossRef]

23. Heydari, S.S.; Mountrakis, G. Effect of classifier selection, reference sample size, reference class distribution and scene heterogeneity in per-pixel classification accuracy using 26 Landsat sites. *Remote Sens. Environ.* **2018**, *204*, 648–658. [CrossRef]

24. Lu, D.; Weng, Q.A. Survey of image classification methods and techniques for improving classification performance. *Int. J. Remote Sens.* **2007**, *28*, 823–870. [CrossRef]

25. Waske, B.; Braun, M. Classifier ensembles for land cover mapping using multitemporal SAR imagery. *ISPRS J. Photogramm. Remote Sens.* **2009**, *64*, 450–457. [CrossRef]

26. Li, C.; Wang, J.; Wang, L.; Hu, L.; Gong, P. Comparison of classification algorithms and training sample sizes in urban land classification with Landsat Thematic Mapper imagery. *Remote Sens.* **2014**, *6*, 964–983. [CrossRef]

27. Abbas, A.W.; Minallh, N.; Ahmad, N.; Abid, S.A.R.; Khan, M.A.A. K-means and ISODATA clustering algorithms for landcover classification using remote sensing. *Sindh Univ. Res. J. (Sci. Ser.)* **2016**, *48*, 315–318.

28. Paola, J.D.; Schowengerdt, R.A. A detailed comparison of backpropagation neural network and maximum-likelihood classifiers for urban land use classification. *IEEE Trans. Geosci. Remote Sens.* **1995**, *33*, 981–996. [CrossRef]

29. Shafri, H.Z.M.; Suhaili, A.; Mansor, S. The performance of maximum likelihood, spectral angle mapper, neural network and decision tree classifiers in hyperspectral image analysis. *J. Comput. Sci.* **2007**, *3*, 419–423. [CrossRef]

30. Santos, J.A.; Ferreira, C.D.; Torres, R.D.S.; Gonalves, M.A.; Lamparelli, R.A.C. A relevance feedback method based on genetic programming for classification of remote sensing images. *Inf. Sci.* **2011**, *181*, 2671–2684. [CrossRef]

31. Ahmad, A.; Quegan, S. Analysis of maximum likelihood classification on multispectral data. *Appl. Math. Sci.* **2012**, *6*, 6425–6436.

32. McRoberts, R.E. A two-step nearest neighbors algorithm using satellite imagery for predicting forest structure within species composition classes. *Remote Sens. Environ.* **2009**, *113*, 532–545. [CrossRef]

33. McRoberts, R.E. Satellite image-based maps: Scientific inference or pretty pictures? *Remote Sens. Environ.* **2011**, *115*, 714–724. [CrossRef]

34. Pal, M.; Mather, P.M. Support vector classifiers for land cover classification. In Proceedings of the Map India Conference, New Delhi, India, 28–31 January 2003.

35. Huang, C.; Davis, L.S.; Townshend, J.R.G. An assessment of support vector machines for land cover classification. *Int. J. Remote Sens.* **2002**, *23*, 725–749. [CrossRef]

36. Bahari, N.I.S.; Ahmad, A.; Aboobaider, B.M. Application of support vector machine for classification of multispectral data. *IOP Conf. Ser. Earth Environ. Sci.* **2014**, *20*. [CrossRef]

37. Balcik, F.B.; Kuzucu, A.K. Determination of land cover/land use using spot 7 data with supervised classification methods. *Int. Arch. Photogramm. Remote Sens. Spat. Inf. Sci.* **2016**, *XLII-2/W1*, 143–146. [CrossRef]

38. Franco-Lopez, H.; Ek, A.R.; Bauer, M.E. Estimation and mapping of forest stand density, volume, and cover type using the k-nearest neighbors method. *Remote Sens. Environ.* **2001**, *77*, 251–274. [CrossRef]

39. Yu, S.; Backer, S.; Scheunders, P. Genetic feature selection combined with composite fuzzy nearest neighbor classifiers for hyperspectral satellite imagery. *Pattern Recognit. Lett.* **2002**, *23*, 183–190. [CrossRef]

40. Lowe, B.; Kulkarni, A. Multispectral image analysis using random forest. *Int. J. Soft Comput. (IJSC)* **2015**, *6*, 14. [CrossRef]

41. Kennedy, R.E.; Yang, Z.; Braaten, J.; Copass, C.; Antonova, N.; Jordan, C.; Nelson, P. Attribution of disturbance change agent from Landsat time-series in support of habitat monitoring in the Puget Sound region, USA. *Remote Sens. Environ.* **2015**, *166*, 271–285. [CrossRef]

42. Basten, K. *Classifying Landsat Terrain Images via Random Forests. Bachelor thesis Computer Science*; Radboud University: Nijmegen, The Netherlands, 2016.

43. Naidoo, L.; Cho, M.A.; Mathieu, R.; Asner, G. Classification of savanna tree species, in the Greater Kruger National Park region, by integrating hyperspectral and LiDAR data in a random forest data mining environment. *ISPRS J. Photogramm. Remote Sens.* **2012**, *69*, 167–179. [CrossRef]

44. Tomppo, E.; Halme, M. Using coarse scale forest variables as ancillary information and weighting of variables in k-NN estimation: A genetic algorithm approach. *Remote Sens. Environ.* **2004**, *92*, 1–20. [CrossRef]

45. Tomppo, E.; Gagliano, C.; De Natale, F.; Katila, M.; McRoberts, R. Predicting categorical forest variables using an improved k-nearest neighbor estimator and Landsat imagery. *Remote Sens. Environ.* **2009**, *113*, 500–517. [CrossRef]

46. Dharamvir. Object oriented model classification of satellite image. *CDQM* **2013**, *16*, 46–54.

47. Machala, M.; Zejdová, L. Forest mapping through object-based image analysis of multispectral and lidar aerial data. *Eur. J. Remote Sens.* **2014**, *47*, 117–131. [CrossRef]

48. Mora, A.; Santos, T.M.A.; Łukasik, S.; Silva, J.M.N.; Falcão, A.J.; Fonseca, J.M.; Ribeiro, R.A. Land cover classification from multispectral data using computational intelligence tools: A comparative study. *Information* **2017**, *8*, 147. [CrossRef]

49. Sowmya, B.; Sheelarani, B. Land cover classification using reformed fuzzy C-means. *Sadhana* **2011**, *36*, 153–165. [CrossRef]

50. Apte, K.S.; Patravali, D.S. Development of back propagation neutral network model for ectracting the feature from a satellite image using curvelet transform. *Int. J. Eng. Res. Gen. Sci.* **2015**, *3*, 226–236.

51. Ma, L.; Li, M.; Ma, X.; Cheng, L.; Du, P.; Liu, Y. A review of supervised object-based land-cover image classification. *ISPRS J. Photogramm. Remote Sens.* **2017**, *130*, 277–293. [CrossRef]

52. Meyfroidt, P.; Lambin, E.F.; Erb, K.H.; Hertel, T.W. Globalization of land use: Distant drivers of land change and geographic displacement of land use. *Curr. Opin. Environ. Sustain.* **2013**, *5*, 438–444. [CrossRef]

53. Bourgoin, C.; Oszwald, J.; Bourgoin, J.; Gond, V.; Blanc, L.; Dessard, H.; Phan, T.V.; Sist, P.; Läderach, P.; Reymondin, L.; et al. Assessing the ecological vulnerability of forest landscape to agricultural frontier expansion in the Central Highlands of Vietnam. *Int. J. Appl. Earth Obs. Geoinf.* **2020**, *84*, 13. [CrossRef]

54. Ha, T.V.; Tuohy, M.; Irwin, M.; Tuan, P.T. Monitoring and mapping rural urbanization and land use changes using Landsat data in the northeast subtropical region of Vietnam. *Egypt. J. Remote Sens. Space Sci.* **2020**, *23*, 11–19. [CrossRef]

55. Drusch, M.; Del Bello, U.; Carlier, S.; Colin, O.; Fernandez, V.; Gascon, F.; Hoersch, B.; Isola, C.; Laberinti, P.; Martimort, P.; et al. Sentinel-2: ESA's optical high-resolution mission for GMES operational services. *Remote Sens. Environ.* **2012**, *120*, 25–36. [CrossRef]

56. Vuolo, F.; Zółtak, M.; Pipitone, C.; Zappa, L.; Wenng, H.; Immitzer, M.; Weiss, M.; Baret, F.; Atzberger, C. Data service platform for Sentinel-2 surface reflectance and value-added products: System use and examples. *Remote Sens.* **2016**, *8*, 938. [CrossRef]

57. Yacouba, D.; Guangdao, H.; Xingping, W. Assessment of land use cover changes using NDVI and DEM in Puer and Simao Counties, Yunnan Province, China. *World Rural Obs.* **2009**, *1*, 1–11.

58. Defries, R.S.; Townshend, J.R.G. NDVI-derived land cover classifications at a global scale. *Int. J. Remote Sens.* **1994**, *15*, 3567–3586. [CrossRef]

59. Pettorelli, N.; Ryan, S.; Mueller, T.; Bunnefeld, N.; Jedrzejewska, B.; Lima, M.; Kausrud, K. The Normalized Difference Vegetation Index (NDVI): Unforeseen successes in animal ecology. *Clim. Res.* **2011**, *46*, 15–27. [CrossRef]

60. Housman, I.W.; Chastain, R.A.; Finco, M.V. An evaluation of forest health insect and disease survey data and satellite-based remote sensing forest change detection methods: Case studies in the United States. *Remote Sens.* **2018**, *10*, 21. [CrossRef]

61. Gorelick, N.; Hancher, M.; Dixon, M.; Ilyushchcnko, S.; Thau, D.; Moore, R. Google earth engine: Planetary-scale geospatial analysis for everyone. *Remote. Sens. Environ.* **2017**, *202*, 18–27. [CrossRef]

62. Housman, I.; Hancher, M.; Stam, C. A quantitative evaluation of cloud and cloud shadow masking algorithms available in Google Earth Engine. Manuscript in preparation. Unpublished work.

63. Roy, D.P.; Li, Z.; Zhang, H.K. Adjustment of sentinel-2 multi-spectral instrument (msi) red-edge band reflectance to nadir BRDF adjusted reflectance (NBAR) and quantification of red-edge band BRDF effects. *Remote Sens.* **2017**, *9*, 1325.

64. Roy, D.P.; Li, J.; Zhang, H.K.; Yan, L.; Huang, H. Examination of sentinel-2A multi-spectral instrument (MSI) reflectance anisotropy and the suitability of a general method to normalize MSI reflectance to nadir BRDF adjusted reflectance. *Remote Sens. Environ.* **2017**, *199*, 25–38. [CrossRef]

65. Soenen, S.A.; Peddle, D.R.; Coburn, C.A. SCS+ C: A modified sun-canopy-sensor topographic correction in forested terrain. *IEEE Trans. Geosci. Remote Sens.* **2005**, *43*, 2148–2159. [CrossRef]

66. Google Earth Engine. Developer's Guide. ImageCollection Reductions. Available online: https://developers. google.com/earth-engine/reducers_image_collection (accessed on 29 October 2017).

67. De Alban, J.D.T.; Connette, G.M.; Oswald, P.; Webb, E.L. Combined landsat and L-band SAR data improves land cover classification and change detection in dynamic tropical landscapes. *Remote Sens.* **2018**, *10*, 306. [CrossRef]

68. Gilat, D.; Hill, T.P. Quantile-locating functions and the distance between the mean and quantiles. *Stat. Neerl.* **1993**, *47*, 279–283. [CrossRef]

69. Google Earth Engine. Developer's Guide. Scale. Available online: https://developers.google.com/earth-engine/scale#scale-of-analysis (accessed on 29 October 2017).

70. Open Geo Blog–Tutorials, Code Snippets and Examples to Handle Spatial Data. Available online: https://mygeoblog.com/ (accessed on 28 October 2018).

71. Teluguntla, P.; Thenkabail, P.S.; Oliphant, A.; Xiong, J.; Gumma, M.K.; Congalton, R.G.; Kamini Yadav, K.; Huete, A. A 30-m landsat-derived cropland extent product of Australia and China using random forest machine learning algorithm on Google Earth Engine cloud computing platform. *ISPRS J. Photogramm. Remote Sens.* **2018**, *144*, 325–340. [CrossRef]

72. Särndal, C.-E.; Swensson, B.; Wretman, J. *Model Assisted Survey Sampling*; Springer: New York, NY, USA, 1992.

73. Chirici, G.; Mura, M.; McInerney, D.; Py, N.; Tomppo, E.O.; Waser, L.T.; McRoberts, R.E. A meta-analysis and review of the literature on the k-Nearest Neighbors technique for forestry applications that use remotely sensed data. *Remote Sens. Environ.* **2016**, *176*, 282–294. [CrossRef]

74. McRoberts, R.E.; Domke, G.M.; Chen, Q.; Næsset, E.; Gobakken, T. Using genetic algorithms to optimize k-Nearest Neighbors configurations for use with airborne laser scanning data. *Remote Sens. Environ.* **2016**, *184*, 387–395. [CrossRef]

75. Cortes, C.; Vapnik, V. Support-vector networks. *Mach. Learn.* **1995**, *20*, 273–297. [CrossRef]

76. Meyer, D.; Leisch, F.; Hornik, K. *Benchmarking Support Vector Machines*; Report Series No. 78; Vienna University of Economics and Business Administration Augasse 2–6, 1090: Wien, Austria, 2002.

77. Knorn, J.; Rabe, A.; Radeloff, V.C.; Kuemmerle, T.; Kozak, J.; Hostert, P. Land cover mapping of large areas using chain classification of neighboring Landsat satellite images. *Remote. Sens. Environ.* **2009**, *113*, 957–964. [CrossRef]

78. Breiman, L. Random forests. *Mach. Learn.* **2001**, *45*, 5–32. [CrossRef]

79. Liaw, A.; Wiener, M. Classification and regression by randomForest. *R News* **2002**, *2*, 18–22.

80. Congalton, R.G.; Green, K. *Assessing the Accuracy of Remotely Sensed Data: Principles and Practices*; Lewis Publishers: Boca Raton, FL, USA, 1999.

81. Cochran, W.G. *Sampling Techniques*, 3rd ed.; Wiley: New York, NY, USA, 1977.

82. Olofsson, P.; Foody, G.M.; Herold, M.; Stehman, S.V.; Woodcock, C.E.; Wulder, M.A. Good practices for estimating area and assessing accuracy of land change. *Remote Sens. Environ.* **2014**, *148*, 42–57. [CrossRef]

83. R Core Team. *R: A Language and Environment for Statistical Computing*; R Foundation for Statistical Computing: Vienna, Austria, 2019; Available online: https://www.R-project.org/ (accessed on 23 March 2020).

84. Qian, Y.; Zhou, W.; Yan, J.; Li, W.; Han, L. Comparing machine learning classifiers for object-based land cover classification using very high resolution imagery. *Remote Sens.* **2015**, *7*, 153–168. [CrossRef]

85. Sutter, J.M.; Kalivas, J.H. Comparison of forward selection, backward elimination, and generalized simulated annealing for variable selection. *Micro. J.* **1993**, *47*, 60–66. [CrossRef]

86. Brown, J.F.; Loveland, T.R.; Ohlen, D.O.; Zhu, Z. The global land-cover characteristics database: The user's perspective. *Photogramm. Eng. Remote Sens.* **1999**, *65*, 1069–1074.

87. Lark, R.M. Components of accuracy of maps with special reference to discriminant analysis on remote sensor data. *Int. J. Remote Sens.* **1995**, *16*, 1461–1480. [CrossRef]

88. Salovaara, K.J.; Thessler, S.; Malik, R.N.; Tuomisto, H. Classification of Amazonian primary rain forest vegetation using Landsat ETM+ satellite imagery. *Remote Sens. Environ.* **2005**, *97*, 39–51. [CrossRef]

89. Foody, G.M. Status of land cover classification accuracy assessment. *Remote Sens. Environ.* **2002**, *80*, 185–201. [CrossRef]
90. Anderson, J.R.; Hardy, E.E.; Roach, J.T.; Witmer, R.E. *A Land Use and Land Cover Classification System for Use with Remote Sensor Data*; U.S. Government Publishing Office: Washington, DC, USA, 1976.
91. Foody, G.M. Harshness in image classification accuracy assessment. *Int. J. Remote Sens.* **2008**, *29*, 3137–3158. [CrossRef]
92. Tomppo, E.; Katila, M.; Makisara, K.; Perasaari, J. *Multi-source National Forest Inventory: Methods and applications*; Springer: Dordrecht, The Netherlands, 2008.
93. Xie, Z.; Chen, Y.; Lu, D.; Li, G.; Chen, E. Classification of land cover, forest, and tree species classes with ziyuan-3 multispectral and stereo data. *Remote Sens.* **2019**, *11*, 164. [CrossRef]

 *remote sensing*

Article

# Predicting Forest Cover in Distinct Ecosystems: The Potential of Multi-Source Sentinel-1 and -2 Data Fusion

Kai Heckel [1,2,*], Marcel Urban [2], Patrick Schratz [3,4], Miguel D. Mahecha [5,6] and Christiane Schmullius [2]

[1] International Max Planck Research School (IMPRS) for Global Biogeochemical Cycles, Max Planck Institute for Biogeochemistry, Hans-Knoell-Str. 10, 07745 Jena, Germany
[2] Department for Earth Observation, Friedrich Schiller University, Grietgasse 6, 07743 Jena, Germany; marcel.urban@uni-jena.de (M.U.); c.schmullius@uni-jena.de (C.S.)
[3] GIScience Group, Friedrich Schiller University, Grietgasse 6, 07743 Jena, Germany; patrick.schratz@uni-jena.de
[4] Department of Statistics, Computational Statistics Group, Ludwig-Maximilian University, 80539 Munich, Germany
[5] Max Planck Institute for Biogeochemistry, Hans-Knoell-Straße 10, 07745 Jena, Germany; mmahecha@bgc-jena.mpg.de
[6] German Centre for Integrative Biodiversity Research (iDiv) Halle-Jena-Leipzig, 04103 Leipzig, Germany
* Correspondence: kai.heckel@uni-jena.de; Tel.: +49-3641-9-48975

Received: 17 December 2019; Accepted: 16 January 2020; Published: 17 January 2020

**Abstract:** The fusion of microwave and optical data sets is expected to provide great potential for the derivation of forest cover around the globe. As Sentinel-1 and Sentinel-2 are now both operating in twin mode, they can provide an unprecedented data source to build dense spatial and temporal high-resolution time series across a variety of wavelengths. This study investigates (i) the ability of the individual sensors and (ii) their joint potential to delineate forest cover for study sites in two highly varied landscapes located in Germany (temperate dense mixed forests) and South Africa (open savanna woody vegetation and forest plantations). We used multi-temporal Sentinel-1 and single time steps of Sentinel-2 data in combination to derive accurate forest/non-forest (FNF) information via machine-learning classifiers. The forest classification accuracies were 90.9% and 93.2% for South Africa and Thuringia, respectively, estimated while using autocorrelation corrected spatial cross-validation (CV) for the fused data set. Sentinel-1 only classifications provided the lowest overall accuracy of 87.5%, while Sentinel-2 based classifications led to higher accuracies of 91.9%. Sentinel-2 short-wave infrared (SWIR) channels, biophysical parameters (Leaf Area Index (LAI), and Fraction of Absorbed Photosynthetically Active Radiation (FAPAR)) and the lower spectrum of the Sentinel-1 synthetic aperture radar (SAR) time series were found to be most distinctive in the detection of forest cover. In contrast to homogenous forests sites, Sentinel-1 time series information improved forest cover predictions in open savanna-like environments with heterogeneous regional features. The presented approach proved to be robust and it displayed the benefit of fusing optical and SAR data at high spatial resolution.

**Keywords:** forest cover; Sentinel-1; Sentinel-2; data fusion; machine-learning; Germany; South Africa; temperate forest; savanna

---

## 1. Introduction

According to the Food and Agriculture Organization of the United Nations (FAO), approximately one-third of global land area is covered by forests, yet exhibiting a decreasing trend since the 1990's [1].

These estimates include decrease in forest area by deforestation and increase by afforestation. Forests are highly vulnerable ecosystems that are not only habitat to a large number of species and the most widely distributed terrestrial type of vegetation, but also act as a key control in the global carbon and water cycle, hence shaping land-atmosphere feedbacks [2–4]. Forests serve as essential sinks for carbon, storing approximately 202–275 PgC [5]. These two account for 82% of the global aboveground biomass carbon (ABC) stores if combined with savanna ecosystems (including woody savanna) [5]. Consequently, land cover changes in these areas are regarded as a major source of emission of greenhouse gases [6,7]. These figures indicate the importance of the monitoring and the related aboveground biomass (AGB) in these areas. The spatial assessment of the amount and related dynamics of woody AGB is crucial to project future developments and assess past changes, not only on the local or national level, but also the global level.

Remote sensing techniques have been applied for decades to foster sustainable forest monitoring. Since the 1970's, this was predominantly accomplished by using optical remote sensing systems due to their more extensive archive, accessibility, as well as straightforward interpretation as compared to microwave data [4,8,9]. Thus, forest cover monitoring using optical data from coarse to fine spatial resolution was conducted on the regional to global scale [10–13]. Nevertheless, the impact of persistent cloud cover/haze over many forested areas, leading to gaps in time series analysis, is a major limitation of optical remote sensing data. This problem can be remediated by the joint use of SAR and optical data [10]. In this context, it could be shown that using both, C- and L-Band SAR together with optical data is capable of significantly improving the forest monitoring results [14–16] and minimizing issues of data continuity [17].

With the start of Sentinel-1 and -2, which are acquiring data in the microwave and optical range of the electromagnetic spectrum, data in high geometrical and temporal resolution became freely available. Recent studies demonstrated the potential of Sentinel-2 data for the distinction of land cover classes and forest types with high accuracy when applying single time steps or multi-date information [18–20]. Similarly, various studies analyzed the suitability of C-Band radar data from Sentinel-1 to investigate land cover [15], forest extent [21], forest change [22], deforestation [23], and woody cover [24]. The fusion of both data sources was reported to increase the classification quality in numerous applications, such as crop [25], forest [26,27], and primary vegetation mapping [28]. The existence of more data sources and fast growing data archives naturally led to an increasing demand for machine-learning approaches that are capable of dealing with high-dimensional multi-source data for forest structure related applications [29,30]. A large number of studies utilized such algorithms in the past to map various land cover related metrics while using multi- and hyper-spectral [31,32] as well as radar data [33]. Similarly, these techniques were used to derive forest/land cover changes [6,34–36] and associated parameters, such as tree species [18,37,38] on different scales, woody cover assessments [39,40], or forest habitats [41]. Multiple studies investigated the potential of Sentinel-1 C-Band SAR data to improve the classification accuracies of optically based approaches of estimating tree cover [42] and characterizing forest ecosystems [17,43]. Further, Sentinel-1 proved to perform well in land cover mapping in heterogeneous landscapes, such as the South African savanna [44].

This study aims at deriving forest cover structures in the study sites in Germany and South Africa while using optical (Sentinel-2) and SAR (Sentinel-1) data in independent and joint approaches while using an innovative CV procedure that takes spatial autocorrelation of the data into consideration.

## 2. Materials

### 2.1. Study Sites

The first study site is located in the federal state of Thuringia and is visualized in Figure 1. Located in the center of Germany, it has an area of approximately 16,200 km$^2$. Land cover predominantly consists of homogeneous coniferous and deciduous forests, as well as agricultural land. The area is characterized by regional climates that can be described as temperate oceanic, exhibiting warm summers with dry periods and cold winters. While the northern part of Thuringia, the 'Thuringian Basin', is one of the driest areas of Germany, rainfall accumulates up to 1500 mm per year in the Thuringian Forest, a mountain range located in the south-east of Thuringia. Precipitation is evenly distributed throughout the year [45].

The second study region, visualized in Figure 2, comprises an area ranging from forest plantations in the Mpumalanga province to the southern parts of Kruger National Park (KNP), covering an area of approximately 19,800 km$^2$. In contrast to the Thuringian study site, this site is highly heterogeneous with regard to land cover, the amount of intra- and interannual rainfall, as well as climatic conditions. While KNP is predominantly characterized by patchy patterns of loose aggregations of vegetation and large portions being covered by bare soil (vegetation growth is heavily dependent on seasonal effects), the elevated plateau in the west of the study area features dense forest structures. According to this, rainfall varies between 500 mm and more than 1000 mm for KNP and the outer areas, respectively [46].

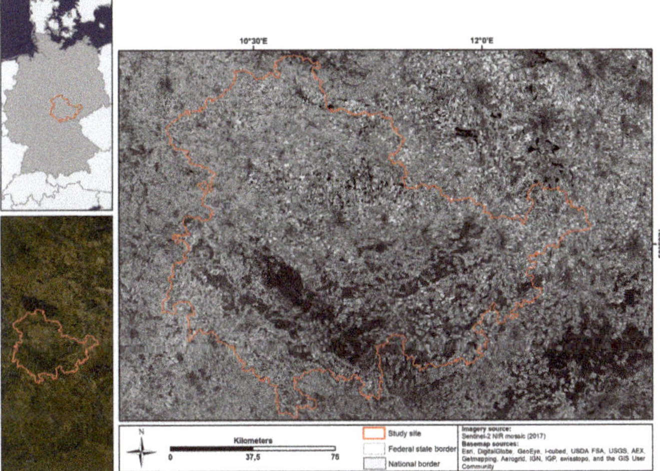

**Figure 1.** Study site in central Germany (Thuringia).

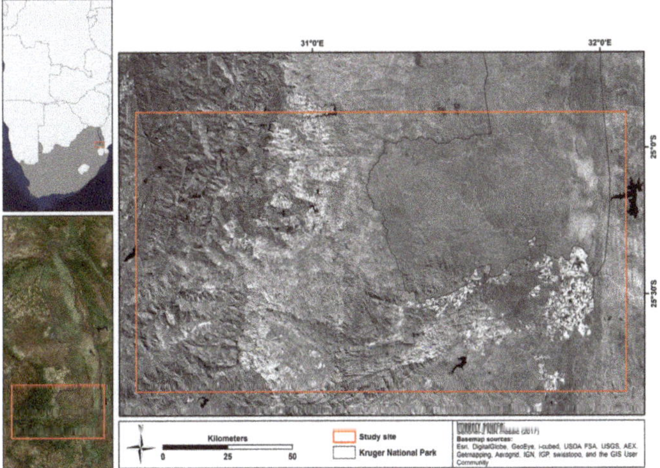

**Figure 2.** Study site in South Africa (southern KNP and forest plantations).

*2.2. Data*

### 2.2.1. Satellite Data

Data from Sentinel-1 and -2 were acquired for the study sites that are shown in Figures 1 and 2. Both data sources were collected in Level 1 format, which required several preprocessing steps prior to image analysis (see Section 3.1). Dual-polarized (VV + VH) Sentinel-1 backscatter intensity and Sentinel-2 data excluding the 60 m bands (B1, B9 + B10) were used in this study. Table 1 provides an overview of the used predictor variables. Data for both study sites were collected multi-temporally. Sentinel-1A data from 2015 to 2017 were used to calculate multi-temporal metrics for both study sites to capture the temporal dynamics of varying land cover types and, thus, compensating for the noticeable noise effects in the C-Band signal [47]. By using temporal features contrary to the application of individual time steps, it is possible to utilize the information of intra-annual trends within the data. To further analyze the impact of seasonality on the classification, these statistics were separated into winter and summer seasons for the SAR time series. As summer period, we selected the months June to September for Thuringia and November to March for the South African study site (predictor suffix = 'sum'). Winter was defined as the duration from December to March and May to September for Thuringia and South Africa, respectively (predictor suffix = 'win'). Only Sentinel-2A scenes with less than 5% cloud coverage were considered for a mosaic for each study site of the dry season 2016. In total, 127 Sentinel-1A, as well as three Sentinel-2A scenes, and 92 Sentinel-1A as well as four Sentinel-2A scenes, were collected for the Thuringia and the South African study site, respectively.

**Table 1.** List of predictor variables from Sentinel-1 and Sentinel-2.

| Sentinel-1 (per Polarization & Season) | Sentinel-2 |
| :---: | :---: |
| minimum | B2 |
| maximum | B3 |
| midhinge | B4 |
| standard deviation | B5 |
| range (95th, 5th percentile) | B6 |
| 5th percentile | B7 |
| 25th percentile | B8 |
| 75th percentile | B8A |
| 95th percentile | B11 |
|  | B12 |
|  | LAI |
|  | FAPAR |

### 2.2.2. Reference Data

Data provided by the State Office for Surveying and Geoinformation of Thuringia served as reference information for the extent of forests in Thuringia. The data originate from a digital land cover model (DLM) and comprise detailed information regarding the extent of forests as well as tree species on a Sentinel sub-pixel level (<10 m). Located south-east of the city of Jena, the 'Roda' reference site (visualized in Figure 3a), which was selected as training and validation subset for the classification algorithm, extends over an area of 261 km$^2$, of which approximately 50% is covered by coniferous (85%) and deciduous (15%) forest. The forest patch size in the Thuringian training and validation site varies between 0.1 ha and 4173 ha with a median of 1.4 ha.

As in any other protected landscape, freely available in situ data are scarce and difficult to acquire. Therefore, for the South African study site, two sets of reference data were combined to characterize the savanna ecosystem of KNP, as well as dense forest plantations west/north-west of the southern KNP (indicated in Figure 3b). Firstly, forest compartment data from York Timbers forestry company were used to define training areas for homogenous forests, as they can also be found within the German study site. For this reference data set, an age and NDVI threshold was applied to filter compartments that were covered by mature forests, and that were not logged during the Sentinel-2 acquisition dates and were also not altered throughout the Sentinel-1 time series period. Secondly, vegetation height metrics that were based on airborne Light Detection and Ranging (LIDAR) measurements were obtained from a canopy height model (CHM), which was made publicly available by Smit et al. [48]. Acquisition dates of this data set were April/May 2010, 2012 and 2014, respectively. For this study, exclusively, the vegetation structure from the last mission date was used to minimize the time gap to the Sentinel-1 acquisitions. The available LIDAR coverage is located on the southern boundary of KNP near Malelane Gate and it comprises an area of 1.8 km by 26 km with a spatial resolution of 2 m. In a next step, the vegetation height was limited to a value of more than 5 m according to FAO's forest definitions [49]. Additionally, we followed FAO's forest cover definition with a tree canopy cover of more than 10 percent and an area of more than 0.5 ha. The size of forest patches in the South African varies significantly from the distribution that can be found in the German study site. While forest patches with an area less than 0.01 ha could be found in the KNP, the forest plantations exhibited patch sizes of up to 9350 ha. The median value for the whole study site is less than 0.1 ha. For both reference sites, we extracted 500,000 data points in a stratified sampling approach to be used for hyperparameter tuning, training, and CV, which were carried out after a spatial partitioning.

**Figure 3.** Reference data sets for both study sites. (a) Thuringia and (b) South Africa.

## 3. Methods

The presented methodology comprises the preprocessing of microwave and optical Sentinel-2 data, acquisition of suitable reference information to characterize strongly different ecosystems as well as the introduction of the concept of spatial autocorrelation to perform tuning and validation procedures while taking spatial dependences in remotely sensed data sets into account. Based on this, forest extents were estimated for Sentinel-1, Sentinel-2, and a fused product, including both sensors. Figure 4 provides the overall workflow.

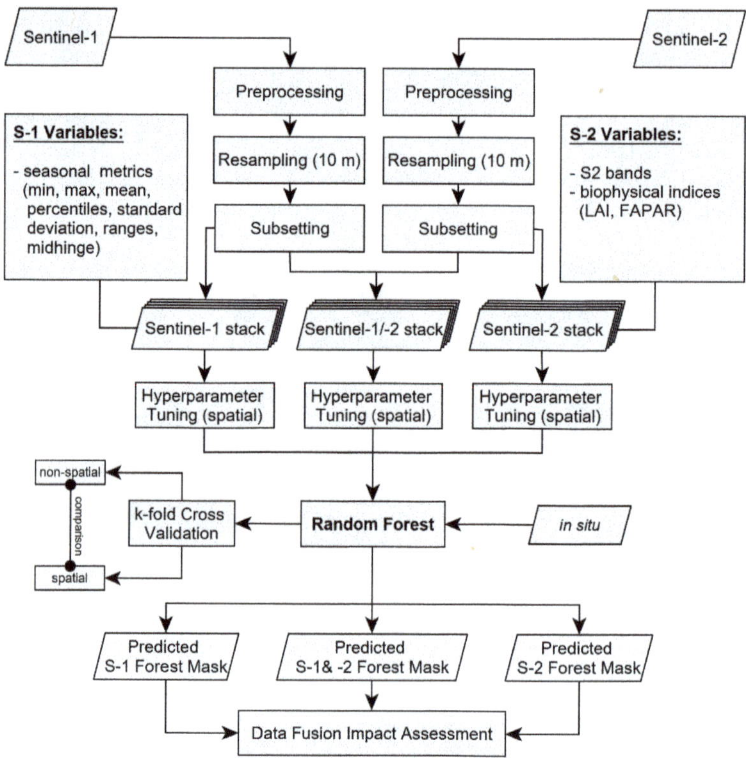


**Figure 4.** Workflow of the forest cover derivation using Sentinel-1 and Sentinel-2.

### 3.1. Preprocessing

Data from both satellites were gathered from ESA's Copernicus Open Access Hub archive. Preprocessing of Sentinel-1 data included multi-looking, geocoding, radiometric calibration, as well as topographic normalization, and it was solely carried out using Gamma routines. Dual-polarized multi-temporal microwave data sets were created at 10 m spatial resolution. Next, Sentinel-2 images were atmospherically, cirrus, and terrain corrected using the Sen2Cor algorithm [50] and Shuttle Radar Topography Mission (SRTM) with a spatial resolution of 1 arc-second ($\approx$30 m). During atmospheric correction, look-up-tables (LUT) for parameters, such as aerosol type and mid-latitude, were automatically chosen based on metadata information. For analysis, Sentinel-2 bands with a resolution of 60 m (channels 1, 9 & 10) were excluded, as their main purpose can be seen within the preprocessing to detect cloud coverage. After preprocessing, multi-temporal Sentinel-1 metrics and biophysical Sentinel-2 indices were calculated and grouped into seasons while using SNAP and R. The latter included the vegetation parameters LAI and FAPAR, which were calculated using the biophysical processor implemented in ESA's SNAP v6.0.0 [51]. As mentioned earlier, the data was split into seasons, with April and October serving as transition months. This split allowed for a more detailed analysis of the importance of predictor variables with respect to seasonality and the related environmental conditions. Finally, predictor variables from both sensors were stacked and subsetted to the respective study areas in R.

### 3.2. Forest Cover Derivation Using Random Forest

#### 3.2.1. Classifier Algorithm Description and Parameter Tuning

For this study, the decision tree classifier random forest (RF) was utilized while using the ranger implementation within the mlr package [52] of the statistical software R [53]. This non-linear, hierarchical ensemble classifier predicts class memberships that are based on the concept of recursive partitioning to create increasingly homogeneous subsets to retrieve a branched network of data splits [54,55]. Out of all possible splits, the predictor variables are selected (randomly by *mtry*) that minimize the Gini impurity (also referred to as 'splitrule'). Throughout the process, the decision tree is produced independently while being controlled by two main parameters, *mtry* and *ntree*. While *mtry* describes the number of predictor variables that are used to split each node, *ntree* defines the number of trees, which are generally characterized by high variance at relatively low bias [56]. The tuning of these two hyperparameters is crucial, as they control not only the accuracy, but also the computation time of the process [57].

By default, most RF implementations define the square root of the number of predictor variables (limited by number of predictor variables) as the *mtry* value and 500 as the sufficient number of trees (*ntree*). Few studies have investigated the impact of *mtry* towards computation time and the resulting model performance [58,59].

As a first step, the hyperparameters that control the performance of the model needed to be adjusted to retrieve the desired results while maintaining a certain computational effort. We used a repeated CV while using 25 repetitions with five folds to ensure the derivation of stable results, leading to consistent accuracies. Further, the procedure was conducted, including the correction of biases possibly introduced by spatial autocorrelation. For the previously mentioned hyperparameters, *mtry* and *ntree* separate feature spaces, retrieved from literature review, were defined, in which best parameter sets were then estimated [38,59]. Tuning of the variables is crucial for allowing the user the investigation of best performing parameter setups and their related computational costs. Feature spaces were limited *a priori* to keep the amount of processing within bounds that meet existent computation resources. Nevertheless, it should be noted that the hyperparameter tuning, theoretically, should optimally not be restricted to fixed value ranges, since this does require expert knowledge of the process as well as the applied training data. While *ntree* was tuned between 50 and 750, the limits of *mtry* were set between 1 and 5. According to Belgiu and Dragut [29], numerous studies revealed that 500 trees are leading to stable accuracies, which implies that the upper boundary is likely to be found in this value range. Additionally, it was found that the number of trees did not tend to be very sensitive to the prediction outcome when applying high-dimensional Sentinel-1/-2 data in combination to derive forest cover [27]. The *mtry* parameter was limited to only a narrow portion of the value domain, which represents the number of predictor variables within the data set. By default, the RF algorithm sets the *mtry* parameter to the square root of the number of input variables [30]. However, studies have shown that the optimal *mtry* value can be found below this value [60]. Therefore, we set the upper boundary of the search space for this hyperparameter lower than the proposed square root of the number of predictors.

#### 3.2.2. Training and Prediction

Once optimal sets of hyperparameters have been defined for both study sites and the individual sensor setups ((i) Sentinel-1 (S-1), (ii) Sentinel-2 (S-2), (iii) Sentinel-1 + Sentinel-2 (S-12)), these were used to perform the training of the model while using the training data for the study site subsets. The trained models were then transferred to the respective study regions described earlier. The derived tuning parameters were also applied for tree cover estimation over the complete study site while assuming that the training data was sufficiently representing the regional landscape diversity. The results were further processed while using a sieving algorithm to meet the selected forest definition, which limits the tree aggregations to a minimum size of 0.5 ha and ten percent tree canopy coverage [49].

### 3.2.3. Importance of Predictor Variables

The analysis of the variable importance was carried out based on all datasets to identify which variables were found to be most useful for the distinction between forest and non-forest in each study site. Several filter methods were tested to assess each predictor variable's value for the classification, in order to find most distinctive input variables for the classification. The gain ratio (GR), which was selected to evaluate variable importance, is an entropy-based algorithm that identifies the weights of discrete attributes based on their correlation with a continuous target attribute. GR was utilized to visualize the contribution of each predictor variable and it represents an extension of information gain G, which calculates the average entropy of a single predictor entity A when a set of observations S is split into subsets of $S_i$. The entropy for data sets with C classes is calculated, as follows:

$$E = \sum_i^C p_i \, \log_2 p_i,\tag{1}$$

where $p_i$ is defined as the probability of the random selection of an element of class i. Consequently, the resulting information exhibits the decrease in entropy of each attribute [61]. The information gain G is then calculated, as given below:

$$G(S, A) = E(S) - E(S, A)\tag{2}$$

G is calculated for every attribute A and the sum of the entropy E with regard to the original set S is then compared to every subset. The predictor entity that maximizes the difference most is selected upon others, being defined as the variable importance within the predictor variable set. However, information gain G is biased whenever the features are branching more complex, as it does not take into account the size of branches as well as their quantity. By using an extension of the previously explained algorithm, the intrinsive information I of each split can be included [61]. The equation of I is given as:

$$I(S, \, A) = -\sum_i \frac{|S_i|}{|S|} \log\!\left(\frac{|S_i|}{|S|}\right)\tag{3}$$

The gain ratio GR then equals:

$$GR(S, \, A) = \frac{G(S, \, A)}{I(S, A)}\tag{4}$$

The results of the calculation of the variable importance are displayed separately for both study sites, respectively, and split up into each specific sensor setup in Section 4.2. Only the seven most important variables were considered and displayed in a combined plot to limit the number of selected predictors.

### 3.2.4. Cross-Validation and Statistical Comparison

A repeated CV was performed using approaches similar to the previously explained tuning algorithm to assess not only the expected classification accuracy, but also to account for spatial dependence in the validation. Within $k$-fold CV the reference information is spatially subdivided into $k$ equally sized partitions of which a single subsample is used for testing and the remaining $k - 1$ portions are used as training information. This process was then repeated $k$ times to make use of all available training data and estimated as an average to estimate a single prediction. Validation in this study was accomplished while using a five-fold approach using 25 repetitions. Following, the average error among all $k$ runs is computed, adding up to 125 RF trials for each sensor setup and study site. A spatial partitioning of the reference data was carried out to account for the existence of spatial autocorrelation, which is omnipresent in remotely sensed data. The term 'spatial portioning' implies the creation of spatially disjoint training and testing partitions while using a $k$-means clustering approach [62]. Typically, CV approaches are performed using random or stratified sampling methods [34,63,64]. However, these do not take the spatial dependence of observations in spatial data sets into account and therefore ignore the bias introduced by non-spatial sampling [65,66]. Thus, commonly adapted CV lacks in consistently generating training and test folds that are independent from each other [64].

A comparison with other FNF products was carried out to identify the quality of our forest classifications. Thus, our results could be compared to completely independent data sets. For this, 10,000 randomly stratified samples were extracted and then used to compute the Jaccard Similarity coefficient (J) between each other and with respect to the reference data sets for each study site [67]. J is commonly used in data science as a tool to measure similarity/dissimilarity in vectors containing binary values and it was found to provide reliable estimates while preventing overoptimistic results [68]. The Jaccard coefficient ranges between 0 and 1, with 0 representing the lowest possible similarity between quantities. This coefficient measures the similarity between two sets of data by dividing the size of the intersection between data set A and B by the size of the union of both sets.

$$J_{A,B} = \frac{|A \cap B|}{|A| + |B| - |A \cap B|} \tag{5}$$

For comparison, three different common FNF products with similar spatial resolution were chosen. Table 2 provides an overview of these data sets.

**Table 2.** Data sets used for similarity analysis.

| | Author | Pixel Size | Period | Data | Study Region |
|---|---|---|---|---|---|
| Landsat FNF | [10] | 30 m | 2000–2018 | Landsat | both |
| ALOS FNF | [69] | 25 m | 2017 | PALSAR-2 | both |
| Copernicus HRL * | [70] | 20 m | 2015 | Sentinel-2, Landsat, SPOT-5, ResourceSat-2 | Thu |
| CCI ** | [71] | 20 m | 2015–2016 | Sentinel-2 | SA |

Notes: * Copernicus High Resolution Layer (HRL)—Forest Type. ** ESA Climate Change Initiative (CCI) S2 Prototype Land Cover Map of Africa.

## 4. Results

### 4.1. Forest Cover Derivation

The final classification of forest cover was carried out using the hyperparameters $ntree = 500$ and $mtry = 1$, which were found to produce reliable results with high accuracy while maintaining a certain computational effort. Table 3, as well as Figure 5, display the accuracies for the CV of the RF based classifications of forested areas in both study sites using Sentinel-1, Sentinel-2 and a fused data product. The overall accuracy was calculated using the median value of all runs of the spatial CV, since this provides a more realistic representation of the actual classification quality, as outliers tended to appear especially in the South African study site.

**Table 3.** Averaged results (median) for each sensor setup from all CV repetitions.

| | Sentinel-1 | Sentinel-2 | Sentinel-1/-2 | |
|---|---|---|---|---|
| **South Africa** | 84.3 | 90.4 | 92.3 | 90.9 |
| **Thuringia** | 90.6 | 93.3 | 93.7 | 93.2 |
| | 87.5 | 91.9 | 93 | Ø |

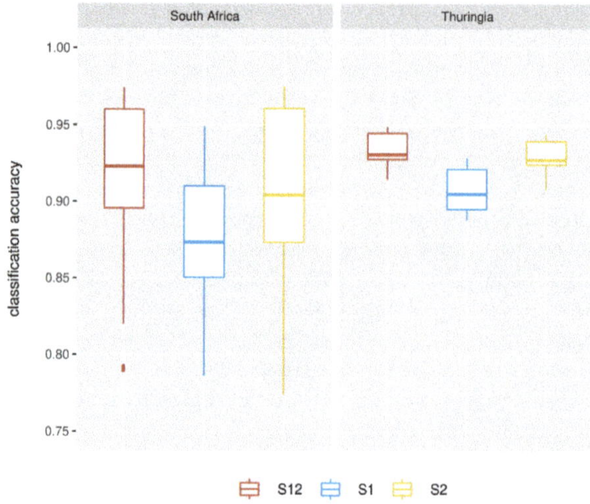

**Figure 5.** Results of CV for both study sites comparing all sensor setups.

Classification accuracies were found to be consistently higher in the German (93.2%) as compared to the South African study site (90.9%) when being averaged over all available sensor setups. Considerably strong variations between these configurations could be observed in both study sites. As visualized in Figure 5, the standard deviation, which was calculated over all CV runs, was significantly larger in the South African study site (0.33), while the results in Thuringia did not differentiate between all CV runs as much (0.012). The difference in the variance visible in the boxplot range can be attributed to the existence of more sufficient training data in the German study site, as well as the appearance of forests in this area, which is much more homogeneous when compared to the South African study site, making it easier for the algorithm to reliably detect forest based sites. It should be noted that the plot is cropped to a lower boundary of 0.75 for visualization reasons as very few values of the Sentinel-1 based classification in South Africa ranged around 0.2. In both study sites, Sentinel-2 tended to outperform Sentinel-1 in the single-sensor approach leading to higher classification accuracies. While the fusion of Sentinel-1 and Sentinel-2 led to the highest classification accuracy in the German site, the potentially lower overall classification accuracy in South Africa also impacted the fused approach.

Figure 6 displays the results of the classifications using all available combinations of sensor setups for the South African study site. Here, Sentinel-1 predicted a greater amount of forest extent in parts of the KNP as well as the area just outside the park (brown) when compared to setups utilizing optical data. It can also be seen that Sentinel-2 only classified few forests/tree aggregations in the reserve area while capturing forest plantations in the west of the KNP very well (dark green). The optical sensor also misclassified agricultural fields that were located south of the Kruger National Park. This was not visible in the multi-temporal SAR classification. Combining both classifications, this was found to positively impact the optical classification. The forest prediction using Sentinel-1 and Sentinel-2 showed a good fit with homogenous forests in the western part of the study area and an underestimation of forest in the Kruger National Park.

In the Thuringian study site, the algorithm achieved higher overall accuracies compared to the South African study area. Results are visualized in Figure 7. Here, it was found that the optical, microwave and the fused FNF classification produced quite similar results, which was also reflected in the accuracies (see Table 3). While the SAR based forest cover map (brown) showed a larger number of noisy and, thus, misclassified pixels, gaps in the optical data caused by cloud cover could be filled by applying data fusion prior to the classification.

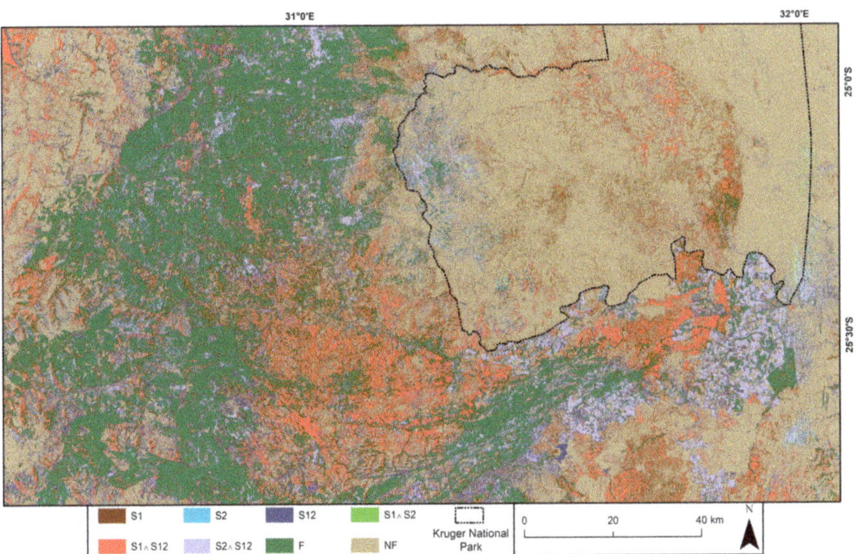

**Figure 6.** FNF map of the South African study site showing pixels being classified by Sentinel-1, Sentinel-2, a fused product (S12) and combinations of these setups; 'F' represents pixels classified as 'forest' by all sensor setups; 'NF' represents the pixels classified as 'non-forest' by all sensor setups.

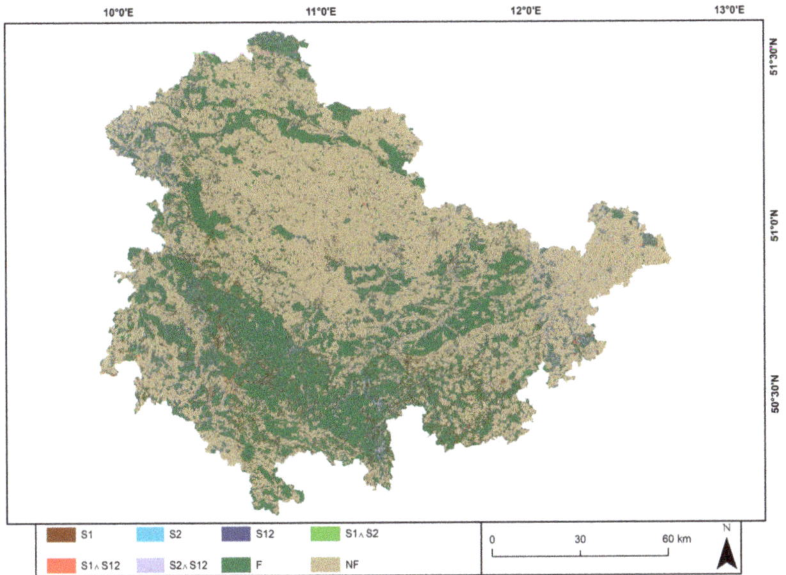

**Figure 7.** FNF map of Thuringia showing pixels being classified by Sentinel-1, Sentinel-2, a fused product (S12) and combinations of these setups; 'F' represents pixels classified as 'forest' by all sensor setups; 'NF' represents the pixels classified as 'non-forest' by all sensor setups.

The discrepancy between varying single sensor setups (microwave vs. optical) was found to be much stronger in the more heterogeneous landscape of the South African study site comprising savanna and forest ecosystems. Here, a sub-classification was performed dividing the study site into two parts, a) the forested areas spreading over from Mpumalanga to the province of Limpopo and b) the savannas of the southern KNP. The results suggested that both sensors produced reliable classifications in the western part of the study site while the savanna part and its vegetation was strongly underestimated by the Sentinel-2 predictors. In contrast, Sentinel-1 C-Band data were able to capture greater portions of tree aggregations in this part of the site. However, the Sentinel-1 setup might led to misclassifications in open savanna, due to limitations of the savanna LIDAR training data set to represent especially lower vegetation (<2 m) and the strong impact of soil moisture, as well as surface roughness. Forest sites with denser canopies were found to be underestimated by the RF classifier when applying SAR data only in both study sites. Here, closed tree canopies force a limited penetration of the C-Band signal. This corresponds to findings of other studies working in this area [39]. Nevertheless, Sentinel-2 outperformed Sentinel-1 only classifications in all study sites in terms of accuracy. This can partly be attributed to the internal variable selection of the RF algorithm that was based on the underlying training data. Consequently, the distinction of FNF in both study sites relied stronger on optical predictor variables, despite the information content stored in the SAR time series.

*4.2. Analysis of Variable Importance in Varying Sensor Setups*

### 4.2.1. Thuringia

Variable importance in both study areas was calculated based on the equations that are given in Section 3.2.3. As visualized in Figure 8, variables that represent the lower range of the SAR time series, such as the minimum or the 25th percentile of the summer season, were found to be the most distinctive predictors for the Sentinel-1 time series in this study site. Here, GR does not vary significantly between the seven most important variables, leading to comparable contributions to the classification result. As the minimum backscatter values throughout the summer months tend to occur towards the end of the growing season for C-Band similar to L-Band SAR data when the signal received from ground surfaces is increased, a better distinction can be found during this period of the year [72]. It was also found that polarization and orientation of the signal being received and emitted do strongly influence the detection of plant structure. Predictors with VH polarization showed to be the most relevant SAR variables for the classification in the German study site. This corresponds to findings of Olesk [22], who found that cross-polarized SAR data from Sentinel-1 is more suitable in most cases when detecting forests as compared to like-polarized microwave radiation.

The analysis of variable importance of the Sentinel-2 classification revealed that the biophysical parameters FAPAR and LAI, which are directly relatable to photosynthetic primary production and activity, exhibited the best differentiation for Thuringian temperate forests [73]. The latter proved to be the most important predictor for forest cover in this study site, which is due to the homogenous tree canopies that can be found in this study site with GR of above 0.3. SWIR (B11, B12), RGB (B2, B3, B4), as well as the shortest red edge band (B5) were also of great importance for the classification performance in the optical setup. Recent studies confirm our findings, which consider the SWIR bands to be among the most important Sentinel-2 channels for vegetation cover mapping [18,74]. The near-infrared (NIR) channels B8 and B8A were found to be only partly capable for a FNF distinction when being compared to the remaining channels of Sentinel-2, which does not correspond with other studies, as these wavelengths tend to be most representative for photosynthetic active vegetation [75].

Analysis of the combined predictor variables from Sentinel-1 and -2 showed that the GR of optical predictors clearly outperformed Sentinel-1 features in terms of variable importance, which was also visible in the final classification. Besides two SAR features (min_VH_sum and p25_VH_sum), optical variables or their derivatives were found to be the most dominant in the classification process and hence the prediction. As Sentinel-2 was favored over the SAR variables in the classification process, this effect could also be seen in the variable importance of the S12 sensor setup (Sentinel-1 and Sentinel-2). Here, no significant differences between the Sentinel-2 variables (SWIR and RGB) with the highest importance were observed, while LAI and FAPAR remained as the most important variables in the optical sensor setup. In general, the variables that were most distinctive in the single-sensor approach of the optical data were also found to be the most important ones for the fused approach.

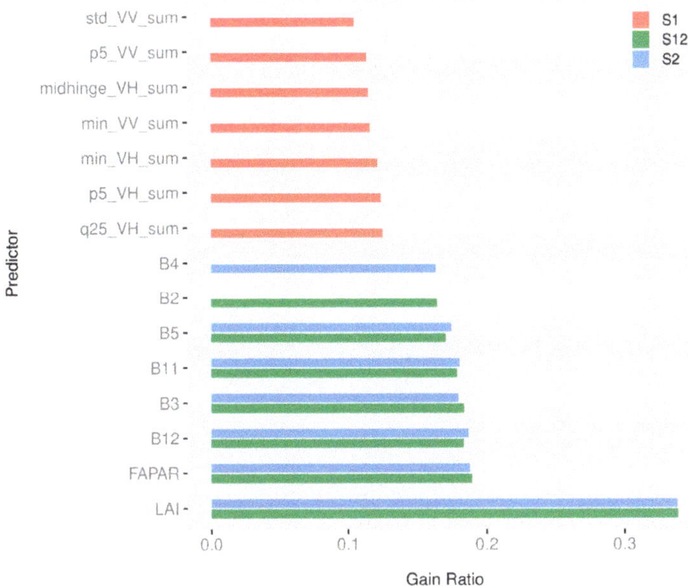

**Figure 8.** Variable importance in the Thuringian study site (seven best predictors).

### 4.2.2. South Africa

Variable importance was separately analyzed for the KNP and the remaining areas of this study area to account for the substantial differences in the appearance of these landscapes due to the great heterogeneity of the South African study site. Figures 9 and 10 display the feature importance for each of the three applied sensor setups in the South African sub study sites.

Similar to the Thuringian study site, the Sentinel-1 features exhibited higher GR values when they were derived from VH-polarized variables in both sub sites. As VH-polarized C-Band data is more sensitive to volume scattering and, therefore, able to monitor vegetation structure it was more important during the classification when compared to co-polarized data. Due to the interaction of cross-polarized predictor variables with shrubs covering large portions of the Kruger National Park, these were found to be more important in the model predictions within the protected savanna ecosystem as compared to their optical counterparts. Between the multi-temporal SAR metrics, differences in GR were relatively small, while variables representing the dry season (winter) ranked higher. This might be due to differences in the vegetation status during the dry season between tree aggregations and their surroundings, thus making it more feasible for the algorithm to detect forests with higher accuracy. Several studies confirm the findings stating that the 'leaf-off' season should be favored for monitoring vegetation structure in savanna ecosystems with SAR information [76–78]. This can be attributed to the increased transparency of deciduous tree canopies and the associated increased penetration depth of the C-Band signal, as well as the decrease of the impact of soil moisture change in this period [75,79]. Consequently, Sentinel-1 was also seen to be more important in the fused approach while using both sensors in combination in the sub site of the open savanna. These results indicate the potential of multi-temporal microwave remote sensing instruments to improve vegetation monitoring in savanna ecosystems.

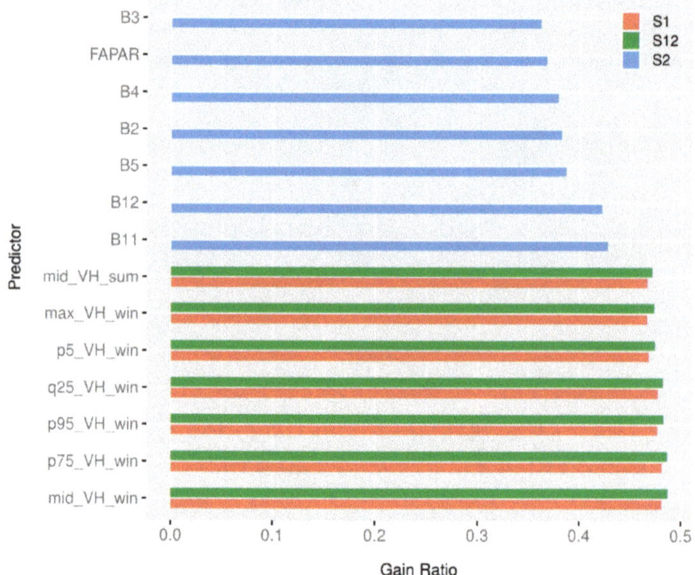

**Figure 9.** Variable importance within the KNP (seven best predictors).

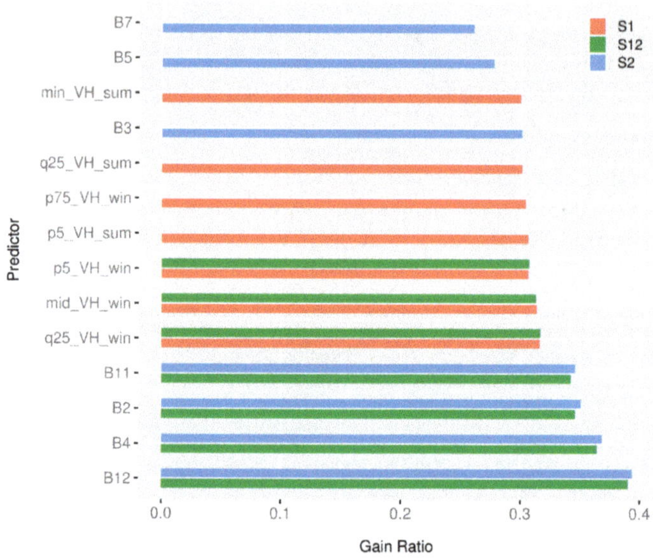

**Figure 10.** Variable importance outside of the KNP (seven best predictors).

Outside the KNP (west), variable importance shifted strongly towards the use of Sentinel-2 predictors by the RF algorithm, as visualized in Figure 10. Similar to the Thuringian study site, the results indicated that the SWIR bands 11 and 12, as well as the channels in the visible bandwidth, are the most important bands for the optical and the fused classification. In contrast, the biophysical parameters were not as distinctive for the classification, as it was observed in Thuringia. This could possibly be explained with the highly heterogeneous composition of land use and land cover in this area. While, dense forest plantations in the west can be represented well by using these indices that can be directly related to vegetation structure, extensive transition regions towards the savanna flora are often characterized by scarcity in terms of plant size and covered area. Information from NIR wavelength were found to be of small importance for the distinction between forest and non-forested areas, which is consistent with the analysis in temperate German forests that was conducted in this and other studies [18]. As this effect could be observed in both study sites, this suggests that the near-infrared channels of Sentinel-2 are only of limited significance for the derivation process of tree cover or tree species mapping when SWIR channels are being implemented in the classification.

Sentinel-1 predictors representing the lower ranges of the SAR time series in VH polarization were also found to be superior to the co-polarized data in terms of variable importance. In contrast, to the protected park area in the east of the South African study site, dry and wet season were found to equally contribute to the model performance. This might also be related to the stability in the forest plantations and their surroundings originating from the weaker seasonal changes of vegetation in this area.

### 4.3. Comparison with Existing FNF Products

In comparison with the chosen reference products, the FNF classifications using Sentinel-1 and Sentinel-2 produced higher J values ($\varnothing = 0.87$ vs. $\varnothing = 0.77$) in both of the study areas. As given in Figure 11, the similarity of a given set of samples of reference with the data sets reveal a distinct relationship between the composition of the study sites and used sensors. While J of classifications based on mostly optical data (S-2, CCI, Landsat) exhibited generally higher values in the Thuringian study site than SAR based ($\varnothing = 0.9$ vs $\varnothing = 0.8$), this trend was found to be reversed in the South African study site ($\varnothing = 0.68$ vs $\varnothing = 0.88$), which exhibits a much higher degree of homogeneity. This indicates the ability of multi-temporal SAR data to detect vegetation in savanna ecosystems adequately representing the intra-annual variability of this complex ecosystem better and, thus, showing that optical data might not be sufficient in this area, at least when few time steps are utilized.

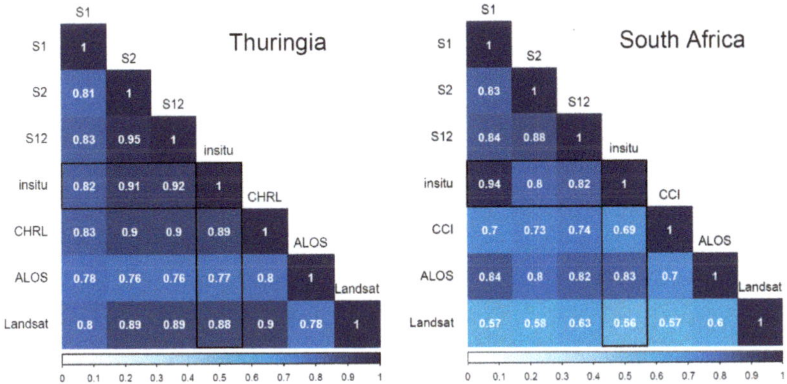


**Figure 11.** J for classifications of all sensor setups (Sentinel-1, Sentinel-2, and Sentinel-1/-2) in the Thuringian and South African site compared with reference data and independent FNF products.

### 4.4. Test of Homogeneity between Classification Distributions

To assess whether the results of the classifications vary significantly from each other, we performed McNemar's test to each individual sensor setup. This non-parametric statistical test is suited to compare the performance of machine-learning-based classifications [80]. The results of McNemar's test indicate that, except for the comparison of classification S-2 vs. S-12 in Thuringia, all of the classifications were considered to significantly differ from each other.

## 5. Discussion

This study investigated the use of multi-source and multi-temporal remote sensing data in varying ecosystems. Firstly, the results revealed that the overall accuracy of the machine-learning based classification while using the RF algorithm increased from using single sensor setups as compared to using Sentinel-1 and Sentinel-2 in a joint approach in both study sites. Combining optical and SAR data led to very high classification accuracies, as illustrated in Table 3. Comparing all available sensor setups, the Sentinel-1 classifications provided the lowest accuracy while still performing reasonably (84.3% to 90.6%) in both study sites. While Thuringia is characterized by relatively homogenous forests, the South African study site comprises forest plantations and savanna ecosystems, thus exhibiting a gradient of increasing heterogeneity from East to West, which also led to great differences in the variable importance for the RF model. In both study sites, the SWIR bands 11 and 12 of Sentinel-2 ranged the highest among the most distinctive optical variables, which other studies also confirmed [18,26]. From Sentinel-1, VH polarized variables were selected as the most important variables in the single sensor and fused data set.

The savanna is a highly variable ecosystem in terms of vegetation composition, as reflected in a non-trivial spectral and spatial appearance, as well as phenology, which drastically changes between dry and wet season [81]. Subsequently, classification results of this study site featured the highest variability in CV-runs over all sensor setups, as visualized in Figure 5.

Further, it could be observed that the SAR data was capable of detecting smaller aggregations of trees in the scarce South African savanna than its optical counterpart, while overestimating tree cover in this area. This can be mainly attributed to the impact of surface roughness to which the C-Band radar is very sensitive, especially in sparsely vegetated areas. This effect was also shown in a study that was conducted in savanna ecosystems using Sentinel-1/-2 data [82]. Due to the heterogeneous training data, which may not represent the complete study area adequately, the accuracy does not fully reflect the positive impact of Sentinel-1 on the savanna classification. Here, the quality of the training and validation data is a major key in obtaining satisfactory and reliable classification results [83]. Thus, a separated variable importance analysis for the South African study site proved the significant role that SAR data plays in the detection of vegetation in open savannas. It should be noted that a separation of the South African study site into two internally more homogenous study areas would potentially lead to an increase in classification accuracy and a strong decrease of the root mean square error.

The high-resolution optical Sentinel-2 data proved to be capable of detecting forests with high accuracy in both study sites, especially in areas with homogenous forest sites. It is important to highlight that Sentinel-2 tends to underestimate savanna vegetation during dry season with the exception of larger tree aggregations, which are often located along river streams. Switching from single time steps to deploying the fully available cloud-free time series of Sentinel-2 data might lead to an improvement to further increase classification accuracy in this ecosystem and, thus, improve the ability to take the strong seasonality of South African savannas into account.

This study further analyzed the impact of sensor fusion of Sentinel-1 and Sentinel-2 to improve forest monitoring in highly variable ecosystems. Similarly to findings of forest monitoring related studies while using optical and radar satellites, the addition of Sentinel-1 data did not significantly improve the overall classification accuracy [26,28]. In both studies, differences between the accuracy of a Sentinel-2 only and fused classification (Sentinel-1 and Sentinel-2) ranged between 1 to 3 %. These results are comparable with our findings. By joining optical and SAR data, the dense Sentinel-1 time series could capture vegetation dynamics in open savanna while being prevented from possible overestimation by Sentinel-2 when vegetation is scarce. The results also indicated that sensor fusion increases the classification accuracy in both study sites averaged over multiple runs of CV. The accuracies show that the individual classifications of sensor setups S-2 and S-12 did not differ as much when compared to the S-1 classification in both study sites. A statistical McNemar test was conducted to check the homogeneity of the distribution of the classifications. The results show that all of the distributions differ (given a significance level of $p = 0.05$ and one degree of freedom) significantly from each other (except for S-2 vs. S-12 in Thuringia), as given in Table 4, with S-1 classifications exhibiting the greatest differences. This indicates the different perception and potential that is provided from the sensors for the varying ecosystems.

**Table 4.** McNemar's test results for varying sensor setups in both study sites. A *p*-value greater than 0.05 defines a difference between two distributions that is not significant and vice versa.

| Product | Thuringia | | South Africa | |
|---|---|---|---|---|
| | $X^2$ | *p*-Value | $X^2$ | *p*-Value |
| S-1 vs. S-2 | 48.3 | <0.05 | 61.6 | <0.05 |
| S-1 vs. S-12 | 66.9 | <0.05 | 57.2 | <0.05 |
| S-2 vs S-12 | 1.4 | 0.23 | 4.8 | <0.05 |

The results of this study demonstrate the ability of machine-learning techniques to produce reliable results from a large number of variables given a relatively low quantity of reference information. Our findings suggest that the RF algorithm favored optical data over multi-temporal SAR data for the detection of forests in the different ecosystems of both study sites. Further, C-Band was found to be a promising data source for the detection of vegetation in dry savanna ecosystems. However, due to the internal variable selection of the RF classifier, this was not acknowledged strong enough within the classification itself. Using in situ data that provides a better representation of the study area might increase the impact of applying a dense Sentinel-1 time series, so that more of the pixels showing strong intra-annual dynamics are being used for training of the model.

## 6. Conclusions

In this study, we investigated the ability of high resolution optical and microwave Sentinel data to derive forest cover in substantially different ecosystems while using machine-learning techniques accounting for the impact of spatial autocorrelation during cross-validation. As study sites, we selected the state of Thuringia in Germany, which is characterized by homogenous dense temperate forests, and an area in South Africa including the southern Kruger National Park, as well as neighboring forest plantations, featuring both homogenous tree aggregations and scarce open savanna vegetation. The results indicated that optical sensors are capable of detecting homogenous tree aggregations with high accuracies while failing at locating large portions of tree cover in open savannas. The addition of multi-temporal microwave information to this data set showed multiple advantages. These are the correction of falsely classified cloud pixels, as well as an improved delineation of small forests in the savanna ecosystem. Thus, our results show that the fusion across wavelengths can lead to classifications with a minimized quantity of misclassifications, while the magnitude is rather small, especially when comparing optical and fused classification. This finding is also reflected in considerably high accuracies of classifications using the joint data sets, which were all cross-validated with multiple repetitions to avoid spatial redundancy and account for outliers. The analysis of variable importance revealed that SWIR and RGB channels range among the most important predictors from Sentinel-2, which corresponds to the findings of other recent publications. The biophysical parameters used in this study (LAI and FAPAR) were found to be useful in detecting forest cover mainly in homogenous temperate environments. As most important multi-temporal Sentinel-1 features, the classifier identified VH-polarized predictors and those representing the lower value range of the time series, such as minimum and the lower percentiles in both study sites. Sentinel-1 variables were not favored strong enough to reveal their full predictive power in some parts of our study sites due to a certain level of 'black box' behavior of the RF algorithm. We further split the South African study site into two parts to reveal their potential impact on the classification, so that the microwave predictors proved their capability to predict tree cover in open savanna ecosystems.

This study demonstrated the beneficial effects of synergistically combining Sentinel-1 and Sentinel-2 to detect forest cover at fine spatial scale. Using CV procedures that account for the existence of spatial dependence within remote sensing data sets our methodology could potentially contribute to improve reliability of activity data (AD) under REDD+ Measurement, Reporting, and Verification. We also applied the robust RF classifier to highly variable ecosystems to examine the robustness of the approach. Future studies could potentially focus on extending from Sentinel-2 time steps to proper time series information to obtain an even better understanding of intra- and inter-annual alterations in the vegetation status. Furthermore, it would be essential to gain more control over how predictor variables are used in machine-learning approaches and, thus, fully reflect their importance in the individual classifications (single sensor) to increase the beneficial effect on the fused models (multi-sensor).

**Author Contributions:** Conceptualization, K.H., M.U., M.D.M. and C.S.; Data curation, K.H. and M.U.; Formal analysis, K.H., M.U. and M.D.M.; Funding acquisition, C.S.; Investigation, K.H.; Methodology, K.H., P.S. and M.D.M.; Software, K.H. and P.S.; Supervision, M.U., M.D.M. and C.S.; Validation, K.H.; Visualization, K.H. and M.D.M.; Writing—original draft, K.H.; Writing—review & editing, K.H., M.U., P.S., M.D.M. and C.S. All authors have read and agreed to the published version of the manuscript.

**Funding:** This research was funded by the International Max Planck Research School for Global Biogeochemical Cycles (IMPRS-gBGC).

**Acknowledgments:** Co-Authors received funding from the German Federal Ministry of Education and Research (BMBF) in the framework of the Science Partnerships for the Assessment of Complex Earth System Processes (SPACES2) under the grant 01LL1701A to D (South African Land Degradation Monitor (SALDi)) as well as from the European Union (EU) Horizon 2020 Research and Innovation Program under the Grant Agreement No. 640176 (BACI—Towards a Biosphere Atmosphere Change Index).

**Conflicts of Interest:** The authors declare no conflict of interest.

## References

1. Food and Agriculture Organization of the United Nations (FAO). *Global Forest Resources Assessment 2015 (FRA 2015)*; FAO: Rome, Italy, 2015.

2. Wulder, M. Optical remote-sensing techniques for the assessment of forest inventory and biophysical parameters. *Prog. Phys. Geogr.* **1998**, *22*, 449–476. [CrossRef]

3. Bonan, G.B. Forests and Climate Change: Forcings, Feedbacks, and the Climate Benefits of Forests. *Science* **2008**, *320*, 1444–1449. [CrossRef] [PubMed]

4. Avitabile, V.; Baccini, A.; Friedl, M.A.; Schmullius, C. Capabilities and limitations of Landsat and land cover data for aboveground woody biomass estimation of Uganda. *Remote Sens. Environ.* **2012**, *117*, 366–380. [CrossRef]

5. Liu, Y.Y.; van Dijk, A.I.J.M.; de Jeu, R.A.M.; Canadell, J.G.; Mccabe, M.F.; Evans, J.P.; Wang, G. Recent reversal in loss of global terrestrial biomass. *Nat. Clim. Chang.* **2015**, 1–5. [CrossRef]

6. Reiche, J.; Verbesselt, J.; Hoekman, D.; Herold, M. Fusing Landsat and SAR time series to detect deforestation in the tropics. *Remote Sens. Environ.* **2015**, *156*, 276–293. [CrossRef]

7. Harris, N.L.; Brown, S.; Hagen, S.C.; Saatchi, S.S.; Petrova, S.; Salas, W.; Hansen, M.C.; Potapov, P.V.; Lotsch, A. Baseline Map of carbon emissions from deforestation in tropical regions. *Science* **2012**, *336*, 1573–1576. [CrossRef]

8. Reiche, J.; Lucas, R.; Mitchell, A.L.; Verbesselt, J.; Hoekman, D.H.; Haarpaintner, J.; Kellndorfer, J.M.; Rosenqvist, A.; Lehmann, E.A.; Woodcock, C.E.; et al. Combining satellite data for better tropical forest monitoring. *Nat. Clim. Chang.* **2016**, *6*, 120–122. [CrossRef]

9. Romijn, E.; Lantican, C.B.; Herold, M.; Lindquist, E.; Ochieng, R.; Wijaya, A.; Murdiyarso, D.; Verchot, L. Assessing change in national forest monitoring capacities of 99 tropical countries. *For. Ecol. Manag.* **2015**, *352*, 109–123. [CrossRef]

10. Hansen, M.C.; Potapov, P.V.; Moore, R.; Hancher, M.; Turubanova, S.A.; Tyukavina, A.; Thau, D.; Stehman, S.V.; Goetz, S.J.; Loveland, T.R.; et al. High-Resolution Global Maps of 21st-Century Forest Cover Change. *Science* **2013**, *342*, 850–854. [CrossRef]

11. Sexton, J.O.; Song, X.-P.; Feng, M.; Noojipady, P.; Anand, A.; Huang, C.; Kim, D.-H.; Collins, K.M.; Channan, S.; DiMiceli, C.; et al. Global, 30-m resolution continuous fields of tree cover: Landsat-based rescaling of MODIS vegetation continuous fields with lidar-based estimates of error. *Int. J. Digit. Earth* **2013**, *6*, 427–448. [CrossRef]

12. Bartholomé, E.; Belward, A.S. A new approach to global land cover mapping from earth observation data. *Int. J. Remote Sens.* **2005**, *26*, 1959–1977. [CrossRef]

13. Loveland, T.R.; Reed, B.C.; Ohlen, D.O.; Brown, J.F.; Zhu, Z.; Yang, L.; Merchant, J.W. Development of a global land cover characteristics database and IGBP DISCover from 1 km AVHRR data. *Int. J. Remote Sens.* **2000**, *21*, 1303–1330. [CrossRef]

14. Lehmann, E.A.; Caccetta, P.; Lowell, K.; Mitchell, A.; Zhou, Z.S.; Held, A.; Milne, T.; Tapley, I. SAR and optical remote sensing: Assessment of complementarity and interoperability in the context of a large-scale operational forest monitoring system. *Remote Sens. Environ.* **2015**, *156*, 335–348. [CrossRef]

15. Balzter, H.; Cole, B.; Thiel, C.; Schmullius, C. Mapping CORINE Land Cover from Sentinel-1A SAR and SRTM Digital Elevation Model Data using Random Forests. *Remote Sens.* **2015**, *7*, 14876–14898. [CrossRef]

16. Erasmi, S.; Twele, A. Regional land cover mapping in the humid tropics using combined optical and SAR satellite data—A case study from Central Sulawesi, Indonesia. *Int. J. Remote Sens.* **2009**, *30*, 2465–2478. [CrossRef]

17. Carrasco, L.; O'Neil, A.W.; Daniel Morton, R.; Rowland, C.S. Evaluating combinations of temporally aggregated Sentinel-1, Sentinel-2 and Landsat 8 for land cover mapping with Google Earth Engine. *Remote Sens.* **2019**, *11*, 288. [CrossRef]

18. Immitzer, M.; Vuolo, F.; Atzberger, C. First Experience with Sentinel-2 Data for Crop and Tree Species Classifications in Central Europe. *Remote Sens.* **2016**, *8*, 166. [CrossRef]

19. Colkesen, I.; Kavzoglu, T. Ensemble-based canonical correlation forest (CCF) for land use and land cover classification using sentinel-2 and Landsat OLI imagery. *Remote Sens. Lett.* **2017**, *8*, 1082–1091. [CrossRef]

20. Belgiu, M.; Csillik, O. Sentinel-2 cropland mapping using pixel-based and object-based time-weighted dynamic time warping analysis. *Remote Sens.* **2018**, *204*, 509–523. [CrossRef]

21. Dostálová, A.; Hollaus, M.; Milenković, M.; Wagner, W. Forest Area Derivation from Sentinel-1 Data. *ISPRS Ann. Photogramm. Remote Sens. Spat. Inform. Sci.* **2016**, *3*, 227–233. [CrossRef]

22. Olesk, A.; Voormansik, K.; Põhjala, M.; Noorma, M. Forest change detection from Sentinel-1 and ALOS-2 satellite images. In Proceedings of the 2015 IEEE 5th Asia-Pacific Conference on Synthetic Aperture Radar (APSAR 2015), Singapore, 1–4 September 2015; pp. 522–527.

23. Reiche, J.; Hamunyela, E.; Verbesselt, J.; Hoekman, D.; Herold, M. Improving near-real time deforestation monitoring in tropical dry forests by combining dense Sentinel-1 time series with Landsat and ALOS-2 PALSAR-2. *Remote Sens. Environ.* **2018**, *204*, 147–161. [CrossRef]

24. Urban, M.; Berger, C.; Heckel, K.; Schratz, P.; Schmullius, C.; Baade, J. Woody Cover Mapping in the Savanna Ecosystem of the Kruger National Park Using Sentinel-1 Time Series. *Koedoe* **2020**. in preparation.

25. Van Tricht, K.; Gobin, A.; Gilliams, S.; Piccard, I. Synergistic Use of Radar Sentinel-1 and Optical Sentinel-2 Imagery for Crop Mapping: A Case Study for Belgium. *Remote Sens.* **2018**, *10*, 1642. [CrossRef]

26. Mercier, A.; Betbeder, J.; Rumiano, F.; Baudry, J.; Gond, V.; Blanc, L.; Bourgoin, C.; Cornu, G.; Cuidad, C.; Marchamalo, M.; et al. Evaluation of Sentinel-1 and 2 Time Series for Land Cover Classification of Forest–Agriculture Mosaics in Temperate and Tropical Landscapes. *Remote Sens.* **2019**, *11*, 979. [CrossRef]

27. Erinjery, J.J.; Singh, M.; Kent, R. Mapping and assessment of vegetation types in the tropical rainforests of the Western Ghats using multispectral Sentinel-2 and SAR Sentinel-1 satellite imagery. *Remote Sens. Environ.* **2018**, *216*, 345–354. [CrossRef]

28. Tavares, P.A.; Beltrão, N.E.S.; Guimarães, U.S.; Teodoro, A.C. Integration of Sentinel-1 and Sentinel-2 for Classification and LULC Mapping in the Urban Area of Belém, Eastern Brazilian Amazon. *Sensors* **2019**, *19*, 1140. [CrossRef] [PubMed]

29. Belgiu, M.; Drăguţ, L. Random forest in remote sensing: A review of applications and future directions. *ISPRS J. Photogramm. Remote Sens.* **2016**, *114*, 24–31. [CrossRef]

30. Gislason, P.O.; Benediktsson, J.A.; Sveinsson, J.R. Random Forests for land cover classification. *Pattern Recognit. Lett.* **2006**, *27*, 294–300. [CrossRef]

31. Lawrence, R.; Wood, S.; Sheley, R. Mapping invasive plants using hyperspectral imagery and Breiman Cutler classifications (Random Forest). *Remote Sens. Environ.* **2006**, *100*, 356–362. [CrossRef]

32. Ghimire, B.; Rogan, J.; Miller, J. Contextual land-cover classification: Incorporating spatial dependence in land-cover classification models using random forests and the Getis statistic. *Remote Sens. Lett.* **2010**, *1*, 45–54. [CrossRef]

33. Urbazaev, M.; Cremer, F.; Migliavacca, M.; Reichstein, M.; Schmullius, C.; Thiel, C. Potential of Multi-Temporal ALOS-2 PALSAR-2 ScanSAR Data for Vegetation Height Estimation in Tropical Forests of Mexico. *Remote Sens.* **2018**, *10*, 1–19. [CrossRef]

34. Baumann, M.; Ozdogan, M.; Kuemmerle, T.; Wendland, K.J.; Esipova, E.; Radeloff, V.C. Using the Landsat record to detect forest-cover changes during and after the collapse of the Soviet Union in the temperate zone of European Russia. *Remote Sens. Environ.* **2012**, *124*, 174–184. [CrossRef]

35. Lv, Z.Y.; Liu, T.F.; Zhang, P.; Benediktsson, J.A.; Lei, T.; Zhang, X.K. Novel Adaptive Histogram Trend Similarity Approach for Land Cover Change Detection by Using Bitemporal Very-High-Resolution Remote Sensing Images. *IEEE Trans. Geosci. Remote Sens.* **2019**, *57*, 1–21. [CrossRef]

36. Desclée, B.; Bogaert, P.; Defourny, P. Forest change detection by statistical object-based method. *Remote Sens. Environ.* **2006**, *102*, 1–11. [CrossRef]

37. Ghosh, A.; Fassnacht, F.E.; Joshi, P.K.; Koch, B. A framework for mapping tree species combining hyperspectral and LiDAR data: Role of selected classifiers and sensor across three spatial scales. *Int. J. Appl. Earth Obs. Geoinf.* **2014**, *26*, 49–63. [CrossRef]

38. Mellor, A.; Haywood, A.; Stone, C.; Jones, S. The Performance of Random Forests in an Operational Setting for Large Area Sclerophyll Forest Classification. *Remote Sens.* **2013**, *5*, 2838–2856. [CrossRef]

39. Main, R.; Mathieu, R.; Kleynhans, W.; Wessels, K.; Naidoo, L.; Asner, G.P. Hyper-Temporal C-Band SAR for Baseline Woody Structural Assessments in Deciduous Savannas. *Remote Sens.* **2016**, *8*, 661. [CrossRef]

40. Naidoo, L.; Mathieu, R.; Main, R.; Kleynhans, W.; Wessels, K.; Asner, G.; Leblon, B. Savannah woody structure modelling and mapping using multi-frequency (X-, C- and L-band) Synthetic Aperture Radar data. *ISPRS J. Photogramm. Remote Sens.* **2015**, *105*, 234–250. [CrossRef]

41. Clerici, N.; Weissteiner, C.J.; Gerard, F. Exploring the use of MODIS NDVI-based phenology indicators for classifying forest general habitat categories. *Remote Sens.* **2012**, *4*, 1781–1803. [CrossRef]

42. Akcay, H.; Kaya, S.; Sertel, E.; Alganci, U. Determination of Olive Trees with Multi-sensor Data Fusion. In Proceedings of the 2019 8th International Conference on Agro-Geoinformatics (Agro-Geoinformatics), Istanbul, Turkey, 16–19 July 2019.

43. Haarpaintner, J.; Davids, C.; Storvold, R.; Johansen, K.; Arnason, K.; Rauste, Y.; Mutanen, T. Boreal forest land cover mapping in Icelarnd and Finland using Sentinel-1A. In Proceedings of the Living Planet, Symposium, Prague, Czech Republic, 9–13 May 2016.

44. Abdikan, S.; Sanli, F.B.; Ustuner, M.; Calò, F. Land cover mapping using sentinel-1 SAR data. In Proceedings of the International Archives of the Photogrammetry, Remote Sensing and Spatial Information Sciences—ISPRS Archives, Prague, Czech Republic, 12–19 July 2016; pp. 757–761.

45. Klima in Thüringen|Thüringer Klimaagentur. Available online: https://www.thueringen.de/th8/klimaagentur/klima/index.aspx (accessed on 16 October 2018).

46. Historical Rain|South African Weather Service. Available online: http://www.weathersa.co.za/home/historicalrain (accessed on 3 July 2019).

47. Gorrab, A.; Zribi, M.; Baghdadi, N.; Mougenot, B.; Fanise, P.; Chabaane, Z.L. Retrieval of Both Soil Moisture and Texture Using TerraSAR-X Images. *Remote Sens.* **2015**, *7*, 10098–10116. [CrossRef]

48. Smit, I.P.J.; Asner, G.P.; Govender, N.; Vaughn, N.R.; van Wilgen, B.W. An examination of the potential efficacy of high- intensity fires for reversing woody encroachment in savannas. *J. Appl. Ecol.* **2016**, *53*, 1623–1633. [CrossRef]

49. Food and Agriculture Organization of the United Nations (FAO). *Global Forest Resources Assessment 2000 (FRA 2000)*; FAO: Rome, Italy, 2001; p. 511.

50. Louis, J.; Debaecker, V.; Pflug, B.; Main-Knorn, M.; Bieniarz, J. Sentinel-2 SEN2COR: L2A Processor for Users. In Proceedings of the 2016 Living Planet, Prague, Czech Republic, 9–13 May 2016.

51. SNAP—ESA Sentinel Application Platform. Available online: http://step.esa.int/main/toolboxes/snap/ (accessed on 20 June 2019).

52. Bischl, B.; Lang, M.; Kotthoff, L.; Schiffner, J.; Richter, J.; Studerus, E.; Casalicchio, G.; Jones, Z. mlr: Machine Learning in R. *J. Mach. Learn. Res.* **2016**, *17*, 1–5.

53. R Core Team. *R: A Language and Environment for Statistical Computing*; R Foundation for Statistical Computing: Vienna, Austria, 2017.

54. Breiman, L.; Friedman, J.; Stone, C.J.; Olshen, R.A. *Classification and Regression Trees*; Chapman & Hall/CRC: Boca Raton, FL, USA, 1984; p. 368.

55. Löw, F.; Conrad, C.; Michel, U. Decision fusion and non-parametric classifiers for land use mapping using multi-temporal RapidEye data. *ISPRS J. Photogramm. Remote Sens.* **2015**, *108*, 191–204. [CrossRef]

56. Breiman, L. Random Forests. *Mach. Learn.* **2001**, *45*, 5–32. [CrossRef]

57. Huang, B.F.F.; Boutros, P.C. The parameter sensitivity of random forests. *BMC Bioinform.* **2016**, *17*, 1–13. [CrossRef] [PubMed]

58. Probst, P.; Bischl, B.; Boulesteix, A.-L. Hyperparameters and Tuning Strategies for Random Forest. *WIREs Data Min. Knowle. Discov.* **2019**, *9*. [CrossRef]

59. Rodriguez-Galiano, V.F.; Ghimire, B.; Rogan, J.; Chica-Olmo, M.; Rigol-Sanchez, J.P. An assessment of the effectiveness of a random forest classifier for land-cover classification. *ISPRS J. Photogramm. Remote Sens.* **2012**, *67*, 93–104. [CrossRef]

60. Schratz, P.; Muenchow, J.; Iturritxa, E.; Richter, J.; Brenning, A. Hyperparameter tuning and performance assessment of statistical and machine-learning algorithms using spatial data. *Ecol. Mod.* **2019**, *406*, 109–120. [CrossRef]

61. Quinlan, J.R. Induction of Decision Trees. *Mach. Learn.* **1986**, *1*, 81–106. [CrossRef]

62. Brenning, A. Spatial cross-validation and bootstrap for the assessment of prediction rules in remote sensing: The R package sperrorest. In Proceedings of the 2012 IEEE International Geoscience and Remote Sensing Symposium, Munich, Germany, 22–27 July 2012; pp. 5372–5375.

63. Wang, J.; Zhao, Y.; Li, C.; Yu, L.; Liu, D.; Gong, P. Mapping global land cover in 2001 and 2010 with spatial-temporal consistency at 250m resolution. *ISPRS J. Photogramm. Remote Sens.* **2015**, *103*, 38–47. [CrossRef]

64. Kuemmerle, T.; Hostert, P.; Radeloff, V.C.; van der Linden, S.; Perzanowski, K.; Kruhlov, I. Cross-border Comparison of Post-socialist Farmland Abandonment in the Carpathians. *Ecosystems* **2008**, *11*, 614–628. [CrossRef]

65. Bui, D.T.; Tuan, T.A.; Klempe, H.; Pradhan, B.; Revhaug, I. Spatial prediction models for shallow landslide hazards: A comparative assessment of the efficacy of support vector machines, artificial neural networks, kernel logistic regression, and logistic model tree. *Landslides* **2016**, *13*, 361–378. [CrossRef]

66. Wollan, A.K.; Bakkestuen, V.; Kauserud, H.; Gulden, G.; Halvorsen, R. Modelling and predicting fungal distribution patterns using herbarium data. *J. Biogeogr.* **2008**, *35*, 2298–2310. [CrossRef]

67. Jaccard, P. Distribution de la flore alpine dans le bassin des Dranses et dans quelques régions voisines. *Bull. Soc. Vaud. Sci. Nat.* **1901**, *37*, 241–272.

68. Wijaya, S.H.; Afendi, F.M.; Batubara, I.; Darusman, L.K.; Altaf-Ul-Amin, M.D.; Kanaya, S. Finding an appropriate equation to measure similarity between binary vectors: Case studies on Indonesian and Japanese herbal medicines. *BMC Bioinform.* **2016**, *17*, 1–19. [CrossRef] [PubMed]

69. Shimada, M.; Itoh, T.; Motooka, T.; Watanabe, M.; Shiraishi, T.; Thapa, R.; Lucas, R. New global forest/non-forest maps from ALOS PALSAR data (2007–2010). *Remote Sens. Environ.* **2014**, *155*, 13–31. [CrossRef]

70. European Environment Agency (EEA). Copernicus Land Monitoring Service—High Resolution Layer Forest. Available online: https://land.copernicus.eu/user-corner/technical-library/hrl-forest\T1\textgreater{} (accessed on 3 July 2019).

71. European Space Agency (ESA). CCI LAND COVER—S2 Prototype Land Cover 20 m Map of Africa 2016. Available online: http://2016africalandcover20m.esrin.esa.int/ (accessed on 1 November 2018).

72. Santoro, M.; Eriksson, L.; Askne, J.; Schmullius, C. Assessment of stand-wise stem volume retrieval in boreal forest from JERS-1 L-band SAR backscatter. *Int. J. Remote Sens.* **2006**, *27*, 3425–3454. [CrossRef]

73. Baret, F.; Guyot, G. Potentials and Limits of Vegetation Indices for LAI and APAR Assessment. *Remote Sens. Environ.* **1991**, *35*, 161–173. [CrossRef]

74. Karasiak, N.; Sheeren, D.; Fauvel, M.; Willm, J.; Dejoux, J.F.; Monteil, C. Mapping Tree Species of Forests in Southwest France using Sentinel 2 Image Time Series. In Proceedings of the 2017 9th International Workshop on the Analysis of Multitemporal Remote Sensing Images (MultiTemp), Brugge, Belgium, 27–29 June 2017; pp. 1–4.

75. Sakowska, K.; Juszczak, R.; Gianelle, D. Remote Sensing of Grassland Biophysical Parameters in the Context of the Sentinel-2 Satellite Mission. *J. Sens.* **2016**. [CrossRef]

76. Mitchard, E.T.A.; Saatchi, S.S.; Lewis, S.L.; Feldpausch, T.R.; Woodhouse, I.H.; Sonké, B.; Rowland, C.; Meir, P. Measuring biomass changes due to woody encroachment and deforestation/degradation in a forest–savanna boundary region of central Africa using multi-temporal L-band radar backscatter. *Remote Sens. Environ.* **2011**, *115*, 2861–2873. [CrossRef]
77. Urbazaev, M.; Thiel, C.; Mathieu, R.; Naidoo, L.; Levick, S.R.; Smit, I.P.J.; Asner, G.P.; Schmullius, C. Assessment of the mapping of fractional woody cover in southern African savannas using multi-temporal and polarimetric ALOS PALSAR L-band images. *Remote Sens. Environ.* **2015**, *166*, 138–153. [CrossRef]
78. Ningthoujam, R.; Balzter, H.; Tansey, K.; Morrison, K.; Johnson, S.; Gerard, F.; George, C.; Malhi, Y.; Burbidge, G.; Doody, S.; et al. Airborne S-band SAR for forest biophysical retrieval in temperate mixed forests of the UK. *Remote Sens.* **2016**, *8*, 609. [CrossRef]
79. Bucini, G.; Hanan, N.P.; Boone, R.B.; Smit, I.P.J.; Saatchi, S.S.; Lefsky, M.A.; Asner, G.P. Woody fractional cover in Kruger National Park, South Africa. In *Ecosystem Function in Savannas*; CRC Press: Boca Raton, FL, USA, 2010; pp. 219–237.
80. McNemar, Q. Note on the sampling error of the difference between correlated proportions or percentages. *Psychometrika* **1947**, *12*, 153–157. [CrossRef]
81. Sankaran, M.; Hanan, N.P.; Scholes, R.J.; Ratnam, J.; Augustine, D.J.; Cade, B.S.; Gignoux, J.; Higgins, S.I.; Le Roux, X.; Ludwig, F.; et al. Determinants of woody cover in African savannas. *Nature* **2005**, *438*, 846–849. [CrossRef] [PubMed]
82. Zhang, W.; Brandt, M.; Wang, Q.; Prishchepov, A.V.; Tucker, C.J.; Li, Y.; Lyu, H.; Fensholt, R. From woody cover to woody canopies: How Sentinel-1 and Sentinel-2 data advance the mapping of woody plants in savannas. *Remote Sens. Environ.* **2019**, *234*, 111465. [CrossRef]
83. Congalton, R.G.; Gu, J.; Yadav, K.; Thenkabail, P.; Ozdogan, M. Global Land Cover Mapping: A Review and Uncertainty Analysis. *Remote Sens.* **2014**, *6*, 12. [CrossRef]

 *remote sensing*

*Article*

# Estimation of Changes of Forest Structural Attributes at Three Different Spatial Aggregation Levels in Northern California using Multitemporal LiDAR


Francisco Mauro [1,*], Martin Ritchie [2], Brian Wing [2,†], Bryce Frank [1], Vicente Monleon [3], Hailemariam Temesgen [1] and Andrew Hudak [4]

[1] Forest Engineering Resources and Management, College of Forestry, Oregon State University, 2150 SW Jefferson Way, Corvallis, OR 97331, USA; bryce.frank@oregonstate.edu (B.F.); temesgen.hailemariam@oregonstate.edu (H.T.)

[2] US Forest Service Pacific Southwest Research Station, 3644 Avtech Parkway, Redding, CA 96002, USA; mritchie@fs.fed.us (M.R.)

[3] US Forest Service Pacific Northwest Research Station, 3200 SW Jefferson Way, Corvallis, OR 97331, USA; vjmonleon@fs.fed.us

[4] US Forest Service, Rocky Mountain Research Station, 1221 S Main St, Moscow, ID 83843, USA; ahudak@fs.fed.us

* Correspondence: francisco.mauro@oregonstate.edu

† Deceased.



Received: 18 March 2019; Accepted: 12 April 2019; Published: 16 April 2019



**Abstract:** Accurate estimates of growth and structural changes are key for forest management tasks such as determination of optimal rotation times, optimal rotation times, site indices and for identifying areas experiencing difficulties to regenerate. Estimation of structural changes, especially for biomass, is also key to quantify greenhouse gas (GHG) emissions/sequestration. We compared two different modeling strategies to estimate changes in V, BA and B, at three different spatial aggregation levels using auxiliary information from two light detection and ranging (LiDAR) flights. The study area is Blacks Mountains Experimental Forest, a ponderosa pine dominated forest in Northern California for which two LiDAR acquisitions separated by six years were available. Analyzed strategies consisted of (1) directly modeling the observed changes as a function of the LiDAR auxiliary information ($\delta$-modeling method) and (2) modeling V, BA and B at two different points in time, including a term to account for the temporal correlation, and then computing the changes as the difference between the predicted values of V, BA and B for time two and time one. We analyzed predictions and measures of uncertainty at three different level of aggregation (i.e., pixels, stands or compartments and the entire study area). Results showed that changes were very weakly correlated with the LiDAR auxiliary information. Both modeling alternatives provided similar results with a better performance of the $\delta$-modeling for the entire study area; however, this method also showed some inconsistencies and seemed to be very prone to extrapolation problems. The $y$-modeling method, which seems to be less prone to extrapolation problems, allows obtaining more outputs that are flexible and can outperform the $\delta$-modeling method at the stand level. The weak correlation between changes in structural attributes and LiDAR auxiliary information indicates that pixel-level maps have very large uncertainties and estimation of change clearly requires some degree of spatial aggregation; additionally, in similar environments, it might be necessary to increase the time lapse between LiDAR acquisitions to obtain reliable estimates of change.

**Keywords:** forest structure change; EBLUP; small area estimation; multitemporal LiDAR and stand-level estimates

## 1. Introduction

Light detection and ranging LiDAR data have been extensively used in forest inventories to provide auxiliary information that is highly correlated with multiple forest structural attributes [1–3]. This strong correlation allows estimating forest structural attributes more efficiently than if only field measurements are available [4]. In addition, the spatially explicit nature of LiDAR enables the mapping of forest attributes at fine resolutions (e.g., [2,5]). Accurate estimates of growth and structural changes are key for forest management as multiple management tasks such as determination of optimal rotation times, calculation of site indexes or the identification of areas experiencing difficulties in regeneration. Estimation of biomass is also key to quantifying greenhouse gas (GHG) emissions/sequestration, to comply with the International Panel on Climate Change (IPCC) reporting and good practice guidelines [6], and to develop a correct appraisal of forest resources for carbon markets. The extensively used area based approach (ABA) [1] provides a way to estimate forest attributes at multiple levels ranging from single pixels to large areas using LiDAR auxiliary information [7]. Availability of repeated LiDAR data acquisitions has opened the door to estimation of changes in forest structural attributes over time (e.g., [8,9]) using the ABA method.

In the ABA, the area under study is covered by a regular grid that will define a population of pixels or grid cells. In this approach, the field plots used to train models and the grid cells are of the same size, typically between 400 m$^2$ and 900 m$^2$. A direct application of predictive models will render predictions for grid units of size too small to be considered of interest for reporting in forest inventories. Areas of interest (AOIs) (i.e., the areas for which estimates are needed) are typically geographic units that can vary in size depending on the particular application. For worldwide inventories or inventories over continents or countries, AOIs are typically administrative or political units such as countries or municipalities. In forest management applications, AOIs are typically stands, compartments or even complete forests or landscapes. All these AOIs require spatial aggregation of grid units. However, validations of predictive models in the ABA literature are typically performed using global metrics of model fit, such as the sample-based root mean square error or bias, that provide average measures of uncertainty for predictions made for pixels or plots. These measures of uncertainty derived from the model fitting stage do not directly translate into measures of uncertainty for predictions for AOIs composed of multiple pixels (i.e., countries, municipalities, forests, stands, etc.). In addition, even when considering single pixels, they are not AOI-specific, as they only provide an average value, across the entire population, of the error that can be expected using a given model.

Thus, it is clear that uncertainty measures used as quality controls in forest inventories need to be made at the AOI-level and change estimation is not an exception. For large areas holding large sample sizes, AOI-specific estimates of means or totals and their measures of uncertainty can be obtained using direct estimators (e.g., [10–12]) that use only use sample data from the AOI under consideration. However, if the AOI sample sizes are not large enough to support direct estimates with reliable precision, then they must be regarded as small areas [13].

Small area estimation (SAE) techniques, especially empirical best linear unbiased predictors (EBLUPs) in combination with the ABA approach have been used to obtain estimates, and their corresponding measures of uncertainty for subpopulations such as municipalities [14], groups or management units [15] and stands [4,14,16,17]. SAE techniques allow correcting the potential bias problems of synthetic predictions (i.e., predictions developed assuming that a general model developed at the population level holds for all subpopulations) and also permit reducing the large variance problems of direct estimators when AOIs sample sizes are small [13]. In addition, while EBLUPs have been extensively used in SAE contexts, they can also be used to produce estimates for subpopulations or AOIs with large sample sizes and preserve important advantages over other methods. First, they allow obtaining model-unbiased estimates and their corresponding measures of uncertainty for all AOIs using a single model that explicitly considers potential variations between AOIs. This is a clear advantage over synthetic methods that assume that a certain relation derived for the entire population holds in all AOIs. A second advantage of EBLUPs is that it is possible to reduce the modeling effort required by direct model-based or model-assisted methods where a model is needed for each AOI. It is thus clear that SAE techniques in combination with LiDAR auxiliary information have potential applications in multiple forest inventories contexts. Unfortunately, to the best of our knowledge, all studies on SAE and forest inventories have focused on estimation of structural attributes at a given point in time, and little is known about: (1) their performance when applied to forest structure change estimation, and (2) about how these techniques compare to other methods used for estimation of changes in AOIs comprising entire populations [10,12,18,19] and especially subpopulations [20].

In this study, we analyzed the two most commonly used strategies to model changes in structural forest attributes using repeated LiDAR acquisitions, and analyzed their performance when used to obtain EBLUPs for AOIs of different size. The first strategy, referred hereafter as the $\delta$-modeling method, considers the change, $\delta$, over the time between LiDAR acquisitions as the model response. The second strategy, which we will call $y$-modeling method, focuses on modeling the structural attributes $y$, and their derived change over time. As a novelty, in the $y$-modeling method, the temporal correlation of both model errors and AOI random effects were taken into account. We considered changes in three structural variables, and AOIs at three different spatial aggregation levels in order to provide insights for future applications where estimates for an entire population and for subpopulations of different sizes are needed. Variables under study are standing volume (V), above ground biomass (B) and basal area (BA) and AOIs subject to analysis are (1) an entire forested area or landscape, (2) subpopulations that in this case are forest stands and (3) pixels as gridded maps are common output in mapping applications.

## 2. Materials and Methods

### 2.1. Study Area

The study area is Blacks Mountains Experimental Forest (BMEF), a 3715 ha forest managed by the United States Forest Service, located northeast of Lassen National Park in northern California, USA (Figure 1). Elevation ranges from 1700 m to 2100 m above sea level. Slopes are gentle (<10%) on the lower parts of the forest and moderate (10%–40%) at higher elevations. Climate is Mediterranean with a certain degree of continentality, with dry summers and wet and cold winters when precipitation is in the form of snow. Average precipitation is 460 mm per year with monthly average temperatures that range from −9 °C to 29 °C. Soils are developed over basalts with depths that range from 1 to 3 m. Ponderosa pine (*Pinus ponderosa Lawson* & C. Lawson) dominated forest occupies the majority of the area. Incense cedar (*Calocedrus decurrens* (Torr.) Florin), white fir (*Abies concolor* (Gordon & Glend) Hildebr) and Jeffrey Pine (*Pinus jeffreyi* Grev & Balf.) are abundant accompanying species. Forest structure is relatively open and the canopy cover varies greatly within the forest (see Figure 1). A more detailed description of the study area can be found in [21,22].

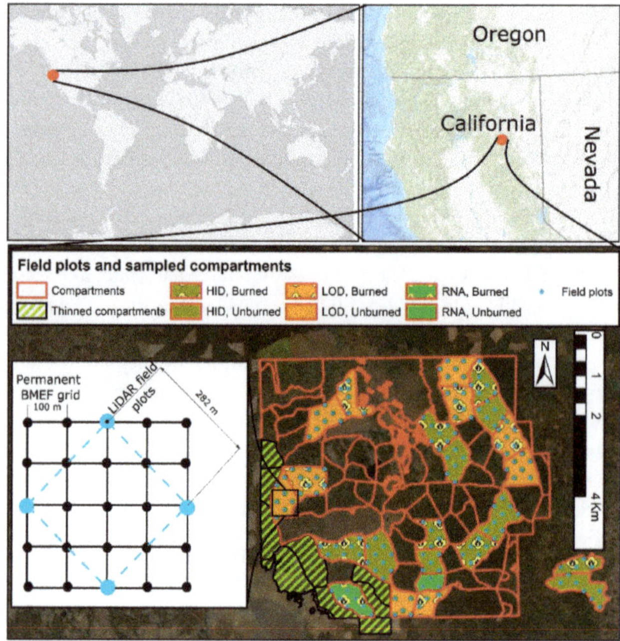


**Figure 1.** Study area location map, delineated stands and field plots, and detailed diagram showing the light detection and ranging LiDAR field plots grid over the permanent Blacks Mountains Experimental Forest (BMEF) grid of permanent makers.

## 2.2. Sampling Design and Field Data

In total, 106 forested stands were delineated in BMEF. Small non-forested patches were masked in the study area and hence were not considered part of the population under study. Out of the 106 forested stands, 24 were selected and sampled in the field. Nine of the remaining 82 unsampled stands were subject to thinning during the period between two available LiDAR acquisitions (i.e., 2009–2015) and all thinning operations were finished by fall 2011. These nine stands are located on the southwestern edge of BMEF and were analyzed separately because the sample of field plots used to train the LiDAR models did not include any stand subject to similar silvicultural interventions (Figure 1). Sampled stands come from a long-term research project initiated in BMEF in 1991 and, excluding the nine thinned stands, were representative of the forest structures and forest management treatments applied in rest of BMEF.

Sampled stands were subject to six different types of treatments resulting from crossing two different factors. The first factor is the structural diversity. It has three levels referred hereafter as low structural diversity (LoD), high structural diversity (HiD) and research natural areas, RNA, or controls. Low structural diversity stands are subject to thinning operations aiming to generate simplified single-strata structures. High diversity stands are subject to thinning where all canopy layers and age groups are preserved, resulting in a multi-storied forest structure with trees of different sizes and ages. Neither the HiD stands nor the LoD stands were subject to thinnings during the period between the two available LiDAR flights. Finally, RNA stands are not subject to any thinning or harvest operation. In total, 10 LoD, 12 HiD and two RNA stands were measured in the field. The second factor under consideration was the presence or absence of prescribed forest fires. Half of the LoD, HiD and RNA stands sampled in the field had been subject to prescribed fires, but only one of the RNA stands was subject to prescribed fires during the period 2009–2015.

A sample of 151, 16 m radius plots (804 m$^2$) were measured in the field during the summer of 2009 and then remeasured during the summer of 2016. All field plots were located on nodes of the 100 m by 100 m grid of monumented markers at BMEF. Coordinates of the makers were determined using traverse methods and survey grade GPS observations and have an accuracy of 15 cm or better (see [23]). For each of the 26 stands selected for sampling, a node of the BMEF 100 m grid was randomly selected and used as a starting node for a 282 m by 282 m grid formed by selecting every other plot of the 100 m grid moving in the diagonal directions. Field measurements were taken on the nodes of the 282 m by 282 m grid (see Figure 1).

Within each field plot all live trees with DBH larger than 9 cm, and all dead standing trees with DBH larger than 12 cm, were stem mapped and measured for DBH and height. Plot basal area (BA) was derived directly from the field measurements. Volume (V) and above ground biomass (B) were computed as the sum of the individual tree volumes and biomasses of all standing trees. Individual tree volumes and biomasses were estimated using species-specific allometric models included in the national volume estimation library (NVEL) and in the national biomass estimation library (NBEL). To account for the one-year difference between acquisition of field measurements in 2016, and the second LiDAR data acquisition obtained in 2015; plot-level values of the variables under analysis were computed for 2015 by linearly interpolating between the values obtained for 2009 and 2016. Finally, for each field plot we computed the change of V, B and BA on a per year basis, as the difference of the plot-level values in 2009 and 2015 divided by 6. For two plots close to the southeastern boundary of the forest, changes in V were extremely large, more than three standard deviations away from the mean value for the change in volume. These anomalous plots were removed from the analysis because such large changes seemed to be derived from edge effects. Plot-level values for 2009, 2016 and per year increments for the period, 2009-2015, for V, B and BA, in the remaining 149 plots are summarized in Table 1.

**Table 1.** Minimum (Min), mean (Mean), standard deviation (Sd), and maximum (Max) of the plot-level values for 2009, 2015 and yearly increments for the period 2009–2015. Values of volume V, basal area BA and biomass B are expressed on a per-hectare basis.

| Variable (Units) | Period | Min | Mean | Sd | Max |
|---|---|---|---|---|---|
| V(m$^3$ ha$^{-1}$) | | 19.87 | 166.93 | 119.66 | 619.43 |
| BA(m$^2$ ha$^{-1}$) | 2009 | 3.81 | 23.43 | 12.02 | 66.54 |
| B(Mg ha $^1$) | | 8.31 | 83.65 | 61.55 | 323.30 |
| V(m$^3$ ha$^{-1}$) | | 17.20 | 175.52 | 117.04 | 644.30 |
| BA(m$^2$ ha$^{-1}$) | 2015 | 3.42 | 25.45 | 12.01 | 67.47 |
| B(Mg ha$^{-1}$) | | 8.34 | 89.38 | 60.29 | 335.03 |
| V(m$^3$ ha$^{-1}$year$^{-1}$) | | −10.89 | 1.43 | 3.88 | 11.19 |
| BA(m$^2$ ha$^{-1}$year$^{-1}$) | Increment 2009–2015 | −0.91 | 0.34 | 0.45 | 1.74 |
| B(Mg ha$^{-1}$year$^{-1}$) | | −5.81 | 0.95 | 1.97 | 5.99 |

For the nine unsampled stands thinned during the period 2009–2015 all thinning operations were completed by fall 2011. In total 427.40 hectares were thinned with prescriptions that varied among stands. Approximately 80% of the area was thinned from below, leaving a residual basal area of 17.22 m$^2$ ha$^{-1}$ to 25.25 m$^2$ ha$^{-1}$. For the remaining 20% of the area, approximately one quarter was not thinned while the other three quarters were thinned to a residual BA that ranged from 6.89 m$^2$ ha$^{-1}$ to 13.77 m$^2$ ha$^{-1}$. Fresh weight of total extractions for the 427.40 hectares subject to thinning was 11,009.38 Mg of logs and 23,164.32 Mg of chipped material.

### 2.3. LiDAR Data Acquisitions

Two LiDAR acquisitions are available for BMEF. The first LiDAR dataset was acquired during the summer of 2009 using a Leica ALS 50 discrete return sensor. Flying altitude was 900 m, side-lap between adjacent flight lines was at least 50% and scanning angle was ±14°. The LiDAR data vendor generated digital terrain models (DTMs) with an accuracy of 15 cm at 95% confidence level. Additional details on the LiDAR data collection for the 2009 acquisition can be found in [24]. The same vendor in the study area performed a second LiDAR acquisition during the summer of 2015 using the same sensor, flying altitude and side-lap specifications. DTMs were also created for 2015 by the vendor.

Four sets of auxiliary variables were considered in this study. The first two sets are composed of 42 LiDAR predictors computed for each acquisition date. Set 1 will represent the predictors for 2009 and Set 2 the predictors for 2015. These predictors are descriptors of the point cloud height distributions and were all relative quantities to avoid introducing noise due to local differences in the point cloud densities of 2009 and 2015 [25]. The third set of predictors, Set 3, was computed as the differences between the 2009 and the 2015 LiDAR predictors. Finally, the fourth set of predictors, Set 4, included the incoming solar radiation computed using the Environmental Systems Research Institute (ESRI) ArcGIS Area Solar Radiation tool [26] with the 2009 digital surface model (DSM) as input; and two treatments: (1) single- or multi-story structural diversity and (2) presence or absence of prescribed fires. All predictors were computed for each field plot and for a grid with a cell size of 805 m² covering the entire BMEF. The cell size matched the field plot size and each cell of the grid was considered a population unit, equivalent to the field plots. Predictors and their corresponding acronyms used in further sections are summarized in Table A1.

### 2.4. AOIs, Target Parameter and Overview of Modelling Strategies

Two different types of subsets of population units will be repeatedly used throughout the manuscript in remaining sections. These subsets and their corresponding notation are: the sample of plots measured in the field, denoted using sub-index $s$ and the target AOIs represented by pixels, denoted using sub-index $\alpha$.

Three different groups of AOIs representing different levels of spatial aggregation were analyzed. The first group represents the largest level of spatial aggregation and represents the entire population under study. Within this group, we considered the set of all sampled stands, SS, and the entire BMEF study area after removing the nine thinned and unsampled stands, SA (i.e., sampled and unsampled but not thinned stands). The second group consists of the 106 forested stands in BMEF. In this group, we considered separately the unsampled and thinned stands (nine stands), unsampled and not thinned stands (73 stands) and sampled and not thinned stands (24 stands). Finally, the third group is the set of all pixels of the LiDAR grid covering the forested area in BMEF.

The main objective of this study was to analyze AOI-specific estimates of the change between 2009 and 2015 for three different structural variables, (V, B and BA). We will use the generic term variable of interest and the letter $y$ to refer to the forest structural variables, and the term target parameter and Greek letter $\Delta$ to refer to the quantities that we seek to estimate. Hereafter, target parameters will always refer to changes over time for the totals of the variables of interest in the considered AOIs, and will be expressed in a per hectare and year basis.

Considering that all pixels have the same area, the target parameter $\Delta_\alpha$ for a generic AOI or subset of population units, $\alpha$, can be expressed as:

$$\Delta_\alpha = \sum_{i=1}^{N_\alpha} K_{y\alpha}(y_{i\alpha15} - y_{i\alpha09}) = \sum_{i=1}^{N_\alpha} K_{\delta\alpha}\delta_{i\alpha}, \tag{1}$$

where $N_\alpha$ is the number of population units (i.e., pixels) in the AOI. The terms $y_{i\alpha15}$ and $y_{i\alpha09}$ respectively represent the value of the variable of interest for 2009 and 2015 for the $i^{th}$ population unit of $\alpha$, and $\delta_{i\alpha}$ is the change, for the $i^{th}$ pixel of $\alpha$, in the variable of interest during the period 2009 to 2015. Finally, for comparability with previous studies, the variables of interest will be expressed in a per unit area basis, and the increments $\delta_{i\alpha}$ will be expressed in a per unit area and year basis. Thus, to ensure that $\Delta_\alpha$ is expressed in the correct units, it is necessary to introduce the factors $K_{y\alpha}$ and $K_{\delta\alpha}$. When $y_{i\alpha15}$

and $y_{i\alpha09}$ are expressed in a per unit area basis $K_{y\alpha} = \frac{1}{6N_\alpha}$ and for $\delta_{i\alpha}$ in a per unit area and year basis $K_{\delta\alpha} = \frac{1}{N_\alpha}$.

We calculated AOI estimates using two different methods. The first, $\delta$-modeling method, for estimation of change uses models similar to those in approach A5 of Poudel et al. [8]. In this approach, the change in a structural variable at the plot/pixel-level ($\delta_{i\alpha}$) is directly modeled as a function of the LiDAR auxiliary variables available for the study area. The second, the $y$-modeling method, uses a modified version of approach A4 of Poudel et al. [8] to obtain AOI-specific estimates of change. Models in this approach jointly relate structural variables ($y_{\alpha15}$ and $y_{\alpha09}$) and LiDAR auxiliary information at a given point in time and account for the correlation between errors obtained for the same plot/pixel at different times. For both methods, variability between stands was accounted by considering them as small areas. Thus, stand-level random effects were included in the models.

*2.5. $\delta$-Modeling Method*

2.5.1. Model $\delta$-modeling Method

Models in the $\delta$-modeling method relate the change (per year) of the variable of interest in a population unit to the auxiliary variables for the population unit. To indicate that these models consider change in the variables of interest directly, model parameters, stand-level random effects and model errors will include the subscript $\delta$. Three different types of auxiliary variables were considered as potential predictors in the $\delta$-modeling method. First, changes in the LiDAR auxiliary variables for the period 2009-2015, Set 3, were considered following Poudel et al. [8] as changes in LiDAR predictors are expected to correlate with growth or changes in forest attributes. Forest structure relates to growth. Thus, the LiDAR auxiliary variables for 2009, Set 1, were also considered as potential predictors that act as proxies for forest structure at the beginning of the period 2009-2015. Finally, the incoming solar radiation and the structural diversity factors and presence of prescribed fires, Set 4, were also considered as potential predictors.

For the $j^{th}$ population unit in the $i^{th}$ stand, models of the $\delta$-modeling method have the form:

$$\delta_{ij} = x_{\delta ij}^t \boldsymbol{\beta}_\delta + v_{\delta i} + \varepsilon_{\delta ij}, \tag{2}$$

where $t$ indicate the transpose operator and $x_{\delta ij}^t$ is a vector of auxiliary variables in which the first element takes the value 1 for the intercept. The term $\boldsymbol{\beta}_\delta$ is a vector of model coefficients where the first element is the intercept of model (2). Selection of auxiliary variables included in the model was performed using the method described in [27]. Stand-level random effects $v_{\delta i}$ are assumed to be independently and identically distributed (i.i.d.) normal random variables $v_{\delta i} \sim N(0, \sigma_{\delta v}^2)$ for all $i = 1, \dots, D$, where $D$ is the total number of stands in the study area. Model errors are i.i.d. normal random variables $\varepsilon_{\delta ij} \sim N(0, \sigma_{\delta \varepsilon}^2)$ independent of the stand-level random effects (i.e., $Cov(\varepsilon_{\delta ij} v_{\delta,k}) = 0$, for all $i$, $j$ and $k$). Models with spatially correlated errors and with non-constant error variances were initially considered but discarded in the model selection stage, as they were not found to be significant (see Section 2.5.3).

For a generic set of population units denoted by subscript $\xi$ (which can represent either $s$ or $\alpha$), the relation in matrix notation between vector of changes of structural variables $\delta_{y\xi}$, and the auxiliary variables included in the model ($X_{\delta\xi}$), is expressed as:

$$\boldsymbol{\delta}_\xi = X_{\delta\xi} \boldsymbol{\beta}_\delta + Z_{\delta\xi} \boldsymbol{v}_\delta + \boldsymbol{\varepsilon}_{\delta\xi}, \tag{3}$$

where $\boldsymbol{\delta}_\xi = (\delta_1, \dots, \delta_{N_\xi})^t$, with $\delta_k$ being the yearly change for the forest structural variable $y$, in the $k^{th}$ unit of $\xi$, and $N_\xi$ is the number of elements in the set $\xi$. The $k^{th}$ element of $\xi$ will be an element of a given stand. To explicitly indicate this membership we will use, when necessary, the sub-indexes $i^{th}$ and $j^{th}$ to respectively indicate the stand and index of the element within the stand. The $k^{th}$ row of the matrix $X_{\delta\xi}$ is $x_{\delta k}^t$. The vector $\boldsymbol{v}_\delta = (v_{\delta 1}, \dots, v_{\delta D})^t$ is a vector of stand-level random effects with variance covariance matrix $G_\delta = \sigma_{\delta v}^2 I_D$, where $I_D$ is the identity matrix of dimension $D$. The matrix $Z_{\xi\delta}$ is a $N_\xi$ x $D$ incidence matrix that describes stand membership for each population unit. The $r^{th}$ row of $Z_{\xi\delta}$ have zeros at all positions except at position $i$, where $i$ is the index of the stand to which the $k^{th}$ unit of $\xi$ belongs. Finally, $\varepsilon_{\xi\delta}$ is a vector of $N_\xi$ model errors with diagonal variance covariance matrix $R_{\xi\delta} = \sigma_{\delta\varepsilon}^2 I_{N_\xi}$.

To simplify the notation, hereafter $\theta_\delta = (\sigma_{\delta\varepsilon}^2, \sigma_{\delta v}^2)$, will represent the vector of variance parameters. The variance covariance matrix of $\delta_\xi$ is:

$$V_{\delta\xi}(\theta_\delta) = Z_{\delta\xi} G_\delta(\theta_\delta) Z_{\delta\xi}^t + R_{\delta\xi}(\theta_\delta). \tag{4}$$

Model (3) is a linear mixed effect model and a special case of the basic unit-level described in [28] (pp. 174).

### 2.5.2. Target Parameter δ-modeling Method

Under model (2) the target parameter (1) for a generic AOI $\alpha$ can be expressed as:

$$\Delta_\alpha = \frac{1}{N_\alpha} \sum_{i=1}^{N_\alpha} \delta_{i\alpha} = \frac{1}{N_\alpha} \mathbf{1}^t \delta_\alpha = \frac{1}{N_\alpha} \mathbf{1}^t (X_{\delta\alpha}\beta_\delta + Z_{\delta\alpha}^t v_\delta + \varepsilon_{\delta\alpha}) = l_{\delta\alpha}^t \beta_\delta + m_{\delta\alpha}^t v_\delta + q_{\delta\alpha}^t \varepsilon_{\delta\alpha}, \tag{5}$$

Thus, the target parameter is a linear target parameter similar to the one considered in [4] where $\mathbf{1}^t$ is a vector of ones and $l_{\delta\alpha}^t = \frac{1}{N_\alpha} \mathbf{1}^t X_{\delta\alpha}$, $m_{\delta\alpha}^t = \frac{1}{N_\alpha} \mathbf{1}^t Z_{\delta\alpha}^t$ and $q_{\delta\alpha}^t = \frac{1}{N_\alpha} \mathbf{1}^t$ are vectors of known constants for the target AOI $\alpha$.

### 2.5.3. Model Selection and Estimator δ-modeling Method

The target parameter $\Delta_\alpha$ was estimated for all considered AOIs using $\hat{\Delta}_\alpha$ the empirical best linear unbiased predictor (EBLUP) described in [29]. For each variable of interest, auxiliary variables included in $X_{\delta\alpha}$ were preselected using the best subset selection procedure described in [29] (pp. 179–180). When models with similar values of model root mean square error or coefficients of determination were compared, the preferred option was to select the model with smallest values of $\sigma_{\delta v}^2$. This criterion is appropriate to minimize the leading term of the AOI specific mean square errors [29] (pp. 176). Pre-selected models considered constant model error variances and no spatial correlation of model errors were fitted using maximum likelihood (ML). In a subsequent stage, models were re-fitted using ML including: (1) an exponential spatial correlation model for the model errors and (2) a non-constant error variance where $\varepsilon_{\delta ij} \sim N(0, \sigma_{\delta\varepsilon}^2 k_{ij}^{2w_\delta})$. The term $k_{ij}$ is the value of the predictor included in the model most correlated to $\delta$ and $w_\delta$ is an additional parameter to account for heteroscedasticity. For all variables, no clear patterns of spatial correlation or non-constant variances were observed, which supports the model form described in Section 2.5.1.

Final estimates $\hat{\theta}_\delta$ of the variance parameters $\theta_\delta$ were obtained using restricted maximum likelihood (REML) with the R [30] package nlme [31]. REML estimates $\hat{\beta}_\delta(\hat{\theta}_\delta)$ of $\beta_\delta$ were functions of the estimated variance parameters (6):

$$\hat{\beta}_\delta(\hat{\theta}_\delta) = \{X_{\delta s}^t \hat{V}_{\delta s}(\hat{\theta}_\delta)^{-1} X_{\delta s}\}^{-1} X_{\delta s}^t \hat{V}_{\delta s}(\hat{\theta}_\delta)^{-1} \delta_s. \tag{6}$$

Matrices $\hat{V}_{\delta s}(\hat{\theta}_\delta)$, $\hat{G}_\delta(\hat{\theta}_\delta)$ and $\hat{R}_{\delta s}(\hat{\theta}_\delta)$ are obtained replacing the estimated variance parameters $\hat{\theta}_\delta$ in $V_{\delta s}(\theta_\delta)$ $G_\delta(\theta_\delta)$ and $R_{\delta s}(\theta_\delta)$, by their REML estimates $\hat{\theta}$. EBLUPs $\hat{\Delta}_\alpha$ are also functions of $\hat{\gamma}$ and are obtained using Equation (7):

$$\hat{\Delta}_\alpha(\hat{\theta}_\delta) = l_{\delta\alpha}^t \hat{\beta}_\delta(\hat{\theta}_\delta) + m_{\delta\alpha}^t \hat{v}_\delta(\hat{\theta}_\delta), \tag{7}$$

where $\hat{v}_\delta(\hat{\boldsymbol{\theta}}_\delta)$ equals:

$$\hat{v}_\delta(\hat{\boldsymbol{\theta}}_\delta) = \hat{\boldsymbol{G}}_\delta(\hat{\boldsymbol{\theta}}_\delta)\boldsymbol{Z}_{\delta s}^t\boldsymbol{V}_{\delta s}(\hat{\boldsymbol{\theta}}_\delta)^{-1}\{\boldsymbol{\delta}_s - \boldsymbol{X}_{\delta\alpha}\hat{\boldsymbol{\beta}}_\delta(\hat{\boldsymbol{\theta}}_\delta)\}. \tag{8}$$

It is important to note that for AOIs in unsampled stands (i.e., pixels in unsampled compartments or the unsampled stands themselves), estimation will be made assuming that the model fit for the sampled stands also holds for the unsampled stands. Under that assumption, $\boldsymbol{m}_{\delta\alpha}^t\hat{v}_\delta(\hat{\boldsymbol{\theta}}_\delta) = 0$ and $\hat{\Delta}_\alpha(\hat{\boldsymbol{\theta}}_\delta) = \boldsymbol{l}_{\delta\alpha}^t\hat{\boldsymbol{\beta}}_\delta(\hat{\boldsymbol{\theta}}_\delta)$ is a synthetic predictor.

### 2.5.4. MSE Estimators for the δ-modeling Method

For all AOIs, the mean squared error of the EBLUP was estimated using the estimator provided by [32] and extended in [4] to account for the fact that AOIs can contain a small number of population units. This estimator is the sum of three components where the last one, $2g_{3,\alpha}(\hat{\boldsymbol{\theta}}_\delta)$, is a bias correction factor:

$$\hat{MSE}\{\hat{\Delta}_{\delta\alpha}(\hat{\boldsymbol{\theta}}_\delta)\} = g_{1\delta\alpha}(\hat{\boldsymbol{\theta}}_\delta) + g_{2\delta\alpha}(\hat{\boldsymbol{\theta}}_\delta) + 2g_{3,\alpha}(\hat{\boldsymbol{\theta}}_\delta), \tag{9}$$

The first term of (9) equals:

$$g_{1\delta\alpha}(\hat{\boldsymbol{\theta}}_\delta) = \boldsymbol{m}_{\delta\alpha}^t\{\hat{\boldsymbol{G}}(\hat{\boldsymbol{\theta}}_\delta) - \hat{\boldsymbol{G}}(\hat{\boldsymbol{\theta}}_\delta)\boldsymbol{Z}_{\delta s}^t\hat{\boldsymbol{V}}_{\delta\alpha}(\hat{\boldsymbol{\theta}}_\delta)^{-1}\boldsymbol{Z}_{\delta s}\hat{\boldsymbol{G}}(\hat{\boldsymbol{\theta}}_\delta)\}\boldsymbol{m}_{\delta\alpha} + \boldsymbol{q}_{\delta\alpha}^t\boldsymbol{R}_{\delta\alpha}(\hat{\boldsymbol{\theta}}_\delta)\boldsymbol{q}_{\delta\alpha}. \tag{10}$$

The second term of (9) is:

$$g_{2\alpha\delta}(\hat{\boldsymbol{\theta}}_\delta) = \boldsymbol{d}_{\delta\alpha}^t\{\boldsymbol{X}_{\delta s}^t\boldsymbol{V}_{\delta s}(\hat{\boldsymbol{\theta}}_\delta)^{-1}\boldsymbol{X}_{\delta s}\}^{-1}\boldsymbol{d}_{\delta\alpha} \tag{11}$$

with $\boldsymbol{d}_{\delta\alpha}^t = \boldsymbol{l}_{\delta\alpha}^t - \boldsymbol{m}_{\delta\alpha}^t\hat{\boldsymbol{G}}(\hat{\boldsymbol{\theta}}_\delta)\boldsymbol{Z}_{\delta s}^t\hat{\boldsymbol{V}}_{\delta s}(\hat{\boldsymbol{\theta}}_\delta)^{-1}\boldsymbol{X}_{\delta s}$. The term $g_{1\delta\alpha}(\hat{\boldsymbol{\theta}}_\delta)$ of $\hat{MSE}\{\hat{\Delta}_{\delta\alpha}(\hat{\boldsymbol{\theta}}_\delta)\}$ accounts for the uncertainty due to the estimation of the random effects while $g_{2\delta\alpha}(\hat{\boldsymbol{\theta}}_\delta)$ accounts for the uncertainty due to estimating $\boldsymbol{\beta}_\delta$.

For model (2), it is possible to compute a bias correction factor for the mean square error estimator, that accounts for the uncertainty due to estimating $\boldsymbol{\theta}_\delta$. This correction factor equals:

$$g_{3\delta\alpha}(\hat{\boldsymbol{\theta}}_\delta) = tr\left\{\left(\left.\frac{\partial \boldsymbol{b}_{\delta\alpha}^t}{\partial \boldsymbol{\theta}_\delta}\right|_{\hat{\boldsymbol{\theta}}_y}\right)\boldsymbol{V}_{\delta s}(\hat{\boldsymbol{\theta}}_\delta)^{-1}\left(\left.\frac{\partial \boldsymbol{b}_{\delta\alpha}^t}{\partial \boldsymbol{\theta}_\delta}\right|_{\hat{\boldsymbol{\theta}}_\delta}\right)^t\overline{\boldsymbol{V}}_{\delta s}(\hat{\boldsymbol{\theta}}_\delta)\right\}, \tag{12}$$

where, $\boldsymbol{b}_{\delta\alpha}^t = \boldsymbol{m}_{\delta\alpha}^t\boldsymbol{G}(\boldsymbol{\theta}_\delta)\boldsymbol{Z}_{\delta s}^t\boldsymbol{V}_{\delta s}(\boldsymbol{\theta}_\delta)^{-1}$ and $\overline{\boldsymbol{H}}_{\delta s}(\hat{\boldsymbol{\theta}})$ is the Fisher information matrix for the fitted model. Explicit formulas for $g_{3\delta\alpha}(\hat{\boldsymbol{\theta}}_\delta)$ are provided [29] (pp. 179–180). This bias correction factor was used as a reference in comparisons with the $y$-modeling method.

All estimators of the mean square errors for AOIs in unsampled stands were made assuming that the model fitted for the sampled stands holds in the unsampled stands. Under this assumption, the leading term $g_{1\delta\alpha}(\hat{\boldsymbol{\theta}}_\delta)$, of $\hat{MSE}\{\hat{\Delta}_{\delta\alpha}(\hat{\boldsymbol{\theta}}_\delta)\}$ will be larger than if the stand containing the AOI was sampled. This occurs because the negative term $\boldsymbol{m}_{\delta\alpha}^t\hat{\boldsymbol{G}}(\hat{\boldsymbol{\theta}}_\delta)\boldsymbol{Z}_{\delta s}^t\hat{\boldsymbol{V}}_{\delta s}(\hat{\boldsymbol{\theta}}_\delta)^{-1}\boldsymbol{Z}_{\delta\alpha}\hat{\boldsymbol{G}}(\hat{\boldsymbol{\theta}}_\delta)\boldsymbol{m}_{\delta\alpha}$ makes the term $g_{1\delta\alpha}(\hat{\boldsymbol{\theta}})$ smaller as the stand sample size increases.

### 2.6. y-Modeling Method

#### 2.6.1. Model y-modeling Method

Models in the $y$-modeling method relate the forest structural variables in a population unit at different points in time with the auxiliary variables for that population unit. To indicate that these models directly consider the variables of interest, model parameters, stand level random effects and model errors will include the subscript $y$. Auxiliary variables considered in the $y$-modeling method include the LiDAR auxiliary variables for 2009 and 2015, (i.e., Set 1 and Set 2, respectively) plus the incoming solar radiation and the factors structural diversity and presence of prescribed fires, Set 4 for both 2009 and 2015. The modeling for the method started obtaining models for the variable of interest for 2009 and models for the variable of interest in 2015.

For given time $t$, the variable of interest in the $j^{th}$ population unit in the $i^{th}$ stand is expressed as:

$$y_{ijt} = x^t_{yijt}\beta_{yt} + u_{yit} + e_{yijt}, \tag{13}$$

where $x^t_{yijt}$ is a vector of auxiliary variables, specific for time $t$, in which the first element takes the value 1. The term $\beta_{yt}$ is a vector of time-specific coefficients with the first element representing the model intercept. The random components of model (13) are the stand-level random effects $u_{yit}$ and the model errors $e_{yijt}$. To account for heteroscedasticity, model errors $e_{yijt}$ were of the form $e_{yijt} = \varepsilon_{yijt}k^{\omega_{yt}}_{ijt}$ with $\varepsilon_{yijt} \sim N(0, \sigma^2_{y\varepsilon t})$; the term $k_{ijt}$, the predictor included in the model for time $t$, is most correlated to $y_t$, and $\omega_{yt}$ is a parameter to model the change in the error variance. The stand-level random effects $u_{yit}$ were assumed to be independently and identically distributed (i.i.d.) normal random variables $u_{yit} \sim N(0, \sigma^2_{yut})$ for all $i = 1, \ldots, D$, where $D$ is the total number of stands in the study area. Model errors were assumed independent of the stand-level random effects (i.e., $Cov(e_{yijt}, u_{ykt}) = 0$, for all $i$, $j$ and $k$). Finally, model errors were consider independent with $Cov(\varepsilon_{yijt}, \varepsilon_{yklt}) = 0$ if $i \neq k$ or $j \neq l$ for both $t = 2009$ and $t = 2015$. Models with spatially correlated errors were initially considered, but discarded for both years in the selection stage as no clear spatial correlation patterns were observed in the residuals. Auxiliary variables included in the model were selected following the same procedure used in the $\delta$-modeling method, using the best subsets selection procedure described in [27].

To account for expected correlations, models for 2009 and 2015 were combined into a single model where stand-level random effects and model errors for 2009 and 2015 were allowed to be time correlated. Then for the $j^{th}$ population unit in the $i^{th}$ stand the two-dimensional vector $y_{ij} = (y_{ij09}, y_{ij15})^t$ of variables of interest was related to the auxiliary variables through the following model:

$$y_{ij} = X_{ij}\beta_y + B_{ij}v_{yi} + e_{yij} \tag{14}$$

with:

$$X_{ij} = \begin{pmatrix} x^t_{yij09} & 0^t_{p15} \\ 0^t_{p09} & x^t_{yijt15} \end{pmatrix}, \beta_y = \begin{pmatrix} \beta_{y09} \\ \beta_{y15} \end{pmatrix}, B_{ij} = \begin{pmatrix} 1 & 1 & 0 \\ 1 & 0 & 1 \end{pmatrix}, v_{yi} = \begin{pmatrix} v_{yi} \\ v_{yi09} \\ v_{yi15} \end{pmatrix}, e_{yij} = \begin{pmatrix} e_{yij09} \\ e_{yij15} \end{pmatrix}, \tag{15}$$

where $0^t_{p2009}$ and $0^t_{p2015}$ are, respectively, row vectors of zeroes of dimensions equal to $x^t_{yij09}$ and $x^t_{yij2015}$.

As with the time-specific models, to account for heteroscedasticity in the combined model, model errors $e_{yijt}$ were of the form $e_{yijt} = \varepsilon_{yijt}k^{\omega_{yt}}_{ijt}$ with $\varepsilon_{yijt} \sim N(0, \sigma^2_{y\varepsilon t})$. The parameters $\omega_{yt}$ were updated when fitting the combined model. Spatial correlation of model errors was not found to be significant when considering each year separately, therefore, no spatial correlation patterns were considered in the combined model. The only source of correlation of model errors present in the combined model was temporal correlation. For a given location, the variables $\varepsilon_{yijtyij09} \sim N(0, \sigma^2_{y\varepsilon09})$ and $\varepsilon_{yij15} \sim N(0, \sigma^2_{y\varepsilon15})$ were allowed to be correlated random variables. The correlation between $\varepsilon_{yij2009}$ and $\varepsilon_{yij2015}$ is $\rho_\varepsilon$ and the variance-covariance matrix of $e_{yij}$ is:

$$Cov\begin{pmatrix} e_{yij09} \\ e_{yij15} \end{pmatrix} = R_{yij} = \begin{pmatrix} \sigma^2_{y\varepsilon09}k^{2\omega_{y09}}_{ij09} & \rho_\varepsilon\sigma_{09}k^{\omega_{y09}}_{ij09}\sigma_{09}k^{\omega_{y15}}_{ij15} \\ \rho_\varepsilon\sigma_{09}k^{\omega_{y09}}_{ij09}\sigma_{09}k^{\omega_{y15}}_{ij15} & \sigma^2_{yv}k^{2\omega_{y15}}_{ij15} \end{pmatrix}. \tag{16}$$

To model correlation between stand-level random effects, three random components $v_{yi}$, $v_{yi2009}$ and $v_{yi2015}$, independent of each other, were considered. These random components had distributions $v_{yi} \sim N(0, \sigma^2_{yv})$, $v_{yi09} \sim N(0, \sigma^2_{yv09})$ and $v_{yi15} \sim N(0, \sigma^2_{yv15})$. Stand-level random effect for a given point, at time $t$, $u_{yit}$, are the sum of a pure stand effect, independent of time $t$, $v_{yi}$, and a time-specific stand

random effect $v_{yi09}$ or $v_{yi15}$. The term $B_{ij}v_{yi} = u_{yi} = \begin{pmatrix} v_{yi} + v_{yi09} \\ v_{yi} + v_{yi15} \end{pmatrix} = \begin{pmatrix} u_{yi05} \\ u_{yi09} \end{pmatrix}$ represents these sums.
The variance covariance matrix of $v_{yi}$ is diagonal, therefore, the variance covariance matrix of $u_{yi}$ is:

$$Cov\begin{pmatrix} u_{yi09} \\ u_{yi15} \end{pmatrix} = G_{yi} = \begin{pmatrix} \sigma_{yv}^2 + \sigma_{yv15}^2 & \sigma_{yv}^2 \\ \sigma_{yv}^2 & \sigma_{yv}^2 + \sigma_{yv15}^2 \end{pmatrix} \tag{17}$$

The fact that the random effect $v_{yi}$ is present for both 2009 and 2015 results in a positive correlation of the terms of $u_{yi}$, with a correlation coefficient $\rho_u = \frac{\sigma_{yv}^2}{(\sigma_{yv}^2 + \sigma_{yv09}^2)(\sigma_{yv}^2 + \sigma_{yv15}^2)}$. In a last step, models with a simpler structure of random effects were fitted and compared to the original models using a likelihood ratio test. Simplified models contained only random effects $v_{yi}$ that did not depend on time (i.e., models did not contain time-specific random effects $v_{yi09}$ and $v_{yi15}$). For simplified models $u_{yi09} = u_{yi15} = v_{yi}$.

For a generic set of population units $\xi$, the combined model can be expressed in matrix notation as:

$$y_\xi = X_{y\xi}\beta_y + Z_{y\xi}v_y + e_{y\xi}, \tag{18}$$

where $y_\xi$, $e_{y\xi}$, and $X_{y\xi}$, are obtained stacking the vectors $y_{ij}$, $e_{yij}$, or the matrices $X_{ij}$ of all units in $\xi$. As no spatial correlation patterns were found, the variance covariance matrix of $e_{y\xi}$ is, $R_{y\xi} = diag_{i,j\in\xi}(R_{yij})$, a block diagonal matrix of dimension $2N_\xi x2N_\xi$ with $2x2$ blocks equal to $R_{yij}$. The vector of stand-level random effects $v = (v_{i1}^t, v_{i2}^t, \ldots, v_{iD}^t)^t$ and the matrix $Z_{y\xi}$ is an incidence matrix of dimension $2N_\xi xD$ for the simplified models and $2N_\xi x3D$ for the models with time-specific random effects. The variance covariance matrix of $y_\xi$ can be expressed as:

$$V_{y\xi}(\theta_y) = Z_{y\xi}G_y(\theta_y)Z_{y\xi}^t + R_{y\xi}(\theta_y). \tag{19}$$

In Equation (19), it is explicitly indicated that matrices $V_{y\xi}(\theta_y)$, $G_y(\theta_y)$ and $R_{y\xi}(\theta_y)$ depend on the vector of variance-covariance parameters $\theta_y = (\sigma_{yv}^2, \sigma_{yv09}^2, \sigma_{yv15}^2, \sigma_{ye09}^2, \sigma_{ye15}^2, \rho_\varepsilon)^t$. For the models with simplified random effects the vector of variance covariance parameters reduces to $\theta_y = (\sigma_{yv}^2, \sigma_{ye09}^2, \sigma_{ye15}^2, \rho_\varepsilon)^t$. Model (18) is a special case of linear mixed effect model with block diagonal covariance structure.

### 2.6.2. Target Parameter *y*-modeling Method

Under model (18) the target parameter (1) for a generic AOI, $\alpha$ is a linear combination of the form:

$$\Delta_\alpha = \frac{1}{6N_\alpha} \sum_{i=1}^{N_\alpha} (y_{i\alpha15} - y_{i\alpha09}) = l_{y\alpha}^t\beta_\delta + m_{y\alpha}^t u_y + q_{y\alpha}^t e_{y\alpha}, \tag{20}$$

where $l_{y\alpha}^t = q_{y\alpha}^t X_{y\alpha}$, $m_{y\alpha}^t = q_{y\alpha}^t Z_{y\alpha}$ and $q_{y\alpha}^t$ are vectors of known constants for the target AOI $\alpha$, with $q_{y\alpha}^t$ a vector of dimension $2N_\alpha$ where the $k^{th}$ element equals $\frac{(-1)^k}{6N_\alpha}$. It is important to remark that for models with a simplified structure of stand random effects, the target parameters do not depend on $u_y$. For these models, $y_{ij15} - y_{ij15} = (x_{yij15}^t\beta_{y15} - x_{yij09}^t\beta_{y09}) + (e_{yij15} - e_{yij09})$, and $u_{yi09} - u_{yi15} = v_{yi} - v_{yi} = 0$. For these type of models, one can expect significant gains in accuracy because it is not necessary to estimate random effects.

### 2.6.3. Estimator *y* modeling Method, and Estimator of the MSE

Model (18) is a linear mixed effects model with block diagonal structure and $\Delta_u$ a linear model parameter; thus, after [29] (pp. 108–110), the EBLUP $\hat{\Delta}_y(\hat{\theta}_y)$ of $\Delta_\alpha$ is:

$$\hat{\Delta}_{y\alpha}(\hat{\theta}_y) = l_{y\alpha}^t\hat{\beta}_y(\hat{\theta}_y) + m_{y\alpha}^t\hat{v}_y(\hat{\theta}_y), \tag{21}$$

where $\hat{\beta}_y(\hat{\theta}_y)$ equals:

$$\hat{\beta}_y(\hat{\theta}_y) = \{X_{ys}^t \hat{V}_{ys}(\hat{\theta}_y)^{-1} X_{ys}\}^{-1} X_{ys}^t \hat{V}_{ys}(\hat{\theta}_y)^{-1} y_s. \tag{22}$$

Matrices $\hat{V}_{ys}(\hat{\theta}_y)$, $\hat{G}_y(\hat{\theta}_y)$ and $\hat{R}_{ys}(\hat{\theta}_y)$ are obtained by replacing the estimated variance parameters $\theta_y$ in $V_{y\xi}(\theta_y)$, $G_y(\theta_y)$ and $R_{y\xi}(\theta_y)$, by their REML estimates $\hat{\theta}_y$. EBLUPs $\hat{\Delta}_\alpha$ are also functions of $\hat{y}$ and are obtained using formula (7), where $\hat{v}_\delta(\hat{\theta}_\delta)$ equals:

$$\hat{v}_y(\hat{\theta}_y) = \hat{G}_y(\hat{\theta}_y) Z_{ys}^t V_{ys}(\hat{\theta}_y)^{-1} \{y_s - X_{y\alpha}\hat{\beta}_y(\hat{\theta}_y)\}. \tag{23}$$

As with the $\delta$-modeling method, estimates for AOIs in unsampled stands were made assuming that the model fit for the sampled stands also applied in the unsampled stands, which leads to $m_{y\alpha}^t \hat{v}_y(\hat{\theta}_y) = 0$ and $\hat{\Delta}_{y\alpha}(\hat{\theta}_y) = l_{y\alpha}^t \hat{\beta}_y(\hat{\theta}_y)$ is a synthetic predictor.

For all AOIs, the estimator of the mean square error of the EBLUP under the $y$-modeling method, $\widehat{MSE}\{\hat{\Delta}_{y\alpha}(\hat{\theta}_y)\}$, is:

$$\widehat{MSE}\{\hat{\Delta}_{y\alpha}(\hat{\theta}_y)\} = g_{1y\alpha}(\hat{\theta}_\delta) + g_{2y\alpha}(\hat{\theta}_y). \tag{24}$$

The terms $g_{1y\alpha}(\hat{\theta}_y)$ and $g_{2y\alpha}(\hat{\theta}_y)$ in (24) are analogous to those in (10) and (11) and have similar interpretation. To compute $g_{1y\alpha}(\hat{\theta}_y)$ and $g_{2y\alpha}(\hat{\theta}_y)$, matrices $\hat{G}_{\delta s}(\hat{\theta}_\delta)$, $\hat{R}_{\delta s}(\hat{\theta}_\delta)$, $\hat{V}_{\delta s}(\hat{\theta}_\delta)$ and $\hat{R}_{\delta\alpha}(\hat{\theta}_\delta)$ must be replaced by $\hat{G}_{ys}(\hat{\theta}_y)$, $\hat{R}_{ys}(\hat{\theta}_y)$, $\hat{V}_{ys}(\hat{\theta}_y)$ and $\hat{R}_{y\alpha}(\hat{\theta}_y)$. For the $y$-modeling method we did not compute the second-order correction factors.

### 2.7. Comparison of Methods

Methods were compared using three different criteria. First, we used general measures of accuracy providing the average error or uncertainty of prediction at the pixel-level (2.7.1); then, we compared methods using AOI-specific estimates and measures of uncertainty (2.7.2). Finally, we assessed the risk of generating biased predictions when using the $\delta$-modeling method and $y$-modeling method in unsampled stands (Section 2.7.3).

#### 2.7.1. General Accuracy Assessment

To compare the $\delta$-modeling method and $y$-modeling method, a fist assessment was made using the cross-validated model mean squared error, *mRMSE*, and the model bias *mBias*:

$$mRMSE = \sqrt{\frac{\sum_{i,j\in s}^{n} (\delta_{ij} - \hat{\delta}_{ij})^2}{n}}, \tag{25}$$

$$mBias = \frac{\sum_{i,j\in s} (\delta_{ij} - \hat{\delta}_{ij})}{n}, \tag{26}$$

where $\delta_{ij}$ is the observed value of change for the $j^{th}$ plot included in the $i^{th}$ sampled stand and $\hat{\delta}_{ij}$ is the predicted value for that plot when model coefficients are obtained removing that plot from the training dataset. For the $y$-modeling method $\hat{\delta}_{ij}$ is obtained using the observed and fitted values of the variable of interest, as $\hat{\delta}_{ij} = \frac{1}{6}(\hat{y}_{ij15} - \hat{y}_{ij09})$ where $\hat{y}_{ij09}$ and $\hat{y}_{ij15}$ are the predictions of $y_{ij09}$ and $y_{ij15}$ are obtained fitting the corresponding $y$-model without the observations for plot $ij$. In addition, we computed *mRMSE* and *mBias* in terms relative to the average changes observed in the sampled plots. These quantities are denoted as $mRRMSE = mRMSE/\hat{\Delta}_f$ and $mRBias = mBias/\hat{\Delta}_f$ where $\hat{\Delta}_f$ is the mean of the changes observed in the field plots.

#### 2.7.2. AOI-specific Comparisons.

For each of the considered areas of interest an estimate by each method (i.e., EBLUPs using either $\hat{\Delta}_{\delta\alpha}(\hat{\theta}_y)$ or $\hat{\Delta}_{y\alpha}(\hat{\theta}_y)$) and their corresponding mean square error estimators (i.e., $\widehat{MSE}\{\hat{\Delta}_{\delta\alpha}(\hat{\theta}_y)\}$ or $\widehat{MSE}\{\hat{\Delta}_{y\alpha}(\hat{\theta}_y)\}$) were available. First, for each AOI and method, we directly compared $\hat{\Delta}_{\delta\alpha}(\hat{\theta}_y)$ and

$\hat{\Delta}_{ya}(\hat{\boldsymbol{\theta}}_y)$, and the square roots of $\hat{MSE}\{\hat{\Delta}_{\delta a}(\hat{\boldsymbol{\theta}}_\delta)\}$ and $\hat{MSE}\{\hat{\Delta}_{ya}(\hat{\boldsymbol{\theta}}_y)\}$. To simplify the notation, we will omit the subscript indicating the target AOI unless it is necessary and refer to $\hat{\Delta}_\delta$ as $\hat{\Delta}_y$. Similarly, after omitting the subscript $\alpha$, the AOI specific root mean square errors will be denoted as:

$$RMSE_\delta = \sqrt{\hat{MSE}\{\hat{\Delta}_{\delta a}(\hat{\boldsymbol{\theta}}_\delta)\}}, \tag{27}$$

$$RMSE_y = \sqrt{\hat{MSE}\{\hat{\Delta}_{ya}(\hat{\boldsymbol{\theta}}_y)\}}, \tag{28}$$

To perform an assessment relative to the predicted values the following coefficient of variation:

$$CV\{\hat{\Delta}_{\delta a}(\hat{\boldsymbol{\theta}}_\delta)\} = \frac{\sqrt{\hat{MSE}\{\hat{\Delta}_{\delta a}(\hat{\boldsymbol{\theta}}_\delta)\}}}{\hat{\Delta}_{\delta a}(\hat{\boldsymbol{\theta}}_\delta)}, \tag{29}$$

$$CV\{\hat{\Delta}_{ya}(\hat{\boldsymbol{\theta}}_y)\} = \frac{\sqrt{\hat{MSE}\{\hat{\Delta}_{ya}(\hat{\boldsymbol{\theta}}_y)\}}}{\hat{\Delta}_{ya}(\hat{\boldsymbol{\theta}}_y)}, \tag{30}$$

was computed for each AOI and method. Finally, for each AOI we compared $CV\{\hat{\Delta}_{\delta a}(\hat{\boldsymbol{\theta}}_y)\}$ to $CV\{\hat{\Delta}_{ya}(\hat{\boldsymbol{\theta}}_y)\}$. To simplify the notation we will refer to these coefficients of variation as $CV_\delta$ and $CV_y$. Finally, for each sampled AOI we computed, using only the field information, the sample mean $\hat{\Delta}_{fa}$ and its standard error:

$$SE_{fa} = \sqrt{\frac{\sum_{i,j\in s}^{n_\alpha}(\delta_{ij} - \hat{\Delta}_{f,\alpha})^2}{(n_\alpha - 1)n_\alpha}}, \tag{31}$$

and its coefficient of variation $CV_{fa}$. In Equation (31), $n_\alpha$ is the number of field plots in the considered AOI and the sub-index $f$ is used to indicate that these quantities are calculated using only field data. Again, to simplify the notation we removed the sub-indexes $\alpha$ unless they were necessary. Finally, the sample mean and the coefficient of variation were then compared to their counterparts (7) and (29) and (21) and (30) obtained by the $\delta$-modeling method and $y$-modeling method, respectively.

### 2.7.3. Extrapolation to Thinned Stands

The fact that thinned stands were not represented in the sample of field plots raises the question of how applicable the models obtained are using either the $\delta$-modeling method or the $y$-modeling method to these stands. Applying the models to these stands involves a degree of extrapolation to a different population and a high risk of producing biased predictions. We assessed this risk by comparing the distributions of the LiDAR predictors included in the models for the $\delta$-modeling method and the $y$-modeling method for the sample of field plots to the distributions of the predictors in: (1) the sampled stands, (2) the unsampled and not thinned stands and (3) unsampled and thinned stands. Within each group (i.e., field plots, FP; sampled stands not thinned, SS; unsampled stands not thinned, UN; and unsampled stands subject to thinning, UT), we estimated density functions for each LiDAR predictor using a Gaussian kernel and a bandwidth determined using Silverman's rule [33]. Note that we considered two AOIs for the largest level of aggregation, the first one is SS and the other one, SA, is the union of SS and UN. We first considered each predictor separately and graphically compared their density functions. Predictors for 2009 and 2015 in the $y$-modeling method were considered separately. For each predictor and group, we computed the area of overlap, AO, with the density function for the field plots which takes value 0 if there is no overlap and value 1 if the distribution of the predictor in the considered group equals the distribution for the sample.

In addition to the area of overlap and aiming to consider all predictors in a given model at once, we calculated $\overline{NT2}$, the average of Mesgaran's novelty index $NT2$ for each model and group [34]. This quantity provides the average Mahalanobis distance from the pixels of the group to the mean of the sample of field plots, and it is expressed in terms relative to the maximum Mahalanobis distance observed in the sample. Values of $\overline{NT2}$ above one indicate that on average pixels in a group are at a distance to the mean of the field plots larger than the distance from the extreme field observation to the mean of the field plots. We also calculated $\overline{NT2}_{mean}$, the average of $NT2$, but using the mean Mahalanobis distance as normalizing constant instead of the maximum. The reference value of one for $\overline{NT2}_{mean}$ indicates that the average Mahalanobis distance from pixels, to the mean of the field plots, is the same as the average of the Mahalanobis distances observed in the sample. Means and variance covariance matrices for computation of Mahalanobis distances are always estimated using the sample of field plots.

## 3. Results

### 3.1. Selected Models δ-modeling Method and y-modeling Method

Selected models for the δ-modeling method included auxiliary variables from Set 1 and Set 3. It was possible to find alternative models including fixed effects for the diversity treatments (i.e., predictors from Set 4) with similar values of *mRMSE* and *mRBias*; however, those models did not improve the model fit. From a practical point of view, models that only depend on the LiDAR variables but do not depend on the structural diversity treatments or the presence/absence of prescribed fires make the models more portable and applicable to stands without needing to know exactly which one of these treatments was applied. Considering that models using the structural diversity and presence of prescribed fires as predictors did not result in important gains in accuracy, we selected models that were not dependent on these treatments (Table 2).

**Table 2.** Summary models for the δ-modeling method. Model coefficients, standard errors of the model coefficients, variance parameters and general metrics for accuracy assessment are provided. Predictor acronyms are explained in Table A1. Coef is the value of the coefficient and Std.Error its corresponding standard error. V indicates volume, BA indicates basal area and B indicates biomass.

| Model | Predictor | Coef | Std. Error | $\hat{\sigma}^2_{\delta v}$ | $\hat{\sigma}^2_{\delta \varepsilon}$ | *mRMSE* | *mRRMSE* | *mBias* | *mRBias* |
|---|---|---|---|---|---|---|---|---|---|
| V(m$^3$ ha$^{-1}$ year$^{-1}$) | Intercept | 1.16 | 0.31 | 0.50 | 10.53 | 3.47 | 241.99% | $-1.83 \times 10^{-4}$ | $-0.01\%$ |
| | δElev_P50$_{15\text{-}09}$ | 1.33 | 0.27 | | | | | | |
| | δPcFstAbv2$_{15\text{-}09}$ | 0.23 | 0.07 | | | | | | |
| BA(m$^2$ ha$^{-1}$ year$^{-1}$) | Intercept | 0.31 | 0.12 | 0.01 | 0.14 | 0.39 | 116.30% | $-8.2 \times 10^{-4}$ | $-0.25\%$ |
| | δPcAllAbv2$_{15\text{-}09}$ | 0.05 | 0.01 | | | | | | |
| | Elev_P75$_{09}$ | $-0.03$ | 0.01 | | | | | | |
| | PcAllAbv2$_{15\text{-}09}$ | 0.02 | <0.01 | | | | | | |
| B(Mg ha$^{-1}$ year$^{-1}$) | Intercept | 1.03 | 0.17 | 0.19 | 2.52 | 1.72 | 180.20% | $-1.09 \times 10^{-3}$ | $-0.11\%$ |
| | δElev_var$_{15\text{-}09}$ | 0.05 | 0.02 | | | | | | |
| | δElev_P50$_{15\text{-}09}$ | 1.03 | 0.20 | | | | | | |
| | δCRR$_{15\text{-}09}$ | $-16.67$ | 6.58 | | | | | | |

For the models in the δ-modeling method, the variance of random the effects, $\hat{\sigma}^2_{\delta v}$, was very small compared to the variance of the model errors, $\hat{\sigma}^2_{\delta \varepsilon}$, (Table 2). This indicated that, in this forest and for these variables, the use of synthetic estimators that do not account for the variability between stands should not cause a strong bias problem.

Models for the y-modeling method showed a pattern similar to that observed for the δ-modeling and only included predictors from Set 1 and Set 2 (Table 3). Errors showed non-constant variance patterns for all variables. The predictor most correlated with the variable of interest (i.e., the predictor used to model the error variance) was the same for 2009 and 2015. For V and B, the variance of model errors was a function of the square of the mean LiDAR elevation (Elev_mean$^2$), and the exponents of the error variance function were very close to those obtained in [4,17,27] for V, and in [27] for B. For BA, variance of model errors was a function of the percentage of first returns above two meters (PcFstAbv2). Based on the results of the likelihood ratio tests, that for all variables resulted in p-values larger than 0.87, simplified models were selected and used for prediction.

**Table 3.** Summary models for the $y$-modeling method. Model coefficients, standard errors of the model coefficients, variance-covariance parameters and general metrics for accuracy assessment are provided. Covariate acronyms are explained in Table A1. Coef is the value of the coefficient and Std.Error its corresponding standard error. Coef is the value of the coefficient and Std.Error its corresponding standard error. V indicates volume, BA indicates basal area and B indicates biomass.

| Model | Year | Covariate | Coef | Std.Error | $\hat{\sigma}^2_{yu}$ | Kijt | $\omega_{yt}$ | $\hat{\sigma}^2_{yt\varepsilon}$ | $\rho_e$ | General Accuracy Metrics for Change Per Hectare and Year | | | |
|---|---|---|---|---|---|---|---|---|---|---|---|---|---|
| | | | | | | | | | | *mRMSE* | *mRRMSE* | *mBIAS* | *mRBias* |
| V(m$^3$ ha$^{-1}$) | 2009 | Intercept | −19.09 | 10.36 | 640.29 | Elev_mean$^2_{09}$ | 0.64 | 3.00 | 0.85 | 3.76 | 262.62% | 0.13 | 9.24% |
| | | Elev_mean$^2_{09}$ | 2.52 | 0.23 | | | | | | | | | |
| | | PcFstAbv2$_{09}$ | 0.63 | 0.05 | | | | | | | | | |
| | 2015 | Intercept | 2.69 | 0.23 | | Elev_mean$^2_{15}$ | 0.61 | 4.17 | | | | | |
| | | Elev_mean$^2_{15}$ | 0.69 | 0.05 | | | | | | | | | |
| | | PcFstAbv2$_{15}$ | −26.30 | 11.10 | | | | | | | | | |
| BA(m$^2$ ha$^{-1}$) | 2009 | Intercept | −0.22 | 1.57 | 7.42 | PcFs-Abv2$_{09}$ | 0.48 | 0.81 | 0.85 | 0.47 | 138.06% | 0.01 | 1.53% |
| | | Elev_P10$_{09}$ | −1.37 | 0.34 | | | | | | | | | |
| | | Elev_P30$_{09}$ | 1.58 | 0.24 | | | | | | | | | |
| | | PcFstAbv2$_{09}$ | 0.51 | 0.03 | | | | | | | | | |
| | 2015 | Intercept | −2.16 | 0.61 | | PcFs-Abv2$_{15}$ | 0.45 | 1.12 | | | | | |
| | | Elev_P10$_{15}$ | 2.56 | 0.51 | | | | | | | | | |
| | | Elev_P20$_{15}$ | 0.57 | 0.03 | | | | | | | | | |
| | | PcFstAbv2$_{15}$ | −0.97 | 1.72 | | | | | | | | | |
| B(Mg ha$^{-1}$) | 2009 | Intercept | −11.86 | 5.19 | 165.69 | Elev_mean$^2_{09}$ | 0.71 | 0.39 | 0.85 | 1.94 | 203.69% | 0.08 | 8.60% |
| | | Elev_mean$^2_{09}$ | 1.19 | 0.11 | | | | | | | | | |
| | | PcFstAbv2$_{09}$ | 0.34 | 0.02 | | | | | | | | | |
| | 2015 | Intercept | 1.31 | 0.12 | | Elev_mean$^2_{15}$ | 0.58 | 1.47 | | | | | |
| | | Elev_mean$^2_{15}$ | 0.37 | 0.03 | | | | | | | | | |
| | | PcFstAbv2$_{15}$ | −14.15 | 5.76 | | | | | | | | | |

### 3.2. General Accuracy Assessment and Comparison of Methods

For all variables and modeling alternatives, values of *mBias* and *mRBias* were orders of magnitude smaller than *mRMSE* and *mRRMSE* (Tables 2 and 3). For all variables and methods, the percentages of explained variance for the change in V, BA and B were low. For the $\delta$-modeling method, models explained 34.38%, 31.37% and 39.04% of the variance of the change in V, BA and B, respectively. For the *y*-modeling method, models explained only 10.65% and 5.37% of V and B, respectively, while for BA the prediction using the *y*-modeling method was not better than the sample mean. In addition, $\delta$-models had values of *mRBias* lower than those obtained for the *y*-models. When instead of the change we considered the forest structural attributes with the *y*-modeling method, percentages of explained variance were 82.16% for V, 82.53% for BA and 82.93% for B. Considering only 2009, the percentage of explained variance for V, BA and B was 81.60%, 83.45%, and 82.84%, respectively. Considering only 2015, the percentage of explained variance for V, BA and B was 82.72%, 81.42% and 82.98%, respectively.

### 3.3. AOI-Specific Estimates

#### 3.3.1. Entire Study Area

Estimates for the sampled stands and for the whole study area using either the $\delta$-modeling method or the *y*-modeling method were consistent with the estimates obtained using only the field information except for BA and B in SA. For the entire study area values of $RMSE_\delta$ tended to be smaller than $RMSE_y$. When considering the sampled stands, SS, approximate confidence intervals computed as $\hat{\Delta}_f \pm 2SE_f$ for the field estimates, and as $\hat{\Delta}_\delta \pm 2RMSE_\delta$ and $\hat{\Delta}_y \pm 2RMSE_y$ for each one of the LiDAR based methods, overlapped for all variables (Table 4) and contained estimates derived from other methods. Differences between the uncertainty of estimates obtained from LiDAR-based methods and the uncertainty of estimates obtained from field-based methods tend to be of small magnitude.

**Table 4.** Average increments of volume V, basal area BA and biomass B in the entire study area excluding the thinned stands (SA) and for the union of the sampled stands (SS). Estimates ($\hat{\Delta}$), root mean square errors ($RMSE$), coefficients of variation ($CV$) and confidence intervals ($CI$) obtained using the $\delta$-modeling method and the *y*-modeling method are compared to estimates ($\hat{\Delta}_f$), standard errors ($SE_f$) coefficients of variation ($CV_f$), and confidence intervals ($CI_f$) using only the field information.

| Variable | Area | $\hat{\Delta}_\delta$ | $RMSE_\delta$ | $CV_\delta$ | $CI_\delta$ | | $\hat{\Delta}_y$ | $RMSE_y$ | $CV_y$ | $CI_y$ | | $\hat{\Delta}_f$ | $SE_f$ | $CV_f$ | $CI_f$ | |
|---|---|---|---|---|---|---|---|---|---|---|---|---|---|---|---|---|
| V(m$^3$ ha$^{-1}$ | SS | 1.66 | 0.27 | 16.29% | 1.12 | 2.20 | 1.95 | 0.32 | 16.48% | 1.31 | 2.60 | 1.43 | 0.32 | 22.21% | 0.80 | 2.07 |
| year$^{-1}$) | SA | 1.67 | 0.30 | 17.98% | 1.07 | 2.27 | 1.98 | 0.29 | 14.67% | 1.40 | 2.56 | | | | | |
| BA(m$^2$ ha$^{-1}$ | SS | 0.36 | 0.03 | 8.68% | 0.30 | 0.42 | 0.37 | 0.04 | 9.93% | 0.30 | 0.45 | 0.34 | 0.04 | 10.87% | 0.26 | 0.41 |
| year$^{-1}$) | SA | 0.42 | 0.04 | 8.41% | 0.35 | 0.49 | 0.44 | 0.04 | 9.61% | 0.35 | 0.52 | | | | | |
| B(Mg ha$^{-1}$ | SS | 1.07 | 0.13 | 12.35% | 0.81 | 1.34 | 1.24 | 0.17 | 13.61% | 0.90 | 1.57 | 0.95 | 0.16 | 16.89% | 0.63 | 1.28 |
| year$^{-1}$) | SA | 1.15 | 0.16 | 13.66% | 0.83 | 1.46 | 1.29 | 0.15 | 11.83% | 0.98 | 1.59 | | | | | |

#### 3.3.2. Stands

Estimated change of V, BA and B in the sampled stands by both the $\delta$-modeling method and the *y*-modeling method agreed with their field-based counterparts in most stands (Figure 2). However, the width of the confidence intervals obtained using the $\delta$-modeling method tended to be larger than the confidence intervals of the estimates derived using the *y*-modeling method (Figure 2).

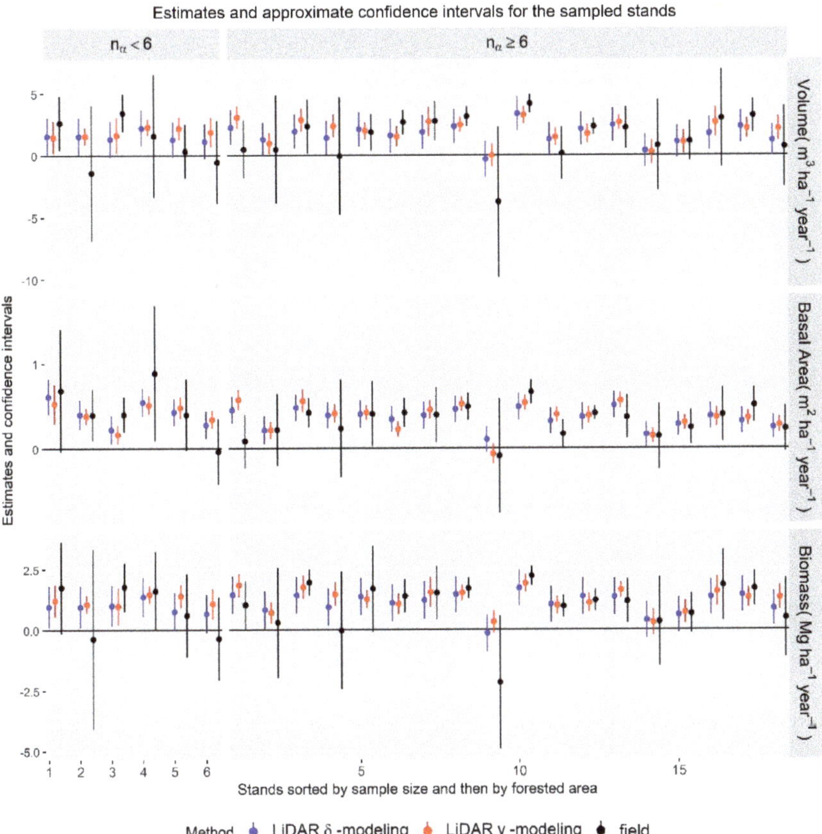



**Figure 2.** Estimates of V, BA and B change for the sampled stands of Blacks Mountains Experimental Forest. LiDAR-derived estimates using the δ-modeling method are indicated by blue dots, LiDAR-derived estimates obtained using the y-modeling method are indicated with red dots and field-based estimates are indicated using black.

For unsampled stands, estimates and confidence intervals had larger variability in stands where the forested area was small (Figure 3). This variability cannot be avoided, and indicates that certain sources of errors cannot be compensated if the number of pixels that are aggregated is low. Finally, for both methods, values of $RMSE_\delta$ and $RMSE_y$ were in the range of 0.25 to 1 m³ ha⁻¹year⁻¹ for V, of 0.02 to 0.15 m² ha⁻¹year⁻¹ for BA and of 0.10 to 0.80 Mg ha⁻¹year⁻¹ for B. However, for B and V, the $RMSE_y$ tended to be smaller than $RMSE_\delta$ while negligible differences between methods were observed for BA (Figures 3 and 4).

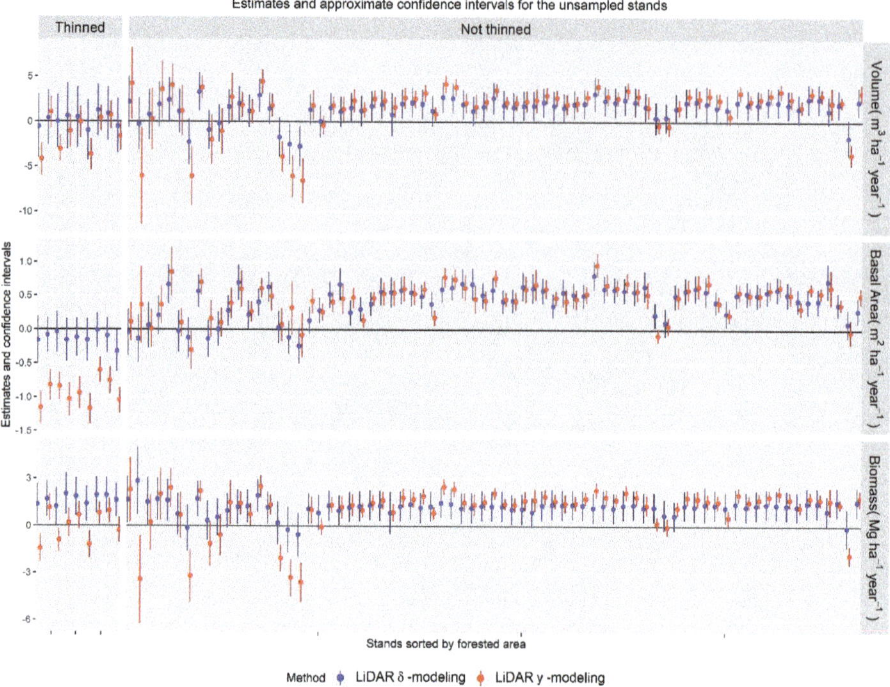



**Figure 3.** Estimates of V, BA and B change for the unsampled stands of Blacks Mountains Experimental Forest. LiDAR-derived estimates using the $\delta$-modeling method are indicated by blue dots and LiDAR-derived estimates obtained using the $y$-modeling method are indicated with red dots. Thinned stands are to the left and non-thinned stands to the right.

For the thinned (and unsampled) stands, differences between $\delta$-modeling method and the $y$-modeling method for BA were large and their confidence intervals did not overlap (Figure 3). For these stands, the estimates for BA using the $\delta$-modeling method tended to indicate almost no changes in BA. Estimates for the thinned stands using the $\delta$-modeling method provided inconsistent results indicating gains in B, and changes close to zero for V and BA. Certain inconsistencies were also observed for stands subject to thinning when using the $y$-modeling method where predictions of the change in V and B were positive for three and five stands respectively. These inconsistencies seem to derive from the fact that the distribution of predictors in Set 3 (i.e., changes in LiDAR predictors) in the thinned stands was rather different to the distribution of these predictors in the sample of field plots, in the sampled stands and in the unsampled and not thinned stands. For predictors of the $y$-modeling method, modeled differences between thinned stands and the remaining groups were of much smaller magnitude. Results for the analysis of the extrapolation risks are presented in detail in Section 3.4.

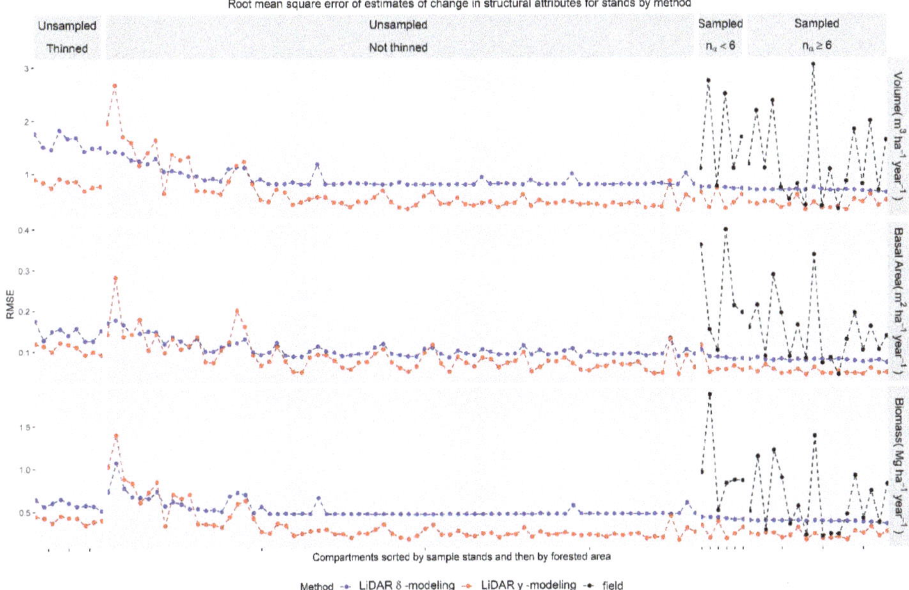


**Figure 4.** Values of $RMSE_\delta$ (blue), $RMSE_y$ (red) and $SE_f$ (black) for the stand-level estimates of V, BA and B.

### 3.3.3. Pixel-level

For both methods, inconsistencies observed at the stand level were observed at the pixel-level, especially the positive predictions of change obtained with the $\delta$-modeling method in the thinned stands (Figure A1). In addition, due to the low correlations of LiDAR predictors with the change in V, BA, and B, predictions at this level have large uncertainties. Mean and median values of $RMSE_\delta$ were 2.30 m³ ha⁻¹ year⁻¹ and 3.30 m³ ha⁻¹ year⁻¹ for V, 0.39 m² ha⁻¹ year⁻¹ and 0.38 m² ha⁻¹ year⁻¹ for BA, and 1.67 Mg ha⁻¹ year⁻¹and 1.65 Mg ha⁻¹ year⁻¹ for B. Mean and median values of $RMSE_y$ were 2.49 m³ ha⁻¹ year⁻¹ and 2.20 m³ ha⁻¹ year⁻¹ for V, 0.48 m² ha⁻¹ year⁻¹ and 0.48 m² ha⁻¹ year⁻¹ for BA and 1.89 Mg ha⁻¹ year⁻¹ and 1.76 Mg ha⁻¹ year⁻¹ for B (Table 5 and Figure A1). Predictions from the $\delta$-modeling method tend to be smoother than predictions from the $y$-modeling method. For all variables, the proportion of pixel-level predictions using the $\delta$-modeling method within the range of values observed for the field plots, was always 99.84% or larger (Figure A2). Considering that these results were obtained in the presence of thinned stands and the relatively small fraction of the forest that was sampled, obtaining less than 0.16% of the predictions outside of the measurement range seems to be a clear sign of over smoothing (see Appendix B).

**Table 5.** Minimum (Min), 5th percentile (p05), mean, median, 95th percentile (p95) and maximum (Max) of $RMSE_\delta$ (27) and $RMSE_y$ (28) for the pixels of the study area.

| Variable | Method | Min | p05 | Mean | Median | p95 | Max |
|---|---|---|---|---|---|---|---|
| V(m³ ha⁻¹ year⁻¹) | $\delta$-modeling method | 0.42 | 0.42 | 2.30 | 3.30 | 3.59 | 9.41 |
| | $y$-modeling method | 0.08 | 0.37 | 2.49 | 2.20 | 6.01 | 32.69 |
| BA(m² ha⁻¹ year⁻¹) | $\delta$-modeling method | 0.38 | 0.38 | 0.39 | 0.38 | 0.40 | 0.59 |
| | $y$-modeling method | 0.11 | 0.30 | 0.48 | 0.48 | 0.64 | 1.47 |
| B(Mg ha⁻¹ year⁻¹) | $\delta$-modeling method | 1.62 | 1.63 | 1.67 | 1.65 | 1.76 | 4.57 |
| | $y$-modeling method | 0.47 | 1.10 | 1.89 | 1.76 | 3.09 | 10.45 |

### 3.4. Extrapolation to Thinned Stands

Estimates of change in B for the thinned stands by both methods were clearly subject to bias problems. The predicted change in B for the total area subject to thinning for the period 2009–2015, using the $\delta$-modeling method was an increase in biomass of 40,469.22 Mg. The predicted change using the $y$-modeling method was a removal of B. However, the predicted removal for the period 2009–2015 was only 1750.29 Mg while the weighted extractions for the thinned stands were orders of magnitude larger. For BA both methods estimated extractions in BA, which is consistent with the fact that these stands were thinned. Estimated changes in BA using the $\delta$-modeling method for the thinned stands ranged from $-0.05$ m$^2$ ha$^{-1}$ to $-1.88$ m$^2$ ha$^{-1}$, which seems to be a very small change in basal area. Estimated changes in BA using the $y$-modeling method ranged from $-3.58$ m$^2$ ha$^{-1}$ to $-7.01$ m$^2$ ha$^{-1}$. An advantage of the $y$-modeling method is that it allows obtaining the values of the structural attributes at a given point in time. Using the $y$-model we estimated BA for the thinned stands for 2015. For those stands where thinning prescriptions dictated leaving a residual BA of 17.22 m$^2$ ha$^{-1}$ to 25.25 m$^2$ ha$^{-1}$, estimated BA for 2015 ranged from 19.87 m$^2$ ha$^{-1}$ to 26.22 m$^2$ ha$^{-1}$, which is in accordance with the thinning prescriptions. For the remaining area subject to thinning the estimated BA for 2015 was 17.64 m2 ha$^{-1}$, while the prescriptions dictated leaving a residual BA ranging from 6.89 m$^2$ ha$^{-1}$ to 13.77 m$^2$ ha$^{-1}$ in 75% of the area and leaving the remaining area untouched. In general, the estimated BA for 2015 are consistent with the prescriptions, which indicates that the $y$-modeling method produces reasonable estimates of BA when extrapolating to the thinned stands. In summary, for the estimation of changes, biases derived from extrapolation seemed to be of larger magnitude for the $\delta$-modeling method although they were also present for the $y$-modeling method.

The extrapolation indexes $\overline{NT2}$ and $\overline{NT2}_{mean}$ showed that predictions in thinned stands, involved a large amount of extrapolation when using the $\delta$-modeling method. For the $y$-modeling method, differences between thinned stands and stands not subject to thinning were of smaller magnitude (Figures 5 and A3). The inspection of the distribution of the LiDAR predictors in the field plots, the sampled and not thinned stands, unsampled and not thinned stands and the unsampled and thinned stands showed similar results for all variables, being the distributions of predictors from Set 3 (i.e., changes in LiDAR predictors for the period 2009–2015) very sensitive to the thinning operations (Figures 6, A4 and A5).

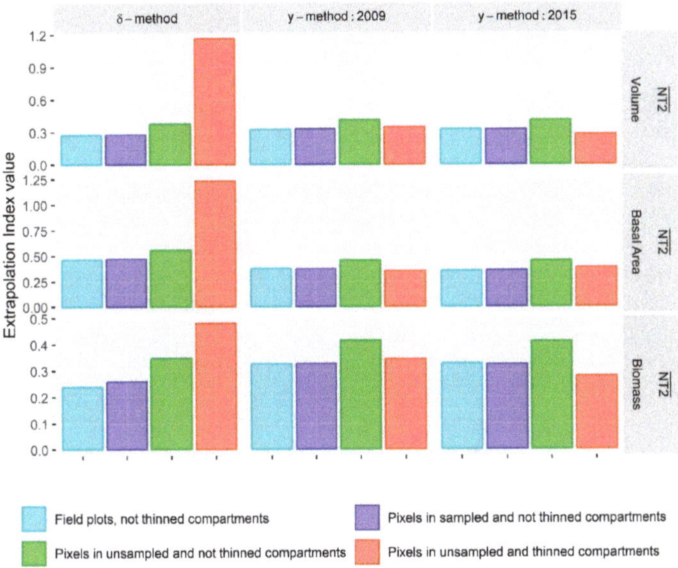



**Figure 5.** Indexes of extrapolation. Average of Mesgaran's novelty index [35], $\overline{NT2}$, for the sampled and not thinned stands (dark blue), unsampled stands not thinned (green) and unsampled and thinned stands (red). The value of this index for the field plots (light blue) provides the baseline value (i.e., the value observed for the sample of field plots).

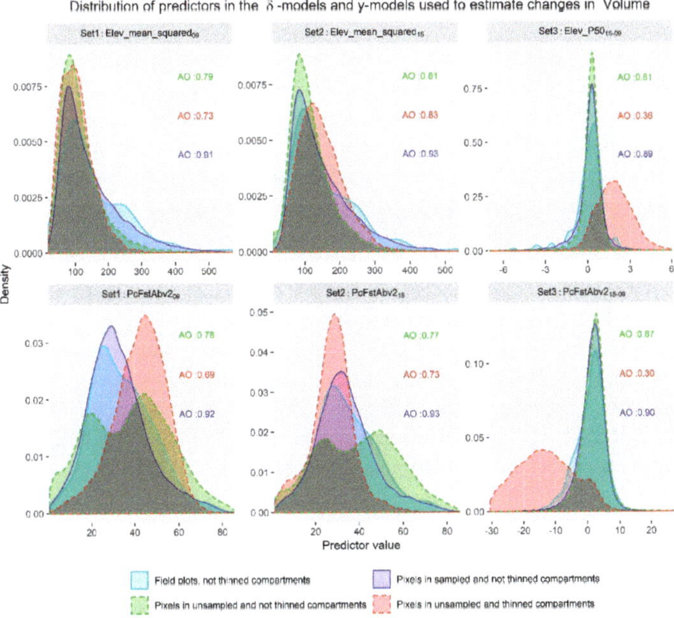



**Figure 6.** Comparison of density functions for the predictors in the models used to estimate changes in Volume using the δ-modeling method and $y$-modeling method in field plots (light blue), sampled and not thinned stands (dark blue), unsampled and not thinned stands (light blue) and unsampled and thinned stands (red). For each group the area of overlap, AO, with the density function for the field plots (green) is provided for each predictor.

## 4. Discusion

### *4.1. General Accuracy Assessment and Comparison of Methods.*

The smallest values of *mRMSE* were obtained using the $\delta$-method, which is consistent with previous results reported by Poudel et al. [8] for V and B in coastal coniferous forest of Western Oregon and by Temesgen et al. [9] for B in spruce-dominated forest of Alaska (Tables 2 and 3). We observed, however, smaller differences between methods. Additionally, as observed in previous studies, (e.g., [8,21,35,36]) where LiDAR auxiliary variables showed a much stronger correlation with structural attributes at a given point in time than with their change.

Values of *mRMSE* for V were 3.47 m$^3$ ha$^{-1}$year$^{-1}$ when using the $\delta$-method and 3.76 m$^3$ ha$^{-1}$ year$^{-1}$ when using the *y*-method. These values are slightly smaller than the *mRMSE* obtained by Poudel et al. [8] using the $\delta$-method (4.74 m$^3$ ha$^{-1}$ year$^{-1}$) and two lidar acquisitions separated in time by five years. For B, *mRMSE* using the $\delta$-method and the *y*-method were, respectively, 1.72 Mg ha$^{-1}$year$^{-1}$ and 1.94 Mg ha$^{-1}$ year$^{-1}$. These values were very close to those reported by Poudel et al. [8] using the $\delta$-method (1.88 Mg ha$^{-1}$ year$^{-1}$) and worse than those reported by Temesgen et al. (1.25 Mg ha$^{-1}$year$^{-1}$ and 1.63 Mg ha$^{-1}$ year$^{-1}$), also using two LiDAR acquisitions separated in time by five years. Values of *RMSE* for BA were similar to those obtained by Næsset and Gobakken [20] in coniferous forest in Norway, using the *y*-method with log-transformed models and two LiDAR acquisitions that were two years apart from each other. In relative terms, for V and B, the values that we obtained for *mRRMSE* were considerably larger than those obtained by Poudel et al. [8]. These differences are due to the fact that observed growth rates in Poudel et al [8] are much higher than we observed at BMEF.

### *4.2. AOI-Specific Estimates*

#### 4.2.1. Entire Study Area

Most studies on estimation of change of structural variables using repeated LiDAR measurements have focused on analyzing indexes of model fit and reported only global measures of accuracy developed at a plot level. There is an important difference between the values of $RMSE_\delta$ and $RMSE_y$ and $mRMSE_\delta$ and $mRMSE_y$ being $mRMSE_\delta$ and $mRMSE_y$ an order of magnitude larger than $RMSE_\delta$ and $RMSE_y$. Model root mean square errors $mRMSE_\delta$ and $mRMSE_y$ provide an average measure of the errors that can occur when predicting a single pixel. For large areas, there will be some level of compensation of overpredicted and underpredicted pixels. Knowing how important that compensation is requires calculating AOI-specific root mean square errors. These AOI-specific measures cannot be directly derived from *mRMSE* because $RMSE_\delta$ and $RMSE_y$ consider factors such as the uncertainty in the estimation of the fixed and random effects that are not accounted for in $mRMSE_\delta$ or $mRMSE_y$. The effect of these factors in $RMSE_\delta$ and $RMSE_y$ can cause that the way two models with similar values of *mRMSE* rank based on this metric could change when attending to *RMSE*. But the most important consequence of the disconnect between $mRMSE_\delta$ and $mRMSE_y$ and AOI-specific measures of uncertainty, is that the former cannot be used as quality controls in LiDAR based inventories.

While numerous studies on estimation of changes using LiDAR rely on global measures of accuracy such as *mRMSE*, exceptions to this trend can be found in the literature [10,12,18–20]. The last four studies used model assisted techniques to derive either landscape or stratum level changes. Reported errors in those studies changed depending on the modeling techniques and study areas, but they all were of similar magnitude for changes in biomass per hectare and year (Table 4). Errors for the methods tested in this study were smaller than those reported by [10], where changes in live carbon stocks in Norway were estimated using generalized regression estimators (GREG). Differences with the errors reported in [10] for carbon, using the same 0.5 biomass to carbon conversion factor, were in the range of 0.12–0.09 Mg ha$^{-1}$ year$^{-1}$. These differences seem to be due to multiple factors such as differences between study areas, changes in live biomass versus changes in standing biomass, time between LiDAR acquisitions and field plot sizes etc. Further investigation is needed to test if the model-based estimators studied here and the GREG estimators in [10] have a similar performance when used under the same conditions.

The study from Magnussen et al. [18] also included model based estimators using the $y$-modeling method. Reported errors were slightly larger than the ones observed here but at the same time smaller in terms relative to the observed mean change. An important result from the comparisons of [18] was the drastic improvement in model accuracy when developing stratum specific models (i.e., a set of model coefficients per stratum) as opposed to a global model for the whole study area. The mixed effect models used in this study can be used in combination with stratification if sample sizes are large. The introduction of stand level random effects allows for certain variability between AOIs that can be applied in situations where AOI sample sizes are limited.

### 4.2.2. Stands

One of the novelties of this study was the analysis of estimates for AOIs with small sample sizes to develop AOI specific models (i.e., stands). While at large scales both LiDAR-based and field-based estimates were very similar and had equivalent accuracies, at the stand-level, LiDAR based estimates, clearly had smaller errors than their field-based counterparts do. Qualitatively, this result for the change in V, BA and B is similar to the results obtained in [15,17] for the structural variables themselves and shows that the LiDAR auxiliary information allows for gains in efficiency when estimating changes in AOIs with small sample sizes. However, due to the low correlation of LiDAR and structural changes, values of $CV_\delta$ and $CV_y$ were oftentimes larger than 50%. These values of $CV$ are larger than those observed for structural variables in similar AOIs in previous studies [4,14,17]. While differences were not of large magnitude $RMSE_y$, tended to be smaller than $RMSE_\delta$. In addition, $RMSE_y$ had a larger variability because errors did not have constant variance. Finally, stand level estimates using the $\delta$-modeling method in the thinned and unsampled stands involved an important degree of extrapolation that can cause inconsistent estimates and severe biases, which indicates that the $\delta$-modeling method is more sensitive to extrapolating than the $y$-modeling method.

### 4.2.3. Pixel-level

For the most detailed level of disaggregation, the magnitude of the errors was very large. This is due to the low correlation between LiDAR auxiliary variables and the change in structural attributes. First-order and second-order texture indexes [37] are auxiliary variables with a promising potential for future research aiming to improve the prediction of structural changes. While for structural variables, maps at the pixel-level can provide a reliable reference about the forest structure; for growth and changes, pixel-level maps like the one in Figure A1 should be taken as mere approximations. They could be used to infer certain trends and patterns, but the high values of $RMSE_\delta$ and $RMSE_y$ show that estimates for a particular location made at the pixel scale can differ significantly with reality. These results clearly indicate that, predictions at such a fine scale are highly unreliable, and it is necessary either to perform some level of spatial aggregation or to increase the lapse between LiDAR acquisitions.

### 4.3. Advantages of Modeling Alternatives

In general, the $\delta$-modeling method was found to be a better alternative to estimate changes for the entire study area than the $y$-modeling method; however, the $y$-modeling method produced better results at the stand-level and also seemed to be advantageous to prevent problems related to extrapolation to values of the covariates outside of those included in model development.

The $\delta$ modeling method offers a faster model developments and fitting, and is significantly simpler than the modeling with the $y$-modeling method, as it is not necessary to consider differences between years and time correlations. The main disadvantage of this method is that it seems to be more prone to extrapolation errors. Predictors from Set 3 are sensitive to intense changes in the forest structure caused for example by thinning (see Figures 5, 6, A4 and A5). The inspection of predictors of alternative models for V and B using this method revealed that inconsistencies of predictions in unthinned stands could be attenuated including more predictors from Set 1. The sensitivity to changes of predictors from Set 3 can be an advantage if all possible changes are correctly represented in the field sample. However, for relatively short periods of time between acquisitions and a low amount of forest operations, changes that are not very frequent in the landscape can be misrepresented or even not included in the sample. Thus, results for areas subject to those changes can be severely biased and inconsistent.

The more complex model development for the $y$-modeling method may be compensated by its ability to produce a richer set of outputs, by its apparently smaller risk of extrapolation and by its more accurate estimates for AOIs with small sample sizes (i.e., stands). In this study we analyzed the performance of the $y$-modeling method when estimating change, but estimates of V, BA and B for all the AOIs in 2009 and 2015 could have been readily obtained using this method. Results from our study also support the idea that the $y$-modeling method has advantages over the $\delta$-modeling method in terms of protection against inconsistent extrapolations. The distributions of predictors from Set 1 and Set 2 in thinned stands were relatively similar to the distributions observed for the sample while the distributions of predictors in Set 3 used in the $\delta$-modeling method, these distributions were rather different (see Figures 5, 6, A4 and A5). The greater similarity between thinned stands and the sample of field plots, for predictors from Set 1 and Set 2, indicates that the effect of thinning, in terms of auxiliary information, can be seen as transition from one situation in 2009 to another in 2015, and both seem to be represented in the field sample (e.g., Figure 5). If structures before and after the thinning (or other changes) are represented in the sample, the need for extrapolation will be limited. Within certain limits, if the sampling design covers all structures present at both points in time, even if there is a particular change from one structure to another that is not represented in the sample, predictions from the $y$-modeling method will not involve large extrapolations.

## 5. Conclusions

The four main conclusions obtained from this study include:

- The change of structural attributes and LiDAR auxiliary information are weakly correlated. This weak correlation seems to more evident in BMEF than in previous studies because of the slower growth in the study area and the relatively short lapse of time between LiDAR acquisitions, which indicates that for future studies in similar areas it might be necessary to increase the time lags between LiDAR flights.
- In general, the $\delta$-modeling method was found to be a slightly more accurate alternative to obtain estimates of change for the whole study area; however, the $y$-modeling method was able to produce better estimates at the stand level. In addition, the $y$-modeling method method also seemed to be less prone to extrapolation problems. This indicates that field campaigns for the $\delta$-modeling method have to be carefully designed while the $y$-modeling method might be less sensitive to certain bias problems.
- Despite the weak correlations with the changes in structural attributes, LiDAR auxiliary information allows obtaining estimates of growth for stands that improve over those derived using only field information.
- The large uncertainty observed for pixel-level predictions indicated that high-resolution maps of growth, generated using LiDAR auxiliary information in similar conditions, should be taken as approximated products.

**Author Contributions:** M.R. and B.W. developed the funding acquisition and data collection procedures. F.M. conceptualized and conducted the analyses and wrote the manuscript daft. M.R. and B.W. also participated in the conceptualization of the study. H.T., V.M., B.F. and A.H. provided significant input for the analyses and throughout the manuscript preparation.

**Funding:** This research received no external funding.

**Acknowledgments:** We would like to thank the USDA Forest Service, Region 5 and Lassen National Forest for assistance in project implementation, and the Guest Editor and two anonymous Reviewers for their constructive comments.

**Conflicts of Interest:** The authors declare no conflict of interest.

## Appendix A

**Table A1.** Sets of candidate predictors used in the study. Predictors included in the models to predict structural changes are highlighted with a boldface font. HiD, LoD and RNA represent the high diversity, low diversity and research natural areas respectively.

| Description Auxiliary Variables Sets 1, 2 and 3 | Acronym | | | Description Auxiliary Variables Set 4 | Acronym |
|---|---|---|---|---|---|
| | Set 1 Year: 2009 | Set 2 Year: 2015 | Set 3, Difference 2015-2009 | | Set 4 |
| Minimum, maximum, mean, mode, standard deviation, variance, coefficient of variation and interquartile range of the distribution of heights of the point cloud. | $Elev\_min_{09}$<br>$Elev\_max_{09}$ | $Elev\_min_{15}$<br>$Elev\_max_{15}$ | $\delta Elev\_min_{15\text{-}09}$<br>$\delta Elev\_max_{15\text{-}09}$ | Incoming solar radiation | Solar_radiation |
| | $Elev\_mean_{09}$<br>**$Elev\_mean^2_{09}$**<br>$Elev\_mode_{09}$ | $Elev\_mean_{15}$<br>**$Elev\_mean^2_{15}$**<br>$Elev\_mode_{15}$ | $\delta Elev\_mean_{15\text{-}09}$<br>$\delta Elev\_mean^2_{15\text{-}09}$<br>$\delta Elev\_mode_{15\text{-}09}$ | Structural diversity, factor with three levels HiD, LoD and RNA. Coded using two dummy variables. RNA reference level. | HiD |
| | $Elev\_stddv_{09}$<br>$Elev\_var_{09}$ | $Elev\_stddv_{15}$<br>$Elev\_var_{15}$ | $\delta Elev\_stddv_{15\text{-}09}$<br>**$\delta Elev\_var_{15\text{-}09}$** | | LoD |
| | $Elev\_CV_{09}$<br>$Elev\_IQ_{09}$<br>$Elev\_AAD_{09}$<br>$Elev\_MADmed_{09}$<br>$Elev\_MADmod_{09}$ | $Elev\_CV_{15}$<br>$Elev\_IQ_{15}$<br>$Elev\_AAD_{15}$<br>$Elev\_MADmed_{15}$<br>$Elev\_MADmod_{15}$ | $\delta Elev\_CV_{15\text{-}09}$<br>$\delta Elev\_IQ_{15\text{-}09}$<br>$\delta Elev\_AAD_{15\text{-}09}$<br>$\delta Elev\_MADmed_{15\text{-}09}$<br>$\delta Elev\_MADmod_{15\text{-}09}$ | Presence absence of prescribed fires. Coded using a dummy variable taking value 1 for stands where prescribed fires are applied and 0 otherwise. | Burned |
| Percentiles of the distribution of heights of the point cloud. | $Elev\_P01_{09}$<br>$Elev\_P05_{09}$<br>**$Elev\_P10_{09}$**<br>$Elev\_P20_{09}$<br>**$Elev\_P30_{09}$**<br>$Elev\_P40_{09}$<br>$Elev\_P50_{09}$<br>$Elev\_P60_{09}$<br>$Elev\_P70_{09}$<br>**$Elev\_P75_{09}$**<br>$Elev\_P80_{09}$<br>$Elev\_P90_{09}$<br>$Elev\_P95_{09}$<br>$Elev\_P99_{09}$ | $Elev\_P01_{15}$<br>$Elev\_P05_{15}$<br>**$Elev\_P10_{15}$**<br>**$Elev\_P20_{15}$**<br>$Elev\_P30_{15}$<br>$Elev\_P40_{15}$<br>$Elev\_P50_{15}$<br>$Elev\_P60_{15}$<br>$Elev\_P70_{15}$<br>$Elev\_P75_{15}$<br>$Elev\_P80_{15}$<br>$Elev\_P90_{15}$<br>$Elev\_P95_{15}$<br>$Elev\_P99_{15}$ | $\delta Elev\_P01_{15\text{-}09}$<br>$\delta Elev\_P05_{15\text{-}09}$<br>$\delta Elev\_P10_{15\text{-}09}$<br>$\delta Elev\_P20_{15\text{-}09}$<br>$\delta Elev\_P30_{15\text{-}09}$<br>$\delta Elev\_P40_{15\text{-}09}$<br>**$\delta Elev\_P50_{15\text{-}09}$**<br>$\delta Elev\_P60_{15\text{-}09}$<br>$\delta Elev\_P70_{15\text{-}09}$<br>$\delta Elev\_P75_{15\text{-}09}$<br>$\delta Elev\_P80_{15\text{-}09}$<br>$\delta Elev\_P90_{15\text{-}09}$<br>$\delta Elev\_P95_{15\text{-}09}$<br>$\delta Elev\_P99_{15\text{-}09}$ | | |
| Canopy relief ratio | $CRR_{09}$ | $CRR_{15}$ | **$\delta CRR_{15\text{-}09}$** | | |
| Percentage of first (Fst) and all (All) returns above 2 m | **$PcFstAbv2_{09}$**<br>**$PcAllAbv2_{09}$** | **$PcFstAbv2_{15}$**<br>$PcAllAbv2_{15}$ | **$\delta PcFstAbv2_{15\text{-}09}$**<br>**$\delta PcAllAbv2_{15\text{-}09}$** | | |
| Ratio all returns above 2 m to first returns | $AllAbv2Fst_{09}$ | $AllAbv2Fst_{15}$ | $\delta AllAbv2Fst_{15\text{-}09}$ | | |
| Percentage of first returns above the mean and mode | $PcFstAbvMean_{09}$<br>$PcFstAbvMode_{09}$ | $PcFstAbvMean_{15}$<br>$PcFstAbvMode_{15}$ | $\delta PcFstAbvMean_{15\text{-}09}$<br>$\delta PcFstAbvMode_{15\text{-}09}$ | | |
| Percentage of all returns above the mean and mode | $PcAllAbvMean_{09}$<br>$PcAllAbvMode_{09}$ | $PcAllAbvMean_{15}$<br>$PcAllAbvMode_{15}$ | $\delta PcAllAbvMean_{15\text{-}09}$<br>$\delta PcAllAbvMode_{15\text{-}09}$ | | |
| Ratio of all returns above the mean and mode to number of first returns | $AllAbvMeanFst_{09}$<br>$AllAbvModeFst_{09}$ | $AllAbvMeanFst_{15}$<br>$AllAbvModeFst_{15}$ | $\delta AllAbvMeanFst_{15\text{-}09}$<br>$\delta AllAbvModeFst_{15\text{-}09}$ | | |
| Proportion of points in the height intervals [0,0.5), [0.5,1), [1,2), [2,4), [4,8) and [8,16) meters. | $Prop0\_05_{09}$<br>$Prop05\_1_{09}$<br>$Prop1\_2_{09}$<br>$Prop2\_4_{09}$<br>$Prop4\_8_{09}$<br>$Prop8\_16_{09}$ | $Prop0\_05_{15}$<br>$Prop05\_1_{15}$<br>$Prop1\_2_{15}$<br>$Prop2\_4_{15}$<br>$Prop4\_8_{15}$<br>$Prop8\_16_{15}$ | $\delta Prop0\_05_{15\text{-}09}$<br>$\delta Prop05\_1_{15\text{-}09}$<br>$\delta Prop1\_2_{15\text{-}09}$<br>$\delta Prop2\_4_{15\text{-}09}$<br>$\delta Prop4\_8_{15\text{-}09}$<br>$\delta Prop8\_16_{15\text{-}09}$ | | |

## Appendix B

Predictions from the $\delta$ modeling method tend to be smoother than predictions from the $y$-modeling method (Figure A1). For all variables, the proportion of pixel-level predictions using the $\delta$-modeling method within the range of values observed for the field plots, was always 99.84% or larger. Considering the presence of thinned stands and the relatively small fraction of the forest that is sampled. Obtaining less than 0.15% of the predictions outside of the measurement range seems to be a clear sign of over smoothing. Predictions using the $y$-modeling method showed a greater variability, especially for BA, and the proportions of predictions inside the range of observed values, $P_y$, were 99.45% for V, 95.82% for BA and 99.29% for B. For BA, pixel-level predictions using the $y$-modeling method were oftentimes negative and of larger magnitude than the changes in BA observed for the plots. However, these pixels represent a small proportion of the total predictions (i.e., 4.02%), and a significant portion of them correspond to pixels in thinned stands. It is important to note that these comparisons of predicted values inform about how similar predictions are by the two analyzed methods and cannot be considered as indicators of accuracy or reliability. For all variables, pixel-level predictions by both methods were strongly correlated with Pearson's correlation coefficients of 0.92, 0.82 and 072 for V, BA and B, respectively (Figure A2). Finally, considering the unsampled and thinned stands, pixel-level predictions obtained by both methods showed the same inconsistencies observed at the stand-level especially for B using the $\delta$-modeling method where only about 4%, of the predicted values were negative (i.e., removals of B). These inconsistencies are clearly due to extrapolations in the thinned stands and are analyzed in more detail in next section.

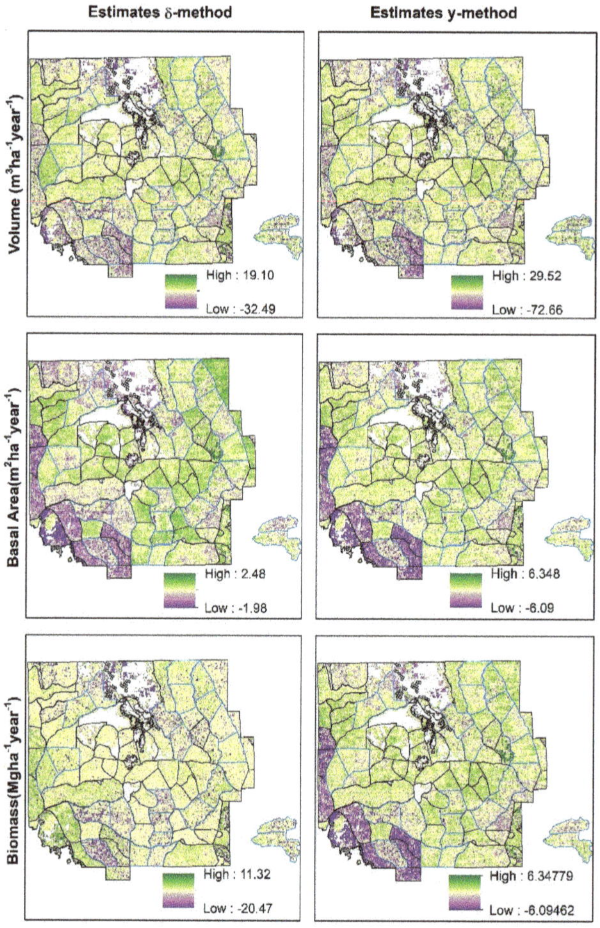


**Figure A1.** Maps of change in V, BA and B and corresponding pixel-level RMSE maps for the $\delta$-modeling method.

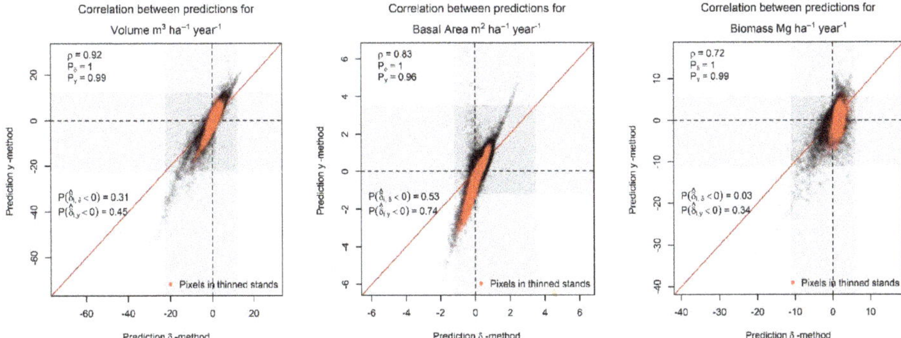

**Figure A2.** Comparison pixel-level predictions for V, BA and B using the δ-modeling method and *y*-modeling method predictions for the unsampled stands subject to thinnings are in red. The range of V, BA and B observed in the sample is indicated by the grey ribbons. The proportions, $P_\delta$ and $P_y$, of predictions within the range of values observed in the sample, and the correlation between predictions from both methods are indicated in the upper left corner. The proportion of pixels in the thinned stands where the δ-modeling method and *y*-modeling method predict losses (i.e., $P(\hat{\delta}_{i,\delta} < 0)$ and $P(\hat{\delta}_{i,y} < 0)$) are indicated on the lower left quadrant of the figure.

## Appendix C

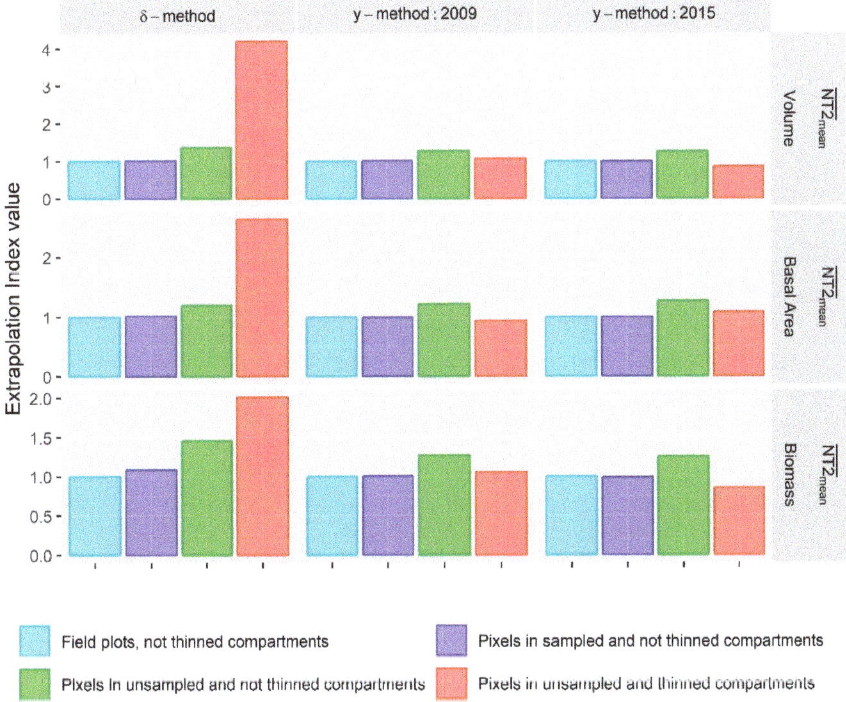

**Figure A3.** Indexes of extrapolation. Average of Mesgaran's novelty index relative to the mean, $\overline{NT2}_{mean}$, for the sampled and not thinned stands (dark blue), unsampled stands not thinned (green) and unsampled and thinned stands (red). The value of this index for the field plots (light blue) provides the baseline value (i.e., the value observed for the sample of field plots).

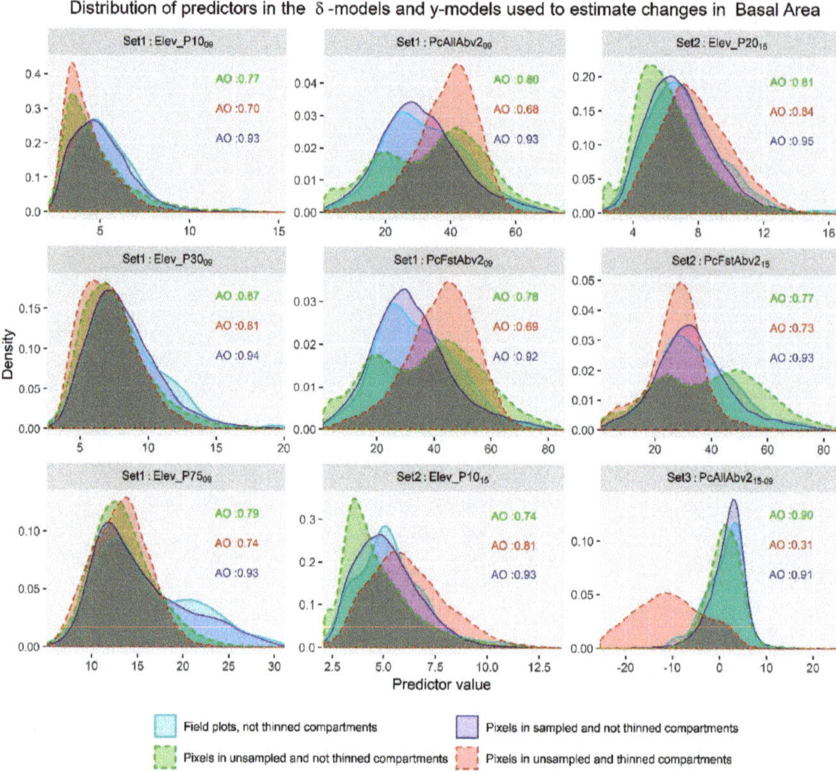



**Figure A4.** Comparison of density functions for the predictors in the models used to estimate changes in Basal Area using the δ-modeling method and *y*-modeling method in field plots (light blue), sampled and not thinned stands (dark blue), unsampled and not thinned stands (light blue) and unsampled and thinned stands (red). For each group the area of overlap, AO, with the density function for the field plots (green) is provided for each predictor.

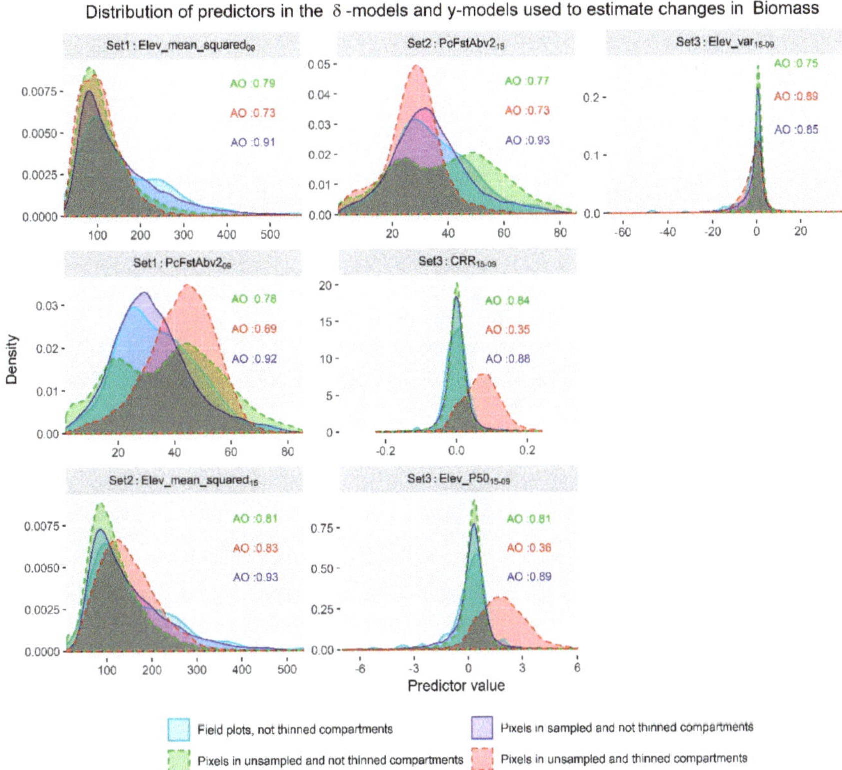


**Figure A5.** Comparison of density functions for the predictors in the models used to estimate changes in Biomass using the δ-modeling method and $y$-modeling method in field plots (light blue), sampled and not thinned stands (dark blue), unsampled and not thinned stands (light blue) and unsampled and thinned stands (red). For each group the area of overlap, AO, with the density function for the field plots (green) is provided for each predictor.

## References


1. Næsset, E. Predicting forest stand characteristics with airborne scanning laser using a practical two-stage procedure and field data. *Remote. Sens. Environ.* **2002**, *80*, 88–99. [CrossRef]
2. Andersen, H.-E.; McGaughey, R.J.; Reutebuch, S.E. Estimating forest canopy fuel parameters using LIDAR data. *Remote. Sens. Environ.* **2005**, *94*, 441–449. [CrossRef]
3. González-Ferreiro, E.; Diéguez-Aranda, U.; Miranda, D. Estimation of stand variables in *Pinus radiata* D. Don plantations using different LiDAR pulse densities. *For. Int. J. For. Res.* **2012**, *85*, 281–292. [CrossRef]
4. Mauro, F.; Molina, I.; García-Abril, A.; Valbuena, R.; Ayuga-Téllez, E. Remote sensing estimates and measures of uncertainty for forest variables at different aggregation levels. *Environmetrics* **2016**, *27*, 225–238. [CrossRef]
5. Valbuena, R.; Packalen, P.; Mehtätalo, L.; García-Abril, A.; Maltamo, M. Characterizing forest structural types and shelterwood dynamics from Lorenz-based indicators predicted by airborne laser scanning. *Can. J. For. Res.* **2013**, *43*, 1063–1074. [CrossRef]
6. Eggleston, H.S.; Buendia, L.; Miwa, K.; Ngara, T.; Tanabe, K. *IPCC Guidelines for National Greenhouse Gas Inventories, Volume 4: Agriculture, Forestry and Other Land Use*; Institute for Global Environmental Strategies: Hayama, Japan, 2006; Volume 4, ISBN 4-88788-032-4.
7. Babcock, C.; Finley, A.O.; Bradford, J.B.; Kolka, R.; Birdsey, R.; Ryan, M.G. LiDAR based prediction of forest biomass using hierarchical models with spatially varying coefficients. *Remote. Sens. Environ.* **2015**, *169*, 113–127. [CrossRef]

8.   Poudel, K.P.; Flewelling, J.W.; Temesgen, H. Predicting Volume and Biomass Change from Multi-Temporal Lidar Sampling and Remeasured Field Inventory Data in Panther Creek Watershed, Oregon, USA. *Forests* **2018**, *9*, 28. [CrossRef]

9.   Temesgen, H.; Strunk, J.; Andersen, H.-E.; Flewelling, J. Evaluating different models to predict biomass increment from multi-temporal lidar sampling and remeasured field inventory data in south-central Alaska. *Math. Comput. For. Nat.-Resour. Sci. (MCFNS)* **2015**, *7*, 66–80.

10.  McRoberts, R.E.; Næsset, E.; Gobakken, T.; Chirici, G.; Condes, S.; Hou, Z.; Saarela, S.; Chen, Q.; Stahl, G.; Walters, B.F. Assessing components of the model-based mean square error estimator for remote sensing assisted forest applications. *Can. J. For. Res.* **2018**, *48*, 642–649. [CrossRef]

11.  Næsset, E.; Gobakken, T.; Solberg, S.; Gregoire, T.G.; Nelson, R.; Ståhl, G.; Weydahl, D. Model-assisted regional forest biomass estimation using LiDAR and InSAR as auxiliary data: A case study from a boreal forest area. *Remote. Sens. Environ.* **2011**, *115*, 3599–3614. [CrossRef]

12.  Massey, A.; Mandallaz, D. Design-based regression estimation of net change for forest inventories. *Can. J. For. Res.* **2015**, *45*, 1775–1784. [CrossRef]

13.  Rao, J.N.K.; Molina, I. Introduction. In *Small Area Estimation*, 2nd ed.; John Wiley & Sons, Inc.: Hoboken, NJ, USA, 2015; pp. 1–8, ISBN 978-1-118-73585-5.

14.  Breidenbach, J.; Astrup, R. Small area estimation of forest attributes in the Norwegian National Forest Inventory. *Eur. J. For. Res.* **2012**, *131*, 1255–1267. [CrossRef]

15.  Mauro, F.; Monleon, V.J.; Temesgen, H.; Ford, K.R. Analysis of area level and unit level models for small area estimation in forest inventories assisted with LiDAR auxiliary information. *PLoS ONE* **2017**, *12*, e0189401. [CrossRef] [PubMed]

16.  Goerndt, M.E.; Monleon, V.J.; Temesgen, H. Small-Area Estimation of County-Level Forest Attributes Using Ground Data and Remote Sensed Auxiliary Information. *For. Sci.* **2013**, *59*, 536–548. [CrossRef]

17.  Breidenbach, J.; Magnussen, S.; Rahlf, J.; Astrup, R. Unit-level and area-level small area estimation under heteroscedasticity using digital aerial photogrammetry data. *Remote. Sens. Environ.* **2018**, *212*, 199–211. [CrossRef]

18.  Magnussen, S.; Næsset, E.; Gobakken, T. LiDAR-supported estimation of change in forest biomass with time-invariant regression models. *Can. J. For. Res.* **2015**, *45*, 1514–1523. [CrossRef]

19.  Næsset, E.; Bollandsås, O.M.; Gobakken, T.; Gregoire, T.G.; Ståhl, G. Model-assisted estimation of change in forest biomass over an 11year period in a sample survey supported by airborne LiDAR: A case study with post-stratification to provide "activity data". *Remote. Sens. Environ.* **2013**, *128*, 299–314. [CrossRef]

20.  Nasset, E.; Gobakken, T. Estimating forest growth using canopy metrics derived from airborne laser scanner data. *Remote. Sens. Environ.* **2005**, *96*, 453–465. [CrossRef]

21.  Ritchie, M.W. Multi-scale reference conditions in an interior pine-dominated landscape in northeastern California. *Ecol. Manag.* **2016**, *378*, 233–243. [CrossRef]

22.  Adams, M.B.; Loughry, L.H.; Plaugher, L.L. *Experimental Forests and Ranges of the USDA Forest Service*; United States Department of Agriculture, Forest Service, Northeastern Research Station: Newton Square, PA, USA, 2008; p. 191.

23.  Oliver, W.W. *Ecological Research at the Blacks Mountain Experimental Forest in Northeastern California*; United States Department of Agriculture, Forest Service, Pacific Southwest Research Station: Albany, CA, USA, 2000; p. 73.

24.  Wing, B.M.; Ritchie, M.W.; Boston, K.; Cohen, W.B.; Olsen, M.J. Individual snag detection using neighborhood attribute filtered airborne lidar data. *Remote. Sens. Environ.* **2015**, *163*, 165–179. [CrossRef]

25.  Hudak, A.T.; Strand, E.K.; Vierling, L.A.; Byrne, J.C.; Eitel, J.U.; Martinuzzi, S.; Falkowski, M.J. Quantifying aboveground forest carbon pools and fluxes from repeat LiDAR surveys. *Remote. Sens. Environ.* **2012**, *123*, 25–40. [CrossRef]

26.  Area Solar Radiation—Help | ArcGIS Desktop. Available online: http://desktop.arcgis.com/en/arcmap/10.6/tools/spatial-analyst-toolbox/area-solar-radiation.htm (accessed on 8 April 2019).

27.  Mauro, F.; Monleon, V.; Temesgen, H.; Ruiz, L. Analysis of spatial correlation in predictive models of forest variables that use LiDAR auxiliary information. *Can. J. For. Res.* **2017**, *47*, 788–799. [CrossRef]

28.  Rao, J.N.K.; Molina, I. *Basic Unit Level Model. In Small Area Estimation*, 2nd ed.; John Wiley & Sons, Inc.: Hoboken, NJ, USA, 2015; pp. 173–234, ISBN 978-1-118-73585-5.

29. Rao, J.; Molina, I. Empirical Best Linear Unbiased Prediction (EBLUP): Theory. In *Small Area Estimation*, 2nd ed.; John Wiley & Sons, Inc.: Hoboken, NJ, USA, 2015; pp. 97–122.

30. R Core Team. *R: A Language and Environment for Statistical Computing*; R Foundation for Statistical Computing: Vienna, Austria, 2018.

31. Pinheiro, J.; Bates, D.; DebRoy, S.; Sarkar, D.; R Core Team. nlme: Linear and Nonlinear Mixed Effects Models. R package version 3.1-137. Available online: https://cran.r-project.org/web/packages/nlme/index.html (accessed on 15 April 2019).

32. Datta, G.S.; Lahiri, P. A unified measure of uncertainty of estimated best linear unbiased predictors in small area estimation problems. *Stat. Sin.* **2000**, *10*, 613–628.

33. Silverman, B.W. *Density Estimation for Statistics and Data Analysis*; Chapman and Hall: Boca Raton, FL, USA, 1986.

34. Mesgaran, M.B.; Cousens, R.D.; Webber, B.L. Here be dragons: A tool for quantifying novelty due to covariate range and correlation change when projecting species distribution models. *Divers. Distrib.* **2014**, *20*, 1147–1159. [CrossRef]

35. Bollandsås, O.M.; Gregoire, T.G.; Næsset, E.; Øyen, B.-H. Detection of biomass change in a Norwegian mountain forest area using small footprint airborne laser scanner data. *Stat. Methods Appl.* **2013**, *22*, 113–129. [CrossRef]

36. Fekety, P.A.; Falkowski, M.J.; Hudak, A.T. Temporal transferability of LiDAR-based imputation of forest inventory attributes. *Can. J. For. Res.* **2014**, *45*, 422–435. [CrossRef]

37. Ozdemir, I.; Donoghue, D.N. Modelling tree size diversity from airborne laser scanning using canopy height models with image texture measures. *For. Ecol. Manag.* **2013**, *295*, 28–37. [CrossRef]

*Review*

# Remote Sensing Support for the Gain-Loss Approach for Greenhouse Gas Inventories

Ronald E. McRoberts [1,2,*], Erik Næsset [3], Christophe Sannier [4], Stephen V. Stehman [5] and Erkki O. Tomppo [6,7]

[1]  Department of Forest Resources, University of Minnesota, Saint Paul, MN 55108, USA
[2]  Raspberry Ridge Analytics, 15111 Elmcrest Avenue North, Hugo, MN 55038, USA
[3]  Faculty of Environmental Sciences and Natural Resource Management, Norwegian University of Life Sciences, P.O. Box 5003, NO-1432 Ås, Norway; erik.naesset@nmbu.no
[4]  Systèmes d'Information à Référence Spatiale, Parc de la Cimaise, 59650 Villeneuve d'Ascq, France; christophe.sannier@sirs-fr.com
[5]  Department of Sustainable Resources Management, State University of New York, College of Environmental Science and Forestry, Syracuse, NY 13210, USA; svstehma@syr.edu
[6]  Department of Electronics and Nanoengineering, School of Electrical Engineering, Aalto University, P.O. Box 11000, 00076 Aalto, Finland; erkki.tomppo@aalto.fi
[7]  Department of Forest Sciences, University of Helsinki, Latokartanonkaari 7, P.O. Box 27, FI-00014 Helsinki, Finland
*  Correspondence: mcrob001@umn.edu

Received: 21 April 2020; Accepted: 8 June 2020; Published: 11 June 2020

**Abstract:** For tropical countries that do not have extensive ground sampling programs such as national forest inventories, the gain-loss approach for greenhouse gas inventories is often used. With the gain-loss approach, emissions and removals are estimated as the product of activity data defined as the areas of human-caused emissions and removals and emissions factors defined as the per unit area responses of carbon stocks for those activities. Remotely sensed imagery and remote sensing-based land use and land use change maps have emerged as crucial information sources for facilitating the statistically rigorous estimation of activity data. Similarly, remote sensing-based biomass maps have been used as sources of auxiliary data for enhancing estimates of emissions and removals factors and as sources of biomass data for remote and inaccessible regions. The current status of statistically rigorous methods for combining ground and remotely sensed data that comply with the good practice guidelines for greenhouse gas inventories of the Intergovernmental Panel on Climate Change is reviewed.

**Keywords:** statistical estimator; IPCC good practice guidelines; activity data; emissions factor; removals factor

---

## 1. Introduction

The Intergovernmental Panel on Climate Change (IPCC) [1] (p.17) identifies five carbon pools to be monitored for forest-related carbon emissions and removals: aboveground biomass, below-ground biomass, litter, dead wood and soil organic carbon. Within the agriculture, forestry and other land use sector, greenhouse gas (GHG) inventories for all these pools are typically conducted using either the stock difference approach or the gain-loss approach [1] (p. 22), [2] (Vol. 4, Chap 2, p. 2.10). With the stock-difference approach, mean annual carbon emissions or removals are estimated as the ratio of the difference in carbon stock estimates at two points in time and the number of intervening years [1] (Chap. 3), [2] (Vol. 4, Chap. 3). For countries with established and comprehensive forest sampling programs, such as national forest inventories (NFI), the stock difference approach is fairly easy to implement. However, for tropical countries that do not have NFIs or that only have a single set of NFI measurements, the gain-loss approach is used more often. With the gain-loss approach, emissions are estimated as differences between additions to and removals from carbon pools. Specifically, emissions are estimated as the product of activity data defined as the areas of "human activity causing emissions and removals" and emissions factors defined as the per unit area responses of carbon stocks for those activities [1] (pp. xvii, 22), [2] (Vol. 1, Chap. 1, Sect. 1.2).

The IPCC characterizes GHG methods for estimating emissions and removals with respect to three tiers or levels of data detail and analytical complexity [1] (p. 19), [3]. Tier 1 is the default method and permits use of default, national, ecological zone, or global emissions factors based on canopy cover reductions, thus making it applicable for any country. However, emissions factors from these sources are subject to considerable uncertainty. Tier 2 is based on the same conceptual structure as Tier 1, but with the expectation that activity data aggregate land use changes between categories and emissions factors are based on national level data. Tier 3 uses models, data from repeated inventories, and fine resolution land use and land use change activity data to produce spatially continuous, sub-national estimates. Gain-loss methods can be implemented at all three tier levels.

Within tiers, the IPCC further documents three approaches for estimating activity data, all involving land use class areas [1] (p. 19). Approach 1 uses aggregations of land use class areas such as would be reported by an NFI, but without regard to the specific geographic locations of those areas of those land use classes; Approach 2 differentiates among land use change classes, but like Approach 1, does not identify the specific locations of individual segments of the classes; and Approach 3 tracks individual land parcels over time.

Regardless of the approach and method, the IPCC specifies two good practice guidelines for GHG inventories: (i) "neither over- nor underestimates so far as can be judged," and (ii) "uncertainties are reduced as far as practicable" [1] (p. 15), [2] (Vol 1. Chap. 1, Sect. 1.2). From a statistical perspective, the satisfaction of these criteria requires use of unbiased estimators and, at minimum, rigorous estimation of uncertainty. In particular, there can be little assurance that uncertainties are reduced until they are first rigorously estimated [4,5].

Considerable recent attention has been devoted to developing and illustrating methods for implementing the gain-loss approach, particularly for estimating activity data, but less so for estimating emissions and removals factors for the aboveground biomass pool [1] (p. 17). The objective of the review is to document and summarize the current role of remote sensing for gain-loss methods for Tier 1 and Tier 2 GHG inventories, with particular emphasis on the two IPCC good practice guidelines. A subordinate objective is to submit statistical guidance developed for the Methods and Guidance Document (MGD) of the Global Forest Observations Initiative to journal-level peer review [1]. As per the MGD, the review focuses on forest land use, particularly forest remaining forest and conversions from and to forest land use. The structure of the review closely follows the structure of the MGD, with an initial focus on estimation of activity data followed by estimation of emissions and removals factors. Within these two categories, the focus is on the statistical estimators, as is the MGD, with the discussion logically proceeding from less complex to more complex.

## 2. Estimating Total Emissions

The ultimate objective is an inference in the form of a confidence interval,

$$\hat{E}^{tot} \pm t \cdot \sqrt{\hat{Var}\left(\hat{E}^{tot}\right)} \tag{1}$$

where $\hat{E}^{tot}$ is the estimate of total emissions and removals, $\hat{Var}\left(\hat{E}^{tot}\right)$ is an estimate of the corresponding variance, and t is a percentile from Student's t-distribution corresponding to the desired confidence and degrees of freedom determined by the sample size and sampling design. With the gain-loss approach, total emissions and removals are estimated as the sum over activities of the products of estimates of activity class areas and corresponding estimates of activity class emissions and removals factors,

$$\hat{E}^{tot} = \sum_{c=1}^{C} \hat{A}^c \cdot \hat{EF}^c \tag{2}$$

where c=1, ... , C indexes activity classes, $\hat{A}^c$ is the estimate of the area for activity class c, and $\hat{EF}^c$ is the estimate of the emission factor for activity class c [1] (pp. xvii, 22), [2] (Vol.1, Chap.1, Sect. 1.2). An estimator of the variance of $\hat{E}^{tot}$ can be formulated using a first-order Taylor series as,

$$\hat{Var}\left(\hat{E}^{tot}\right) = \sum_{c=1}^{C}\left[\left(\hat{A}^c\right)^2\cdot\hat{Var}\left(\hat{EF}^c\right) + 2\cdot\hat{A}^c\cdot\hat{EF}^c\cdot\hat{Cov}\left(\hat{A}^c,\hat{EF}^c\right) + \left(\hat{EF}^c\right)^2\cdot\hat{Var}\left(\hat{A}^c\right)\right] \tag{3}$$

Reference [6] (Appendix). If activity areas and emission factors are estimated independently, then $\hat{Cov}\left(\hat{A}^c,\hat{EF}^c\right) = 0$ may be assumed. The technical challenge with the gain-loss method is to formulate the activity data estimators, $\hat{A}^c$ and $\hat{Var}\left(\hat{A}^c\right)$, and the estimators for the emissions and removals factors, $\hat{EF}^c$ and $\hat{Var}\left(\hat{EF}^c\right)$. Multiple country-level examples of the use of the gain-loss method for estimating total emissions and removals can be found in the country Fact Sheets at the REDD+ Web Platform [7], although rigorous uncertainty estimation is not always documented. Illustrations that include uncertainty assessments and that document estimation details can be found in [1] (p. 158) and [6].

## 3. Activity Data

The estimation of activity data typically entails three components: reference data, auxiliary data, and statistical estimators. When estimating activity data, auxiliary data are typically acquired from external sources, so that the primary purpose of reference data is to serve as the basis for estimates. However, when estimating emissions and removals factors (Section 4), auxiliary data are often developed by the user, so that reference data may have multiple purposes, including as training data for constructing remote sensing-based maps, as map validation data, and as a source of data for calculating estimates. For estimating activity data, reference data can be obtained from both ground and remote sensing sources. The primary purpose of auxiliary data is to enhance or improve estimates based on reference data, most often to increase precision. Auxiliary data, like reference data, can be obtained from both ground and remote sensing sources.

### 3.1. Data

### 3.1.1. Reference Data

For the estimation of activity data, the reference data are the most accurate information on land use change classes available and are typically acquired as measurements of ground plots, or more commonly as image interpretations. To facilitate design-based inference, reference data are most often acquired using a probability sampling design [8,9] (Section 3.2).

Regardless of the estimator used with reference data (Section 3.2), observation errors and other uncertainties in reference data tend to induce bias into estimators [10–13]. Sun et al. [14] reported that manual interpretations of Google Earth and other fine resolution imagery by three trained interpreters were not as reliable as ground observations for seven land cover classes in Central Asia. Thus, for most applications, ground reference data are considered to be preferable, although Foody [10,11] argues that even ground reference data may be subject to error. However, the acquisition of probability samples of ground reference data for remote and inaccessible regions may be prohibitively expensive, if not logistically infeasible. The common alternative is to acquire reference data in the form of visual image interpretations, albeit with the caveat that such reference data are of greater quality than any auxiliary map data (Section 3.1.2), with respect to factors such as resolution and accuracy [1] (pp. 125, 139) [8,15,16].

Mowrer and Congalton [17] characterize reference data with non-negligible uncertainty as imperfect reference data. Næsset [18,19] reported that interpretations of crown coverage for structurally homogenous Norwegian boreal forests differed statistically significantly among interpreters and among different times of year for the same interpreter. Further, interpretations of broad tree species groups by 12 professional interpreters using stereo aerial photography produced only 31–79% agreement with field reference data. These results were consistent with the results of a review of more than 10 studies from the Nordic countries from the 1970s and 1980s. An additional finding was, as might be expected, that interpretations were less accurate for more complex forests. For five trained interpreters of videography, Powell et al. [20] reported interpreter disagreement of almost 30% for five land cover classes in the Brazilian Amazon. Pengra et al. [21] examined duplicate interpretations of land-cover reference class labels obtained by well-trained interpreters from a probability subsample of 2900 pixels (30 m × 30 m) of the United States of America. For the 8-class land cover legend, the overall agreement between interpreters was 88%, but agreement by class ranged from 46% for the disturbed class to 94% for water, with more prevalent classes having greater agreement than rare classes. Thus, reference data in the form of visual interpretations of remotely sensed data, even by well-trained professional interpreters, could be subject to substantial interpreter disagreement and error.

McRoberts et al. [13] used a combination of photo interpretations by multiple professional interpreters and simulation data to assess the effects of imperfect reference data on the bias and precision of estimators of land cover class proportions, to characterize conditions that affect the magnitudes of bias and precision, and to develop a variance estimator that incorporates the effects of interpreter error. Several relevant conclusions were drawn from the study. First, estimator bias resulting from interpreter error is greater for greater inequality in areas of land cover classes, greater for smaller map and interpreter accuracies, greater for fewer interpreters, and greater for greater correlations among interpretations for different interpreters. For some scenarios, seven or more interpreters were necessary to reduce biases to acceptable levels. Second, failure to incorporate the effects of interpreter error into variance estimators led to underestimates of standard errors by factors as great as 2.0.

The important lesson from these studies is that imperfect reference data induce bias into statistical estimators, sometimes substantial bias despite only small errors [10–13], and lead to non-compliance with the first IPCC good practice guideline. In addition, failure to incorporate the uncertainty associated with imperfect reference data into statistical variance estimators leads to under-estimation of variances and non-compliance with the second IPCC good practice guideline.

Multiple strategies for dealing with imperfect reference data may be considered. First, greater numbers of interpreters reduce the effects of imperfect reference data [13]. Second, prior to operational interpretation, interpreters can calibrate their interpretations, with respect to known field conditions and/or to each other [22] (p. 82). Third, in the absence of unanimous interpretations, interpreters may discuss the specific sample units and agree on a consensus interpretation. Finally, instead of using majority interpretations leading to categorical reference observations (e.g., 0 for non-forest, 1 for forest), continuous reference observations in the form of the proportions of forest interpretations among interpreters for the same sample unit are possible.

### 3.1.2. Auxiliary Data

For purposes of estimating activity data, auxiliary data are secondary, i.e., the reference data are the primary source of information on which estimates are based, whereas the auxiliary data are used only to improve the estimation process by increasing the precision of estimates. Because activity data are estimates of change in areas of land use classes, both the reference and auxiliary data are closely related to change. In particular, auxiliary data are often in the form of a map or spatial product whose classes reflect or can be aggregated to reflect land use change classes related to human activities, such as deforestation, reforestation, afforestation, and forest remaining forest, including the special case of forest degradation, which entails the substantial loss of biomass, but no conversion from forest to other land use.

Although the scope of this review does not include techniques and methods for constructing remote sensing-based maps, regional and global land use and land use change maps can be used as auxiliary data when estimating activity data. The most widely known example is the Global Forest Change dataset [23], a percentage tree cover map for the year 2000, in which trees are defined as vegetation taller than 5m in height at a 30-m pixel size. Annual forest gain and loss data from 2001 to 2019 are also included. Forest loss is a binary layer (1: loss, 0: no loss), and is understood as complete or the comprehensive removal of forest cover and "defined as a stand-replacement disturbance, or a change from a forest to non-forest state." Forest gain is the complete or comprehensive recovery of forest cover and "defined as the inverse of loss, or a non-forest to forest change entirely within the study period". Examples of uses of the Global Forest Change dataset for estimating land cover class areas include Sannier et al. [24], Næsset et al. [25], and McRoberts et al. [26]. In addition, the Global Forest Change dataset can be adapted to match national forest definitions [24,25,27]. Although few examples appear in the literature, maps for this purpose can be produced locally, and are typically more accurate than global maps [22,28]. McRoberts et al. [26] demonstrate how data from multiple maps can be combined to produce a new, more accurate map.

A potentially important issue is that while activity data are defined in terms of land use change, auxiliary data in the form of maps based on satellite spectral data inevitably depict land cover change. Further, land cover change does not necessarily correspond to land use change. For example, forest land that has been completely harvested loses its forest cover, but retains its forest land use. For use with the stratified (Section 3.2.3) and model-assisted (Section 3.2.4) estimators, auxiliary data in the form of land cover change maps in lieu of land use change maps do not induce bias into the estimators, but rather just reduce precision.

### 3.2. Statistical Estimators

#### 3.2.1. Design-Based Inference

Design-based inference, also characterized as probability-based inference, is based on three assumptions. First, a probability sample incorporating some form of randomization is used and constitutes the basis for validity. Second, apart from negligible observation or measurement error, each population unit is assumed to have one, and only one, possible value. Third, the selection of population units into the sample is based on positive and known probabilities of selection. Familiar sampling designs include simple random (SRS), systematic (SYS), stratified (STR), multi-phase and multi-stage sampling designs. The design-based simple expansion (Section 3.2.2), stratified (Section 3.2.3), post-stratified (Section 3.2.4) and model-assisted (Section 3.2.5) estimators herein considered for the estimation of activity data are either unbiased or asymptotically unbiased. The uncertainty estimation for these estimators entails comparing observations to their corresponding means or model predictions. Of importance, for statistical estimators to be unbiased, they must be consistent with the probability sampling design used to collect the reference data, i.e., the known probabilities of selection must be incorporated into the estimators. For example, the simple expansion estimator, Equation (6), is generally biased if used with reference data collected with a stratified sampling design.

If the reference data are categorical, then for land use change class, c, a new variable is defined, such that for each reference sample unit, $y_i^c = 1$ if class c is observed, or $y_i^c = 0$ if any class other than c is observed. If the reference data are continuous, such as differences in proportion forest cover at two times, then for each land use change class, c, the new variable takes on a value in the interval [0,1] corresponding to the difference in proportions of the reference sample unit in class c. Letting $\mu^c$ be the proportion of the area of interest in class c, the estimator of the area of class c is,

$$\hat{A}^c = A^{tot} \cdot \hat{\mu}^c \qquad (4)$$

with

$$SE\left(\hat{A}^c\right) = A^{tot} \cdot SE(\hat{\mu}^c) \tag{5}$$

where $A^{tot}$ is the total area which is assumed to be without error. The activity data challenge, then, is to formulate the estimators, $\hat{\mu}^c$ and $SE(\hat{\mu}^c)$.

### 3.2.2. Simple Expansion Estimator

The simple expansion (EXP) estimator of $\mu^c$ is,

$$\hat{\mu}^c_{EXP} = \frac{1}{n} \sum_{i=1}^{n} y_i^c, \tag{6}$$

with

$$SE\left(\hat{\mu}^c_{EXP}\right) = \sqrt{\frac{1}{n \cdot (n-1)} \sum_{i=1}^{n} \left(y_i^c - \hat{\mu}^c_{EXP}\right)^2} \tag{7}$$

where n is the sample size. Of importance, the EXP estimators use no auxiliary data, which further establishes that the reference data are the primary source of information on which estimates are based.

### 3.2.3. Stratified Estimators

Stratified (STR) estimation uses a map or similar spatial product to increase the precision of estimates. If the map classes are the same as the land use change activity classes, an intuitive estimator of $\mu^c$ is simply the sum of the areas of all map units classified as activity class c. However, this estimator, characterized as pixel-counting, is biased, because it does not account for classification errors and, therefore, does not comply with the first IPCC good practice guideline pertaining to avoiding under- and over-estimation; this approach should be avoided. The reason for the bias is that due to classification error, the map class of interest may include some units that are not actually of the activity class of interest and, similarly, other map classes may include some map units that are of the activity class of interest.

For the purposes of estimating activity data, STR estimators assume that reference data are acquired using a STR sampling design, guided by an activity class map, for which the classes serve as strata. An advantage of the combination of stratified sampling and estimation is that within-stratum sample sizes can be controlled and selected for the primary purpose of more precisely estimating the areas of key land use change classes such as deforestation. In particular, when the class of interest is small, SRS and SYS sampling designs generally do not produce sample sizes sufficiently large enough to satisfy precision requirements. STR designs facilitate the allocation of greater sample sizes for smaller strata corresponding to land use change classes, and smaller sample sizes for larger strata corresponding to stable land use classes.

Cochran [29] (p. 134) recommends no more than 6–8 strata. Särndal et al. [30] (pp. 267, 407) recommend minimum within-stratum sample sizes of 10–20; Cochran [29] (p. 134) recommends minimum within-stratum sample sizes of 20; and for temperate forest inventories, Westfall et al. [31] recommend within-stratum sample sizes of at least 20.

The STR estimator of the proportion of the total area in class c is,

$$\hat{\mu}^c_{STR} = \sum_{h=1}^{H} w_h \cdot \hat{\mu}^c_h \tag{8}$$

with

$$SE\left(\hat{\mu}^c_{STR}\right) = \sqrt{\sum_{h=1}^{H} w_h^2 \cdot \frac{\hat{\sigma}_h^2}{n_h}} \tag{9}$$

where h = 1, ... , H indexes the strata, $w_h$ is the stratum weight calculated as the proportion of the population in the *h*th stratum, and $n_h$ is the sample size for the *h*th stratum,

$$\hat{\mu}_h^c = \frac{1}{n_h} \sum_{i=1}^{n_h} y_{hi}^c, \tag{10}$$

$$\hat{\sigma}_h^2 = \frac{1}{n_h - 1} \sum_{i=1}^{n_h} \left( y_{hi}^c - \hat{\mu}_h^c \right)^2, \tag{11}$$

and $y_{hi}^c$ is the observation for the *i*th sample reference unit in the *h*th stratum. An additional advantage of the combination of stratified sampling and estimation is that the SEs are typically smaller than for the simple expansion estimators.

For many forests, both natural change and human-induced disturbance occur most frequently in the vicinity of the forest/non-forest boundary. In addition, classification uncertainty for most maps is greatest at boundaries between map classes. Finally, because of both classification error and geo-reference errors, the erroneous assignment of ground plots to strata inconsistent with their observations is often most severe near map class boundaries. Some of the adverse effects of these phenomena on precision can be alleviated by constructing an additional small stratum with corresponding small stratum weight along class boundaries. Thus, heterogeneous ground plots found near the boundaries can be confined to small strata with small stratum weights, thereby having less detrimental effects on precision. For a stratification based on a forest/non-forest map, McRoberts et al. [32] constructed a forest edge stratum consisting of a multi-pixel buffer on the forest side of the forest/non-forest map boundary and a non-forest edge stratum consisting of a multi-pixel buffer on the non-forest side of the boundary. The effect of the two additional strata was to decrease the SEs of the estimated forest area by as much as 12%.

These buffer strata have an additional beneficial effect. Activity data are often required at relatively short intervals, perhaps as frequently as annually or biennially. For such short reporting intervals, area change may be small. For the purposes of precisely estimating area change, STR designs based on change maps commonly entail using a small change stratum with a small stratum weight, but a disproportionately large sample size; similarly, no-change strata would have large stratum weights, but with disproportionately small sample sizes. A risk associated with this approach is that only a few sample units with change observations erroneously assigned to a no-change stratum can both greatly inflate standard errors and induce a much greater range in estimates of the mean, even though the estimator remains unbiased. Buffer strata, as previously described, can be used to alleviate at least some of these adverse effects [33].

### 3.2.4. Post-Stratified Estimators

Even though reference data may have been acquired using an SRS or SYS sampling design, some of the benefits of stratified estimation can still be realized using the post-stratified (PSTR) estimators. The primary difference between the combination of STR sampling and stratified estimation and the combination of SRS or SYS sampling and post-stratified estimation is that with the former combination, the map-based stratification guides the sampling, whereas with the latter combination, the map has no influence on the sampling. Further, whereas within-stratum sample sizes are considered fixed with the STR estimators, they are considered random with the PSTR estimators. The PSTR estimator of the class proportion is the same as for the STR estimator of Equation (8), i.e., $\hat{\mu}_{PSTR}^c = \hat{\mu}_{STR}^c$. However, the PSTR variance estimator has a slightly different form to accommodate the random within-stratum sample sizes,

$$SE\left(\hat{\mu}_{PSTR}^c\right) = \sqrt{\sum_{h=1}^{n_h} \left[ w_h \cdot \frac{\hat{\sigma}_h^2}{n} + (1 - w_h) \cdot \frac{\hat{\sigma}_h^2}{n^2} \right]^2} \tag{12}$$

Some researchers consider the variance and SE to be conditional on the sample, in which case the estimator of Equation (9) can be used instead of Equation (12) [34] (pp. 152–156). The same recommendations regarding the maximum number of strata and the minimum within-stratum sample sizes for stratified estimation pertain for post-stratified estimation.

### 3.2.5. Model-Assisted Estimators

The model-assisted (MA) estimators can be used with any probability sampling designs. The first component of model-assisted estimators of the proportion, $\mu^c$, is formulated as the synthetic (SYN) estimator,

$$\hat{\mu}^c_{SYN} = \frac{1}{N} \sum_{i=1}^{N} \hat{y}^c_i,$$ (13)

where N is the total number of population units and $\hat{y}^c_i$ is a prediction for the *i*th population unit. However, the SYN estimator may be biased because of prediction error. For SRS and SYS designs, the bias can be estimated as,

$$\hat{Bias}\left(\hat{\mu}^c_{SYN}\right) = \frac{1}{n} \sum_{i=1}^{n} \left(\hat{y}^c_i - y^c_i\right).$$ (14)

The asymptotically unbiased model-assisted estimator of the proportion of the total area is then,

$$\hat{\mu}^c_{MA} = \hat{\mu}^c_{SYN} - \hat{Bias}\left(\hat{\mu}^c_{SYN}\right)$$
$$= \frac{1}{N} \sum_{i=1}^{N} \hat{y}^c_i - \frac{1}{n} \sum_{i=1}^{n} \left(\hat{y}^c_i - y^c_i\right)$$ (15)

with standard error,

$$SE\left(\hat{\mu}^c_{MA}\right) = \sqrt{\frac{1}{n \cdot (n-1)} \cdot \sum_{i=1}^{n} \left(\varepsilon_i - \bar{\varepsilon}\right)^2},$$ (16)

where

$$\varepsilon_i = y^c_i - \hat{y}^c_i \text{ and } \bar{\varepsilon} = \frac{1}{n} \sum_{i=1}^{n} \varepsilon_i.$$ (17)

When the prediction technique used to obtain $\hat{y}^c_i$ is formulated and calibrated using data external to the area of interest, or otherwise does not use observations of *y* from the sample, the estimator is characterized as the difference (DIF) estimator, whereas if the prediction technique is calibrated using data internal to the area of the interest, the estimator is characterized as the generalized regression (GREG) estimator [35]. For sampling designs other than SRS and SYS, Equations (14)–(16) must be revised to accommodate the features of the designs (e.g., [25]).

Representative examples, as opposed to an exhaustive list of tropical applications which are sparse, of uses of the STR, PSTR, DIF and GREG estimators for estimating land cover and/or land use area and area change are reported in Table 1. For categorical response variables, the STR and PSTR estimators are generally expected to produce greater precision than the model-assisted estimators [26,36], although Stehman [36] notes three exceptions: (i) large overall accuracies, (ii) small true proportions, and (iii) small reference sets. For continuous response variables, the model-assisted estimators are generally more precise than the STR and PSTR estimators. Exceptions are when the PSTR estimator is formulated as a model-assisted estimator [16], and possibly when the map resolution is much coarser than the resolution of the reference data [37].

**Table 1.** Representative examples of land cover class area and area change estimation.

| Parameter | Response Variable | Estimator * | References |
|---|---|---|---|
| Area | Class | STR | Stehman [36] |
| | | PSTR | McRoberts et al. [6,26,32,38], Stehman [16,36] |
| | | DIF | Stehman [36], Khan et al. [39] |
| | | GREG | Gallego [40], Stehman [16], Vibrans et al. [41], McRoberts et al. [26,38] |
| | Proportion | STR | Stehman [42] |
| | | PSTR | McRoberts [43] |
| | | DIF | McRoberts [43] |
| | | GREG | Sannier et al. [24,44] |
| Area change | Class | STR | Olofsson et al. [8,9,33], Ying et al. [45] |
| | | PSTR | McRoberts et al. [6], Næsset et al. [46] |
| | | DIF | |
| | | GREG | Stehman [16], Næsset et al. [46], McRoberts and Walters [47], McRoberts et al. [48] |
| | Proportion | STR | Claggett et al. [49], Tyukavina et al. [50] |
| | | PSTR | Pickering et al. [51] |
| | | DIF | Zimmerman et al. [52] |
| | | GREG | Stehman [16], Sannier et al. [24,44] |

* STR: Stratified, PSTR: Post-stratified, DIF: Difference, GREG: Generalized regression.

### 3.3. Time Series Analyses

With the increasing availability of free Landsat and Sentinel 2 spectral data at regular temporal intervals over many years, opportunities for tracking land cover and land use changes over time are greatly enhanced [3]. Two applications are attracting attention. First, trajectory analyses track metrics for individual pixels over time for purposes of characterizing phenomena, such as gradual forest degradation, abrupt disturbances such as harvest followed by recovery, and dates of change [53–60]. The second application entails estimating differences or trends in forest area or forest area change at multi-year intervals, or perhaps even annually. With this approach, a remote sensing-assisted estimate of area or area change and an associated SE are calculated, as per Section 3.2, for each temporal point of interest.

For both applications, compliance with the second IPCC good practice guideline related to rigorous uncertainty assessment presents challenges. For the first application, for which dates of individual pixel-level changes are estimated, very large numbers of tests of significance will inevitably lead to large numbers of false positives and false negatives. For the second application which focuses on multi-year trends of annual differences, an intuitive approach would be to use a regression or similar prediction technique for estimating trends or an ANOVA technique followed by a simultaneous multiple inference method for determining which, if any, temporal estimates differ statistically significantly from other temporal estimates. A complicating factor, however, is that both regression and ANOVA techniques assume that the response variable, in this case forest area change, is observed with at most negligible uncertainty. For this kind of trend analysis, this assumption is not satisfied, because the standard errors associated with the individual change estimates are likely non-negligible. Special techniques similar to hybrid inference would be necessary to accommodate and account for the uncertainty in the temporal area or area change estimates [35,61–63].

## 4. Emissions and Removals Factors

Emissions and removals factors, by definition, represent carbon change per unit area. Estimates of these factors can be obtained from three primary sources: published default values [64], ground biomass observations that are converted to carbon, and remote sensing-based biomass maps from which estimates can be converted to carbon. For this review, the focus is remote sensing-assisted estimation of biomass change per unit area, particularly aboveground, live tree biomass change per unit area.

The state-of-the-science for estimating biomass change as a precursor to estimating emissions and removals factors is considerably less mature than for estimating activity data. To estimate biomass change, reference data for two dates are required, although an exception may be for complete deforestation, for which there is no remaining biomass at the second time. Two approaches for estimating change can be used, the indirect approach and the direct approach. With the indirect approach, total or mean biomass per unit area is estimated for an activity as the difference in two biomass estimates, one based on a sample acquired at time 1 and the second based on a sample acquired at time 2, usually for a different set of sample unit locations, because otherwise, the direct approach would be used. With the direct approach, biomass change is estimated directly, using biomass change observations as reference data. Of necessity, the direct approach requires ground level observations of biomass for the same locations at the two times, a requirement that is currently difficult to satisfy for many tropical applications. Thus, at the current time, the indirect approach is more commonly used.

For estimating either biomass or biomass change, remote sensing-based maps are used for three purposes: first, as auxiliary data to enhance design-based estimates using probability samples of ground reference data; second, for direct estimation when constructed using either probability or non-probability samples of ground reference data; and, third, as sources of reference data. Multiple biomass maps are available for the first and third purposes: a circa year 2001, 250-m, MODIS-based map for the USA [65]; a global, circa year 2005, 1-ha GLAS, ALOS, and Landsat-based map constructed by NASA's Jet Propulsion Laboratory [66]; the year 2010, 1-ha, lidar and SAR-based GlobBiomass map [67]; a circa 2007-2008, 500-m, GLAS and MODIS-based, pan-tropical map [68]; and the European Space Agency's year 2017 global datasets of AGB [69]. Global, regional, or even large area biomass change maps are not commonly available, at least partially as a result of the lack of ground biomass change observations. The emphasis of the sections that follow relates to the use of maps to estimate biomass, although the estimation of biomass change is completely analogous.

### 4.1. Probability Samples of Ground Reference Data

Remote sensing-based forest attribute maps, not just biomass or biomass change maps, can be used as sources of auxiliary data for increasing the precision of ground-based estimates of both biomass and biomass change. If observations of biomass or biomass change are obtained using a probability sampling design, any of the design-based STR, PSTR, or model-assisted estimators described in Section 3.2 can be used, albeit with $\mu^c$ denoting mean biomass or mean biomass change rather than an area proportion.

The statistical forms of the STR and PSTR estimators are the same as described for estimating activity data. For both estimators, the primary purpose of the maps is to serve as a basis for stratifications, which, in turn, are used to increase precision. For both these estimators, the map values are aggregated into a small number of contiguous classes that serve as strata [6,35,70]. The map-based stratifications can also be used to guide sample allocations to strata.

For use with the model-assisted estimators, the primary purpose of the maps is to serve as a source of predictions. The forms of both model-assisted estimators are the same as described for estimating activity data in Equations (13)–(16). Although many of the early applications used linear regression models, additional prediction techniques have been used, including nonlinear models [36,71,72]; k-nearest neighbors [73,74], and random forests [73,75].

Representative examples, as opposed to an exhaustive list of tropical applications which are sparse, of the uses of maps as auxiliary data for the estimation of biomass and biomass change, are reported in Table 2.

**Table 2.** Representative examples of biomass and biomass change estimation.

| Parameter | Estimator * | References |
|---|---|---|
| Biomass (volume) | STR | Næsset et al. [76,77] |
| | PSTR | McRoberts et al. [6,37,69], Tomppo et al. [74] (p. 56) |
| | DIF | McRoberts et al. [37], Næsset et al. [77] |
| | GREG | Næsset et al. [25,76,77], Poorazimy et al. [71] |
| Biomass (volume) change | STR | Næsset et al. [78] |
| | PSTR | McRoberts et al. [7,48], Næsset et al. [46] |
| | DIF | |
| | GREG | Gregoire et al. [79], McRoberts et al. [6,48], Næsset et al. [46,78], Esteban et al. [75] |

* STR: Stratified, PSTR: Post-stratified, DIF: Difference, GREG: Generalized regression.

*4.2. Non-Probability Samples of Ground Reference Data*

Although probability samples of ground data may not be available to support design-based inferential methods, ground observations and measurements may still be available from other sources, such as long-term research plots, local forest management plots, and pre-harvest plots. If conditions associated with these plots conform to the features of activities of interest such as forest-remaining-forest, pre-harvest as a form of deforestation, or thinning treatments as a form of degradation, they can be used for estimating emissions and removals factors. Two key challenges are associated with the use of data from these kinds of plots. First, differences in data features acquired from these kinds of plots must be reconciled before the ground data can be combined with remotely sensed data, and second, the lack of a consistent probability sampling design means that less familiar and more complex model-based inferential methods must be used instead of design-based inferential methods.

4.2.1. Data-Related Challenges

The data-related challenges result from the different plot configurations and different measurement protocols inevitably used for the different sources of reference data. In general, smaller plots tend to have more extreme per unit area observations than larger plots that are at least marginally more representative of the entire population. Thus, combining data for mixed size plots runs the risk of skewing relationships with remotely sensed data toward the data from the smaller, less representative plots. For modeling applications, multiple studies have shown the advantages of larger plots with smaller area to perimeter ratios that minimize edge effects [78,80,81]. Although no studies evaluating the effects of constructing models using data for mixtures of small and large plots are known, the effects are expected to be a form of heteroscedasticity and possibly model predictions skewed toward the data for the smaller plots. Differences in measurement protocols include, but are not limited to, differences in minimum diameters of trees to be measured and minimum height at maturity. For otherwise comparable plot configurations, plot-level biomass would be larger for smaller minimum diameters. Moreover, plots with trees that only marginally satisfy a small minimum height criterion might not even be measured if the criterion was larger.

If data for plots with different configurations and measurement protocols are to be combined, some form of data harmonization is necessary. For example, the smallest among multiple plot radii can be selected, and data for the now smaller plot can be recalculated. Similarly, the largest among multiple minimum diameters can be selected. The issue of harmonization of national forest inventories in Europe has received considerable attention, including the development of useful harmonization methods. Although developed for temperate forests, these methods are likely also applicable for tropical forests [82,83].

### 4.2.2. Model-Based Inference

Model-based inference, also characterized as model-dependent inference, relies on a quite different set of underlying assumptions than the more familiar design-based inference [72]. First, validity is based on correct model specification rather than a probability sample. Second, an entire distribution of possible values is assumed for each population unit, rather than just a single value. Third, randomization is via the realized observations from the distributions characterizing the population units selected for the sample rather than the sampling design. An important consequence of the first assumption is that model-based inference does not require probability samples of reference data for constructing the model. Although probability samples may be used and, in fact, may be preferable, purposive and other non-probability samples may also produce entirely valid model-based inferences [30] (p. 534). The absence of a requirement for a probability sample means that model-based inference can be used for applications for which design-based inference is not possible, such as when a non-probability sample is used to construct the model.

The model-based estimator of the mean is simply the synthetic estimator of Equation (13). As noted in Section 3.2.5, the synthetic estimator may be biased. The model-based SE of the estimate requires three sources of uncertainty information: (1) variances and covariances among population unit predictions due to sampling variability associated with the training data used to construct the map, (2) residual uncertainty in the form of differences between the map's training data and the corresponding map unit values, and (3) spatial correlation among the residuals. Of these three sources, the effects of sampling variability are often the greatest. Examples of model-based inference relevant for greenhouse gas inventories include McRoberts et al. [72] and Saarela et al. [84].

### 4.3. Remote Sensing-Based Maps as Sources of Reference Data

In the absence of ground data, large area, regional, or global biomass maps can be used as sources of reference data for estimating biomass. However, because the maps consist of map unit predictions subject to uncertainty, compliance with the IPCC good practice guidelines requires special considerations. In particular, the map must be validated, if not in its entirety, then preferably for a validation unit coincident with the area of interest and/or perhaps for a sample of validation units. Validation consists of a statistically rigorous test of the hypothesis of no difference between the map-based estimate and an estimate obtained using independent (ind) data of the form,

$$t = \frac{\hat{\mu}_{map} - \hat{\mu}_{ind}}{\sqrt{\hat{Var}(\hat{\mu}_{map}) + \hat{Var}(\hat{\mu}_{ind})}}. \tag{18}$$

where t follows Student's t-distribution with degrees of freedom determined by the sum of the sample sizes for the two variance estimates. Although these independent data serve as reference data for this analysis, for this discussion, they continue to be characterized only as independent data to avoid confusion with the reference data to be acquired from the map. Four estimates are therefore required: (1) the estimate based on the independent data, $\hat{\mu}_{ind}$; (2) the variance of the independent data estimate, $\hat{Var}(\hat{\mu}_{ind})$; (3) the map-based estimate for the validation unit, $\hat{\mu}_{map}$; and (4) the variance of the map-based estimate, $\hat{Var}(\hat{\mu}_{map})$.

### 4.3.1. Reference Data Estimates

McRoberts et al. [37] demonstrate two inferential approaches for acquiring the independent data necessary for the comparison of Equation (18). The first approach uses a sample of the independent data and auxiliary data to calculate an estimate for the validation unit that can be compared to the map estimate. If the independent data are obtained from a probability sample, one of the design-based estimators of Section 3.2 can be used, for which the map that is subject to validation can be used as auxiliary data. If the independent data are not obtained using a probability sampling design, the model-based estimators described in Section 4.2.2. and by McRoberts et al. [37] can be used.

The second approach uses a local map of greater quality than the biomass map as the source of independent data. Greater quality is with respect to factors such as finer resolution, use of additional auxiliary data, and use of local training data. The local map is sampled to obtain independent data using any convenient sampling intensity and sampling design, although probability sampling designs greatly simplify estimation. Because the local map also consists of predictions rather than observations, the uncertainty in the local map must be incorporated into the estimation of overall uncertainty using hybrid inference [35,61–63].

### 4.3.2. Map-Based Estimates

The map-based mean for Equation (18) is simply the mean over all map units obtained using the SYN estimator of Equation (13). Because the data used to construct the map are typically not available, estimation of the SE for the map-based estimate can be particularly challenging and typically entails model-based inferential methods, as described in Section 4.2.2. However, unlike Section 4.2.2, where the remote sensing-based map is constructed by the user, global maps used as sources of reference data are generally not constructed by the user. Thus, information on the sources of uncertainty necessary for calculation of the SE must be provided by the map authors. Unfortunately, however, they rarely provide such information, presumably because they do not know it is necessary or they do not know how to produce it. McRoberts et al. [37] describe some approximations and bounds on the uncertainty, but the validation process would be greatly facilitated if map authors were cognizant of these requirements.

### 5. Summary

Remotely sensed imagery and land use change maps are crucial for the statistically rigorous estimation of activity data. The post-stratified estimators are appropriate and effective for use with land use change maps and data obtained from an existing probability ground sampling program. When such ground data are unavailable or cannot be readily obtained, the combination of land use change maps, land use change interpretations of stratified samples from imagery, and the stratified estimators have been demonstrated to be particularly efficient. When sample unit data are continuous, such as plot- or pixel-level proportions of land use change, the model-assisted estimators can be used. In general, the design-based stratified, post-stratified and model-assisted estimators are more frequently used and are easier to use than model-based estimators.

Methods for estimating emissions and removals factors are much less mature than methods for estimating activity data, likely because the former rely much more heavily on ground data which may not be available. When probability-based samples of ground data are available or can be acquired, remote sensing-based biomass maps can be used as sources of auxiliary data for increasing the precision of estimates of emissions and removals factors obtained using any of the design-based stratified, post-stratified, and model-assisted estimators.

When some ground data are available, but were not acquired using probability sampling designs and/or not with similar plot configurations and measurement protocols, model-based inferential methods can be used to estimate biomass or biomass change. If the ground data are acquired from multiple sources, a degree of data harmonization will inevitably be necessary.

When sufficient ground data are not available or cannot be acquired, biomass maps can be used in lieu of ground data. However, the required statistical methods are more complex, because the uncertainty in the maps must be accommodated to comply with the IPCC good practice guidelines for greenhouse gas inventories. These methods typically require model-based inferential methods that are difficult to implement, apart from substantial meta-data regarding sources and effects of uncertainty (Section 4.2.2) that must be provided by the map authors.

Additional research should focus in several areas: (1) the documentation of actual tropical applications, (2) the development of rigorous uncertainty assessment methods for use with time series methods, and (3) greater attention by map authors on providing meta-data for global maps to facilitate uncertainty estimation.

**Author Contributions:** Conceptualization: R.E.M., E.N., C.S., S.V.S.; writing—original draft preparation, R.E.M.; writing—review and editing: R.E.M., E.N., C.S., S.V.S., E.O.T. All authors have read and agreed to the published version of the manuscript.

**Funding:** This research received no external funding.

**Acknowledgments:** The authors thank two anonymous reviewers for their constructive comments.

**Conflicts of Interest:** The authors declare no conflict of interest.

## References


1. GFOI. *Integration of Remote-Sensing and Ground-Based Observations for Estimation of Emissions and Removals of Greenhouse Gases in Forests: Methods and Guidance from the Global Forest Observations Initiative, Ed. 2.0;* Food and Agriculture Organization: Rome, Italy, 2016; Available online: https://www.reddcompass.org/frontpage (accessed on 8 June 2020).

2. Eggleston, H.S.; Buendia, L.; Miwa, K.; Ngara, T.; Tanabe, K. (Eds.) *2006 IPCC Guidelines for National Greenhouse Gas Inventories, Volume 4: Agriculture, Forestry and Other Land Use;* Institute for Global Environmental Strategies: Hayama, Japan, 2006; Available online: http://www.ipcc-nggip.iges.or.jp/public/2006gl/index.html (accessed on 8 June 2020).

3. Bucki, M.; Cuypers, D.; Mayaux, P.; Achard, F.; Estreguil, C.; Grassi, G. Assessing REDDC performance of countries with low monitoring capacities: The matrix approach. *Environ. Res. Lett.* **2012**, *7*, 014031. [CrossRef]

4. Gregoire, T.G.; Næsset, E.; McRoberts, R.E.; Ståhl, G.; Andersen, H.-E.; Gobakken, T.; Ene, L.; Nelson, R. Statistical rigor in lidar-assisted estimation of aboveground forest biomass. *Remote Sens. Environ.* **2016**, *173*, 98–108. [CrossRef]

5. Birigazzi, L.; Gregoire, T.G.; Finegold, Y.; Golec, R.C.D.; Sandker, E.; Donegan, E.; Gamarra, J.G.P. Data quality reporting: Good practice for transparent estimates from forestand land cover surveys. *Environ. Sci. Policy* **2019**, *96*, 85–94. [CrossRef]

6. McRoberts, R.E.; Næsset, E.; Gobakken, T. Comparing the stock-change and gain–loss approaches for estimating forest carbon emissions for the aboveground biomass pool. *Can. J. For. Res.* **2018**, *48*, 1535–1542. [CrossRef]

7. REDD+ Web Platform. Available online: https://redd.unfccc.int/fact-sheets.html (accessed on 8 June 2020).

8. Olofsson, P.; Foody, G.M.; Stehman, S.V.; Woodcock, C.E. Making better use of accuracy data in land change studies: Estimating accuracy and area and quantifying uncertainty using stratified estimation. *Remote Sens. Environ.* **2013**, *129*, 122–131. [CrossRef]

9. Olofsson, P.; Foody, G.M.; Herold, M.; Stehman, S.V.; Woodcock, C.E.; Wulder, M.A. Good practices for estimating area and assessing accuracy of land change. *Remote Sens. Environ.* **2014**, *148*, 42–57. [CrossRef]

10. Foody, G.M. The impact of imperfect ground reference data on the accuracy of land cover change estimation. *Int. J. Remote Sens.* **2009**, *30*, 3275–3281. [CrossRef]

11. Foody, G.M. Assessing the accuracy of land cover change with imperfect ground reference data. *Remote Sens. Environ.* **2010**, *14*, 2271–2285. [CrossRef]

12. Foody, G.M. Ground reference data error and the mis-estimation of the area of land cover change as a function of its abundance. *Remote Sens. Lett.* **2013**, *4*, 8. [CrossRef]

13. McRoberts, R.E.; Stehman, S.V.; Liknes, G.C.; Næsset, E.; Sannier, C.; Walters, B.F. The effects of imperfect reference data on remote sensing-assisted estimators of land cover class proportions. *Isprs J. Photogramm. Remote Sens.* **2018**, *142*, 292–300. [CrossRef]

14. Sun, B.; Chen, X.; Zhou, Q. Analyzing the uncertainties of ground validation for remote sensing land cover mapping in the era of big geographic data. In *Spatial Data Handling in Big Data Era. Advances in Geographic Information Science;* Zhou, C., Su, F., Harvey, F., Xu, J., Eds.; Springer: Singapore, 2017; pp. 31–38.

15. Mannel, S.; Price, M.; Hua, D. A method to obtain large quantities of reference data. *Int. J. Remote Sens.* **2006**, *27*, 623–627. [CrossRef]

16. Stehman, S.V. Model-assisted estimation as a unifying framework for estimating the area of land cover and land-cover change from remote sensing. *Remote Sens. Environ.* **2009**, *113*, 2455–2462. [CrossRef]

17. Mowrer, H.T.; Congalton, R.G. (Eds.) *Quantifying Spatial Uncertainty in Natural Resources: Theory and Applications for GIS and Remote Sensing*; Sleeping Bear Press: Ann Arbor, MI, USA, 2000.

18. Næsset, E. The effect of season upon registrations of stand mean height, crown closure and tree species on aerial photos. *Commun. Skogforsk* **1991**, *44*, 1–28.

19. Næsset, E. The effect of scale, type of film and focal length upon interpretation of tree species mixture on aerial photos. *Commun. Skogforsk* **1992**, *45*, 1–28.

20. Powell, R.L.; Matzke, N.; de Souza, D.; Clark, M.; Numata, I.; Hess, L.L.; Roberts, D.A. Sources of error in accuracy assessment of thematic land-cover maps in the Brazilian Amazon. *Remote Sens. Environ.* **2004**, *90*, 221–234. [CrossRef]

21. Pengra, B.P.; Stehman, S.V.; Horton, J.A.; Dockter, D.J.; Schroeder, T.A.; Yang, Z.; Cohen, W.B.; Healey, S.P.; Loveland, T.R. Quality control and assessment of interpreter consistency of annual land cover reference data in an operational national monitoring program. *Remote Sens. Environ.* **2020**, *238*, 111261. [CrossRef]

22. Guyana Forestry Commission. Guyana REDD+ Monitoring Reporting & Verification System (MRVS) Interim Measures Report 01 October 2010–31 December 2011 Version 1, 15 June 2012. Available online: http://occguyana.org/lcds/index.php/documents/reports/national/guyana-mrvs-interim-measures-reports-1/61-guyana-forestry-commission-guyana-redd-monitoring-reporting-verification-system-mrvs-interim-measures-report-01-october-2010-31-december-2011-version-3-26-july-2012/file (accessed on 8 June 2020).

23. Hansen, M.C.; Potapov, P.V.; Moore, R.; Hancher, M.; Turubanova, S.A.; Tyukavina, A.; Thau, D.; Stehman, S.V.; Goetz, S.J.; Loveland, T.R.; et al. High-Resolution global maps of 21st-century forest cover change. *Science* **2013**, *342*, 850–853. [CrossRef]

24. Sannier, C.; McRoberts, R.E.; Fichet, L.-V. Suitability of Global Forest Change data to report forest cover estimates at national level in Gabon. *Remote Sens. Environ.* **2016**, *173*, 326–338. [CrossRef]

25. Næsset, E.; Ørka, H.O.; Solberg, S.; Bollandsås, O.M.; Hansen, E.H.; Mauya, E.; Zahabu, E.; Malimbwi, R.; Chamuya, N.; Olsson, H.; et al. Mapping and estimating forest area and aboveground biomass in miombo woodlands in Tanzania using data from airborne laser scanning, TanDEM-X, RapidEye, and global forest maps: A comparison of estimated precision. *Remote Sens. Environ.* **2016**, *175*, 282–300. [CrossRef]

26. McRoberts, R.E.; Vibrans, A.C.; Sannier, C.; Næsset, E.; Hansen, M.C.; Walters, B.F.; Lingner, D.V. Methods for evaluating the utilities of local and global maps for increasing the precision of estimates of subtropical forest area. *Can. J. For. Res.* **2016**, *46*, 924–932. [CrossRef]

27. Mermoz, S.; Bouvet, A.; Le Toan, T.; Herold, M. Impacts of the forest definitions adopted by African countries on carbon conservation. *Environ. Res. Lett.* **2018**, *13*, 104014. [CrossRef]

28. De Wasseige, C.; Flynn, J.; Louppe, D.; Hiol Hiol, F.; Mayaux, P. (Eds.) *The Forests of the Congo Basin–State of the Forest*; Weyrich: Neufchâateu, Belgium, 2013.

29. Cochran, W.G. *Sampling Techniques*, 3rd ed.; Wiley: New York, NY, USA, 1977.

30. Särndal, C.-E.; Swensson, B.; Wretman, J. *Model Assisted Survey Sampling*; Springer: New York, NY, USA, 1992.

31. Westfall, J.A.; Patterson, P.L.; Coulston, J.W. Post-stratified estimation: Within-strata and total sample size recommendations. *Can. J. For. Res.* **2011**, *41*, 1130–1139. [CrossRef]

32. McRoberts, R.E.; Wendt, D.G.; Nelson, M.D.; Hansen, M.H. Using a land cover classification based on satellite imagery to improve the precision of forest inventory area estimates. *Remote Sens. Environ.* **2002**, *81*, 36–44. [CrossRef]

33. Olofsson, P.; Arévalo, A.; Espejo, A.B.; Green, C.; Lindquist, E.; McRoberts, R.E.; Sanz, M.J. Mitigating the effects of omission errors on area and area change estimates. *Remote Sens. Environ.* **2020**, *236*, 111492. [CrossRef]

34. Gregoire, T.; Valentine, H. *Sampling Strategies for Natural Resources and the Environment*; Chapman & Hall/CRC: Boca Raton, FL, USA, 2008.

35. Ståhl, G.; Saarela, S.; Schnell, S.; Holm, S.; Breidenbach, J.; Healey, S.P.; Patterson, P.L.; Magnussen, S.; Næsset, E.; McRoberts, R.E.; et al. Use of models in large-area forest surveys: Comparing model-assisted, model-based and hybrid estimation. *For. Ecosyst.* **2016**, *3*, 5. [CrossRef]

36. Stehman, S.V. Estimating area from an accuracy assessment error matrix. *Remote Sens. Environ.* **2013**, *132*, 202–211. [CrossRef]

37. McRoberts, R.E.; Nasset, E.; Saatchi, S.; Liknes, G.C.; Walters, B.F.; Chen, Q. Local validation of global biomass maps. *Int. J. Appl. Earth Obs. Geoinf.* **2019**, *83*, 101931. [CrossRef]

38. McRoberts, R.E. Post-classification approaches to estimating change in forest area using remotely sensed auxiliary data. *Remote Sens. Environ.* **2014**, *151*, 149–156. [CrossRef]

39. Khan, A.; Hansen, M.C.; Potapov, P.V.; Adusei, B.; Pickens, A.; Krylov, A.; Stehman, S.V. Evaluating Landsat and RapidEye data for winter wheat mapping and area estimation in Punjab, Pakistan. *Remote Sens.* **2018**, *10*, 489. [CrossRef]

40. Gallego, F.J. Remote sensing and land cover area estimation. *Int. J. Remote Sens.* **2004**, *25*, 3019–3047. [CrossRef]

41. Vibrans, A.C.; McRoberts, R.E.; Moser, P.; Nicoletti, A.L. Using satellite image-based maps and ground inventory data to estimate the area of the remaining Atlantic forest in the Brazilian state of Santa Catarina. *Remote Sens. Environ.* **2013**, *130*, 87–95. [CrossRef]

42. Stehman, S.V. Estimating area and map accuracy for stratified random sampling when the strata are different from the map classes. *Int. J. Remote Sens.* **2014**, *35*, 4923–4939. [CrossRef]

43. McRoberts, R.E. Probability- and model-based approaches to inference for proportion forest using satellite imagery as ancillary data. *Remote Sens. Environ.* **2010**, *114*, 1017–1025. [CrossRef]

44. Sannier, C.; McRoberts, R.E.; Fichet, L.-V.; Makaga, E.M.K. Using the regression estimator with Landsat data to estimate proportion forest cover and net proportion deforestation in Gabon. *Remote Sens. Environ.* **2014**, *151*, 138–148. [CrossRef]

45. Ying, Q.; Hansen, M.C.; Potapov, P.V.; Tyukavina, A.; Wang, L.; Stehman, S.V.; Moore, R.; Hancher, M. Global bare ground gain from 2000 to 2012 using Landsat imagery. *Remote Sens. Environ.* **2017**, *194*, 161–176. [CrossRef]

46. Næsset, E.; Bollandsås, O.M.; Gobakken, T.; Gregoire, T.G.; Ståhl, G. Model-assisted estimation of change in forest biomass over an 11 year period in a sample survey supported by airborne LiDAR: A case study with post-stratification to provide "activity data". *Remote Sens. Environ.* **2013**, *128*, 299–314. [CrossRef]

47. McRoberts, R.E.; Walters, B.F. Statistical inference for remote sensing-based estimates of net deforestation. *Remote Sens. Environ.* **2012**, *124*, 394–401. [CrossRef]

48. McRoberts, R.E.; Næsset, E.; Gobakken, T.; Bollandsås, O.M. Indirect and direct estimation of forest biomass change using forest inventory and airborne laser scanning data. *Remote Sens. Environ.* **2015**, *164*, 36–42. [CrossRef]

49. Claggett, P.R.; Okay, J.A.; Stehman, S.V. Monitoring regional riparian forest cover change using stratified sampling and multiresolution imagery. *J. Am. Water Resour. Assoc.* **2010**, *46*, 334–343. [CrossRef]

50. Tyukavina, A.; Hansen, M.C.; Potapov, P.V.; Parker, D.; Okpa, C.; Stehman, S.V.; Kommareddy, I.; Turubanova, S. Congo Basin forest loss dominated by increasing smallholder clearing. *Sci. Adv.* **2018**, *4*, eaat2993. [CrossRef]

51. Pickering, J.; Stehman, S.V.; Tyukavina, A.; Potapov, P.; Watt, P.; Jantz, S.M.; Bholanath, P.; Hansen, M.C. Quantifying the trade-off between cost and precision in estimating area of forest loss and degradation using probability sampling in Guyana. *Remote Sens. Environ.* **2019**, *221*, 122–135. [CrossRef]

52. Zimmerman, P.L.; Housman, I.W.; Perry, C.H.; Chastain, R.A.; Webb, J.B.; Finco, M.V. An accuracy assessment of forest disturbance mapping in the western Great Lakes. *Remote Sens. Environ.* **2013**, *128*, 176–185. [CrossRef]

53. Kennedy, R.E.; Cohen, W.B.; Schroeder, T.A. Trajectory-based change detection for automated characterization of forest disturbance dynamics. *Remote Sens. Environ.* **2007**, *110*, 370–386. [CrossRef]

54. Cohen, W.B.; Yang, Z.; Kennedy, R. Detecting trends in forest disturbance and recovery using yearly Landsat time series: 2. TimeSync—Tools for calibration and validation. *Remote Sens. Environ.* **2010**, *114*, 2199–2924. [CrossRef]

55. Verbesselt, J.; Hyndman, R.; Newnham, G.; Culvenor, D. Detecting trend and seasonal changes in satellite image time series. *Remote Sens. Environ.* **2010**, *114*, 106–115. [CrossRef]

56. Zhu, Z.; Woodcock, C.E. Continuous change detection and classification of land cover using all available Landsat data. *Remote Sens. Environ.* **2014**, *144*, 152–171. [CrossRef]

57. Zhu, Z.; Woodcock, C.E.; Olofsson, P. Continuous monitoring of forest disturbance using all available Landsat imagery. *Remote Sens. Environ.* **2012**, *122*, 75–91. [CrossRef]

58. Reiche, J.; de Bruin, S.; Hoekman, D.; Verbesselt, J.; Herold, M. A Bayesian approach to combine Landsat and ALOS PALSAR time series for near real-time deforestation detection. *Remote Sens.* **2015**, *7*, 4973–4996. [CrossRef]

59. Reiche, J.; Verbesselt, J.; Hoekman, D.; Herold, M. Fusing Landsat and SAR time series to detect deforestation in the tropics. *Remote Sens. Environ.* **2015**, *156*, 276–293. [CrossRef]

60. Arevalo, P.; Woodcock, C.E.; Olofsson, P. Continuous monitoring of land change activities and post-disturbance dynamics from Landsat time series: A test methodology for REDD+ reporting. *Remote Sens. Environ.* **2019**, *238*, 111051. [CrossRef]

61. Corona, P.; Fattorini, L.; Franceschi, S.; Scrinzi, G.; Torresan, C. Estimation of standing wood volume in forest compartments by exploiting airborne laser scanning information: Model-based, design-based, and hybrid perspectives. *Can. J. For. Res.* **2014**, *4*, 1303–1311. [CrossRef]

62. McRoberts, R.E.; Chen, Q.; Domke, G.M.; Ståhl, G.; Saarela, S.; Westfall, J.A. Hybrid estimators for mean aboveground carbon per unit area. *For. Ecol. Manag.* **2016**, *378*, 44–56. [CrossRef]

63. Condés, S.; McRoberts, R.E. Updating national forest inventory estimates of growing stock volume using hybrid inference. *For. Ecol. Manag.* **2017**, *400*, 48–57. [CrossRef]

64. IPCC. User Manual, Database on Greenhouse as Emission Factors. Version 3.0. 2018. Available online: https://www.ipcc-nggip.iges.or.jp/EFDB/documents/EFDB_User_Manual.pdf (accessed on 8 June 2020).

65. Blackard, J.A.; Finco, M.V.; Helmer, E.H.; Holden, G.R.; Hoppus, M.L.; Jacobs, D.M.; Lister, A.J.; Moisen, G.G.; Nelson, M.D.; Riemann, R.; et al. Mapping, U.S. forest biomass using national forest inventory data and moderate resolution information. *Remote Sens. Environ.* **2008**, *112*, 1658–1677. [CrossRef]

66. Saatchi, S.S.; Harris, N.L.; Brown, S.; Lefsky, M.; Mitchard, E.T.; Salas, W.; Zutta, B.R.; Buermann, W.; Lewis, S.L.; Hagen, S.; et al. Benchmark map of forest carbon stocks in tropical regions across three continents. *Proc. Natl. Acad. Sci. USA* **2011**, *108*, 9899–9904. [CrossRef]

67. Santoro, M.; Cartus, O.; Mermoz, S.; Bouvet, A.; Le Toan, T.; Carvalhais, N.; Rozendaal, D.; Herold, M.; Avitabile, V.; Quegan, S.; et al. A detailed portrait of the forest aboveground biomass pool for the year 2010 obtained from multiple remote sensing observations. *Geophys. Res. Abstr.* **2018**, *20*, EGU2018-18932.

68. Baccini, A.; Goetz, S.J.; Walker, W.S.; Laporte, N.T.; Sun, M.; Sulla-Menashe, D.; Hackler, J.; Beck, P.S.A.; Dubayah, R.; Friedl, M.A.; et al. Estimated carbon dioxide emissions from tropical deforestation improved by carbon-density maps. *Nat. Clim. Chang.* **2012**, *2*, 182–185. [CrossRef]

69. Santoro, M.; Cartus, O. ESA Biomass Climate Change Initiative (Biomass_cci): Global datasets of forest above-ground biomass for the year 2017, v1. *Cent. Environ. Data Anal.* **2019**. [CrossRef]

70. McRoberts, R.E.; Gobakken, T.; Næsset, E. Post-stratified estimation of forest area and growing stock volume using lidar-based stratifications. *Remote Sens. Environ.* **2012**, *125*, 157–166. [CrossRef]

71. Särndal, C.E. The calibration approach in survey theory and practice. *Surv. Methodol.* **2007**, *33*, 99–119.

72. McRoberts, R.E.; Næsset, E.; Gobakken, T. Inference for lidar-assisted estimation of forest growing stock volume. *Remote Sens. Environ.* **2013**, *128*, 268–275. [CrossRef]

73. Poorazimy, M.; Shataee, S.; McRoberts, R.E.; Mohammadi, J. Integrating airborne laser scanning data, space-borne radar data and digital aerial imagery to estimate aboveground carbon stock in Hyrcanian forests, Iran. *Remote Sens. Environ.* **2020**, in press. [CrossRef]

74. Tomppo, E.; Haakana, M.; Katia, M.; Peräsaari, J. *Multi-Source National Forest Inventory-Methods and Applications*; Managing Forest Ecosystems; Springer: Dordrecht, The Netherlands, 2008.

75. Esteban, J.; McRoberts, R.E.; Fernández-Landa, A.; Tomé, J.L.; Næsset, E. Estimating forest volume and biomass and their changes Using random forests and remotely sensed data. *Remote Sens.* **2019**, *11*, 1994. [CrossRef]

76. Næsset, E.; Gobakken, T.; Solberg, S.; Gregoire, T.G.; Nelson, R.; Ståhl, G.; Weydahl, D. Model-assisted regional forest biomass estimation using LiDAR and InSAR as auxiliary data: A case study from a boreal forest area. *Remote Sens. Environ.* **2011**, *115*, 3599–3614. [CrossRef]

77. Næsset, E.; McRoberts, R.E.; Pekkarinen, A.; Saatchi, S.; Santoro, M.; Trier, O.D.; Zahabu, E.; Gobakken, T. Use of local and global maps of forest canopy height and aboveground biomass to enhance local estimates of biomass in miombo woodlands in Tanzania. *Int. J. Appl. Earth Obs. Geoinf.* **2020**, in press.

78. Næsset, E.; Bollandsås, O.M.; Gobakken, T.; Solberg, S.; McRoberts, R.E. The effects of field plot size on model-assisted estimation of aboveground biomass change using multitemporal interferometric SAR and airborne laser scanning data. *Remote Sens. Environ.* **2015**, *168*, 252–264. [CrossRef]
79. Gregoire, T.G.; Ståhl, G.; Næsset, E.; Gobakken, T.; Nelson, R.; Holm, S. Model-assisted estimation of biomass in a LiDAR sample survey in Hedmark county, Norway. *Can. J. For. Res.* **2011**, *41*, 83–95. [CrossRef]
80. Mauya, E.W.; Hansen, E.H.; Gobakken, T.; Bollandsås, O.M.; Malimbwi, R.E.; Næsset, E. Effects of field plot size on the prediction accuracy of aboveground biomass in airborne laser scanning-assisted inventories in tropical rain forests of Tanzania. *Carbon Balance Manag.* **2015**, *10*, 10. [CrossRef]
81. Tomppo, E.; Kuusinen, N.; Mäkisara, K.; Katila, M.; McRoberts, R.E. Effects of field plot configurations on the uncertainties of ALS-assisted forest resource estimates. *Scand. J. For. Res.* **2017**, *32*, 488–500. [CrossRef]
82. McRoberts, R.E.; Tomppo, T.; Schadauer, K.; Vidal, C.; Ståhl, G.; Chirici, G.; Lanz, A.; Cienciala, A.; Winter, S.; Smith, W.B. Harmonizing national forest inventories. *J. For.* **2009**, *107*, 179–187. [CrossRef]
83. Tomppo, E.; Gschwantner, T.; Lawrence, M.; McRoberts, R.E. (Eds.) *National Forest Inventories: Pathways to Common Reporting*; Springer: Heidelberg, Germany, 2010.
84. Saarela, S.; Schnell, S.; Tuominen, S.; Balázs, A.; Hyyppä, J.; Grafström, A. Effects of positional errors in model-assisted and model-based estimation of growing stock volume. *Remote Sens. Environ.* **2016**, *172*, 101–108. [CrossRef]

MDPI
St. Alban-Anlage 66
4052 Basel
Switzerland
Tel. +41 61 683 77 34
Fax +41 61 302 89 18
www.mdpi.com

*Remote Sensing* Editorial Office
E-mail: remotesensing@mdpi.com
www.mdpi.com/journal/remotesensing

www.ingramcontent.com/pod-product-compliance
Lightning Source LLC
LaVergne TN
LVHW071549170726
843515LV00008B/2252